# Analytical Supercritical Fluid Extraction Techniques

# Analytical Supercritical Fluid Extraction Techniques

Edited by

E.D. RAMSEY
School of Applied Sciences
University of Glamorgan
Pontypridd
UK

KLUWER ACADEMIC PUBLISHERS
DORDRECHT / BOSTON / LONDON

Library of Congress Cataloging Card Number: 98-67006

---

Published by Kluwer Academic Publishers,
P.O. Box 17, 3300 AA Dordrecht, The Netherlands.

Sold and distributed in North, Central and South America
by Kluwer Academic Publishers,
101 Philip Drive, Norwell, MA 02061, U.S.A.

In all other countries, sold and distributed
by Kluwer Academic Publishers,
P.O. Box 322, 3300 AH Dordrecht, The Netherlands.

*Printed on acid-free paper*



Printed in Great Britain

# Contents

# Contributors


| | |
|---|---|
| **J.W. Algaier** | Isco Inc., PO Box 5347, 4700 Superior Street, Lincoln, NE 68504, USA |
| **I.G.M. Anderson** | British American Tobacco, Regents Park Road, Millbrook, Southampton SO15 8TL, UK |
| **R. Babecki** | School of Applied Sciences, University of Glamorgan, Pontypridd, Mid Galmorgan CF37 1DL, UK |
| **K.D. Bartle** | School of Chemistry, University of Leeds, Leeds LS2 9JT, UK |
| **J.M. Bayona** | Department of Environmental Chemistry, Centro de Investigacion y Desarrollo, Jordi Girona, 18-26-E-08034 Barcelona, Spain |
| **M.D. Burford** | Unilever Research, Port Sunlight Laboratory, Quarry Road East, Bebington, Wirral L63 3JW, UK |
| **A.A. Clifford** | School of Chemistry, University of Leeds, Leeds LS2 9JT, UK |
| **N.J. Cotton** | Smith and Nephew, Group Research Center, York Science Park, Heslington, York YO1 5DF, UK |
| **G.R. Davies** | Isco Inc., PO Box 5347, 4700 Superior Street, Lincoln, NE 68504, USA |
| **J.R. Dean** | Department of Chemical and Life Sciences, University of Northumbria at Newcastle, Ellison Building, Newcastle upon Tyne NE1 8ST, UK |
| **D.I. James** | Nicolet Instruments Ltd, Budbrooke Road, Warwick CV34 5XH, UK |
| **M.I. Littlewood** | Nicolet Instruments Ltd, Budbrooke Road, Warwick CV34 5XH, UK |

**D.C. Messer** Isco Inc., PO Box 5347, 4700 Superior Street, Lincoln, NE 68504, USA

**B. Minty** School of Applied Sciences, University of Glamorgan, Pontypridd, Mid Galmorgan CF37 1DL, UK

**P.M. O'Brien** Nicolet Instruments Ltd, Budbrooke Road, Warwick CV34 5XH, UK

**C.G. Prange** Isco Inc., PO Box 5347, 4700 Superior Street, Lincoln, NE 68504, USA

**E.D. Ramsey** School of Applied Sciences, University of Glamorgan, Pontypridd, Mid Galmorgan CF37 1DL, UK

**M.W. Raynor**[1] Department of Chemistry and Applied Chemistry, University of Natal, Durban 4041, South Africa

**A.T. Rees** Nycomed Amersham, Cardiff Laboratories, Forest Farm, Whitchurch, Cardiff CF4 8YD, UK

**A.C. Rosselli** Isco Inc., PO Box 5347, 4700 Superior Street, Lincoln, NE 68504, USA

**N. Saim** Department of Chemistry, Faculty of Physical and Applied Sciences, Universiti Kebangsaan, 43650 UKM Bangi, Selangor, Malaysia

**H.J. Vandenburg** School of Chemistry, University of Leeds, Leeds LS2 9JT, UK

[1] Now at Matheson Gas Products, Advanced Technology Center, 1861 Lefthand Circle, Longmont, CO 80501, USA

# Preface

Since the late 1980s supercritical fluid extraction (SFE) has attracted considerable attention as a sample preparation procedure. The successful implementation of this technique can lead to improved sample throughput, more efficient recovery of analytes, cleaner extracts, economic replacement of halogenated solvents and a high level of automation compared with conventional sample preparation procedures. The present text was conceived as an update of *Supercritical Fluid Extraction and its Use in Chromatographic Sample Preparation*, edited by Dr. S.A. Westwood, which largely focused on the on-line combination of SFE with chromatographic techniques. However, in keeping with current trends, this book has also been expanded to provide more details of off-line SFE, with newer developments being described in separate chapters. The topics described within this text are illustrated with many 'state-of-the-art' applications, and each chapter provides a comprehensive list of references.

The first chapter deals with the basic principles of SFE, discussing the properties of supercritical fluids, factors affecting the kinetics of extraction and modelling of SFE. Chapter 2 is devoted to the essential aspects of SFE instrumentation, describing the features and benefits of various instrument configurations, automation and future developments. Off-line SFE of solid matrices is covered in Chapter 3, which provides important details concerning sample preparation, *in situ* chemical derivatisation, extract cleanup procedures, high-temperature SFE, extraction of metals and methods for optimising SFE experimental parameters. Techniques involving SFE of liquid matrices form the subject of Chapter 4 which deals with relevant instrument considerations for such applications. Other topics covered in this chapter include factors affecting the choice between direct and indirect liquid SFE procedures, *in situ* sample derivatisation, modifications to liquid samples to promote analyte extraction efficiencies, recovery of metal ions from aqueous media, enzymes and inverse SFE. The next three chapters are devoted to the on-line coupling of SFE with gas chromatography (GC), capillary and packed column supercritical fluid chromatography (SFC), with the emphasis being placed on practical considerations for the selection of the best techniques for different applications and sample matrices. The on-line combination of SFE with high-performance liquid chromatography (HPLC) remains largely unexplored; reasons for this form the subject of Chapter 8, which also reviews off-line SFE as a sample preparation procedure for HPLC. The applications cited within this chapter serve to dispel any notion that SFE is applicable only to analytes which are amenable to GC and SFC. The on-line combination of SFE with spectroscopic techniques and mass spectrometry are covered in Chapter 9, which describes how these procedures offer great potential for the rapid confirmation or quantitation of target analytes along with the provision of structural information for unknown species. Insofar as all current sample preparation techniques have limitations which prevent their universal application, the final chapter describes the principles and applications of microwave-assisted and accelerated solvent extraction as emerging alternatives to SFE. For the convenience of the reader, an appendix which contains pressure conversion scales and supercritical fluid carbon dioxide density tables appear at the end of the book.

E.D. Ramsey
Pontypridd
April 1998

# Abbreviations

| | |
|---|---|
| AA | acetic anhydride |
| AAS | atomic absorption spectroscopy |
| AC | $N^4$-acetylsulphamethazine |
| AES | atomic emission spectroscopy |
| ANOVA | analysis of variance |
| APCI | atmospheric pressure chemical ionisation |
| APE | alcohol phenol ethoxylate |
| ASE | accelerated solvent extraction |
| AVR | automated variable restrictor |
| BEC | Bond Elute Certify |
| BHC | benzene hexachloride |
| BHT | 2,6-ditertiarybutyl-4-methylphenol/butylated hydroxytoluene |
| BSTFA | *N*,*O*-bis(trimethylsilyl)trifluoracetamide |
| BTEX | benzene, toluene, ethylbenzene, xylene |
| CBs | chlorinated benzenes |
| CC | cryogenic collection |
| CI | chemical ionisation |
| CID | collision-induced dissociation |
| CPTH | 3-chloro-*p*-toluidine hydrochloride |
| cSFC | capillary supercritical fluid chromatography |
| DAD | photodiode-array detector |
| DBCP | 1,2-dibromo-3-chloropropane |
| DBDTC | dibutyldithiocarbamate |
| DCM | dichloromethane |
| DDD | dichlorodiphenyldichloroethane |
| DDE | dichlorodiphenyldichloroethylene |
| DDT | dichlorodiphenyltrichloroethane |
| DDVP | dichlorvos |
| DEDTC | diethyldithiocarbamate |
| DEHP | di(2-ethylhexyl) phthalate |
| DES | diethylstibestrol and desaminosulphamethazine |
| DEX | dexamethasone |
| DHA | docosahexanoic acid |
| DHTDMAC | dihardenetallowdimethylammonium chloride |
| DIMP | diisopropyl methylphosphonate |
| DIP | direct insertion probe |

| | |
|---|---|
| DMHA | dimethylhexylamine |
| DTDMAC | ditallowdimethylammonium chloride |
| ECD | electron capture detection |
| EI | electron ionisation |
| ELISA | enzyme-linked immunosorbent assay |
| EPA | eicosapentaenoic acid/(US) Environmental Protection Agency |
| ESE | enhanced solvent extraction |
| ESI | electospray ionisation |
| FAMES | fatty acid methyl esters |
| FDDC | bis(trifluoroethyl)dithiocarbamate |
| FID | flame ionisation detection |
| FOD | 2,2-dimethyl-6,6,7,7,8,8,8-heptafluoro-3,5-octanedione |
| %FOY | percentage finish on yarn |
| FPD | flame photomeric detection |
| FTIR | Fourier transform infra-red spectroscopy |
| GC | gas chromatography |
| GPC | gel permeation chromatography |
| GR | $N^4$-glucuronylsulphamethazine |
| GSR | Gram–Schmidt reconstructed (chromatogram) |
| HAPA | halogenated aromatic phenoxy derivative of an aliphatic alkane |
| HAU | halogenated derivative of urea |
| HCB | hexachlorobenzene |
| HCH | hexachlorohexane/hexachlorocyclohexane |
| HDCP | high-density crystalline polymer |
| HFA | hexafluoroacetylacetone |
| HPLC | high-pressure (or high-performance) liquid chromatography |
| HPMC | hydroxypropyl methylcellulose |
| HTSFE | high-temperature SFE |
| i.d. | inner diameter |
| IPA | isopropyl alcohol |
| LC | liquid chromatography |
| LDPE | low-density polyethylene |
| LLE | liquid–liquid extraction |
| MAE | microwave-assisted extraction |
| MBC | carbendazim |
| MDP | medroxyprogesterone |
| MEBOH | mebeverine alcohol |
| MEKC | micellar electrokinetic chromatography |
| MGA | melengestrol acetate |
| MI | methyl iodide |
| MOC | methoxychlor |
| MSD | mass selective detector |

| | |
|---|---|
| MTOA | methyltrioctylammonium chloride |
| N4 | $N^4$-acetylsulphamethoxazole |
| NIST | National Institute of Standards and Technology |
| NMR | nuclear magnetic resonance spectroscopy |
| NNA | neutron activation of analysis |
| NPD | nitrogen phosphorus specific detector |
| OCP | organochlorine pesticide |
| o.d. | outer diameter |
| ODS | octadecylsilyl-modified silica |
| OPP | organophosphate pesticide |
| PAC | polycyclic aromatic compound |
| PAH | polycyclic aromatic hydrocarbon/polynuclear aromatic hydrocarbon |
| PBT | poly(1,4-butylene terephthalate) |
| PCB | polychlorinated biphenyl |
| PCCD | polychlorinated dibenzo-*p*-dioxins |
| PDTC | pyrollidinedithiocarbamate |
| PEEK | poly(ether-ether-ketone) |
| PET | poly(ethylene terephthalate) |
| PFBBr | pentaflourobenzylbromide |
| PFE | pressurised fluid extraction |
| PTFE | poly(tetrafluoroethene) |
| PTV | programmed temperature vaporiser |
| PUF | polyurethane foam |
| RPD | radiofrequency plasma detector |
| RSD | relative standard deviation |
| SDB | styrene–divinyl benzene |
| SDM | sulphadimethoxine |
| SFC | supercritical fluid chromatography |
| SFDE | supercritical fluid derivatisation extraction |
| SFE | supercritical fluid extraction |
| SFR | supercritical fluid reactor |
| SIM | single ion monitoring |
| SMI | sulphamethizole |
| SMOP | sulphamethoxypyridazine |
| SMOZ | sulphamethoxazole |
| SMR | sulphamerazine |
| SMZ | sulphamethazine |
| $SO_4$ | $N^4$-sulphatosulphamethazine |
| SPA | solid phase absorption |
| SPE | solid phase extraction |
| SQX | sulphaquinoxaline |
| SRM | standard reference material |
| TACA | triamcinolone acetonide |

| | |
|---|---|
| TAM | trialkylamine |
| TBA | tetrabutylammonium hydroxide/cation |
| TBOH | trenbolone |
| TBP | tributyl phosphate |
| TBPO | tri-*n*-butylphosphine oxide |
| TBZ | thiabendazole |
| TCP | 2,4,6-trichlorophenol |
| TEA | triethylamine and thermal energy analysis |
| TEPP | tetraethylpyrophosphate |
| TFA | triflouroacetic acid |
| TGA | thermal gravimetric analysis |
| THA | tetrahexylammonium hydrogen sulphate |
| THAB | tetrahexylammonium bromide |
| THF | tetrahydrofuran |
| THPAB | tetraheptylammonium bromide |
| TIC | total ion chromatogram |
| TID | thermionic detection |
| TLC | thin-layer chromatography |
| TMAOH | tetramethylammonium hydroxide |
| TMPA | trimethylphenylammonium hydroxide |
| TOPO | tri-*n*-octylphosphine oxide |
| TPH | total petroleum hydrocarbon |
| TPPO | triphenylphosphine oxide |
| TTA | thienoyltrifuoroacetylacetone |
| ZER | zeranol |
| 2,4-D | 2,4-dichlorophenoxyacetic acid |
| 2,4,5-T | 2,4,5-trichlorophenoxyacetic acid |

# 1 Introduction to supercritical fluid extraction in analytical science

A.A. CLIFFORD

## 1.1 Introduction

Supercritical fluid extraction (SFE) is becoming an important tool in analytical science and has seen rapid development in the past few years. Manufacturers are now producing instrumentation designed for the routine application of the technique. It has the advantages, compared with liquid extraction, that

- it is usually less expensive in terms of laboratory time;
- the solvent is easier to remove;
- pressure (as well as temperature and the nature of the solvent) can be used to select, to some extent, the compounds to be extracted;
- carbon dioxide is available, to be used as a pure or modified solvent, with its convenient critical temperature, its cheapness and non-toxicity.

This book describes the principles and methods available for those considering using the technique for their analytical problems. This first chapter explains the basic principles of SFE, and starts with a short introduction to supercritical fluids and their properties.

From the viewpoint of methodology, SFE is often classified as off-line or on-line. In off-line SFE the sample is subjected to a flow of fluid, usually at constant temperature and pressure, and the extract or, in the case of a kinetic experiment, a series of samples is collected at regular time intervals from the eluting fluid after depressurizing, by passing it through a solvent for example. These samples are analysed later. In on-line SFE the SFE instrument is coupled directly to the analytical instrument, as in SFE–gas chromatography (SFE–GC) for example. Typically, the sample is extracted by a flowing stream of fluid at a particular temperature and pressure for a certain length of time and the extract deposited, after depressurizing, on the front of a GC column. The extraction is then stopped while chromatographic analysis is carried out. Apart from possible convenience and time-saving, on-line SFE has the advantage that all of the extract can be analysed, whereas in off-line SFE the extracted material is trapped in, say, 1 ml of solvent and only a portion of this is used for further analysis, by injection into a GC for example. This can give rise to improvements in sensitivity.

## 1.2 Pure and modified supercritical fluids

A pure supercritical fluid is a substance above its critical temperature and pressure. Above its critical temperature it does not condense or evaporate to form a liquid or a gas but is a fluid, with properties changing continuously from gas-like to liquid-like as the pressure increases. This allows extraction to be selective to some extent. Figure 1.1 shows the phase diagram (schematic) of a single substance. The line between the liquid and gas regions is the gas–liquid coexistence curve, which is a graph of vapour pressure versus temperature. As we move upwards along this curve, the density of the liquid phase decreases as a result of thermal expansion, and the density of the gas phase increases as a result of the increase in pressure. At the critical point, the densities (and other properties) of both phases become identical and the distinction between gas and liquid disappears. The hatched area shows the temperature–pressure region usually described as a supercritical fluid.

The temperature and pressure coordinates of the critical point are the critical temperature, $T_c$, and critical pressure, $p_c$. Table 1.1 shows the critical parameters of some compounds useful as supercritical fluids [1]. One compound, $CO_2$, has so far been the most widely used, because of its convenient critical temperature, cheapness, non-explosive character and non-toxicity. Because the molecule is non-polar it is classified as a non-polar solvent,

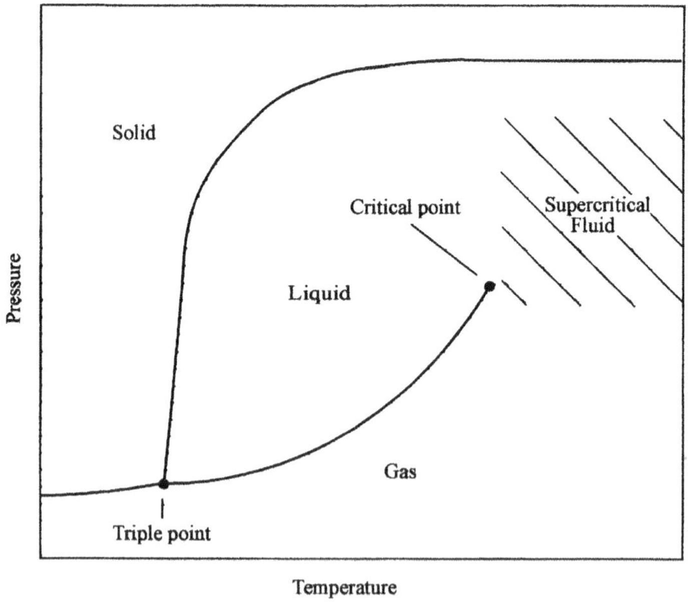


**Figure 1.1** A schematic representation of the phase diagram of a single substance, showing the supercritical fluid region as a hatched area.

**Table 1.1** Substances useful as supercritical fluids. Source: ref. 1

| | $T_c$ (K) | $p_c$ (bar) | $Z_c$ | $\omega$ |
|---|---|---|---|---|
| Carbon dioxide | 304 | 74 | 0.274 | 0.225 |
| Ethane | 305 | 49 | 0.285 | 0.099 |
| Ethene | 282 | 50 | 0.280 | 0.089 |
| Propane | 370 | 43 | 0.281 | 0.153 |
| Xenon | 290 | 58 | 0.287 | 0 |
| Ammonia | 406 | 114 | 0.244 | 0.250 |
| Nitrous oxide | 310 | 72 | 0.274 | 0.165 |
| Fluoroform | 299 | 49 | 0.259 | 0.260 |

Note: $T_c$ = critical temperature; $p_c$ = critical pressure; $Z_c$ = critical compression factor; $\omega$ = acentric factor.

although it has some limited affinity with polar solutes because of its large molecular quadrupole. Thus pure $CO_2$ can be used for many large organic solute molecules even if they have some polar character.

For the extraction and chromatography of more polar molecules, it is common to add modifiers or entrainers, such as the lower alcohols, to $CO_2$, usually in small quantities. Other properties can also be imparted to $CO_2$ by modifiers, such as decreased polarity, aromaticity, chirality and the ability to complex metal ion compounds. In such cases it is important to be aware of the modifier–$CO_2$ phase diagram to ensure that the solvent is in one phase. For example for methanol–$CO_2$ at 50°C there is only one phase above 95 bar whatever the composition, but below this pressure two phases can occur. The phase diagram for a binary mixture, such as methanol–$CO_2$, can be represented by a three-dimensional figure, whose axes are pressure, $p$, temperature, $T$, and mole fraction, $x$. At a particular temperature a cross-section through such a diagram is a two-dimensional $x$–$p$ plot, of which an example is given for methanol–$CO_2$ at 50°C in Figure 1.2 based on data published by Brunner *et al.* [2]. At very low pressures (which are not of importance in SFE) a single gaseous phase exists at all compositions, which are mixtures of $CO_2$ and methanol vapour. At intermediate pressures, both gaseous and liquid phases can occur, dependent on composition. At high mole fractions of $CO_2$ the mixture is gaseous, at high methanol concentrations it is liquid and at intermediate compositions both phases exist. The liquid + gas region reaches a maximum in pressure at the critical point (C) for this particular temperature. Consider what happens to a mixture of the critical composition at a pressure below the critical pressure (where it will be in two phases) as the pressure is raised. The liquid will dissolve more $CO_2$, the gas will solvate more methanol and the gas will increase in density more rapidly than the liquid. Eventually, at the critical point, the compositions and densities of the two phases will become identical. Thus above the critical pressure only one supercritical fluid phase will exist. (One should mention that at very much higher pressures, of no concern in SFE, other phases such as solids can occur.)

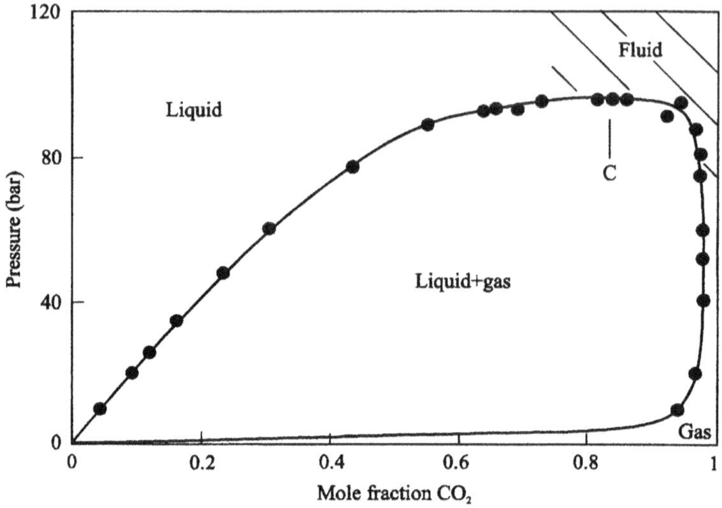


**Figure 1.2** The phase diagram of a methanol–$CO_2$ mixture at 50°C. C = point at which liquid + gas mixture reaches maximum pressure. Source: ref. 2.

Thus to be under truly supercritical fluid conditions the pressure needs to be above the critical pressure of the mixture for the particular temperature. However, in the context of SFE, where the proportion of modifier is often small, part of the gaseous phase is often considered as 'supercritical' as the pure gaseous component is above its critical pressure and temperature. Hence the hatched area in the figure is that usually loosely called 'supercritical'. It should be mentioned that, for both pure fluids and mixtures, many of the advantages of a supercritical fluid are possessed by liquids which are just subcritical, and these are used in industrial processes, for example in the extraction of hops. The term 'near critical' is used to describe both situations and is preferred by some people. And again, although SFE is normally carried out by a one-phase fluid, because of possible experimental problems and inconsistent results it is possible that a two-phase extraction may have an advantage in terms of the agitation of the matrix to be extracted.

SFE (and also supercritical fluid chromatography, SFC) take advantage of the fact that a supercritical fluid can have properties intermediate between those of a liquid and a gas and that these properties can be controlled by pressure. Table 1.2 shows some rather approximate typical values of important properties: density (this is related to solvating power) [3], viscosity (related to flow rates) [4] and diffusion coefficients (related to mass transfer within the fluid) [5]. One property advantage for SFE is that solubilities, and particularly the relative solubilities of two compounds, can be controlled via both pressure and temperature, making extraction selective to a limited extent. Other advantages are the relatively easy removal of the solvent and the

**Table 1.2.** The density, $\rho$ [3], and viscosity, $\eta$ [4], of carbon dioxide and the diffusion coefficient for naphthalene in carbon dioxide, $D$ [5], under gas, supercritical and liquid conditions

| | $\rho$ (kg m$^{-3}$) | $\eta$ (μPa s) | $D$ (m$^2$ s$^{-1}$) |
|---|---|---|---|
| Gas (313 K, 1 bar) | 2 | 16 | $5.1 \times 10^{-6}$ |
| Supercritical (313 K, 100 bar) | 632 | 17 | $1.4 \times 10^{-8}$ |
| Liquid (300 K, 500 bar) | 1029 | 133 | $8.7 \times 10^{-9}$ |

facilitation of mass transfer in the extracting fluid owing to the higher diffusion coefficients compared with those of liquids. The disadvantage of using a supercritical fluid is that high-pressure technology is involved. Although SFE and SFC are the two areas where supercritical fluids have been widely exploited, research into the use of these fluids in other areas, such as preparative SFC, chemical reactions, recrystallization and electrochemistry, is proceeding.

## 1.3 Density of a supercritical fluid

A supercritical fluid changes from being gas-like to liquid-like as the pressure is increased, and its thermodynamic properties change in the same way. Close to the critical temperature, this change occurs rapidly over a small pressure range. The most familiar property is the density, and its behaviour is illustrated in Figure 1.3. This shows three density–pressure isotherms, and at

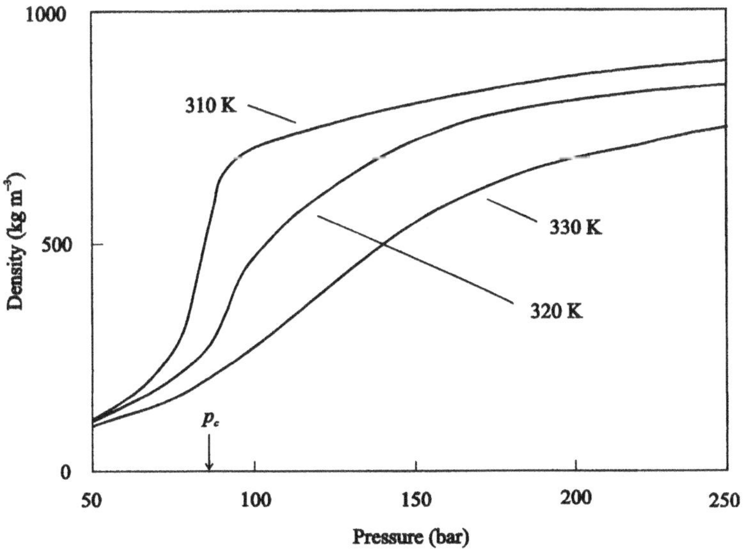


**Figure 1.3** Density–pressure isotherms for carbon dioxide.

the lowest temperature, 6 K above the critical temperature, the density change is seen to increase rapidly at around the critical pressure. As the temperature is raised the change is less dramatic and moves to higher pressures. One consequence is that it is difficult to control the density near the critical temperature and, as many effects are correlated with the density, control of experiments and processes can be difficult. Other properties, such as enthalpy, also show these dramatic changes near the critical temperature.

The behaviour of density, as well as all other thermodynamic functions, as a function of pressure and temperature can be predicted by an equation of state. Some of these have an analytical form, but the most accurate equations are complex numerical forms that have been obtained by intelligent fitting of a wide range of thermodynamic data, such as is carried out at the International Union of Pure and Applied Chemistry Thermodynamic Tables Project Centre at Imperial College in London. They have carried out a study for a number of gases suitable as supercritical fluids and, in particular, for carbon dioxide [3]. A more recent equation of state for carbon dioxide is that published by Span and Wagner [6]. For many other purposes, however, adequate predictions can be made by using a simpler analytical equation. A large number of more complex and realistic equations of state have been proposed and an example of these is now discussed, that of Peng and Robinson [7], which is chosen because of its wide application in the field of supercritical fluids. The Peng–Robinson equation is one of a family of cubic equations of state developed from that of van der Waals, which for a one-component fluid is given by

$$p = \frac{RT}{V-b} - \frac{a}{V^2} \tag{1.1}$$

where $a$ and $b$ are constants known as the van der Waals parameters. The equation is an adaptation of the perfect-gas equation of state in which the volume has been reduced by $b$, the so-called excluded volume, to allow for the physical size of the molecules, and the pressure has been reduced by $a/V^2$, to allow for the effect of attraction between the molecules. Above the critical temperature and outside the liquid + gas region, the equation reproduces qualitatively the pressure–volume–temperature behaviour of a fluid. It is described as a cubic equation of state because, when multiplied throughout by $V^2(V-b)$ to remove both denominators, the equation contains a term in $V^3$.

For the Peng–Robinson equation the second term in the van der Waals equation is modified by making the parameter $a$ a function of temperature and including $b$ in the denominator:

$$p = \frac{RT}{V-b} - \frac{a(T)}{V^2 + 2Vb - b^2} \tag{1.2}$$

By using the fact that at the critical point the first and second derivatives of pressure with respect to volume are zero, the following relationships are obtained, when $a$ and $b$ are calculated from the critical temperature and pressure:

$$a(T_c) = \frac{0.45724R^2T_c^2}{p_c} \quad (1.3)$$

and

$$b = \frac{0.07780RT_c}{p_c} \quad (1.4)$$

By the same method $V_c$, the critical molar volume, is calculated to be $3.9514b$ and thus $Z_c = p_cV_c/RT_c = 0.3074$. This can be compared with experimental values, shown in Table 1.1. It is closer to these values than the theoretical values obtained from most other equations of state, although it is still 11% away from the experimental value for carbon dioxide. Hence the Peng–Robinson equation is used in supercritical studies.

The variation of $a$ with $T$ was obtained by Peng and Robinson by fitting to experimental hydrocarbon vapour pressures and obtaining the relationship

$$a(T) = a(T_c)\{1 + (0.37464 + 1.54226\omega - 0.26992\omega^2)[1 - (T/T_c)^{1/2}]\}^2 \quad (1.5)$$

which introduces the acentric factor, $\omega$, into the equation. Without it, the equation would predict the same vapour pressure curve for all substances in terms of reduced pressure, $p/p_c$, versus reduced temperature, $T/T_c$. This is found to be approximately the case for many substances whose molecules are spherically symmetric and it is also found that their vapour pressure falls to approximately $0.1p_c$ when the temperature falls to $0.7T_c$. For most fluids, especially those with non-spherically symmetric molecules, the vapour pressure falls more rapidly than this. Asymmetric molecules in a liquid rotate more freely as the temperature rises, and for this to happen they must move farther apart on average. When this happens their intermolecular binding energy is reduced and they pass more easily into the gas phase. Thus the vapour pressure will rise more rapidly with temperature for asymmetric molecules than for spherically symmetric molecules. Polar molecules will also lose attractive potential energy as the temperature rises as their orientation becomes more random and this will cause a more rapid change in vapour pressure with temperature. This will be especially true when hydrogen bonding is involved.

To quantify these effects an acentric factor, $\omega$, was defined by Pitzer [8] as

$$\omega = -\log\left[\frac{p(T = 0.7T_c)}{p_c}\right]^{-1} \quad (1.6)$$

Thus for spherically symmetrical molecules, where $p(T = 0.7T_c) \simeq 0.1p_c$, such as xenon, $\omega$ is essentially zero and for methane it is small, at 0.011. Values for some other substances are shown in Table 1.1.

## 1.4 Viscosity and diffusion

At low pressures, below one atmosphere, the (dynamic) viscosity, $\eta$, of a gas is approximately constant, but thereafter rises with pressure in a similar way to density, $\rho$. However, the dependencies of density and viscosity on pressure at constant temperature are not conformal. Of interest therefore is the kinematic viscosity, $\kappa = \eta/\rho$, calculated by my colleagues and me [9], which is illustrated in Figure 1.4. At constant temperature, kinematic viscosity falls from high values at low pressure until the critical density and then rises slightly. As well as illustrating the comparative behaviour of dynamic viscosity and density, the kinematic viscosity is proportional to the pressure drop through a non-turbulent system for a given mass flow rate. For a uniform capillary column of radius $a$, with gas flowing through at a given mass flow rate of $m$, the pressure variation with length $l$ along the column is given by

$$\frac{\mathrm{d}p}{\mathrm{d}l} = -\left(\frac{8m}{\pi a^4}\right)\left(\frac{\eta}{\rho}\right) \tag{1.7}$$

A comprehensive correlation for the viscosity of carbon dioxide has been published [4].

Table 1.2 shows typical values for the density and viscosity of a gas, supercritical fluid and liquid, taking carbon dioxide as an example. Using the example given the viscosity of a supercritical fluid is much closer to that of a gas than that of a liquid. Thus pressure drops through supercritical

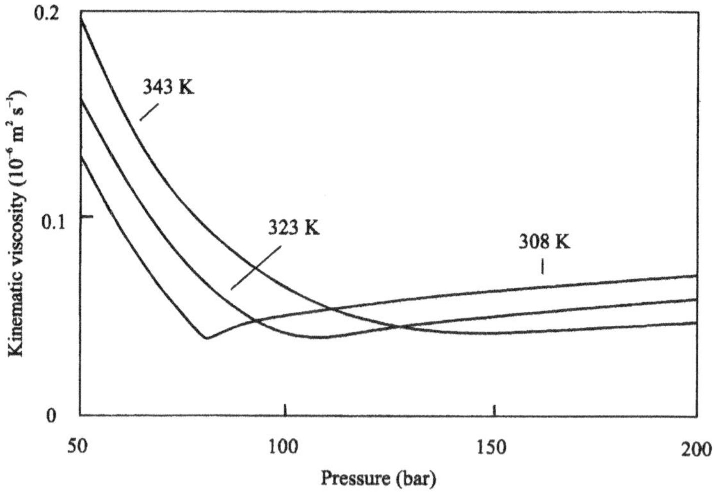


**Figure 1.4** Isotherms for the kinematic viscosity, $\kappa$ (equal to the dynamic viscosity, $\eta$, divided by the density, $\rho$) for carbon dioxide.

extraction apparatus are less than those for the equivalent liquid processes, which is advantageous.

Diffusion coefficients, also shown in Table 1.2, for naphthalene in carbon dioxide, are higher in a supercritical fluid than they are in a liquid. They are approximately inversely related to the fluid density [5]. The advantage shown in Table 1.2 is seen not to be so great and the main diffusional advantage lies in the fact that typical supercritical solvents have smaller molecules than do typical liquid solvents. The diffusion coefficient for naphthalene in a typical liquid would be closer to $1 \times 10^{-9}\ m^2\ s^{-1}$. Thus diffusion coefficients in supercritical fluid experiments and processes are typically an order of magnitude higher than they are in a liquid medium. This has the advantage of faster transport in extraction.

## 1.5 Solubility in a supercritical fluid

The behaviour, at constant temperature, of the solubility of a substance in a supercritical fluid, in terms of mole fraction, is illustrated schematically in Figure 1.5. When the pressure is close to zero only the solute is present as vapour and the mole fraction of solute is unity. There is then an initial fall almost to zero at very low pressures as the solvent is added and the solute is diluted without being much solvated. After staying close to zero there is then a rise in solubility at around the critical density of the fluid, that is, when the density is rising rapidly with pressure. This rise is due to solvation arising from attractive forces between the solvent and solute molecules. Thereafter the solubility may exhibit a fall, represented by the dashed line. If this occurs, it is because at higher pressures the system is becoming

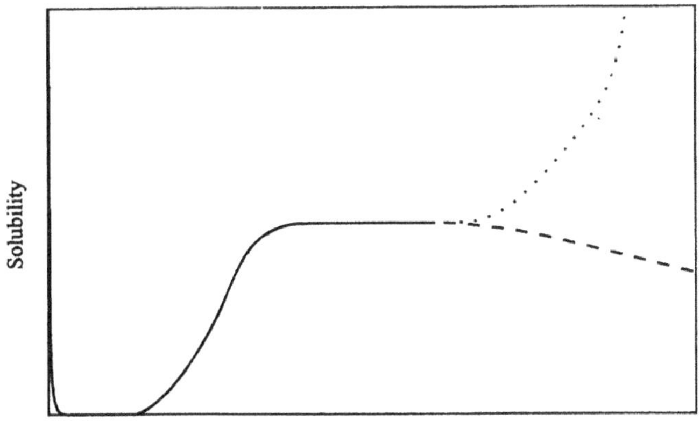


**Figure 1.5** A schematic illustration of the behaviour of solubility in a supercritical fluid. A description of the curves is given in the text, section 1.5.

compressed and repulsive solute–solvent interactions are important. The solute can be said to be 'squeezed out' of the solvent. Alternatively, a rise may occur, as represented by the dotted line. This happens if there is a critical line present at high pressures at the temperature of the isotherm and the solubility will rise towards it. The rising type of curve is a feature of smaller more volatile molecules and higher temperatures and *vice versa*. All situations between the two curves occur.

Correlation of supercritical fluid solubility data is not straightforward. All the features shown in Figure 1.5 can be reproduced qualitatively by any equation of state. For quantitative fitting more refined equations of state are useful in certain regions, and of these the Peng–Robinson equation has been the most widely used. However, even this equation is not successful in fitting all the data at all pressures and temperatures. A further problem is that the parameters necessary for using the equation of state, such as the critical temperature and pressure of the solute and its vapour pressure and acentric factor, are not always available. This problem has been discussed by Johnston *et al.* [10]. They came to the conclusion that a cruder empirical correlation with density is the best available route for most compounds.

## 1.6 Factors affecting supercritical fluid extraction

Extraction by a supercritical (or any) fluid is never complete in finite time but can be considered to be successful in a given time, for analytical extractions, on the basis of the accuracy required. SFE is relatively rapid initially, but there then follows a long tail in the curve of percentage extracted versus time. In a typical situation 50% is extracted in 10 minutes, but it may be 100 minutes before 99% is extracted. It is not correct, therefore, to assume that extraction is completed if it has been carried out for two consecutive equal periods of time and the second period produces only a tenth of the compound extracted in the first period. It is necessary for every application to carry out an experimental long extraction and study the results by the methods given below.

The process of extraction can be considered to involve the three factors shown in the following SFE triangle:

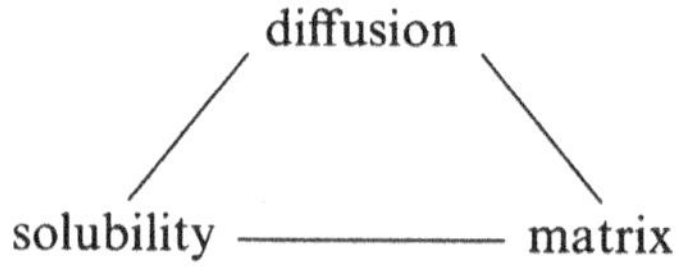


First, the solute must be sufficiently soluble in the supercritical fluid. If this is not the case it will be revealed by interpretation of the kinetic recovery curve, as will be shown below. If solubility is insufficient the situation may be improved by adding a modifier to the fluid, as described earlier (section 1.2).

Second, the solute must be transported sufficiently rapidly, by diffusion or otherwise, from the interior of the matrix in which it is contained. The latter 'diffusion' process may be normal diffusion of the solute or it may involve diffusion in the fluid through pores in the matrix. The time-scale for diffusion will depend on the diffusion coefficient and the shape and dimensions of the matrix or matrix particles. Of these the shortest dimension is of great importance, as the times depend on the square of its value. Values for this quantity of 1 mm or preferably less are usually necessary.

Third, the analyte must be released by the matrix. This last process may involve desorption from a matrix site, passage through a cell wall or escape from a cage formed by polymer chains. It can be slow and in some cases it appears that part of the substance being extracted is locked into the structure of the matrix. An example is the SFE of additives and lower oligomers from polymers, which can give much lower results than obtained by dissolving the polymer in a solvent, or using liquid extraction at higher temperatures, which swells the polymer to a greater extent. Thus SFE will not always give the total amount of a compound in a sample, only the amount 'extractable' under particular SFE conditions. It may be that the latter is of interest, for example if one is concerned with migration of additives from polymers into foodstuffs, but if the total amounts are required SFE may not be applicable in some cases. Preliminary experiments and comparisons with other methods are necessary. The process can be strongly temperature-dependent and thus higher temperatures may improve the situation. The addition of modifiers may often reduce the matrix effect; in fact modifiers are often more important in this respect than in enhancing solubility. The mechanism is thought to involve interactions with surfaces.

Another problem in SFE is the presence of water. Water is not very soluble in carbon dioxide and it can 'mask' the analytes to be recovered. The rate of extraction may sometimes be equal to the rate of water removal. Addition of diatomaceous earth, anhydrous magnesium sulphate or another drying agent to the sample matrix may help. Modifiers such as methanol which improve water solubility are another solution.

The initial step in the SFE process will be the entry of fluid material into the matrix. This may be the ingress of fluid into the pores of a plant matrix or between soil particles. The miscibility of nitrogen and oxygen with carbon dioxide under pressure means that penetration is rapid. Another situation is the absorption of the fluid into a polymer, which causes swelling and consequently enhances extraction. An example where this is revealed to be the case is given below. This first step of fluid entry is not thought to be a rate-determining step in SFE.

Figure 1.6 shows examples of the types of curves of recovery versus time that can be obtained in SFE. Curve (a) is a typical curve obtained when the process is controlled by diffusion. When matrix effects are significant the results may have the form of curve (b). Curve (c) is an example of

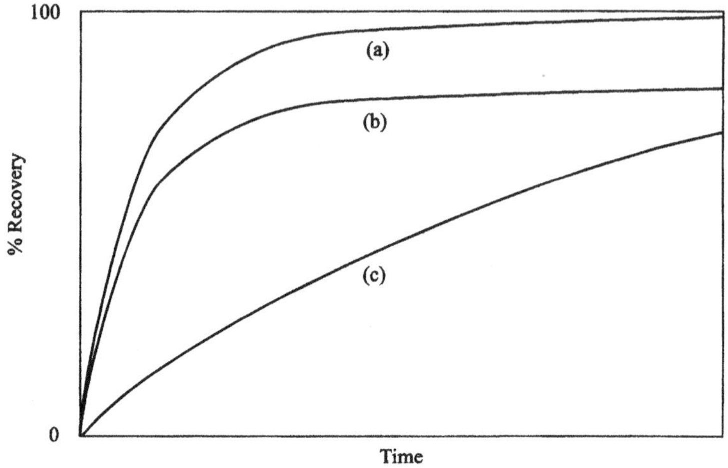


**Figure 1.6** Examples of recovery curves: (a) a typical diffusion-controlled curve; (b) a curve showing significant matrix effects; (c) a curve of a poorly soluble analyte.

recovery behaviour when the extracted analyte is not very soluble in the extracted fluid.

## 1.7 Modelling of supercritical fluid extraction

A series of models developed by my colleagues and me have been used for interpreting the results of SFE on a small scale [11–15]. Four steps are considered in these models:

1. rapid fluid entry into the matrix;
2. a reversible release process such as desorption from matrix sites or penetration of a biological membrane;
3. transport, by diffusion or otherwise, to the edge of a matrix particle;
4. removal by solvation in the fluid.

Figure 1.7 illustrates steps 2–4 in the process. Step 1 is considered to be too fast to affect the kinetics of recovery significantly. In the next two sections, a model is described in which steps 2 and 4 are also considered fast, and so transport out of the matrix is the rate determining step. This will occur when there are no significant matrix effects and the solubility of the extracted substance is very high. In later sections situations are considered where solubility and later matrix effects are involved. These various situations are initially explained by avoiding much of the inherent mathematics. Fuller descriptions of the derivation of the relevant equations are given at the end of the chapter.

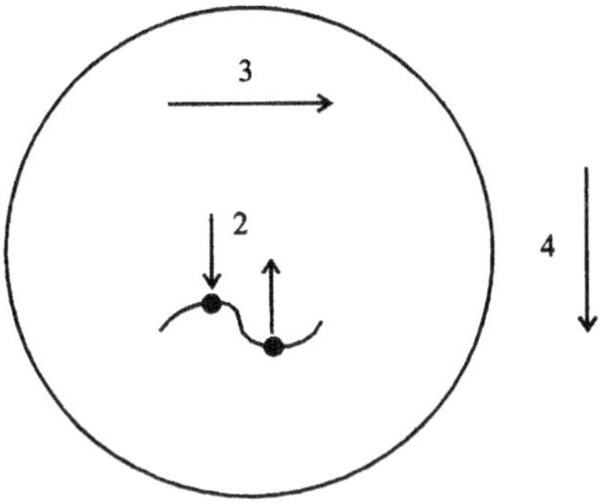


**Figure 1.7** Steps in the supercritical fluid extraction process: 1. rapid fluid entry into the matrix (not shown); 2. a reversible release process such as desorption from matrix sites or penetration of a biological membrane; 3. transport, by diffusion or otherwise, to the edge of a matrix particle; 4. removal by solvation in the fluid.

## 1.8 Continuous dynamic supercritical fluid extraction controlled by diffusion

We now consider the extraction of a matrix in a continuous flow of fluid, which is fast enough for the concentration of a particular solute to be well below its solvation limit and where there are no matrix effects. The rate-determining process is therefore the rate of transport out of the matrix. Most practical examples of extraction are complex, but it is found that simple models can account for the main behavioural features and lead to methods of treatment for the results of SFE. For these simple theoretical models, we assume an effective diffusion coefficient, $D$, and a particular geometry for the matrix and solve the appropriate differential equation (the Fourier equation) with assumed boundary conditions. The latter are that the compound is initially uniformly distributed within the matrix and that as soon as extraction begins the concentration of compound at the matrix surfaces is zero (corresponding to no solubility limitation). The solutions of the Fourier equation for various geometries are given by Carslaw and Jaeger [16], in the context of heat conduction (where the same equation applies) and also by Crank [17], who has translated Carslaw's equations into diffusion notation. Two simple geometries will be discussed here: those of a sphere, which will be applied to extraction of spherical particles as well as irregularly shaped powdered particles; and those of a slab with two infinite dimensions, which will be applied to pieces of thin film.

The solution for a sphere, described as the hot-ball model because of the analogy of the mathematical solutions with those for a hot spherical object being dropped into cold water, is explained in more detail elsewhere [11]. If the mass of solute in the matrix is $m_0$ initially and $m$ after a given time, a plot of $\ln(m/m_0)$ versus time has the form given by Figure 1.8. It is characterized by a relatively rapid fall onto a linear portion, corresponding to an

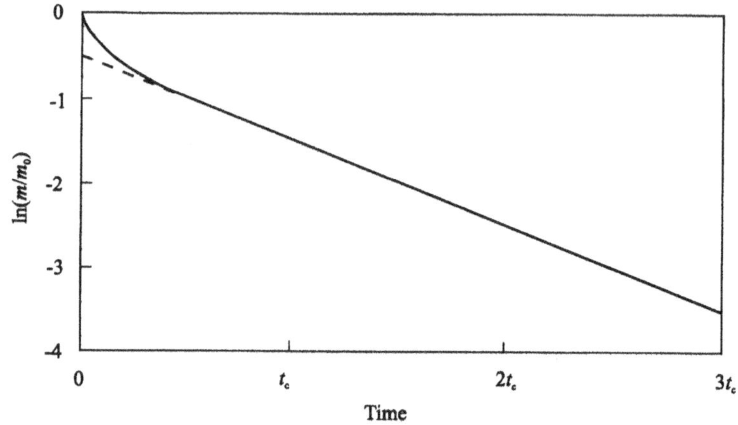


**Figure 1.8** Theoretical curve for the dynamic supercritical fluid extraction of a sphere, where extraction is controlled by diffusion. $m$ = mass of solute in the matrix; $m_0$ = initial mass of solute in the matrix; $t_c$ = characteristic time.

exponential 'tail'. The physical explanation of the form of the curve is that the initial portion is extraction, principally out of the outer parts of the sphere, which establishes a smooth concentration profile across each particle, peaking at the centre and falling to zero at the surface. When this has happened, the extraction becomes an exponential decay.

The curve is characterized by two parameters: a characteristic time, $t_c$, and the intercept of the linear portion, $-I$, which has the value $-0.5$ (actually $-0.4977$) for the sphere. The slope of the linear portion is $-1/t_c$ and the linear portion begins at approximately $0.5t_c$; $t_c$ is theoretically related to the effective diffusion coefficient out of the matrix, $D$, and the radius of the sphere, $a$, by the equation

$$t_c = \frac{a^2}{\pi^2 D} \tag{1.8}$$

The value of the effective diffusion coefficient will usually not be known, although its order of magnitude may be commented on. Most measurements published for $D$ are for true diffusion and for small molecules in relatively mobile solvents, as described by Tyrrell and Harris [18], and $D$ is of the order of $10^{-9}\,m^2\,s^{-1}$. For systems of interest to SFE, $D$ will be between one order (for oils) and four orders (for solids) of magnitude below this value. For example, values for various solutes in polymers have been given which are of the order of $10^{-11}$ and $10^{-12}$. Equation (1.8) shows a squared dependence on $a$ and rationalizes the commonsense rule that for rapid extraction matrix particles must be small. This may be achieved for solids by crushing or grinding and for liquids by coating on a finely divided substrate or spraying or mechanical agitation. For solid matrix particles with a value of $a$ of the order of 0.1 mm, typical values of $t_c$ are between 10 and 100 minutes.

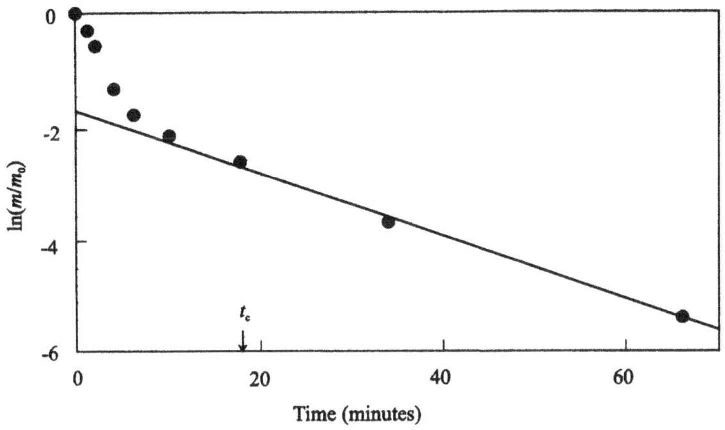


**Figure 1.9** Continuous extraction of 1,8-cineole from crushed, dried rosemary with $CO_2$ at 50°C and 400 bar. $m$ = mass of solute in the matrix; $m_0$ = initial mass of solute in the matrix; $t_c$ = characteristic time.

Figure 1.9 shows some experimental results for the extraction of 1,8-cineole from crushed rosemary at 50°C using $CO_2$ at 400 bar. Extraction was continued almost until exhaustion to allow the calculation of values of $m$ and $m_0$. Similar curves are obtained for the extraction of five other major compounds from rosemary (α-pinene, camphor, camphene, borneol and bornyl acetate) and also for several other types of system [11]. The experimental results are consistent with the theoretical curve in that the points are close to a straight line after a time of approximately $0.5t_c$; $t_c$ has a value of about 18 minutes in this case, which is obtained from the slope of the straight-line portion (it is the time taken for the line to fall one log unit). However, the curve differs from the theoretical curve of Figure 1.8 in two respects. First, the intercept, $I$, is greater, and this is discussed in the next paragraph. Second, the curve does not fall as steeply from zero, and this is thought to be a result of the effect of solubility limitation, which is discussed in section 1.8.

In general, the value of $I$ depends on the particle shape and size distribution (in particular the surface-to-volume ratio for shape) and also the distribution of solute within the matrix particles (i.e. whether the solute is primarily located near the surface or in the interior of the particle). For a model system of spheres of the same size, with uniform solute concentration, it is 0.5. For real systems values of $c.$ 2 are common and prediction of the values is not really possible. Thus usually values of $t_c$ and $I$ can only be obtained by experiment. A small-scale dynamic extraction followed by the application of an appropriate analytical technique is therefore an important preliminary study in designing a routine quantitative analytical procedure.

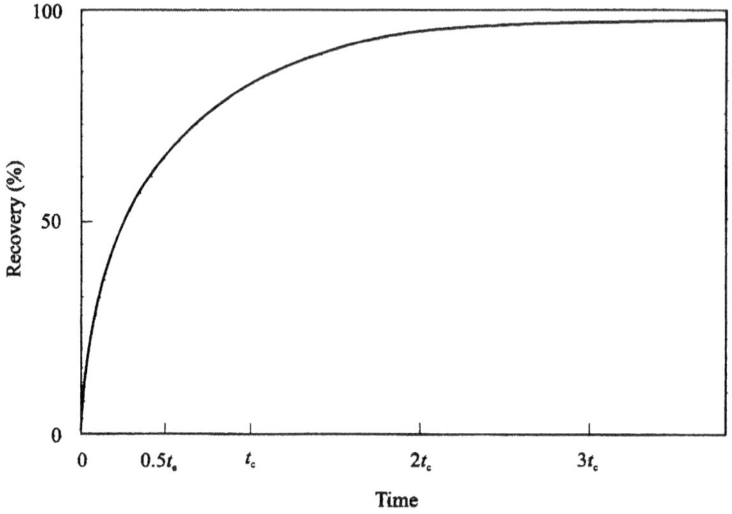


**Figure 1.10** Theoretical recovery curve for the dynamic supercritical fluid extraction of a sphere, where extraction is controlled by diffusion. $t_c$ = characteristic time.

The information in Figures 1.8 and 1.9 can also be given in terms of percentage extraction versus time, and this is shown in Figure 1.10 for a sphere. As can be seen the majority is extracted in a time of $0.5t_c$ (63%). Another 14% is extracted in the next period of $0.5t_c$ and thereafter there is a long tail and it is $4.8t_c$ before 99% is extracted.

Although the spherical model is adaptable to the irregular geometry of matrix particles, for extraction from a thin film of well-defined geometry a separate, though similar, study of a suitable model is desirable. In this case our model would be that of an infinite slab of thickness $L$, on the basis that the surface dimensions of the film are far larger than this thickness. It is then necessary, as before, to solve the diffusion equation for the system with appropriate boundary conditions, and the appropriate solutions are again given by Carslaw and Jaeger [16]. Adaptation of the published solutions leads to the curve of $\ln(m/m_0)$ versus time shown in Figure 1.11. The curve is similar to that for a sphere, with the curve falling more steeply initially, and later becoming approximately linear, with a slope of $-1/t_c$, where, in the case of an infinite slab,

$$t_c = \frac{L^2}{\pi^2 D} \tag{1.9}$$

However, it falls more rapidly onto the straight portion than does the equivalent curve for a sphere, i.e. after a time of approximately $0.25t_c$. Extrapolation of the linear portion of the curve to the $t = 0$ axis gives an intercept of $-0.2100$, i.e. $I = 0.2100$, compared with a value for the sphere of 0.4977.

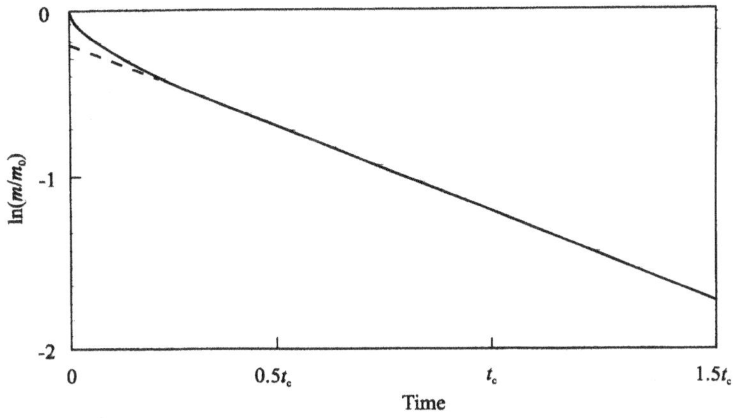


**Figure 1.11** Theoretical curve for the dynamic supercritical fluid extraction of a film, where extraction is controlled by diffusion. $m$ = mass of solute in the matrix; $m_0$ = initial mass of solute in the matrix; $t_c$ = characteristic time.

Qualitatively, the theoretical curve of percentage extracted versus time for an infinite slab is similar to that for a sphere and exhibits the same long tail. Some 37% of the material is extracted during an initial period of $0.25t_c$. The time required to extract 99% of the material, however, is $4.4t_c$, i.e. 17 times the time needed to extract the first 37%.

Figure 1.12 shows some experimental results for extraction from polymer film [12]. The sample was a film of poly(ethylene terephthalate) (PET), 1.2 mm in thickness; extraction was carried out at 70°C with $CO_2$ at 400 bar and results shown for the extraction of the cyclic trimer of ethylene

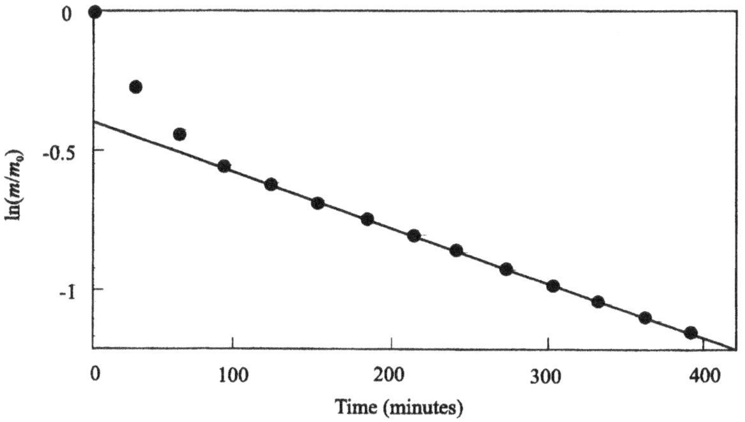


**Figure 1.12** Continuous dynamic extraction of the cyclic trimer from poly(ethylene terephthalate) with $CO_2$ at 70°C and 400 bar. $m$ = mass of solute in the matrix; $m_0$ = initial mass of solute in the matrix.

terephthalate. Figure 1.12 is a curve of the form of Figure 1.11 with a steeper portion falling onto a straight line after approximately 125 minutes. The slope of the straight-line portion gives the result that $t_c = 506$ minutes. Thus the straight line appears to set in at $0.25t_c$, in agreement with the model. However, the value of $I$, at 0.39, is above the theoretical value of 0.21 (similar to what was obtained in the studies using the spherical model). Here the geometry is well known and another explanation must be sought. A plausible explanation in this case is that a higher proportion of the oligomer near the surface is extractable under the conditions used. (It should be mentioned that the amount of cyclic trimer extractable under these conditions is considerably below that obtained by more rigorous extraction methods: an example of the existence of 'extractable' and 'non-extractable' material in SFE.) From the slope obtained from Figure 1.12 and the thickness of the film, a value for the diffusion coefficient of the cyclic trimer in PET at 70°C can be obtained from the results to be $2.1 \times 10^{-13}\,m^2\,s^{-1}$. No literature value is available, but the result has the correct order of magnitude, by comparison with other diffusion coefficients in polymers quoted by Mills [19].

In the case of the spherical model, the occurrence of an intercept below that of the theoretical value indicates either non-uniform distribution of extractable compound or irregular particle shape. In the case of extraction from a film of known geometry, the latter is the only possibility, and so in this case it is worthwhile to investigate the effect of non-uniform distribution on the theoretical results. A model distribution is required for such an investigation, and one of the simplest available for this purpose is an exponential fall-off in concentration from each surface. This is of the form $c_0 \exp(-x/\alpha)$, where $c_0$ is the concentration at the surface, $x$ the perpendicular distance in from the surface and $\alpha$ a distance parameter giving the characteristic distance of the exponential fall-off. Figure 1.13 shows the concentration profile schematically. The detailed equations have been published [12] and are given

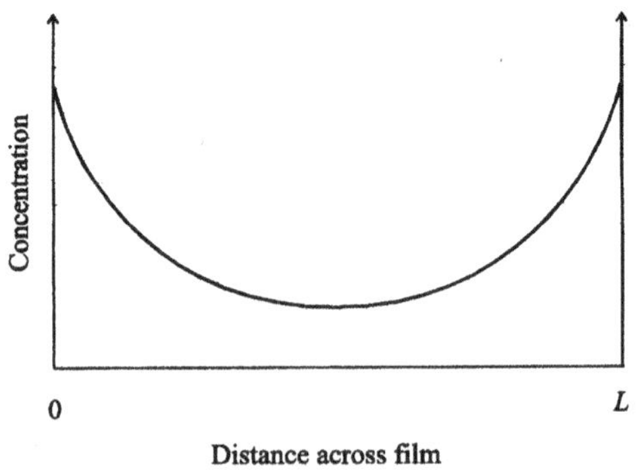


**Figure 1.13** Model concentration profile across a film.

**Table 1.3** Values for the intercept, $I$, for extractions from a film with a non-uniform initial solute distribution for various values of the ratio of the distance parameter for the distribution, $\alpha$, and the thickness of the film, $L$

| $\alpha/L$ | $I$ |
|---|---|
| $\infty$ | 0.2100 |
| 1 | 0.2277 |
| 0.5 | 0.2779 |
| 0.3 | 0.3820 |
| 0.1 | 1.0103 |
| 0.05 | 1.6338 |
| 0.01 | 3.2199 |
| 0.005 | 3.9120 |
| 0.001 | 5.5215 |

briefly at the end of the chapter; here it is sufficient to assert that the solutions are of the general form of Figure 1.11, but with the intercept becoming lower as $\alpha$ becomes smaller, that is, as the concentration falls off more rapidly from the surface. Table 1.3 give the values of $I$ expected for various values of $\alpha/L$. The value of 0.39 obtained with the results of Figure 1.12 is seen to correspond to a value for $\alpha/L$ of about 0.3, indicating that the concentration of extractable analyte has fallen to about 20% of its surface value in the centre of the film. Of course, the precise profile in the experimental film does not have to be of precisely the exponential form, but the analysis indicates the extent of the predominance of extractable compound near the surface. It may be worth repeating that the total cyclic trimer is probably uniformly distributed during manufacture, and the intercept value is indicating only that the compound near the surface is more extractable.

## 1.9 Continuous dynamic supercritical fluid extraction controlled by both diffusion and solvation

Of the four steps in SFE (sections 1.6 and 1.7) steps 3 and 4 are now both considered to be rate determining [13]. So far it has been assumed that the solubility of the solute in the supercritical fluid is essentially infinite and transport out of the matrix has controlled the rate of extraction. In this section extraction out of a sphere of radius $a$ is assumed to be controlled by two effects: transport through the sphere by diffusion or otherwise; and partition between the sphere and the fluid at its surface. As before, transport will be quantified by an effective diffusion coefficient, $D$. Partition is important at the surface of a sphere and is quantified by the partition coefficient, defined as a ratio of concentration, $K$, of the solute between the supercritical fluid and the material of the sphere. The appropriate equations are obtained by solving the diffusion equation within a sphere, subject to the boundary condition at its surface determined by partition and flux at the surface as described in more detail in section 1.11. The important parameters in determining this boundary condition and the relative importance of the two rate-determining steps are $a$, $D$ and $K$, as previously defined, and also $F$, the volume rate of flow of the fluid, and $A$, the surface area of all of the spheres. It is convenient to define a combined parameter, $h$, which is defined by $h = FK/AD$.

The larger the value of $ha$, the more important transport is in determining the rate of extraction, whilst for smaller values of $ha$ solvation in the fluid and removal by the fluid flow becomes more rate-determining. Adaptation of the appropriate solutions for heat conduction equations [16] gives, after some manipulation, equations for $\ln(m/m_0)$ as a function of time [$t_c$, given by equation (1.8)], which are plotted in Figure 1.14 for various values of $ha$.

When $ha$ is large, this is because $K$ is large and $D$ is small according to its definition. Diffusion is then the slow and important step and this is shown in the lowest curve in Figure 1.14. This curve is identical to that shown in Figure 1.8. As $ha$ decreases, both the slope and the intercept of the straight-line

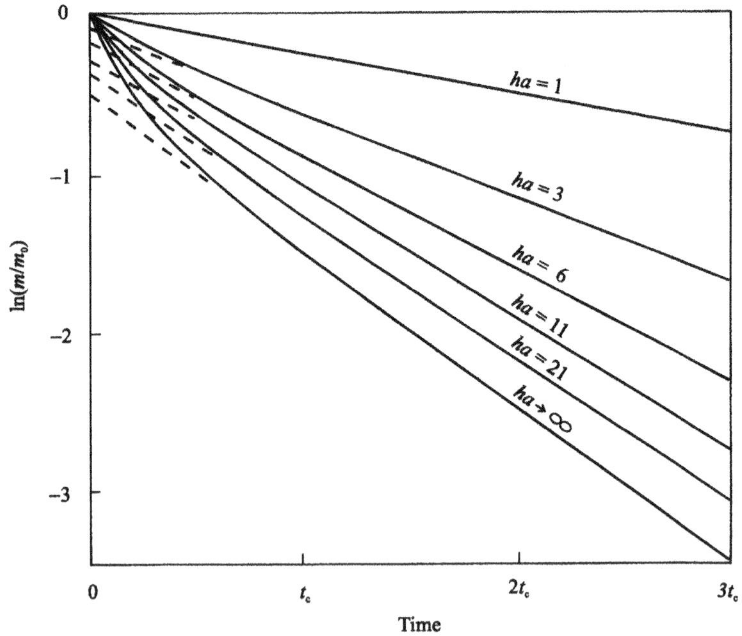


**Figure 1.14.** Theoretical curves for supercritical fluid extraction of a sphere, including solvation effects, for different values of the parameter $ha$. $m$ = mass of solute in the matrix; $m_0$ = initial mass of solute in the matrix; $t_c$ = characteristic time. $h = FK/AD$; $F$ = volume rate of flow in the liquid; $K$ = ratio of concentration of the solute between the supercritical fluid and the sphere; $A$ = surface area of all the spheres; $D$ = effective diffusion coefficient; $a$ = radius of the sphere.

**Table 1.4** Parameters for the spherical model, including both transport and solvation

| *ha* | $\lambda_1$ | *I* | Recovery after $0.3t_c$ (%) |
|---|---|---|---|
| $\infty$ | 3.1416 | 0.4997 | 63 |
| 21 | 2.9930 | 0.3731 | 57 |
| 11 | 2.8628 | 0.2884 | 52 |
| 6 | 2.6537 | 0.1887 | 46 |
| 3 | 2.2889 | 0.0866 | 36 |
| 1 | 1.5708 | 0.0146 | 23 |

Note: $h = FK/AD$; $F$ = volume rate of flow of the fluid; $K$ = ratio of concentration of the solute between the supercritical fluid and the material of the sphere; $A$ = surface area of all the spheres; $D$ = effective diffusion coefficient; $a$ = radius of sphere; $I$ = intercept of the linear portion of the graph of $\ln(m/m_0)$ versus time; $m$ = mass of solute in the matrix; $m_0$ = intial mass of solute in the matrix; for $\lambda_1$ refer to text, section 1.12.4.

portion of the curve decrease. Values of the intercept, showing this more quantitatively, are given in Table 1.4. When *ha* is very small, corresponding to poor partition into the fluid and rapid diffusion, SFE behaves exponentially and the plot of $\ln(m/m_0)$ versus time becomes a straight line. The curve for $ha = 1$ can be seen to be close to this condition. For $ha \rightarrow 0$ the curve is given by

$$\frac{m}{m_0} = \exp\left\{-\left(\frac{KF}{V}\right)t\right\} \tag{1.10}$$

where $V$ is the volume of the matrix. In this situation, only partition is important in controlling extraction, which is first order, with the rate coefficient being determined by the product of the partition coefficient and the ratio of the volume flow rate of the fluid to the volume of the matrix being extracted.

The intermediate situation is illustrated in Figure 1.15, which shows how the concentration profile changes during extraction. Initially [Figure 1.15(a)], it is constant across the sphere. Passage to the profile shown in Figure 1.15(b) corresponds to the non-linear portion of the curves in Figure 1.14. Once this profile is established, it reduces in size but maintains the same shape, as shown in Figure 1.15(c), during the final exponential decay. If *ha* is large, the vertical portion of the profiles in parts (b) and (c) are very small and the non-exponential part of the extraction curve is more important. If *ha* is small, the curved portion of the profiles in parts (b) and (c) are very flat and the whole extraction curve is exponential.

Plots of percentage recovery versus time, drawn from the same equations, are shown in Figure 1.16 for various values of *ha*. For $ha = 1$, representing limitation by partitioning into the fluid, a slow recovery of exponential form is obtained. As *ha* is increased, the rate of recovery rises and the form changes to that of diffusion control, similar to that shown in Figure

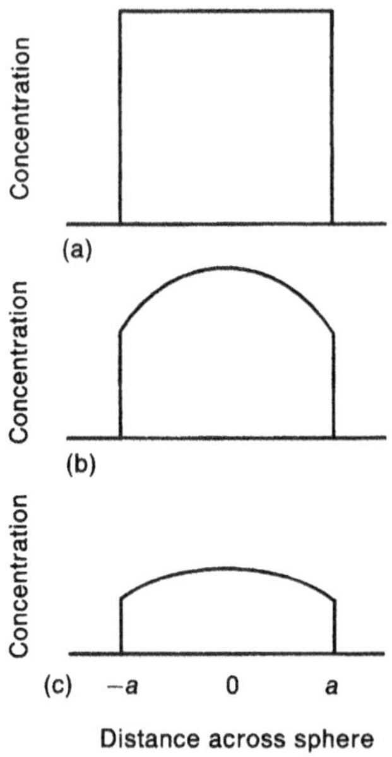


**Figure 1.15** Concentration profiles across a sphere of radius $a$ during supercritical fluid extraction involving transport and solvation effects. Parts (a)–(c) are described in the text, section 1.9.

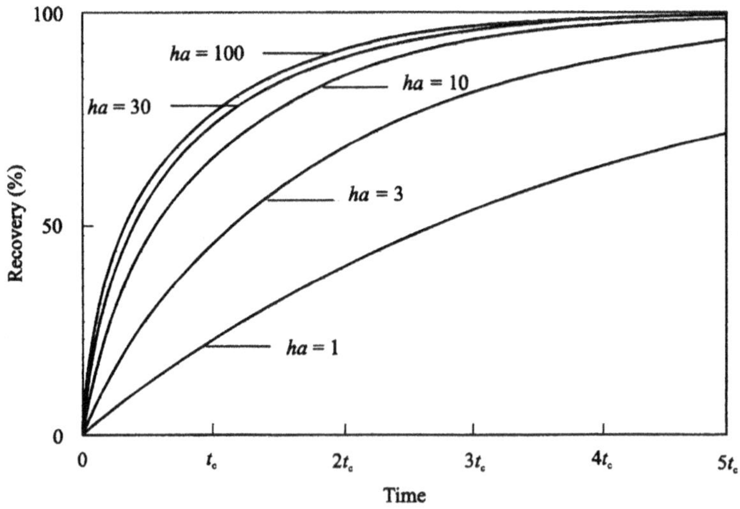


**Figure 1.16** Plots of the percentage recovery during supercritical fluid extraction of a sphere as a function of time for different values of $ha$. $h = FK/AD$; $F$ = volume rate of flow in the liquid; $K$ = ratio of concentration of the solute between the supercritical fluid and the sphere; $A$ = surface area of all the spheres; $D$ = effective diffusion coefficient; $a$ = radius of the sphere; $t_c$ = characteristic time.

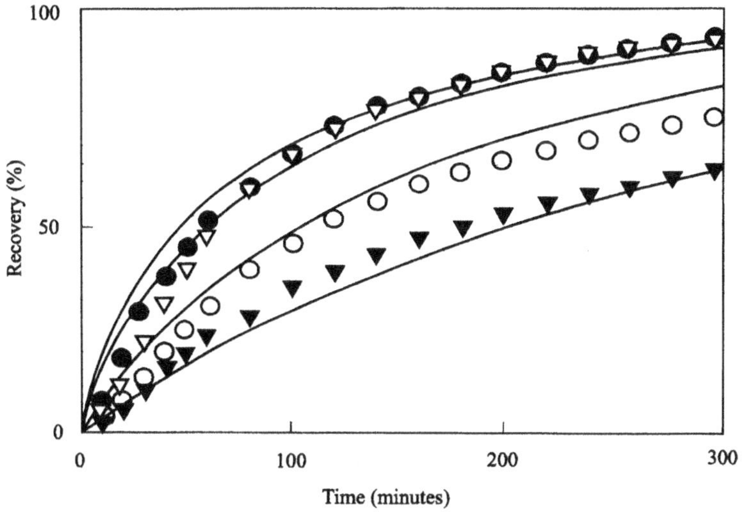


**Figure 1.17** Comparisons of experimental data and model predictions (continuous lines) for supercritical fluid extraction of *m*-xylene and *p*-xylene from polystyrene beads at various flow rates: ▼ = 0.1 ml min$^{-1}$; ○ = 0.25 ml min$^{-1}$; ▽ = 0.70 ml min$^{-1}$; ● = 1.25 ml min$^{-1}$.

1.10. However, raising *ha*, by increasing solubility or flow rate, has diminishing returns, because when diffusion control takes over, increases in *ha* have little effect. Thus the curves for $ha = 30$ and $ha = 100$ are very similar. The curves are plotted versus time in terms of $t_c$ and the relationship to real time is given by the parameter $D/a^2$ using equation (1.8). Thus, if experimental data are fitted to the theoretical curves, the two parameters *ha* and $D/a^2$ are obtained.

If the flow rate is varied at constant pressure and temperature for SFE from a polymer, $D/a^2$ is expected to remain constant whereas *ha* is expected to rise in proportion to the volume flow rate, *F*. Data for the SFE of the combined amounts of *m*-xylene and *p*-xylene from polystyrene beads, varying in size from 0.18 mm to 2.0 mm diameter, for various flow rates [14] were fitted to the appropriate equations; the comparison is shown in Figure 1.17. (The flow rates were measured as liquid $CO_2$ at the pump but will be proportional to the fluid flow rate in the extraction cell.) For all the theoretical curves values of $D/a^2 = 0.0009$ and $ha = 16$ ($F$ ml$^{-1}$ min$^{-1}$) were used. Thus only two parameters were used to fit the curves, and there is qualitative agreement, bearing in mind that the sample did not consist of spheres of uniform size as strictly required by the theory.

If the pressure is varied at constant flow rate and temperature, both $D/a^2$ and *ha* are expected to change. Thus the recovery curves must be fitted for individual pressures and this has been done for the extraction of Irgafos 168 (tris-(2,4-di-*tert*-butyl) phosphite) from polypropylene at various pressures (Figure 1.18). The particles were irregular spheres of diameter

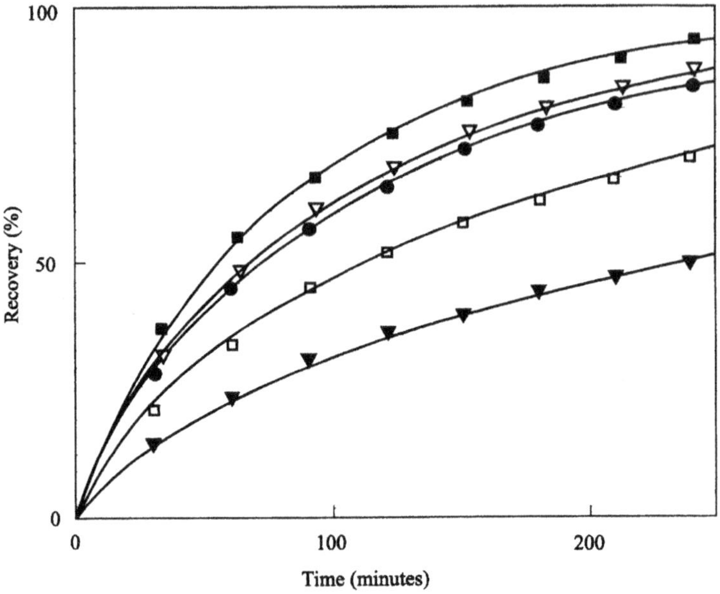


**Figure 1.18** Comparisons of experimental data and model predictions (continuous lines) for supercritical fluid extraction of Irgafos 168 [tris-(2,4-di-*tert*-butyl) phosphite] from polypropylene at various pressures: ▼ = 75 bar; □ = 105 bar; ● = 175 bar; ▽ = 200 bar; ■ = 400 bar.

$0.8 \pm 0.2$ mm and extraction was carried out at 45°C with pure $CO_2$ at a flow rate of 7 ml $s^{-1}$, measured with a bubble flow meter at 20°C and 1 bar [14]. Fitting is now much better and the parameters obtained from the fitting are given in Table 1.5. The values of *ha* are also shown in Figure 1.19, plotted against pressure, and can be seen to have the same form as a solubility curve (Figure 1.5). This is to be expected as $K$ is proportional to solubility as will be $h$. The values of $D/a^2$ in Table 1.5 also rise with pressure and this is explained by the higher absorption of the supercritical fluid substance at

**Table 1.5** Values of the parameters obtained by fitting the data shown in Figure 1.18

| Pressure (bar) | $D/a^2 \times 10^5$ ($s^{-1}$) | *ha* |
|---|---|---|
| 75 | 21 | 3.2 |
| 105 | 48 | 5.8 |
| 175 | 90 | 7.3 |
| 200 | 100 | 8.1 |
| 400 | 160 | 8.2 |

Note: $h = FK/AD$; $F$ = volume rate of flow in the liquid; $K$ = ratio of concentration of the fluid between the supercritical fluid and the sphere; $A$ = surface area of all the spheres; $D$ = effective diffusion coefficient; $a$ = radius of sphere.

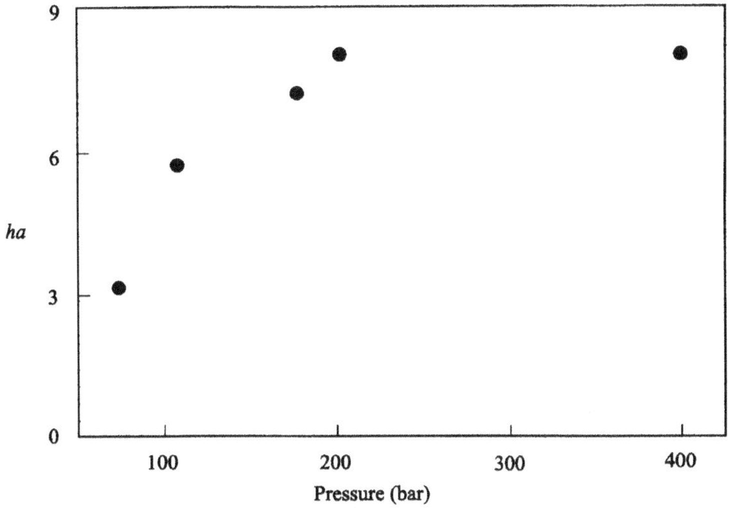


**Figure 1.19** Values of the parameter *ha* obtained by analysis of the data in Figure 1.18. $h = FK/AD$; $F$ = volume rate of flow in the liquid; $K$ = ratio of concentration of the solute between the supercritical fluid and the sphere; $A$ = surface area of all the spheres; $D$ = effective diffusion coefficient; $a$ = radius of the sphere.

higher pressures, causing the polymer to swell, raising the diffusion coefficient. Thus, with polymers, increasing the pressure can be beneficial to SFE, even above pressures where the solubility is no longer rising.

The effect of pressure on SFE, because of its influence on solubility, is well known. It is most obvious if extractions are carried out for a particular time. Table 1.4 gives the percentage recovery, predicted by the model for a period of $0.3t_c$, for various values of *ha*, which is proportional to solubility. Although the relationship is by no means linear, there is a correlation between *ha* and therefore solubility with the amount extracted. Figure 1.20 shows the solubility of atrazine, predicted by the Peng–Robinson equation of state, as a function of pressure, and the experimental percentage recovery of atrazine from soil, also as a function of pressure [20]. The SFE was carried out at 80°C for 15 min using pure $CO_2$ at a constant flow rate of $5\,ml\,s^{-1}$ measured with a bubble flow meter at 20°C and 1 bar. This is an example of a so-called pressure threshold curve for SFE.

## 1.10 Continuous dynamic supercritical fluid extraction controlled by diffusion, solvation and matrix effects

Of the three factors which are thought to control SFE, that of matrix effects is the least well understood. Although matrix effects in SFE are inherently complex and many effects may be invoked, it is useful to compare the predictions

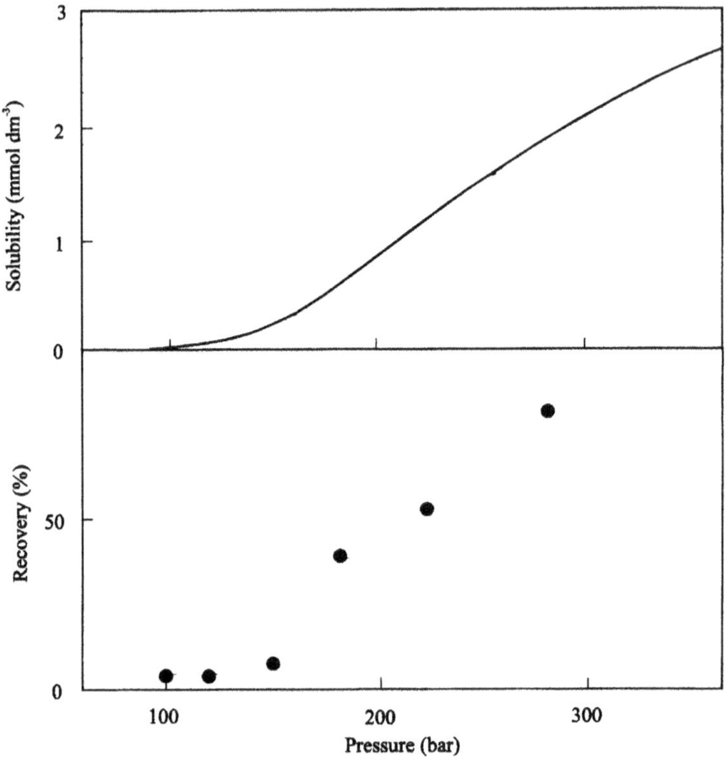


**Figure 1.20** Percentage recovery of atrazine from soil by supercritical fluid extraction with $CO_2$ at different pressures after 15 minutes at 80°C and constant flow rate, compared with predicted solubility at the same temperature.

of a relatively simple model with experiment and demonstrate which features of SFE the model will predict and which other features it will not explain. The model can then be used as a basis for further development.

The outstanding feature of matrix effects in SFE is that in some experiments although extraction is carried out until very little further solute is emerging and the extraction appears to be complete not all the solute has been removed. This can be seen by comparison of yields with extraction by liquids or by SFE using other fluids or higher temperatures. The matrix thus appears to be preventing the release of some of the solute. The extractions, which appear to be approaching a final recovery of less than 100%, are still, in fact, slowly rising, although this is not always observed because the amounts being extracted at later times are below detection limits. This can be demonstrated by carrying out extractions for an abnormally long period. Some of the results for the extraction of polyaromatic hydrocarbons from contaminated soil are shown in Figure 1.21. SFE was carried out with pure $CO_2$ at 55°C and a flow rate of 0.9 ml min$^{-1}$ [15]. The figure of 100%

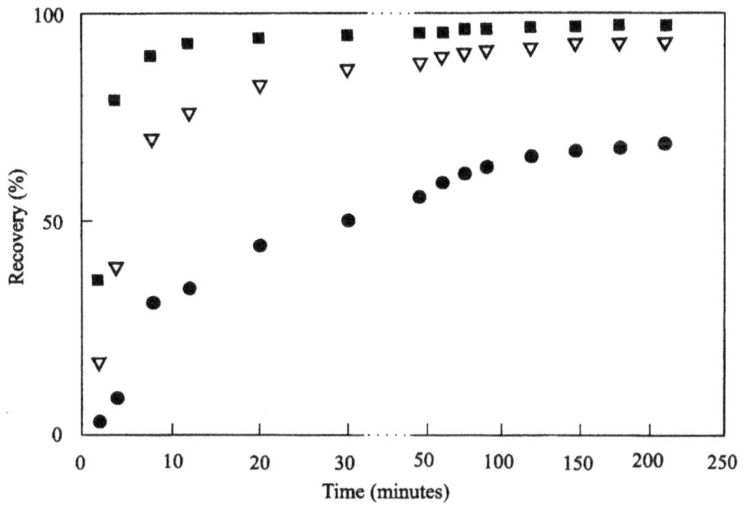


**Figure 1.21** Supercritical fluid extraction of chrysene (■), benzo[*b*]fluoranthene plus benzo[*k*]fluoranthene (▽); indeno[1,2,3-*cd*]pyrene (●) from contaminated soil.

recovery is based on the sum of two extractions plus the amount recovered by 14 h of ultrasonication of the SFE residue in methylene chloride. These curves show the following features. First, there is an initial slower extraction at very short times. This is not very obvious from all the results as the initial extraction period is outside the timescale for this effect, but it is quite visible for the extraction of indeno[1,2,3-*cd*]pyrene from contaminated soil and some other curves show vestiges of this effect. This cannot be due to experimental start-up effects, as these would be the same for all compounds. There then follows a more rapid extraction phase which ceases often well below 100% recovery. Last, there is a much slower final extraction phase heading towards 100% recovery.

A model has been developed [15] for a spherical matrix particle of radius $a$, and within it the solute is considered to be in both the adsorbed state and the free state, with concentrations depending on time and position within the particle. The terms 'adsorbed' and 'desorbed' are used as examples of a more general situation of the molecules being bound and released. It is assumed that the solute is totally adsorbed initially and that its concentration is uniform throughout the particle. The following four processes are then considered to occur:

- fluid entry is assumed to be rapid and begins the reversible release process;
- the reversible release process, such as desorption and adsorption, or alternatively the penetration of a barrier such as a cell wall, is described by the first-order rate coefficients $k_1$ (e.g. desorption) and $k_2$ (e.g. readsorption);

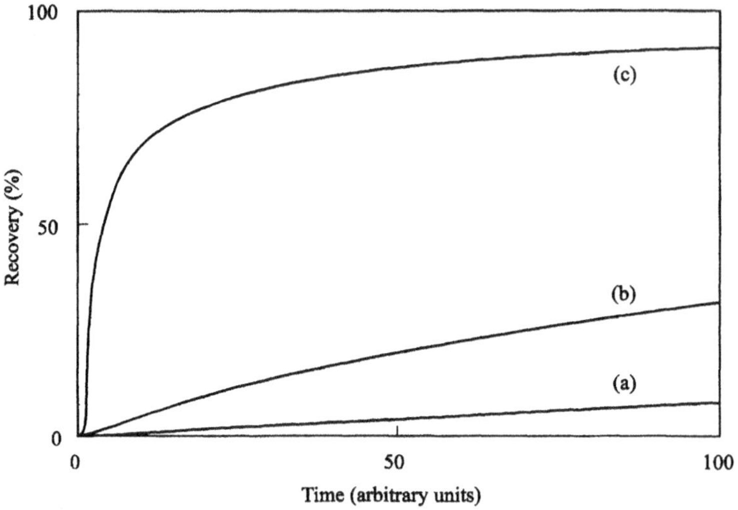


**Figure 1.22** Predicted curves of percentage recovered versus time obtained from the model which includes reversible release, transport and solvation. The input parameters for the calculation were: $k_1 = 10$, $k_2 = 30$, $D/a^2 = 0.1$ for all curves, and $ha = 1$, 10 and tends to infinity for curves (a), (b) and (c), respectively.

- transport through the matrix particle may be by normal diffusion or by diffusion through the fluid in channels in the matrix, or some other process; it is nevertheless modelled as diffusion and given an effective diffusion coefficient, $D$; adsorption and desorption will be occurring during this transport process;
- removal by the solvent then occurs; it is described, as before, by the parameter $h$.

Appropriate equations are then obtained to give a prediction of the recovery as a function of time in terms of the input parameters to the model: the rate coefficients $k_1$, $k_2$ and $(D/a^2)$, which are in units of inverse time, and the dimensionless parameter $ha$, which is proportional to the solubility. Some predictions from the model are shown in Figure 1.22. The input parameters for the calculation were: $k_1 = 10$, $k_2 = 30$, $D/a^2 = 0.1$ for all curves, and $ha = 1$, 10 and tends to infinity for curves (a), (b) and (c), respectively. The units of time are the same as in the input rate coefficients. The curves show the kinetic features of dynamic SFE which have been attributed to the effect of the matrix: the slow initial extraction; a more rapid extraction phase; and a slow final phase, which can be so slow that it appears that extraction is complete when only a fraction of the solute present in the matrix has been recovered.

Investigation of the model equations in detail leads to an appreciation of the physical processes occurring during the three phases of the extraction process. The separation of the process into the three phases, as described

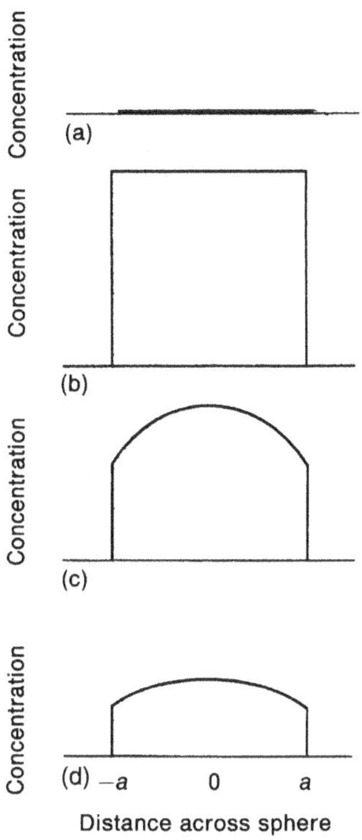


**Figure 1.23** Concentration profiles across a sphere of radius $a$ during supercritical fluid extraction involving matrix, transport and solvation effects. Parts (a)–(d) are described in the text, section 1.10.

below, is only approximate and uses the schematic description of the development of concentration profiles across the spherical model particle during extraction (Figure 1.23). In this diagram a uniform initial distribution of solute is assumed. The three phases of extraction are described in terms of this diagram as follows. It has been assumed that initially the solute is adsorbed or otherwise held in the matrix. The initial rate of extraction is therefore zero and only builds up as the solute is released from the matrix. This initial phase is more pronounced if the rate of release, $k_1$, is slow. During this phase, the concentration of free solute builds up to a steady state, illustrated by the transition from situations illustrated in Figure 1.23(a) to (b). Thus the initial slow phase corresponds to the attainment of a steady-state concentration of mobile solute molecules. Transport of the solute through the particle will occur only if there is a concentration gradient. Initially, therefore, extraction takes place only from the edge of the particles. This will erode the concentration profile at the edge of the particle, promoting transport from further inside it. The concentration profile will develop to that illustrated in Figure 1.23(c). The rate of this process will be high, initially equal to the rate of release, $k_1$. The second rapid phase corresponds to the establishment of a smooth concentration profile falling towards the edge of the particle. Once this concentration profile has been established it will decay in value, but will keep its shape, as evident from the transition from Figure 1.23(c) to (d). This decay will be exponential and its rate will be determined by solubility (which controls the concentration at the particle edge), diffusion and the equilibrium between bound and free species and thus may be slow. The final slow phase corresponds to the exponential decay of the established concentration profile.

## 1.11 Extrapolation of continuous extraction results

For all models, and in practice, extraction becomes exponential after the initial period. This opens up the possibility of using extrapolation to obtain quantitative analytical information in a shorter time than would be required for exhaustive extraction. If extraction is carried out at least as long as the initial non-exponential period to obtain an extracted mass $m_1$, followed by extraction over two subsequent equal time periods to obtain masses $m_2$ and $m_3$, respectively, then it can be readily shown that $m_0$, the total mass in the sample, is given by

$$m_0 = m_1 + \frac{m_2^2}{m_2 - m_3} \tag{1.11}$$

It can be seen that if the value of $m_3$ is found to be very small and can be considered zero, that is, the extraction is almost complete after the second time period, the equation simplifies to

$$m_0 = m_1 + m_2 \tag{1.12}$$

as would be expected. If not, equation (1.11) may be used to obtain $m_0$, provided the difference between $m_2$ and $m_3$ is large enough compared with the errors in the two quantities. This is not too serious a problem, as usually the last term in equation (1.11) is small compared with $m_1$. Two polymer examples are given below, one involving pellets (nominally spheres) and the other a film.

In the case of extraction from polymers, there is an advantage in working with the original (rather than ground) sample pellets, as there is a danger, suggested by some of the experiments carried out by my colleagues and me, that the results are affected by the grinding process (perhaps by loss of solute or a change in its extractability). However, a fairly exhaustive extraction of polymer pellets of a few millimetres in diameter is likely to take 80 h. The extrapolation procedure was therefore investigated for this type of system. Table 1.6 gives data for the extraction of 2,6-ditertiarybutyl-4-methylphenol (BHT) from standard polypropylene cylinders of *c.* 3 mm in length and diameter, with additive concentrations known to within 1% m/m [11]. Although extraction was carried out for 8 h with only 57% of the additive

**Table 1.6** Extrapolation to obtain final quantities in the extraction of 2,6-ditertiarybutyl-4-methylphenol (BHT) (0.2% m/m) from 178.4 mg of standard polypropylene pellets using pure $CO_2$ at 50°C and at 400 bar

| Extraction time (min) | Weight extracted (μg) | Cumulative time (min) | Weight extracted (μg) |
|---|---|---|---|
| 0–20 | 7.1 | – | – |
| 20–60 | 25.0 | – | – |
| 60–120 | 45.7 | – | – |
| 120–180 | 36.8 | – | – |
| 180–240 | 26.8 | 0–240 | 141.4 |
| 240–300 | 16.4 | – | – |
| 300–360 | 17.1 | 240–360 | 33.5 |
| 360–480 | 27.8 | 360–480 | 27.8 |
| Total | 202.7 | – | – |
| Given total | 356.8 | – | – |
| Total from equation (1.11) | 338.3 | – | – |

Percentage difference between given total and equation (1.11): −5.2

extracted, an estimate of the final amount was made using equation (1.11) and was *c.* 5.2% below the given value. From the form of the curve of $\ln(m/m_0)$ versus time for this system, and calculations from the known diffusion coefficient for BHT in high-density polyethylene, it can be deduced that the linear portion of the curve is not well achieved in the first extraction period. Thus the model is assuming the tail is falling off more rapidly than is in fact the case, hence the low result. If a better result were desired this could be obtained by making the sacrifice of a longer extraction time. It appears in this case that the great majority, if not all, of the additive is extractable by SFE under the conditions used, perhaps helped by the fact that the BHT molecule is a small one.

The second example is the extraction of cyclic trimer from the PET films illustrated in Figure 1.12 [12]. As the extraction was considered to be by no means exhaustive, the extrapolation procedure was used initially to estimate the total extractable oligomer. These calculations are shown in Table 1.7, in which three different sets of times were used. These gave results of reasonable agreement, with an average amount of 190 ± 5 μg in the 2.739 g PET sample, or $(6.9 \times 10^{-3})$% m/m. This is a considerably smaller percentage of oligomer than obtained by other methods. One explanation is that much of the oligomer is locked in and unextractable under the conditions of the extraction experiments.

## 1.12 Derivations and discussions of model equations

Detailed discussions of the appropriate equations in each section have been avoided hitherto, and these details are now given. The models are for

**Table 1.7** Extrapolation to obtain final quantities in the extraction of cyclic trimer from poly(ethylene terephthalate) (PET) film by means of pure $CO_2$ at 70°C and 400 bar

| Extraction time (min) | Weight extracted (μg) | Predicted weight (μg)[a] |
|---|---|---|
| 0–150 | 93.88 | |
| 150–270 | 19.08 | |
| 270–390 | 15.44 | 193.88 |
| 0–210 | 103.47 | |
| 210–300 | 13.61 | |
| 300–390 | 11.32 | 184.36 |
| 0–270 | 112.96 | |
| 270–330 | 8.14 | |
| 330–390 | 7.30 | 191.91 |

[a] Results are cumulative and are calaculated from equation (1.11) in text. The total extracted experimentally was, in each case, 128.40 μg.

continuous dynamic extraction and, unless stated otherwise, are for uniform initial concentration across the matrix of the solute to be extracted.

### *1.12.1 Extraction from a sphere controlled by transport only*

The model consists of a solid sphere of radius $a$ with a uniform initial concentration of a dissolved compound that is immersed into a fluid in which a zero concentration of the compound is maintained. Adaptation of the published solutions for the differential equation (the Fourier equation) with appropriate boundary conditions leads to the following equation for the ratio of the mass, $m$, of extractable compound that remains in the matrix sphere after extraction for time, $t$, to that of the initial mass of extractable compound, $m_0$:

$$\frac{m}{m_0} = \frac{6}{\pi^2}\sum_{n=1}^{\infty}\frac{1}{n^2}\exp\left(-\frac{n^2\pi^2 DT}{a^2}\right) \tag{1.13}$$

where $n$ is an integer and $D$ the diffusion coefficient of the compound in the material of the sphere. Equation (1.13) may be simplified by defining, as was done earlier in equation (1.8), a quantity $t_c$, which is a characteristic time for the extraction, to give, after expanding the summation:

$$\frac{m}{m_0} = \frac{6}{\pi^2}\left[\exp(-t/t_c) + \frac{1}{4}\exp\left(-\frac{4t}{t_c}\right) + \frac{1}{9}\exp\left(-\frac{9t}{t_c}\right) + \ldots\right] \tag{1.14}$$

The solution is thus a sum of exponential decays in which at longer times the later (more rapidly decaying) terms will decrease in importance and the first exponential term in square brackets will become dominant. This can be seen again if the natural logarithm of this equation is taken, after factorizing the term $\exp(-t/t_c)$ from the square bracket, to obtain:

$$\ln\left(\frac{m}{m_0}\right) = \ln\left(\frac{6}{\pi^2}\right) - \frac{t}{t_c} + \ln\left[1 + \frac{1}{4}\exp\left(-\frac{3t}{t_c}\right) + \frac{1}{9}\exp\left(-\frac{8t}{t_c}\right) + \ldots\right] \quad (1.15)$$

The term $\ln(6/\pi^2)$ is equal to $-0.4977$, the final term in this equation equals 0.4977 at $t = 0$ and so, as required, at $t = 0$ $\ln(m/m_0)$ is also equal to zero. A plot of $\ln(m/m_0)$ versus time therefore tends to become linear at longer times, when the last term in equation (1.15) tends to zero, and $\ln(m/m_0)$ is given approximately by:

$$\ln(m/m_0) = -0.4977 - \frac{t}{t_c} \quad (1.16)$$

Figure 1.8 is a plot of equation (1.15), and the straight-line portion, which is continued to the $t = 0$ line as a dashed line, is a plot of equation (1.16).

### *1.12.2 Extraction from a film controlled by transport only*

Here the model is a rectangular slab of thickness $L$, whose other dimensions are infinite. As before, we assume initial uniform distribution of the compound to be extracted and a diffusion coefficient of $D$. In this case, as the slab is infinite, we must consider $m$ to be the amount of the compound in a section of the slab, of given area, at time $t$, and $m_0$ to be the amount in the same section at $t = 0$. Adaptation of the appropriate solutions gives:

$$\frac{m}{m_0} = \frac{8}{\pi^2}\sum_{n=0}^{\infty}\frac{1}{(2n+1)^2}\exp\left[-\frac{1}{L^2}(2n+1)^2\pi^2 Dt\right] \quad (1.17)$$

where $n$ is again an integer. As before, equation (1.17) may be simplified by defining, in equation (1.9) above, a characteristic time for the extraction, $t_c$, and we obtain:

$$\frac{m}{m_0} = \frac{8}{\pi^2}\left[\exp\left(-\frac{t}{t_c}\right) + \frac{1}{9}\exp\left(-\frac{9t}{t_c}\right) + \frac{1}{25}\exp\left(-\frac{25t}{t_c}\right) + \ldots\right] \quad (1.18)$$

This is a similar sum of exponential decays, with the first exponential term in the square brackets becoming dominant at longer times, and this happens more rapidly than in the case of the spherical model. This is a feature of the lower surface-to-volume ratio $1/L$ for the infinite slab (neglecting the edges), as compared with $3/a$ for the sphere. This equation gives, after factorization of $\exp(t/t_c)$ from the square brackets, taking natural logarithms and substituting the numerical value of $\ln(8/\pi^2)$:

$$\ln\left(\frac{m}{m_0}\right) = -0.2100 - \frac{t}{t_c} + \ln\left[1 + \frac{1}{9}\exp\left(-\frac{8t}{t_c}\right) + \frac{1}{25}\exp\left(-\frac{24t}{t_c}\right) + \ldots\right] \quad (1.19)$$

A plot of $\ln(m/m_0)$ versus time therefore again becomes linear at longer times and the nature of the equations are such that this linear portion is reached more rapidly than in the case of a sphere. The approximate equation for this model at longer times is

$$\ln\left(\frac{m}{m_0}\right) = -0.2100 - \frac{t}{t_c} \tag{1.20}$$

Figure 1.11 is a plot of equation (1.19), and the straight-line portion, which is continued to the $t = 0$ axis as a dashed line, is a plot of equation (1.20).

### *1.12.3 Extraction from a film, with non-uniform concentration distribution, controlled by transport only*

Here the extractable compound in an infinite slab is considered to be distributed not uniformly but with concentration falling off exponentially from the surface, as illustrated schematically in Figure 1.13. The concentration is a function of the distance $x$ along a coordinate perpendicular to the slab surface. If the slab surfaces are located at $x = 0$ and $x = L$, then the concentration is initially given by:

$$c(x, 0) = c_0\left[\exp\left(-\frac{x}{\alpha}\right) + \exp\left(-\frac{(L - x)}{\alpha}\right)\right] \tag{1.21}$$

where $\alpha$ is a distance parameter. The average concentration in the slab, $c_{av}$, obtained by integrating this expression with respect to $x$ between 0 and $L$, and dividing by $L$ is initially:

$$c_{av}(0) = \left(\frac{2c_0}{L}\right)\left[1 - \exp\left(\frac{L}{\alpha}\right)\right] \tag{1.22}$$

Adaptation of the appropriate heat conduction equation [16] gives the following integral equation for determining the concentration profile at a later time from the initial concentration profile, when the surfaces of the infinite slab are maintained at zero concentration by continuous extraction unlimited by solubility:

$$c(x, t) = \left(\frac{2}{L}\right)\sum_{n=1}^{\infty}\sin\left(\frac{n\pi x}{L}\right)\exp\left(-\frac{n^2 t}{t_c}\right)\int_0^L c(x', 0)\sin\left(\frac{n\pi x'}{L}\right)\mathrm{d}x' \tag{1.23}$$

where $t_c$ is defined by equation (1.9) and $n$ is an integer. Substitution of $c(x, 0)$ from equation (1.21) and integration leads to an expression for $c(x, t)$. This expression is then further integrated with respect to $x$ and divided by $L$ to give the average concentration across the slab at $t$, as follows:

$$c_{av}(t) = 8c_0\left[1 + \exp\left(\frac{L}{\alpha}\right)\right]\sum_{n=0}^{\infty}\left[\left(\frac{L}{\alpha}\right)^2 + \pi^2(2n+1)^2\right]^{-1}\exp\left[-\frac{(2n+1)^2 t}{t_c}\right] \tag{1.24}$$

As it is true that

$$\frac{m}{m_0} = \frac{c_{\mathrm{av}}(t)}{c_{\mathrm{av}}(0)} \tag{1.25}$$

$m/m_0$ can be obtained by dividing equation (1.24) by equation (1.22) to be given by:

$$\frac{m}{m_0} = \frac{4\alpha}{L}\left[1+\exp\left(-\frac{L}{\alpha}\right)\right]\left[1-\exp\left(-\frac{L}{\alpha}\right)\right]^{-1}\sum_{n=0}^{\infty}\left[1+\frac{\alpha^2\pi^2(2n+1)^2}{L^2}\right]^{-1} \times \exp\left[-\frac{(2n+1)^2 t}{t_{\mathrm{c}}}\right] \tag{1.26}$$

This equation is of the same type as equation (1.17), derived for a uniform initial concentration distribution, with a sum of exponential decays of which the first term becomes dominant at longer times, that is, $\ln(m/m_0)$ becomes linear with time at longer times. Also it can be readily shown that in the limit of very large values of $\alpha/L$, which correspond to uniform distribution of concentration initially, equation (1.26) becomes identical to equation (1.17), as required. However, for finite values of $\alpha/L$, and especially for values less than unity, the higher terms in the sum of exponentials fall off less rapidly in importance than was the case for the uniform initial distribution. This results in the initial rapid extraction being of greater importance and in a lowering of the intercept of the linear portion.

The value of the intercept, $-I$, may be found from the natural logarithm of the coefficient of the first term in the series ($n = 0$), that is:

$$-I = \ln\left\{\frac{4\alpha}{L}\left[1+\exp\left(-\frac{L}{\alpha}\right)\right]\left[1-\exp\left(-\frac{L}{\alpha}\right)\right]^{-1}\left[1+\left(\frac{\alpha\pi}{L}\right)^2\right]^{-1}\right\} \tag{1.27}$$

Some values of $I$ for various values of $\alpha/L$ are given in Table 1.3.

### 1.12.4 *Extraction from a sphere controlled by transport and solvation*

In previous models it has been assumed that the concentration of solute in the fluid adjacent to the matrix surfaces is zero and consequently the concentration of solute at the surfaces of the matrix particles or films is also zero. The equations were obtained by using $c(0, t)$ as a boundary condition during extraction. We now assume that the concentration of solute in the fluid, $c_{\mathrm{f}}$, is finite and is determined by the rate of loss of solute out of the matrix and the volume rate of flow of fluid, $F$. The appropriate equation is:

$$-AD\left(\frac{\mathrm{d}c}{\mathrm{d}x}\right)_{x=0} = Fc_{\mathrm{f}} \tag{1.28}$$

where $A$ is the surface area of the matrix and $x$ is the distance perpendicular to the surface, increasing to zero at the surface. The term on the left-hand side of equation (1.28) is therefore the rate of loss of solute from the matrix. This will be equal at the steady state, which is assumed to be rapidly established, to the rate of transport of solute away from the matrix, which is given by the right-hand side of the equation.

It is now assumed that the concentration in the fluid is proportional to the concentration in the matrix at the surface, which for a sphere will be at $a$, the proportionality constant being the partition coefficient, $K$, and thus

$$c_f = Kc(a, t) \tag{1.29}$$

Combining equations (1.28) and (1.29) gives

$$-\left(\frac{dc}{dx}\right)_{x=0} = hc(a, t) \tag{1.30}$$

where

$$h = \frac{KF}{AD} \tag{1.31}$$

Equation (1.30) is the boundary condition that will be used in conjunction with the diffusion equation to obtain the equations for solubility limitation. It can be seen that if $h$ becomes very large, compared with the inverse of the distances involved, then $c(0, t)$ becomes close to zero and the simpler models developed earlier apply. This occurs as expected, from equation (1.31), if the partition coefficient (solubility in the fluid) and flow rate are high and the diffusion coefficient, i.e. the rate of diffusion out of the matrix, is low.

Adaptation, as before, of the appropriate heat conduction equations corresponding to this problem [16], which correspond to Newtonian heat loss, gives, after integration over the sphere to obtain the average concentration and after some algebraic manipulation, the following equation for extraction from a sphere with uniform initial concentration distribution:

$$\frac{m}{m_0} = 6\sum_{n=1}^{\infty}\left(\frac{ha}{\lambda_n}\right)^2 [ha(ha-1) + \lambda_n^2]^{-1} \exp\left[-\left(\frac{\lambda_n}{\pi}\right)^2 \frac{t}{t_c}\right] \tag{1.32}$$

where $\lambda_n$ are the roots of the equation

$$\lambda \cot(\lambda) = 1 - ha \tag{1.33}$$

These roots are tabulated or are available from computer library routines and, in the limit of large values of $ha$, are given by

$$\lambda_n = n\pi \tag{1.34}$$

Substitution of these values of $\lambda_n$ for very large values of $ha$ into equation (1.32) gives equation (1.13), i.e. that of the hot-ball model, the sphere in a fluid with zero solute concentration, as expected.

Plots of values of $\ln(m/m_0)$, obtained from equation (1.32), are shown in Figure 1.14 for various values of $ha$. The curve for $ha$ tending towards infinity corresponds to the hot-ball model with zero concentration of solute in the fluid, as previously shown in Figure 1.8. In all cases the curves become linear at longer times, as the terms for $n = 1$ become dominant. As $ha$ falls both the slope and the intercept of the linear section fall. At long times $\ln(m/m_0)$ is given approximately by

$$\ln\left(\frac{m}{m_0}\right) = \ln\left\{6\left(\frac{ha}{\lambda_1}\right)^2 [ha(ha-1)+\lambda_1^2]^{-1}\right\} - \left(\frac{\lambda_1}{\pi}\right)^2 \frac{t}{t_c} \tag{1.35}$$

The slopes of the linear portion of the graphs are thus equal to $-(\lambda_1/\pi)^2/t_c$ and the intercepts $-I$ are given by

$$-I = \ln\left\{6\left(\frac{ha}{\lambda_1}\right)^2 [ha(ha-1)+\lambda_1^2]^{-1}\right\} \tag{1.36}$$

Values of $\lambda_1$ and $I$ are given in Table 1.4 for the values of $ha$ which were used in the plots of Figure 1.14. It can be seen that both these quantities fall as $ha$ falls, i.e. as the effect of solvation limitation increases. The effect of solubility limitation is thus to reduce the high rate at the beginning of the extraction, to reduce the slope of the linear portion and to move the linear portion upwards on the graph. The extent of these effects increases as the pressure falls and the solubility decreases. The model is consistent with some of the experimental results described earlier and shown in Figure 1.20.

### *1.12.5 Extraction from a film controlled by transport and solvation*

Using a similar appropriate boundary condition and adapting the appropriate heat conduction equations [16], which in this case correspond to Newtonian heat loss from a slab, gives the following equation for extraction from an infinite slab with uniform initial concentration distribution:

$$\frac{m}{m_0} = 8\sum_{n=0}^{\infty}\left(\frac{hL}{\lambda_n}\right)^2 [hL(hL+2)+\lambda_n^2] \exp\left[-\left(\frac{\lambda_n}{\pi}\right)^2 \frac{t}{t_c}\right] \tag{1.37}$$

where $t_c$ is given by equation (1.9) and $\lambda_n$ are the roots of the equation

$$\lambda \tan\left(\frac{\lambda}{2}\right) = hL \tag{1.38}$$

Again these roots are tabulated or are available from computer library routines and, in the limit of large values of $hL$, are given by

$$\lambda_n = (2n-1)\pi \tag{1.39}$$

Substitution of these values of $\lambda_n$ for very large values of $ha$ into equation (1.37) gives equation (1.17), as expected. Equation (1.37) will not be discussed further, except to say that it gives a series of curves of the general form of Figure 1.14.

### *1.12.6 Extraction from a sphere controlled by transport, solvation and matrix effects*

As before, the model consists of a spherical matrix particle of radius $a$, and any point in the particle is described by its distance, $r$, from the centre. We consider concentrations of the solute in both the adsorbed and free states, $c'$ and $c$, respectively, which are a function of $r$ and $t$. It is assumed that the solute is totally adsorbed initially and that its concentration is uniform throughout the particle at $c_0$. This is expressed by boundary conditions:

$$c'(r,0) = c_0 \tag{1.40}$$

$$c(r,0) = 0 \tag{1.41}$$

As solvation is being taken into account, the boundary condition expressed by equation (1.30), with $h$ defined by equation (1.31), also applies. The equations for the rates of change of the two concentrations are:

$$\frac{\partial c'}{\partial t} = -k_1 c' + k_2 c \tag{1.42}$$

$$\frac{\partial c}{\partial t} = +k_1 c' - k_2 c + D\left\{\left(\frac{\partial^2 c}{\partial r^2}\right) + \left(\frac{2}{r}\right)\left(\frac{\partial c}{\partial r}\right)\right\} \tag{1.43}$$

Both equations contain terms for the rates of adsorption and desorption and equation (1.43) also contains a term for the diffusion of the desorbed solute. The differential equations (1.42) and (1.43) are now solved in conjunction with the boundary conditions: equations (1.30), (1.40) and (1.41). This can be done by the method of Laplace transforms. After integration over the particle, an expression for the fraction of solute remaining after time $t$ is found to be

$$\frac{m}{m_0} = \sum_{n=1}^{\infty} C_n \left[\frac{\exp(-p_n t)}{p_n} - \frac{\exp(-q_n t)}{q_n}\right] \tag{1.44}$$

The equations for the rate coefficients for these decays are

$$p_n = k_1 + k_2 + \frac{D\lambda_n^2}{a^2} - \left[\left(k_1 + k_2 + \frac{D\lambda_n^2}{a^2}\right)^2 - \frac{4k_1 D\lambda_n^2}{a^2}\right]^{1/2} \tag{1.45}$$

and

$$q_n = k_1 + k_2 + \frac{D\lambda_n^2}{a^2} + \left[\left(k_1 + k_2 + \frac{D\lambda_n^2}{a^2}\right)^2 - \frac{4k_1 D\lambda_n^2}{a^2}\right]^{1/2} \tag{1.46}$$

where $\lambda_n$ are the roots of equation (1.33) as before. The coefficients $C_n$ in equation (1.44) are given by

$$C_n = 6k_1 h^2 D\left\{[\lambda_n^2 + ha(ha-1)]\left[\left(k_1 + k_2 + \frac{D\lambda_n^2}{a^2}\right)^2 - \frac{4k_1 D\lambda_n^2}{a^2}\right]^{1/2}\right\}^{-1} \tag{1.47}$$

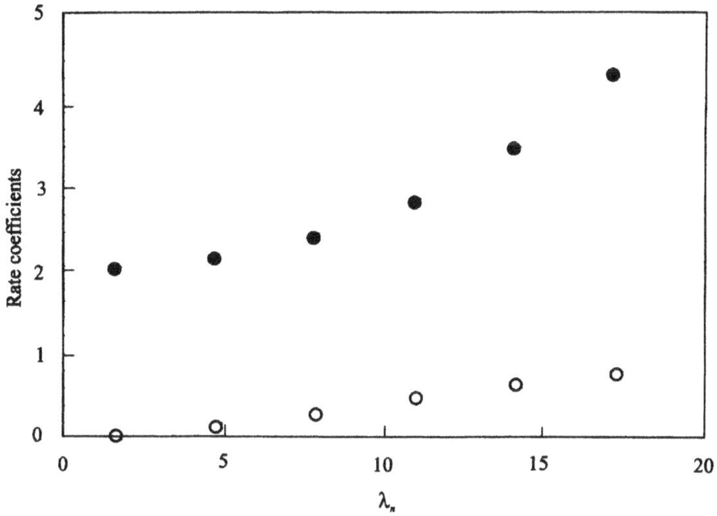


**Figure 1.24** Values for the rate coefficients for the first six terms of equation (1.44) in text (section 1.12.6), plotted against the first six values of $\lambda_n$, which are roots of equation (1.33). $\bigcirc = p_n$, $\bullet = q_n$. Input parameters: $k_1 = 1$; $k_2 = 1$; $D/a^2 = 0.01$; $ha = 1$; the units of all input and output rate coefficients are the same.

Equation (1.43) and its subsidiary equations (1.33) and (1.45) to (1.47) give a prediction of the fraction remaining in the matrix after extraction for a given time $t$ in terms of the input parameters to the model: the rate coefficients $k_1$, $k_2$ and $D/a^2$, which are in units of inverse time, and the dimensionless parameter $ha$, which is proportional to the solubility. Equation (1.43) contains two infinite series of exponential decays, one of which is negative. The negative series is relatively fast; inspection of equations (1.45) and (1.46) shows that $q_n$ must lie between $k_1 + k_2$ and infinity, whereas $p_n$ must lie between zero and $k_1$. Figure 1.24 is a diagram showing the first few members of each series for some representative input parameters: $k_1 = 1$, $k_2 = 1$, $(D/a^2) = 0.01$ and $ha = 1$, the units of all input and output rate coefficients being the same.

If the input parameters are such that $p_1$ is much smaller than $k_1$, it can be shown that $p_1$ is given approximately by

$$p_1 = 3ha\left(\frac{k_1}{k_1 + k_2}\right)\frac{D}{a^2} \tag{1.48}$$

Thus if SFE is being carried out under adverse conditions such that the solubility and thus $ha$ is small, the readsorption rate coefficient is greater than that for desorption, and the transport rate $D/a^2$ is small, all these factors combine to reduce the rate of extraction at long times. Furthermore, as discussed previously, a large fraction of the solute will extract at this rate. The quantities multiplying the exponential decays are $C_n/p_n$ and $C_n/q_n$, respectively, which means that the more rapidly decaying terms in $q_i$ are numerically smaller than the terms in $p_i$, and that the initial rate of extraction will be zero, as on differentiation the pairs of terms will cancel as $t$ tends to zero. $C_n$ falls as $n$ increases, causing the series to converge. For lower values of $ha$ (lower solubility) this convergence is more rapid, so that for compounds with poor solubility only the $C_1$ terms are important. Figure 1.22 shows some curves predicted from the model. The input parameters for the calculation were: $k_1 = 10$, $k_2 = 30$, $D/a^2 = 0.1$ for all curves, and $ha = 1$, 10 and tends to infinity for curves (a), (b) and (c), respectively. The units of time are the same as in the input rate coefficients. They show the three-phase characteristics of a slow initial extraction (the rate of extraction being zero at $t = 0$ as discussed above), a more rapid rise and then a slow-down in the final rate, as explained by equation (1.48).

### 1.12.7 *Extraction from a sphere controlled by transport, solvation and matrix effects, with non-uniform initial concentration*

Analogously to the earlier treatment of a film, the compound, adsorbed initially, is considered to be distributed not uniformly but with concentration falling off exponentially from the surface, that is:

$$c'(r,0) = c_0 \exp\left(\frac{r-a}{\alpha}\right) \tag{1.49}$$

where $\alpha$ is a distance parameter giving the steepness of the fall-off in concentration. This equation then replaces the boundary condition for $c'$ in equation (1.40). The change in boundary condition has no effect on the form of equation (1.44) or on the values of $p_n$ and $q_n$. The coefficients $C_n$ in equation (1.44) are, however, changed and this equation becomes:

$$\frac{m}{m_0} = \sum_{i=1}^{\infty} C'_n \left[\frac{\exp(-p_n t)}{p_n} - \frac{\exp(-q_n t)}{q_n}\right] \tag{1.50}$$

where $C'_n$ is given in terms of the original $C_n$ by

$$\begin{aligned} C'_n = C_n \lambda_n^3 \Bigg\{ & \left[\left(\frac{\lambda_n \alpha}{a}\right)^2 + 1\right]\left[\lambda_n + \left(\frac{\lambda_n \alpha}{a}\right)(ha - 1)\right. \\ & + \left.\left[\left(\frac{\lambda_n \alpha}{a}\right)^2 - 1\right]\frac{\lambda_n \alpha}{a}\right\}\left\{3ha\left[\left(\frac{\lambda_n \alpha}{a}\right)^2 + 1\right]\lambda_n^2 - \frac{2\lambda_n^2 \alpha}{a}\right. \\ & + 2\left(\frac{\lambda_n \alpha}{a}\right)^2\left[1 - \exp\left(-\frac{a}{\alpha}\right)\right]\Bigg\}^{-1} \end{aligned} \tag{1.51}$$

As $\alpha/a$ tends to infinity, representing uniform distribution, equation (1.51) becomes $C'_n = C_n$, as required. For finite values of $\alpha/a$, the smaller its value the more important are the faster decay terms in equation (1.50).

Here an exponential distribution has been considered, although equations for other distributions may be readily obtained by the same methods. The model does allow qualitative discussion of material concentrated near or away from the surface to be made.

### *1.12.8 Extrapolation using the models*

For all models the equations approximated to at longer times have the form of

$$m = K \exp\left(-\frac{t}{\tau}\right) \tag{1.52}$$

for $m$, the amount of solute remaining after time $t$, where $K$ and $\tau$ are constants for a particular extraction model [equal to $m_0 \exp(-0.4977)$ for a sphere and to $m_0 \exp(-0.2100)$ for an infinite slab, taking on more complex values for other models]. For the extrapolation procedure as described above, we require an initial extraction of an amount $m_1$, in a period of $t = 0 - t_1$, by which time the extraction should have approximately reached the linear portion of the curve of $\ln(m/m_0)$ versus time. There should then be two subsequent extractions, in equal periods, of amounts $m_2$ and $m_3$, these periods terminating at $t_2$ and $t_3$, respectively, that is:

$$t_3 - t_2 = t_2 - t_1 \tag{1.53}$$

On substitution of the appropriate values at the end of each of the three extraction periods into equation (1.52) we obtain:

$$m_0 - m_1 = K \exp\left(-\frac{t_1}{\tau}\right) \tag{1.54}$$

$$m_0 - m_1 - m_2 = K \exp\left(-\frac{t_2}{\tau}\right) \tag{1.55}$$

and

$$m_0 - m_1 - m_2 - m_3 = K \exp\left(-\frac{t_3}{\tau}\right) \tag{1.56}$$

After dividing equation (1.54) by equation (1.55) and also dividing equation (1.55) by equation (1.56), the following two equations are obtained:

$$\frac{m_0 - m_1}{m_0 - m_1 - m_2} = K \exp\left(\frac{t_2 - t_1}{\tau}\right) \tag{1.57}$$

$$\frac{m_0 - m_1 - m_2}{m_0 - m_1 - m_2 - m_3} = K \exp\left(\frac{t_3 - t_2}{\tau}\right) \tag{1.58}$$

From equation (1.53), the right-hand sides of equations (1.57) and (1.58) are identical and so the left-hand sides can be equated. After algebraic manipulation, we obtain equation (1.11) (section 1.11).

## References


1. Reid, R.C., Prausnitz, J.M. and Polling, B.E. (1986) *The Properties of Gases and Liquids*, McGraw-Hill, New York.
2. Brunner, E., Hultenschmidt, W. and Schlichtharle, J. (1987) Phase equilibria in five binary mixtures. *Journal of Chemical Thermodynamics*, **19**, 273–91.
3. Angus, S., Armstrong, B. and de Reuck, K.M. (1976) *International Thermodynamic Tables of the Fluid State: Volume 3, Carbon Dioxide*, Pergamon Press, Oxford.
4. Vesovic, V., Wakeham, W.A., Olchowy, G.A. *et al.* (1990) The transport properties of carbon dioxide. *Journal of Physical and Chemical Reference Data*, **19**, 763–808.
5. Clifford, A.A. and Coleby, S.E. (1991) Diffusion of a solute in dilute solution in a supercritical fluid. *Proceedings of the Royal Society of London*, **A433**, 63–79.
6. Span, R. and Wagner, W. (1996) A new equation of state for carbon dioxide covering the fluid region from the triple-point temperature to 1100 K at pressures up to 800 MPa. *Journal of Physical and Chemical Reference Data*, **25**, 1509–96.
7. Peng, D.Y. and Robinson, D.B. (1976) A new two-constant equation of state. *Industrial and Engineering Chemistry Fundamentals*, **15**, 59–64.
8. Pitzer, K.S. (1955) The volumetric and thermodynamic properties of fluids II. Compressibility factor, vapour pressure and entropy of vaporisation. *Journal of the American Chemical Society*, **77**, 3433–40.
9. Bartle, K.D., Boddington, T., Clifford, A.A. and Shilstone, G.F. (1989) Effect of pressure drop across the column on average densities and capacity factors in supercritical fluid chromatography. *Journal of Chromatography*, **471**, 347–55.
10. Johnston, K.P., Peck, D.G. and Kim, S. (1989) Modelling supercritical fluids: how predictive is it? *Industrial and Engineering Chemistry Research*, **28**, 1115–25.
11. Bartle, K.D., Clifford, A.A., Hawthorne, S.B. *et al.* (1990) A model for dynamic extraction using a supercritical fluid. *Journal of Supercritical Fluids*, **3**, 143–9.
12. Bartle, K.D., Boddington, T., Clifford, A.A. *et al.* (1991) Supercritical fluid extraction and chromatography of oligomers from poly(ethylene terephthalate) films. *Analytical Chemistry*, **63**, 2371–7.
13. Bartle, K.D., Boddington, T., Clifford, A.A. and Hawthorne, S.B. (1992) The effect of solubility on the kinetics of supercritical fluid extraction. *Journal of Supercritical Fluids*, **5**, 207–12.
14. Clifford, A.A., Bartle, K.D. and Zhu, S. (1995) Supercritical fluid extraction of polymers: theoretical explanation of pressure and flow-rate effects. *Analytical Proceedings*, **32**, 227–30.
15. Clifford A.A., Burford, M.D., Hawthorne, S.B. *et al.* (1995) The effect of the matrix on the kinetics of supercritical fluid extraction. *Journal of the Chemical Society Faraday Transactions*, **91**, 1333–8.
16. Carslaw, H.S. and Jaeger, J.C. (1959) *Conduction of Heat in Solids*, Clarendon Press, Oxford.
17. Crank, J. (1975) *The Mathematics of Diffusion*, Clarendon Press, Oxford.
18. Tyrrell, H.J.V. and Harris, K.R. (1984) *Diffusion in Liquids*, Butterworth, London.
19. Mills, N.J. (1986) *Plastics – Microstructure, Properties and Applications*, Edward Arnold, London.
20. Ashraf, S., Bartle, K.D., Clifford, A.A. *et al.* (1992) Prediction of the conditions for supercritical fluid extraction of atrazine from soil. *Analyst*, **117**, 1697–1700.

# 2 Supercritical fluid extraction instrumentation

D.C. MESSER, G.R. DAVIES, A.C. ROSSELLI, C.G. PRANGE AND J.W. ALGAIER

## 2.1 Introduction

Since the late 1980s, supercritical fluid extraction (SFE) instrumentation has evolved from a chemical engineering research tool into large-scale commercial applications (e.g. the extraction of caffeine from coffee beans) and finally into the analytical laboratory as an efficient preparation method for a wide range of chemical compounds from a variety of samples. This chapter reviews SFE instrumentation as it applies to the analytical laboratory.

All SFE systems consist of five basic components, with each component being an integral part of a fully functional system. Analytical SFE systems are usually defined by the ability to extract samples ranging in sizes up to a litre. SFE systems with sample capacities accepting greater than a litre fall into the categories of either pilot or process scale. Although system components between analytical and process scale SFE change only slightly, the focus of this work is analytical SFE. Generally, an analytical SFE system consists of a source of fluid, a pump, a heated extraction or sample vessel, a restrictor and a collection or trapping device (Figure 2.1). Before examining each of the SFE components, clarification must be made of some common terms and techniques used in SFE.

## 2.2 Analyte and matrix

Two commonly used terms in SFE are analyte and matrix. No matter how complex the sample, the target compounds to be extracted, collected and analysed are termed 'analytes'. The material from which the analytes are extracted is termed the 'matrix'. The matrix may range from soils, polymers, prepared foods or even biological tissues.

## 2.3 Modifier addition

Termed 'entrainer' in early research papers [1], the term 'modifier' refers to the addition of a solvent to the supercritical fluid. Often modifiers are used

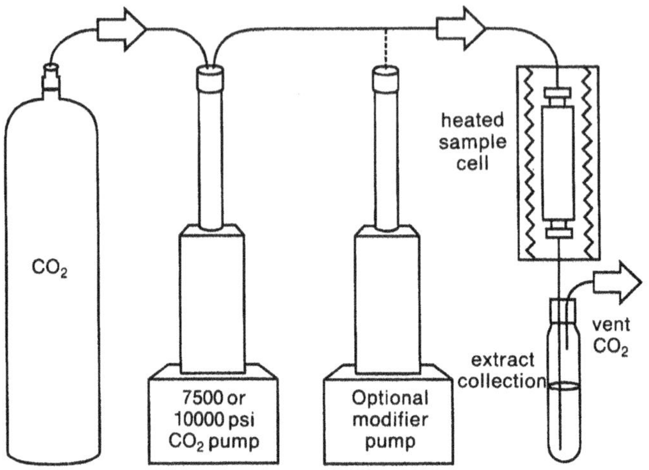


**Figure 2.1** Schematic of a basic supercritical fluid extraction system.

for the extraction of polar compounds. Examples of modifiers include dichloromethane, methanol and ethanol.

## 2.4 On-line and off-line supercritical fluid extraction

Both on-line and off-line systems share the same SFE components. On-line SFE couples the extraction with an analytical technique such as supercritical fluid chromatography (SFC), gas chromatography (GC), Fourier transform infrared spectroscopy (FTIR) or high-pressure liquid chromatography (HPLC).

In off-line SFE the extracted analyte is collected independently by means of a solid phase or liquid phase trap. An analytical instrument such as GC, SFC or HPLC is then used to analyse the extract.

## 2.5 Supercritical fluid delivery

Supercritical fluid delivery at constant pressures ranging from 10 to 70 MPa and at sufficient volumes, generally up to 500 ml of supercritical fluid (SF) per extraction, are requirements in analytical SFE. Unlike liquid chromatography pumps, SFE pumps are required to deliver compressible liquids under varying pressures and flow rates. Moreover, in contrast to the centrifuge and reciprocating pumps used in process scale SFE [2], analytical SFE pumps must be designed to fit the space limitations of the typical analysis laboratory, perform at reasonable noise levels, stabilize rapidly and provide for reproducible SF delivery. With these requirements, commercial or self-built analytical SFE systems have adopted three types of pumps: syringe, reciprocating and pneumatic amplifier. Each type, of course, has advantages and disadvantages, which are described below.

### *2.5.1 Syringe pumps*

The use of syringe pumps in SFE evolved from research and development in SFC. In early SFC work the Varian 8500 syringe pump was modified to allow for pump pressure control [3, 4]. The successful use of syringe pumps led to the develoment of a commercial SFC system followed by their usage in SFE [5]. Since their inital use there has been a considerable number of improvements in syringe pump technology [6]. For example, microprocessor control along with improved pressure-sensing transducers result in greater accuracy and precision for constant-pressure SF delivery. Rapid pressurization and electronic stabilization controls allow for quick set-up. Mechanical improvements, such as the triple ball-nut screw drive, result in smoother operation and quieter pumps. Figure 2.2 shows some of these improvements in a typical syringe pump design. Of particular importance is the seal for the piston which compresses the SF and ensures leak-proof operation. Seal material

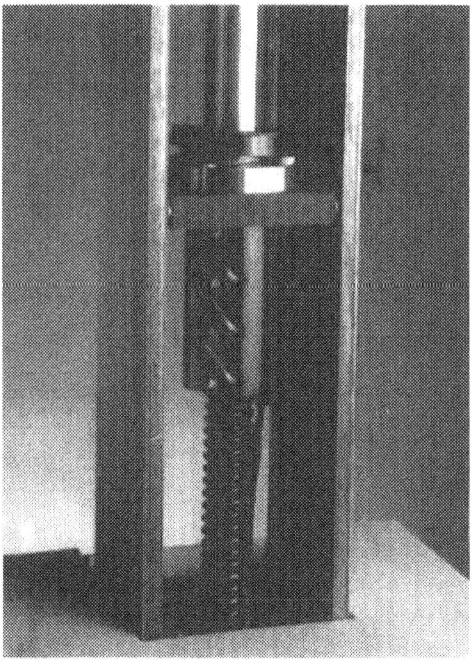

**Figure 2.2** Triple recirculating ball nut in a high-pressure syringe pump drive.

must not only function throughout a wide pressure range but also be resistant to the solvent or SF used and must not leach compounds into the fluid, which would be disastrous for post-SFE trace analysis procedures. To meet these criteria, a graphite-filled teflon seal is generally employed.

An often cited advantage of syringe pumps is the wide range of flow rates possible, from 1.0 μl/min to 90 ml/min during constant pressure mode. In SFE the high flow rate capability of the syringe pump ensures rapid pressurization control for accelerated preparation of initial extraction conditions. Also, the low flow rate capability provides optimal constant pressure stability control, ensuring pressure maintenance under static and low-flow applications. Although not strictly necessary, better refilling of syringe pumps is achieved by cooling the pump heads. An additional advantage is that syringe pumps do not require the use of helium headspace $CO_2$ supply cylinders, the use of which can result in non-reproducible SFE results [7].

The often cited disadvantage of syringe pumps is the limited delivery volume. One manufacturer (Isco Inc., Lincoln, NE) markets a large-volume syringe pump that can deliver a maximum of 266 ml liquid at a maximum pressure of 7500 psi. Delivery volume limitation can also be overcome by using two syringe pumps equipped with continuous-flow connections (Figure 2.3). Whilst one pump is providing SF the other can refill for uninterrupted flow delivery of unlimited volumes.

**Figure 2.3** Dual syringe pump system configured for continuous-flow delivery of $CO_2$ with alternating refill and delivery.

Another important aspect of SFE is modifier addition to the supercritical fluid. In many SFE applications modifier addition results in more efficient extractions. One example is the extraction of lipids from meat and eggs in which phospholipid extraction requires alcohol-modified supercritical $CO_2$ [8, 9]. In syringe pump systems, modifier addition is performed on the high-pressure side of the pumps in a static mixing 'T' prior to the extraction vessel. Syringe pump control systems ensure proper modifier addition by first pressurizing the extraction system with supercritical fluid only, followed by modifier pump pressurization. When the modifier pump pressure equals the system pressure, a check valve opens into the mixing 'T' and extraction begins. Since both pumps are now at equal pressure, the amount or ratio of modifier to supercritical fluid is microprocessor controlled by varying the syringe pump flow rate. The precision with which the syringe pump can supply modifiers, even at low flow rates, facilitates optimization studies and reproducible SFE performance.

### 2.5.2 *Reciprocating piston pumps*

Reciprocating piston pumps, which were originally used for liquid delivery in HPLC, have been adapted for use in SFC, followed by SFE [10–13]. Since these pumps were designed for room-temperature liquid delivery, modifications are required for SFC and SFE operation, the chief of these being the requirement for cooling the pump heads down to *c.* 0°C in order to enhance pump efficiency and to avoid cavitation or vapor lock, which may occur when the pump piston reverses direction to draw fluid into the piston chamber. Since HPLC reciprocating pumps generally have a chamber volume of only 50–100 μl, the pump must repeatedly fill and refill in order to deliver fluid flows of millilitres per minute. With each compression stroke, heat is generated. Low boiling point organic solvents and supercritical fluids, which are typically gases at room temperature, will vaporize when drawn into a hot pump head. Cavitation and pump inefficiency results under these conditions. To avoid this problem, pump heads are cooled by means of three methods:

- $CO_2$ split-flow stream cooling;
- thermal electric contact cooling [14];
- fluid flow cooling, which also requires a separate recirculating pump [15].

In addition, the use of a siphon tube in the cylinder containing the eventual supercritical fluid allows liquid rather than gas to be drawn into the pump head from the bottom of the storage cylinder. Another problem encountered with the reciprocating pumps are pressure pulsations [12] which require the use of a pulse dampener. One advantage of the reciprocating pump, of course, is the ability to pump continuously without the need for a second pump.

For modifier addition, however, a second dedicated pump is required with a high-pressure mixing 'T' or low dead volume magnetic mixer. This was earlier thought to be avoidable by using premixed supply cylinders which contained a specified amount of modifier with the presupercritical fluid. However, with this approach consistency of delivering a fixed amount of modifier is a problem, resulting in varying modifier/SF ratios with tank or cylinder usage [16]. In sum, the low-cost availability of reciprocating pumps coupled with the relative ease with which they can be adapted to deliver liquid $CO_2$ has resulted in their use in commercial SFE systems.

### *2.5.3 Pneumatic amplifier pumps*

The pneumatic amplifier pump was designed to produce a controlled output pressure for pumping compressible fluids. One application of these pumps is high-pressure slurry packing of HPLC columns. These pumps work by applying gas pressure (usually air) to the low-pressure side of a large-surface-area piston which displaces fluid on the opposite small-surface-area, high-pressure side. Pressure programming control is achieved through the use of stepper motors [17]. Although these pumps have been used in commercial SFE systems [18] they are generally less quiet in operation than syringe or reciprocating pumps.

All three types of pumps – syringe, reciprocating and pneumatic amplifier – are used in commercial and self-built SFE systems. Several reports, such as that by Greibrokk and co-workers [13] in 1986, dealing with the delivery of supercritical fluids suggest preference for one type of pump. Since this aspect has been the subject of considerable engineering advancement along with ongoing development, the pros and cons of each pumping system should be carefully evaluated to match the needs of individual SFE applications.

## 2.6 Extraction vessels

Early SFE vessels or sample cartridges were simply stainless-steel HPLC column bodies which could withstand and contain SFE pressures (up to 70 MPa). A 5–10 μm porous stainless-steel frit fitted at both ends retained the sample. Later, a wrench-free sealing mechanism was developed by Keystone Scientific (Bellefonte, PA) which allowed for easier use in non-commercial extractors.

Commercial SFE sample vessels are adaptations of stainless-steel tubes with wrench-free fittings (Figure 2.4). A sample vessel patented by Isco [19] introduced a unique design, in which the vessel or sample cartridge is a thin-walled material (Figure 2.5). This design allows for equal pressurization of the inside and outside of the cartridge (Figure 2.6). Thus, it is the extraction chamber which must contain the high pressures, not the cartridge. Not only

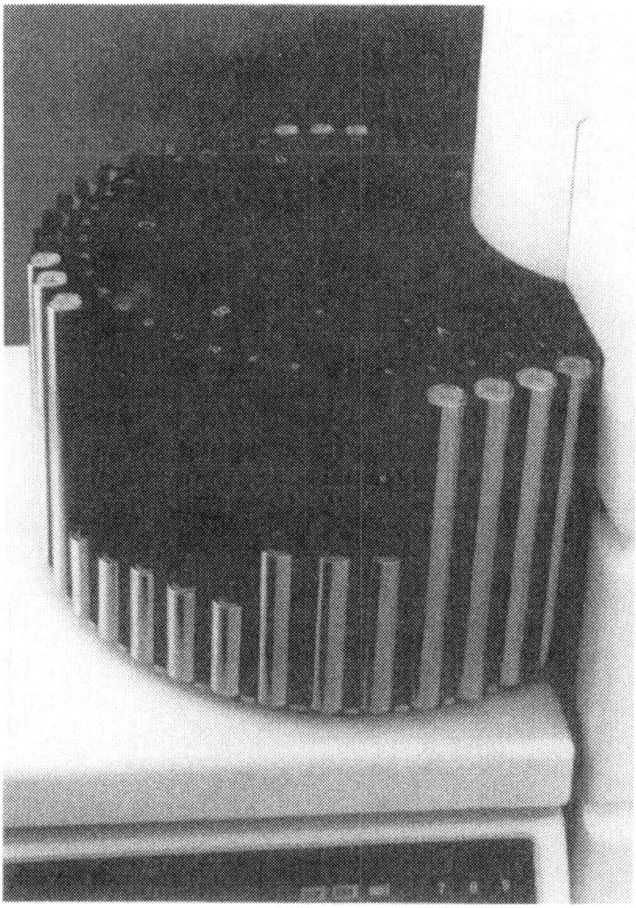

**Figure 2.4** Conventional supercritical fluid extraction sample vessels consist of stainless-steel tubes with press-in frits and polished ends for sealing surfaces (shown here) or with threaded end caps and provision for connection of tube fittings.

does this dramatically lower the production costs of the cartridge but also it permits the use of various materials in its construction. These include stainless-steel, aluminium, poly(ether-ether-ketone) (PEEK) and a high-density crystalline polymer (HDCP). For SFE applications involving temperatures greater than 100°C, PEEK cartridges are unsuitable because of material breakdown and leaching of compounds into the SF $CO_2$. The aluminium and HDCP cartridges, being lighter in weight than stainless-steel, are ideally suited for applications in which weight loss of the sample is important (e.g. fat or oil extraction).

Typical analytical SFE vessels range in volume from 0.5 to 50 ml. The sample is usually retained by porous stainless-steel frits with pore sizes of

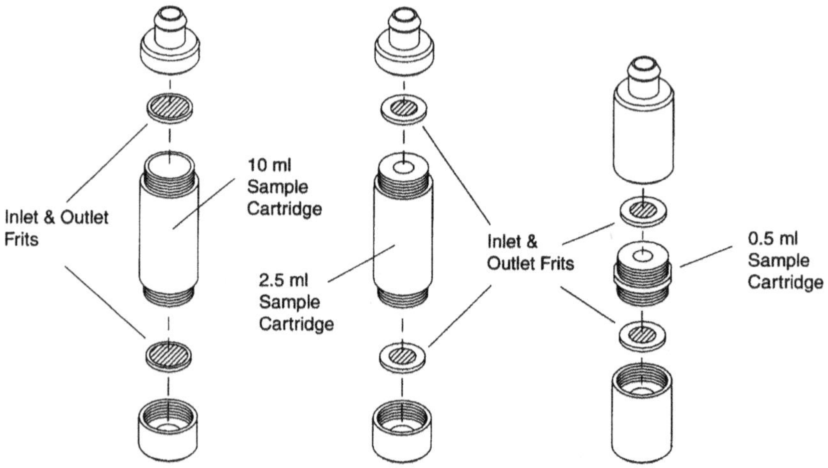


**Figure 2.5** Isco SFX sample cartridges have internal volumes of 0.5 ml, 2.5 ml, or 10 ml with identical outer dimensions to fit a patented pressure-equalizing extraction chamber.

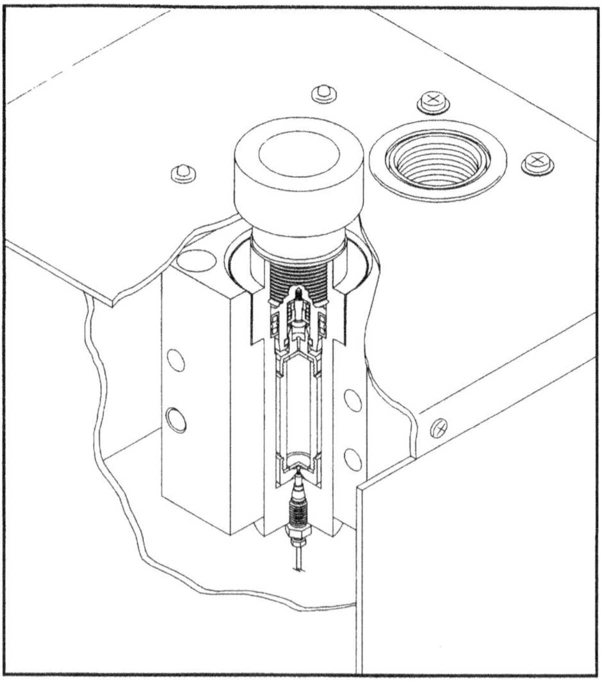

**Figure 2.6** In Isco SFX extractors, the sample cartridge is loaded into a closely fitted pressure chamber. The space surrounding the cartridge is statically pressurized with $CO_2$ to equalize the pressure of $CO_2$ flowing through the sample inside the cartridge. This design enables the use of lightweight, disposable cartridges.

2–5 μm. In general, it is suggested that the SFE vessel volume should exceed that of the sample by 10% [20]. The shape of the extraction vessel (long and narrow, or short and fat), and its orientation (horizontal or vertical) have been studied by Langenfeld *et al.* [21]. They concluded that the extraction rates were virtually identical under the variations tested. Overall, sample vessels used by the operator will be dependent on those available from the manufacturer of the SFE instrument.

No matter what type of SFE vessel is used, safety is always a concern in the laboratory. For example, although nitrous oxide ($N_2O$) has been successfully used for several SFE applications its use should be avoided. It is an oxidant that can react explosively with organic matrices. A graphic example has been reported [20] in which a vessel rated at 10 000 psi ruptured during the decaffeination of coffee using $N_2O$. It was estimated that the vessel burst once the pressure exceeded 80 000 psi. Warnings concerning the safety of $N_2O$ in SFE have been published [22].

## 2.7 Supercritical fluid extraction flow-control devices and restrictors

Flow-control devices, which are commonly known as 'restrictors', function primarily as maintainers of back pressure for the entire SFE system. In any SFE system, an optimal restrictor should possess the following characteristics:

- rugged construction to prevent plugging and analyte carry-over;
- provide constant, even flow for a variety of modified and unmodified extraction fluids;
- provide a ready interface with the trap or analyte collection device;
- function at the temperature ranges necessary to optimize analyte trapping.

In present commercial and self-built SFE systems, restrictors are of two general types: fixed and variable flow. Both types are described below.

### *2.7.1 Fixed-flow restrictors*

The initial restrictors employed in SFE included crimped stainless-steel tubes [23] as well as restrictors developed for SFC technology and transferred to SFE. The flow rate allowed by a fixed-flow restrictor depends upon several factors, including type of fluid, pressure and temperature and the individual physical characteristics of the restrictor. The restrictors previously and currently utilized in SFC include linear, tapered, integral and frit restrictors. However, given the unique analytical demands of SFE, SFC flow restrictors typically perform less than optimally in SFE.

*Linear flow restrictors in supercritical fluid extraction applications*

Although several other materials have been utilized, fused silica capillaries were amongst the first and most commonly used linear flow restrictors

**Figure 2.7** Linear capillary restrictor, showing capillary inner diameter (i.d.).

used in SFE. These restrictors are fabricated from appropriate lengths of capillary tubing of small internal diameter (i.d.), usually less than 50 μm (Figure 2.7). Fused silica restrictors are economical, reliably produced by the user and quick to replace [24–28]. The flow rate is dependent upon the i.d. and length of the restrictor. Unfortunately, fused silica linear flow restrictors become brittle with the use of certain modified fluids, such as methanol-modified $CO_2$. This results in a short-lived restrictor, which necessitates frequent replacement.

All linear restrictors are prone to plugging, which results from loss in analyte solubility as the density of the SF drops during its decompression along the length of the restrictor. Cooling due to the Joule–Thompson effect as the fluid expands can further exacerbate this plugging. Although restrictor heating reduces plugging, fused silica linear flow restrictors prove difficult to heat thoroughly and consistently [29].

The commercial development of coaxial heated stainless-steel restrictors [30] (Figure 2.8) has greatly increased the range of SFE applications using this type of restrictor. This is attributed to the elimination of Joule–Thompson cooling and a heating arrangment which ensures that the SF decompresses at the very tip of the restrictor. Also, the restrictor life-time is not shortened by the use of modified SFs.

*Tapered restrictor*

Tapered restrictors were developed for use in SFC [31] (Figure 2.9) to reduce the length of the decompression zone thus minimizing analyte deposition and associated plugging [25, 27]. Generally, these restrictors are manually produced by flame heating fused silica capillary tubing and gently pulling to produce a tapered end with inner diameters of 1–5 μm. These restrictors are difficult to manufacture reproducibly and are very fragile [32]. Although Chester *et al.* [31] have experimented with robotics to improve production and reproducibility, tapered restrictor fragility continues to be a problem.

*Integral restrictors*

Integral restrictors [33–35] are generally produced by heating the tip of a fused silica capillary tube until occluded (Figure 2.10(a)). The excess silica is abraded until the desired orifice is achieved (Figure 2.10(b)). Considerable skill is required to manufacture reproducible integral restrictors by this technique. An alternative and apparently much easier method of producing integral restrictors involves the use of a microfusion device [36]. With this approach the end of a fused silica capillary tube is locked into position between two discharge electrodes, mounted on the stage of a ×50 microscope. The electrodes

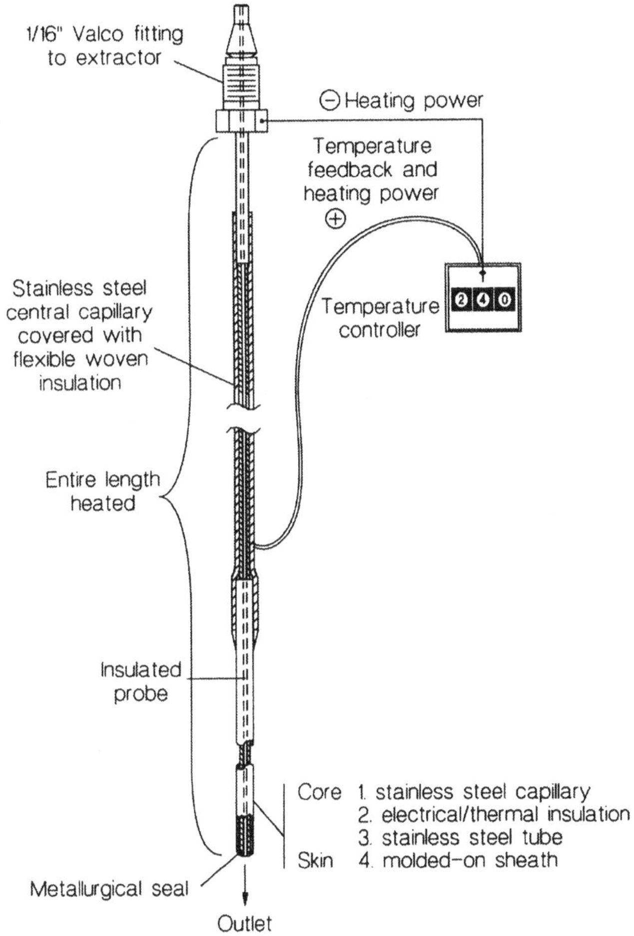


**Figure 2.8** Coaxial heated stainless-steel (linear capillary) restrictor.

can be arced in a controlled manner (in short bursts towards the end of the process) such that the end of the fused silica capillary tube or column becomes almost completely imploded. The integral restrictor is durable and has a short flow path for decompression, therefore it is less prone to plugging.

*Frit restrictors*

Frit restrictors [37–39] were developed employing the theory of 'many paths' for supercritical fluid decompression which uses a porous frit packing to produce a back pressure (Figure 2.11). Because of the shortened decompression zone, frit restrictors are less prone to plugging. Unfortunately, they are relatively expensive and difficult to manufacture compared with other fused silica restrictors.

**Figure 2.9** Tapered restrictor.

**Figure 2.10** Integral restrictor: (a) silica capillary melted to occlude tip; (b) tip abraded to produce orifice of suitable size.

### *2.7.2 Variable-flow restrictors*

Variable-flow restrictors have a distinct advantage over fixed-flow restrictors by eliminating the interdependence between flow rate and pressure–temperature parameters. Back-pressure regulators, previously used in SFC, are inappropriate for use in SFE because of excessive dead volume. These early back-pressure regulators preceded the development of low dead volume, manually controlled heated restrictors. Manufacturers of SFE instrumentation have produced several iterations of variable restrictors from those which were manually regulated to today's computer-controlled designs. SFE technology continues to advance with the development and evolution of computer-regulated, electronically controlled restrictors with optimized collection interfaces. Below is a short developmental perspective of variable restrictors.

*Early Suprex manual restrictor*

Early designs of a manual variable restrictor (Suprex) [39] regulated the flow rate by manually reducing the i.d. of PEEK tubing with a clamp-like device at a point just prior to decompression (Figure 2.12). Although certainly not ideal, this restrictor was a significant improvement over earlier fixed restrictors. Plugging was minimal as the decompression zone was short, and if partial plugging did occur the obstruction usually could be cleared by opening the restrictor. The device, however, was difficult to heat and the PEEK tubing lost its elasticity with repeated use. On the plus side, the restrictor could be rebuilt easily by the operator so commercial use was both cost effective and user friendly.

*Hewlett-Packard variable restrictor*

Hewlett-Packard integrated a heated automatic variable restrictor [40] and solid phase trap into their SFE system. This restrictor is an axially driven,

**Figure 2.11** Frit restrictor.

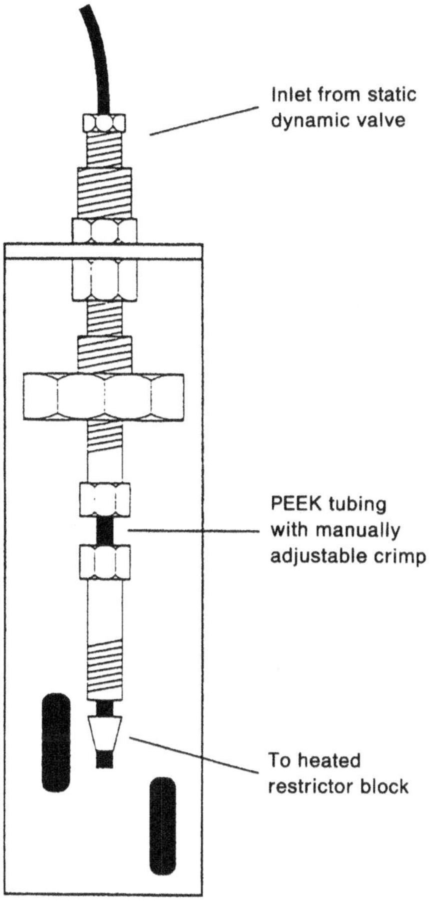


**Figure 2.12** Manually variable restrictor based on compression crimping of poly(ether-ether-ketone) (PEEK) tubing.

low dead volume solenoid valve, in which flow is computer regulated through an orifice and diaphragm design. The solenoid valve controls the flow rate by impinging the diaphragm into and out of an expansion nozzle. Placed before a solid phase trap this restrictor is reliable and allows the changeover of different extract traps.

*Isco variable restrictors*

Isco developed a manually adjustable restrictor [41] optimized for use with solvent collection (Figure 2.13). The restrictor consists of a long, thin stainless-steel needle surrounded by a coaxially heated (150°C maximum) stainless-steel barrel. The barrel is then covered with a chemically inert, high-temperature polymer which acts as an electrical and thermal insulator.

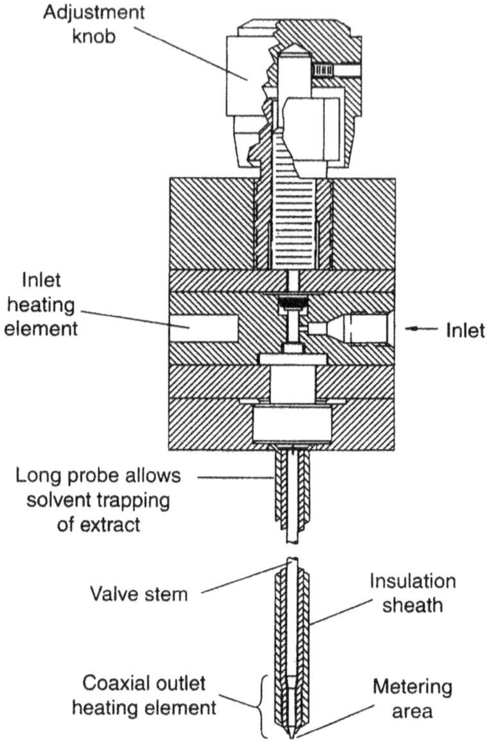


**Figure 2.13** Manually adjustable restrictor designed for direct liquid analyte trapping in a solvent-filled vial.

This design is ideal for liquid trapping as the extended barrel tolerates immersion in common organic solvents. Because the entire length of the restrictor is heated, the fluid remains in the supercritical state, with decompression occurring at the very tip of the restrictor, maximizing analyte transfer to the collection solvent. Moreover, the temperature of the restrictor block and barrel are individually controlled, allowing for optimal settings. With the flow rate regulated by adjusting the knob at the top of the restrictor, a wide flow range is possible.

Isco later evolved a fully automated supercritical fluid extractor, integrating the concept of the manual variable restrictor (Figure 2.14). Successful incorporation into an automated SFE instrument (Isco SFX 3560) [42] allows for a computer-controlled version of the restrictor utilizing a stepping motor to regulate flow rate, with the commercial advantage of hands-free operation.

*Suprex automated restrictor*

An automated variable restrictor (AVR) was also marketed by Suprex [43]. This design concept is based on the principle of a modified needle valve

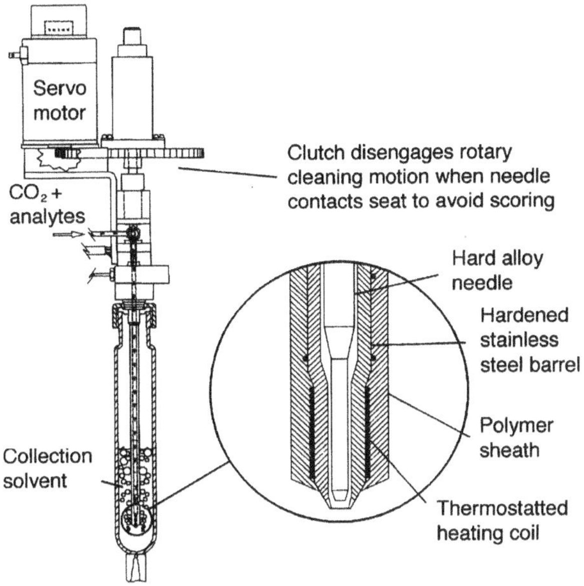


**Figure 2.14** Automated restrictor designed for direct liquid analyte trapping.

(Figure 2.15), which replaces the traditional tapered needle with a blunt stem. The blunt stem, controlled by a computer-driven electric motor, regulates the flow of supercritical fluid. The heated AVR is the most versatile variable restrictor as it interfaces with both solid phase and liquid traps. Featured on several Suprex commercial instrument models, the AVR has the additional feature of being the most user serviceable of all the variable restrictors described.

### 2.7.3 *Summary*

Since the late 1980s, flow-control devices (restrictors) used with supercritical fluid extractors have progressed from primitive, unheated, crimped stainless-steel tubing to heated, zero or low dead volume control valves. Several of the manual and computer-controlled valves have been optimized for use with analyte collection apparatus. Improvement in restrictor design has resulted in an increase in the successful use of SFE for a vast variety of analytes and matrices. Consequently, since optimization of analyte collection is interfaced with efficient recoveries, the reproducibility of SFE methods has significantly improved. Further work and improvements will no doubt focus on maintenance-free, automated variable restrictors for both on-line and off-line collection modes.

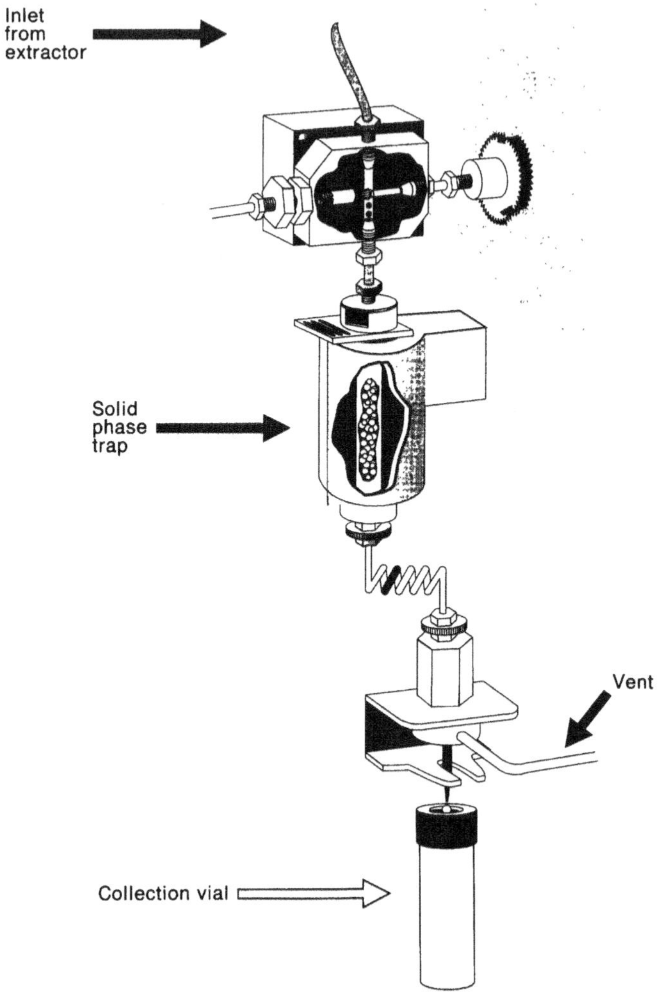


**Figure 2.15** Suprex automated restrictor.

## 2.8 Supercritical fluid extraction collection modes

After efficiently extracting the analytes of interest from the sample matrix it is crucial to any analytical method to collect quantitatively and reproducibly the analytes for further analysis. SFE is no different from any other analytical sample preparation technique in this respect. However, since SFE is more often than not considered a preparative technique, quantitative collection is frequently overlooked. The result of this is, of course, misinterpretation of the data. Quantitative collection or trapping of supercritical fluid extracted compounds is a highly significant aspect of SFE. There have been examples where poor extraction efficiency was thought to be the source of an SFE problem. When carefully and properly analysed, a poor or inefficient collection technique was the true cause. It is imperative that SFE users and researchers fully understand the advantages and disadvantages of each collection method. By the same token a basic understanding of the chemical and physical attributes of both the analyte and the trapping method will lead to correct decision making and successful collection. Below is a summary of SFE collection modes (see also sections 3.3 and 4.2.5). These modes can be divided into two collection methods. The first is off-line analysis, in which the entire analyte is collected and manually transported to an analysis instrument (Figure 2.16). The second is on-line SFE, in which the analyte is directly

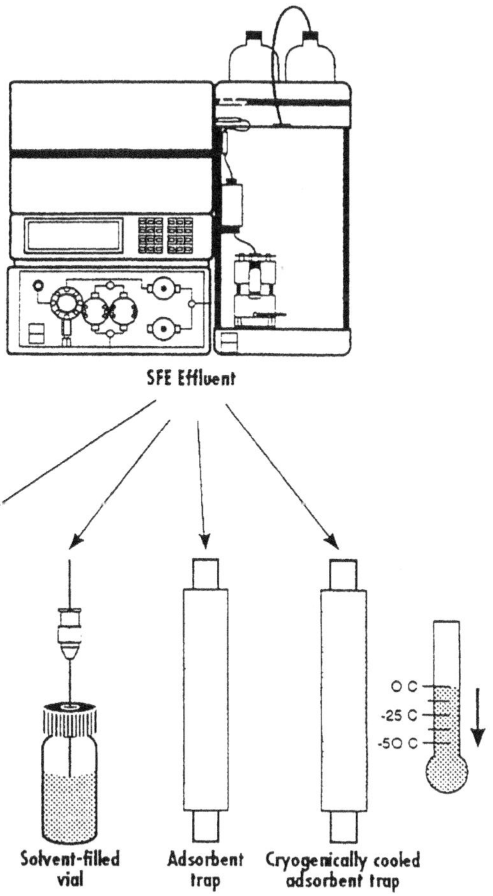


**Figure 2.16** Supercritical fluid extraction (SFE) collection modes for off-line analyte trapping.

collected to the input of an analysis instrument without user intervention. For the latter method, on-line SFE, only a brief outline will follow since these methods are fully explained in the following chapters. Whether off-line or on-line SFE is employed, the basis of collection is, in effect, to change the fluid conditions to force analyte equilibrium to favor the collection media over solvation in the supercritical fluid. To accomplish this, the temperature or pressure or both are lowered to subcritical conditions in which the analyte is more soluble in the trap solvent rather than the now subcritical fluid or gas. Alternatively, the analyte depressurizes from the SF solution onto a solid support. Relatively simple experimental designs involving internal standards and reference standards can be used to establish collection efficiency. A complete understanding of this will allow for optimization of collection and will lead to successful extraction protocols.

### *2.8.1 Off-line liquid trapping*

Liquid trapping or collection is when the SFE restrictor output is allowed to depressurize in a small volume (generally *c.* 5 ml) of an organic solvent. The selection of this solvent must be such that the analyte is reasonably to highly soluble. Other parameters such as temperature and pressure must also be considered. Langenfeld and co-workers [21] examined collection solvent parameters for collection efficiencies of semivolatile pollutants, concluding that by controlling the collection solvent temperature at 5°C one can significantly improve the collection efficiency. Burford *et al.* [44] continued this work to demonstrate that the level of solvent and the depth of the restrictor in the collection solvent could be adjusted to improve collection efficiency, demonstrating the physical parameters of solvent collection. Solvent collection has been successful in the quantitative trapping of a variety of analytes from chlorinated hydrocarbons [45] to thiosulfinates from garlic and onion [46]. Considerable research has been directed towards the study of various parameters of solvent trapping, such as collection efficiencies between open-flask and closed-flask containers [47] and collection efficiencies as a function of single and mixed solvents [48]. Solvent collection was considered a method of choice for only non-volatile and semivolatile compounds, until Jameson [49] successfully demonstrated pressurized liquid collection. This method, in which the restrictor depressurizes in a solvent trap that is under 30 psi pressure, controls the bubble size and allows for a more efficient transfer of the analyte to the collection fluid. Thus this method improves on the solvent collection of lower molecular weight semivolatiles and volatile compounds, allowing for solvent collection of analytes such as flavors and fragrances.

### *2.8.2 Off-line solid phase collection*

Solid phase collection or sorbent trapping was considered an alternative trapping method to liquid trapping. Evaluations and comparisons of both methods have established that each method has preferred niches in analyte collection [50, 51]. Newer SFE systems offer both collection types to optimize available choices for SFE applications. Volatile and low molecular weight semivolatiles, such as flavor components, are best collected by means of solid phase trapping. Here the supercritical fluid passes over a bed of small particles (40–100 μm diameter). The collection beads are cooled, in some cases down to −50°C for optimal trapping. Once the extraction is complete the trap is rinsed with organic solvent to elute the analyte quantitatively into a vial. Comparisons of sorbent trap types such as stainless-steel beads, silica particles, glass beads and octadecylsilyl ($C_{18}$) modified silica have been studied [52–55]. Again, trapping conditions varied with analyte type from collection of a drug from animal feed [56] to polycyclic aromatic hydrocarbons (PAHs) from marine sediment [57], demonstrating the versatility of this collection method.

### *2.8.3 Off-line solventless collection*

In some applications it is possible to collect an SFE analyte quantitatively into an empty vial. One application is the collection of fat from food products. Here the restrictor is placed in a vial containing only glass wool. The fat is collected efficiently on glass wool, which, once dried to a constant weight, can be used gravimetrically to determine the percentage of fat [58]. In a different application, Miller *et al.* [59] demonstrated solventless collection of polychlorinated biphenyls (PCBs), PAHs, gasoline and diesel fuel.

### *2.8.4 On-line collection modes*

Various successful on-line methods have been reported. These methods directly couple the SFE output with a trapping method which provides direct input of the analytes into an analytical instrument. Various reviews have documented the success of on-line SFE techniques [60–62]. The principal advantage of on-line SFE is the quantitative transfer of extracted analytes to the analysis instrument. In addition, reduction of contamination levels in on-line SFE have been reported [63]. Hawthorne *et al.* [64] demonstrated the effectiveness of SFE–GC for the extraction and analysis of flavor and fragrance compounds. An account describing the coupling of SFE and GC was later published [65]. On-line SFE–GC forms the subject of Chapter 5.

Other successful SFE coupling methods include on-line SFE–fluorescence detection [66], on-line SFE–FTIR [67–69] and on-line SFE–SFC in the analysis of vitamins [70] and in the analysis of polymer additives [71]. As with all transfer methods, the techniques of quantitative transfer require evaluation. Ashraf-Khorassani *et al.* [72] examined SFE–SFC coupling involving a unique intermediate cryogenic collection procedure.

On the negative side, however, it is possible to co-extract undesirable compounds. Should these interfere with the analysis instrument costly repair or shortened column life may result. It is therefore recommended to ascertain the full range of extracted components by using off-line SFE prior to the use of on-line SFE.

### *2.8.5 Summary*

In conclusion, SFE collection methods must be carefully examined to ensure quantitative and reproducible trapping of the analytes under study. There is now a wealth of information in the scientific literature. Besides some excellent reviews on SFE [60–62], there is an SFE research paper database, *SF Searcher: SFE Database*, which may be obtained from Isco Inc. (4700 Superior Street, Lincoln, NE 68505, USA). This database contains pertinent SFE references from about 1981 to the present.

## 2.9 Automation of supercritical fluid extraction

As soon as the first commercial SFE instruments were available many users expressed the need for systems that could extract numbers of samples greater than two, with minimal attention and manipulation. Early researchers of SFE technology found that, although SFE was indeed much faster than conventional liquid extraction techniques for single samples, considerable operator labor was required to run multiple samples, limiting the number of samples that could be extracted in a normal eight-hour working day. The need for increased sample throughput was especially noticeable as SFE moved from the laboratories of pioneering researchers, with a minimal sample load requirement, to quality assessment or control and testing laboratories which routinely analyse multiple batches of samples. The SFE market place was thus requiring an automated system which would operate overnight, increase productivity, needed little operator intervention and required minimal maintenance with short down times. Unfortunately, technological obstacles hindered development. Chief among these was the development of a reliable automated flow restrictor [73]. With considerable development, the first commercial automatic flow restrictor was introduced in 1990 by Hewlett-Packard on their model 7680A system. Although this system had only a single-sample capacity it laid the groundwork for later developments in automated SFE. In the review below, the evolution of automated SFE systems is divided into two distinct categories, each with advantages and disadvantages.

### *2.9.1 Parallel supercritical fluid extraction systems*

In the parallel approach, three or more samples are extracted simultaneously, achieving some level of automation in the process. These early SFE systems, such as the Suprex SFE/50M and the Dionex 723, which could simultaneously extract four or eight samples, respectively, both used fixed-flow restrictors. Later, Suprex incorporated the DuraFlow® restrictor, which permitted the user to vary the extraction flow rate manually. Although a significant improvement over fixed restrictors, automatic plug clearance was not present. The development of a practical automated system was thus delayed until the general availability of an automated attendant-free restrictor. With the advent of the automatic restrictor, first integrated into the HP 7680A SFE system, practical parallel systems became theoretically feasible. However, these were not in fact produced until later, when the LECO Corporation (St Joseph, MI, USA) introduced the RFA-100 Rapid Fat Analyzer (later FA-100) in 1995. The reason for this development gap was that many manufacturers chose to use automatic restrictors in sequential rather than parallel systems, mainly because of the cost and complexity of producing parallel systems. The development of automated systems thus progressed down the sequential rather than parallel path until the introduction of the RFA-100.

The RFA-100, capable of performing three simultaneous extractions, consists of a 'master' module (single-piston reciprocating pump, self-contained refrigeration unit and microprocessor) (Figure 2.17) which can control as many as two additional, 'slave', modules, to provide a total of nine simultaneous extractions. Although this system provides good throughput it requires the operator to change extraction vessels manually at the end of the extraction, rendering it incapable of unattended operation for extended

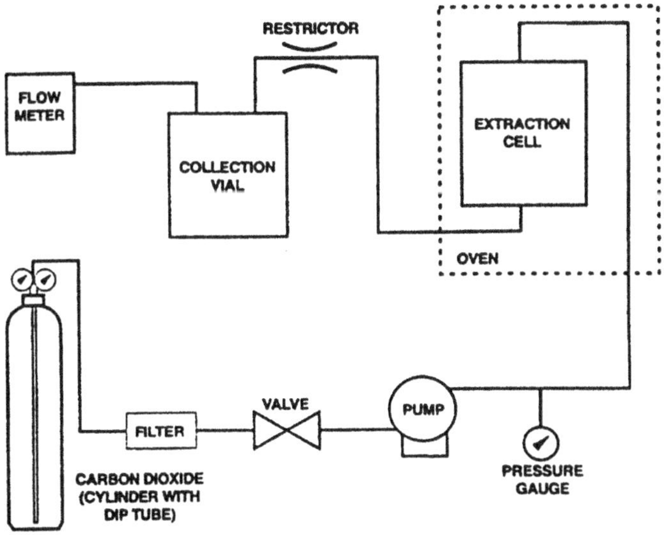


**Figure 2.17** LECO FA-100 parallel extractor, showing the flow of the three channels. Reproduced with permission from LECO Corporation, 3000 Lakeview Avenue, St Joseph, MI 49085-2396, USA.

periods. This may not be a problem for some plants which operate around the clock but does not meet the needs of the method development or research laboratories, which require a system capable of extracting under different conditions without constant attention. These latter requirements are thus best met by sequential rather than parallel automated systems. In addition, the RFA-100 is by design limited to a range of SFE applications as analyte (fat or oil) collection is accomplished by trapping into vials containing glass wool. The levels of fat extracted are determined gravimetrically.

### *2.9.2 Sequential supercritical fluid extraction systems*

The first practical automated sequential SFE system was the HP 7680T, introduced in 1992. An extension of the HP 7680A (see Figure 4.8), its design includes the addition of a sample turret (hence the 'T' designation) to increase its load to eight sample vessels or 'thimbles', each having a sample capacity of up to 7 ml. The maximum pressure and temperature at which extractions can be performed is 386 atm and 150°C, respectively. In use, the loaded stainless-steel thimbles are sealed with finger-tight end caps containing integral porous stainless-steel frits (see Figure 4.4). These are placed in the turret, which rotates to the appropriate position. A piston or ram rises up from underneath the turret and conveys the sample thimble into the oven where the extraction begins. Analytes are collected on a solid phase trap which is automatically washed by a solvent (or solvents) into GC-type autosampler vials mounted on a carousel. The entire system is controlled by computer. The system is able to run a maximum of eight samples which, given a typical extraction time of 30–45 min, cannot be used during overnight operations. This situation was indirectly addressed by the later addition of the 'bridge', which allowed the HP 7680T to be interfaced with a GC or HPLC system via an autosampler. Here the autosampler arm acts as a bridge to transfer vials from the SFE rack to the autosampler rack on the chromatograph. This provided an additional level of analytical automation but did not improve the SFE throughput since only eight samples could be extracted before operator intervention. The pumping system is a dual-piston reciprocating pump derived from the HP 1080 HPLC. With all things considered, the HP 7680T was the first system that could realistically be called 'automated'.

The next system commercially introduced in late 1992 was the Suprex AutoPrep-44 (AP-44) (Figure 2.18). The AP-44 has a capacity of 44 extraction vessels, with a maximum volume of 10 ml each, and operates analogously to the HP 7680T. The notable exception is the automated restrictor or solid phase analyte collection features, which are housed in separate modules. In this way, the collection module or AccuTrap®, can be used by other Suprex SFE systems. Another difference between the HP system and the Suprex system is that the AP-44 utilizes an on-board microprocessor rather than direct computer control.

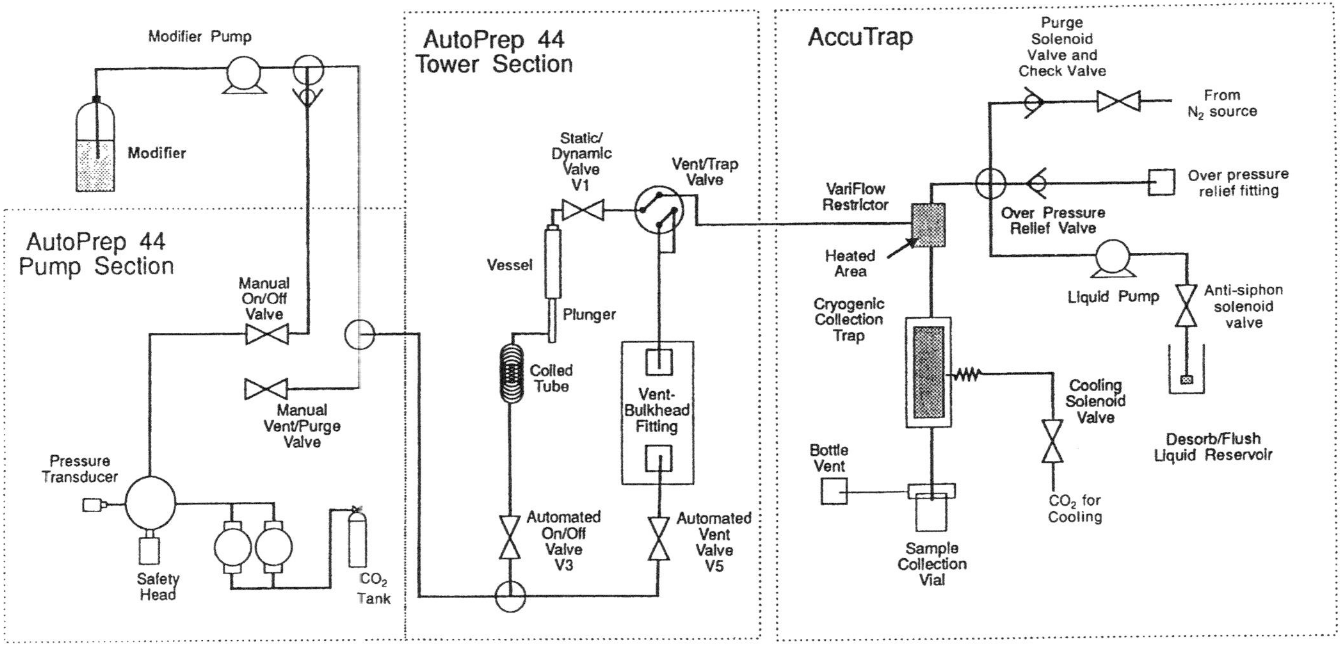


**Figure 2.18** Suprex AutoPrep 44 automated extractor flow diagram.

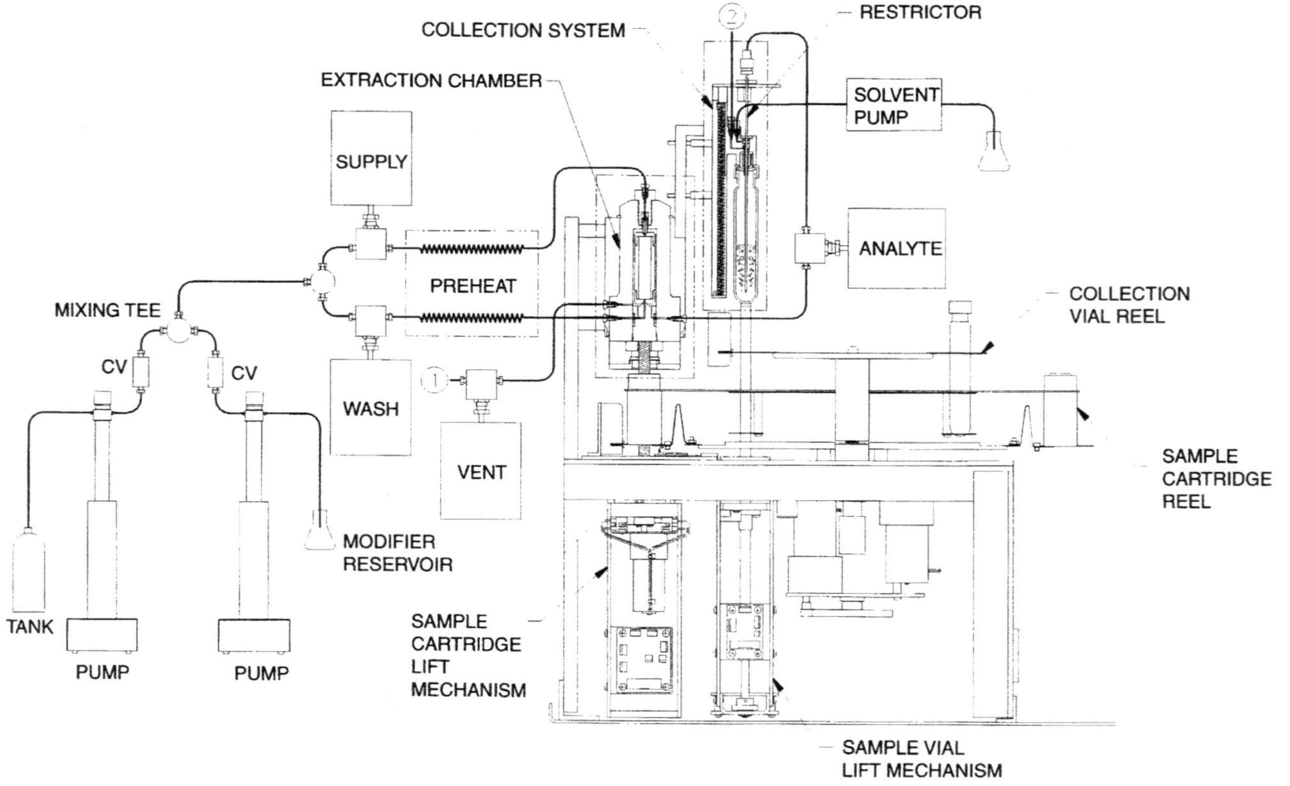


**Figure 2.19** Schematic of the Isco SFX 3560 automated extractor. CV = check valve.

The pump on the AP-44 is specifically designed for compressible fluids such as liquid carbon dioxide, rather than being a modified HPLC pump. This dual-piston reciprocating pump uses ultrahard ceramic cylinder liners to reduce piston seal wear and acts as an insulating medium to eliminate the requirement for ancillary cooling. In this regard it succeeded, but if cooling is to be entirely dispensed with the extraction fluid (usually carbon dioxide) has to be supplied in cylinders with a head pressure of helium of 1500–2000 psi in order to maintain pumping efficiency. The restrictor and its control also differ from the HP SFE systems in that the restrictor controls the extraction fluid flow while the pump maintains the set pressure. With the HP SFE systems the flow rate is controlled by the pump, with the restrictor controlling the pressure. Each vendor claims advantages in their design. Both systems represent significant technological advances.

In 1993, Isco introduced the SFX 3560, an automated 24-sample SFE system (Figure 2.19). This system differs from the two automated systems above in several important design features, one of which is the pumping system, the SFX 3560 using syringe pumps instead of reciprocating pumps. The SFX 3560 also uses the proprietary pressure-equalizing cartridge system, where the sample is contained in a relatively thin-walled cartridge which in turn is placed inside a larger pressure vessel. Since the pressure inside and outside the cartridge is the same, the need for heavy thick-walled extraction vessels is eliminated, allowing for the use of inert polymeric cartridges which are inexpensive, easy to use and quickly cooled [19].

Another important difference in the design of the SFX 3560 is the mode of analyte collection. Whereas HP and Suprex have developed cooled solid-phase trapping, Isco uses solvent collection, in which analytes transfer to a suitable collection solvent after passing through the restrictor. This collection method is best suited to the collection of less volatile analytes, although, with experience, relatively volatile molecules can be trapped.

### *2.9.3 Summary*

It is interesting to note the chronology of events in the development of automated sequential systems and parallel systems. It is unknown if either approach will prevail in the future. Given the diversity of needs in analytical SFE, it is most likely that both types will fill different niches. The most probable future developments in automated SFE systems will evolve from current systems leading to better adaptation for routine use.

## 2.10 Future developments

Many advancements in the field of SFE have occurred during the 1990s, with several more on the horizon. One major advancement in SFE instrumentation includes the introduction of totally automated SFE systems, which have increased the productivity of both research and quality control laboratories. Other likely advancements in SFE instrumentation are discussed below.

### *2.10.1 Supercritical fluid extraction in the production environment*

The use of SFE for process monitoring and control within industry offers significant potential especially when interfaced on-line to analytical instruments such as infra-red spectrometers. Instruments designed for the production environment will have to be rugged and simple to use. Possible features needed for this environment include: bar-code reading of samples for chain of custody and an integrated balance for gravimetric analysis. Since production environments often require high sample throughput per working shift the further development of automated parallel SFE systems rather than sequential systems should afford advantages in this respect.

### *2.10.2 Field portable systems*

With the increasing need for rapid environmental screening tests, the development of field portable SFE seems a logical development. With slight modifications, some SFE systems can be mounted on a cart and operated using a gas generator. Once the extraction is complete, either an ELISA assay or field portable GC analysis may follow.

### *2.10.3 Pressurized fluid extraction*

Some SFE manufacturers have modified their SFE instruments to work with both supercritical fluids and liquid solvents for enhanced solvent extraction (ESE). Instruments of the future will be designed to work across the entire spectrum of gases and solvents, hence the advent of the pressurized fluid extraction (PFE) system.

## References


1. Dooley, K.M., Kao, C.-P., Gambrell, R.P. and Knopf, F.C. (1987) The use of entrainers in the supercritical extraction of soils contaminated with hazardous organics. *Industrial Engineering Chemical Research*, **26**, 2058–62.
2. Vance, S.W. (1994) Supercritical fluid service, in *Supercritical Fluid Processing of Food and Biomaterials* (ed. S.S.H. Rizvi), Blackie Academic & Professional, London, pp. 93–102.
3. Van Lenten, F.J. and Rothman, L.D. (1976) Pressure control of a liquid chromatograph pump. *Analytical Chemistry*, **48**, 1430–2.
4. Peaden, P.A., Fjeldsted, J.C., Lee, M.L. *et al.* (1982) Instrumental aspects of capillary supercritical fluid chromatography. *Analytical Chemistry*, **54**, 1090–3.
5. Allington, R.W. (1983) *US Patent*, 4,422,942.
6. Allington, R.W. and Jones, J.N. (1989) *US Patent*, 4,882,063.

7. Kordikowski, A., Robertson, D.G. and Poliakoff, M. (1996) Acoustic determination of the helium content of carbon dioxide from He pressure cylinders and FT-IR studies of the density of the resulting supercritical $CO_2$: implications for reproducibility in supercritical experiments. *Analytical Chemistry*, **68**, 4436–40.
8. Wehling, R.L., Froning, G.W., Cuppett, S.L. and Niemann, L. (1992) Extraction of cholesterol and other lipids from dehydrated beef using supercritical carbon. *Journal of Agricultural Food Chemistry*, **40**, 1204–7.
9. Froning, G.W., Wehling, R.L., Cuppett, S.L. *et al.* (1990) Extraction of cholesterol and other lipids from dried egg yolk using supercritical carbon dioxide. *Journal of Food Science*, **55**, 95–8.
10. Gere, D.R., Board, R. and McMinigill, D. (1982) Supercritical fluid chromatography with small particle diameter packed columns. *Analytical Chemistry*, **54**, 736–40.
11. Lauer, H.H., McManigill, D. and Board, R.D. (1983) Mobile-phase transport properties of liquefied gases in near-critical and supercritical fluid chromatography. *Analytical Chemistry*, **55**, 1370–5.
12. Greibrokk, T., Blilie, A.L., Johansen, E.J. and Lundanes, E. (1984) New system for delivery of the mobile phase in supercritical fluid chromatography. *Analytical Chemistry*, **56**, 2681–4.
13. Greibrokk, T., Doehl, J., Farbrot, A. and Iversen, B. (1986) Mobile phase delivery in supercritical fluid chromatography. *Journal of Chromatography*, **371**, 145–52.
14. Koebler, I. and Williams, G.P. (1993) *US Patent*, 5,193,991.
15. Wright, B.W. and Fruchter, J.S. (1992) *US Patent*, 5,087,360.
16. Schweighardt, F.K. and Mathias, P.M. (1993) Impact of phase equilibria on the behavior of cylinder stored $CO_2$–modifier mixtures used as supercritical fluids. *Journal of Chromatographic Science*, **31**, 207–11.
17. Pariente, G.L., Pentoney, S.L., Griffiths, P.R. and Shafer, K.H. (1987) Computer controlled pneumatic amplifier pump for supercritical fluid chromatography and extractions. *Analytical Chemistry*, **59**, 808–13.
18. Durst, K., Porer, N.L., Riches, R.A. *et al.* (1992) *International Patent*, WO92/06058.
19. Allington, R.W., Clay, D.L., Jameson, D.G. and Winter, R.R. (1992) *US Patent*, 5,132,014.
20. Taylor, L.T. (1996) *Supercritical Fluid Extraction*, John Wiley, New York.
21. Langenfeld, J.J., Burford, M.D., Hawthorne, S.B. and Miller, D.J. (1992) Effects of collection solvent parameters and extraction cell geometry on SFE efficiencies. *Journal of Chromatography*, **594**, 297–307.
22. Raynie, D.E. (1993) Warning concerning the use of nitrous oxide in supercritical fluid extractions. *Analytical Chemistry*, **65**, 3127–8.
23. Smith, R.D., Fjeldsted, J.C. and Lee, M.L. (1983) Direct fluid injection interface for capillary supercritical fluid chromatography–mass spectrometry, *Journal of Chromatography*, **247**, 231–43.
24. Fjeldsted, J.C., Kong, R.C. and Lee, M.L. (1983) Capillary supercritical fluid chromatography with conventional flame detectors. *Journal of Chromatography*, **279**, 449–55.
25. Smith, R.D. and Udseth, H.R. (1983) Mass spectrometry with direct supercritical fluid injection. *Analytical Chemistry*, **55**, 2266–72.
26. Pinkston, J.D. and Henstschel, R.T. (1993) Evaluation of flow restrictors for open-tubular supercritical fluid chromatography at pressures up to 560 atm. *Journal of High Resolution Chromatography*, **16**, 269–74.
27. Cortes, H.J. (1992) Developments in multidimensional separation systems. *Journal of Chromatography*, **626**, 3–23.
28. Yang, Y., Hawthorne, S.B. and Miller, D.J. (1995) Estimating flow-rates for sub- and supercritical fluid extractions with linear restrictors. *Journal of Chromatography*, **690**, 131–9.
29. Hawthorne, S.B., Miller, D.J., Burford, M.D. and Macomber, J. (1993) Construction of a robust stainless steel-clad fused silica restrictor for use in supercritical fluid extraction. *Journal of Chromatography*, **648**, 445–9.
30. Burford, M.D., Hawthorne, S.B., Miller, D.J. and Braggins, T. (1992) Comparison of methods to prevent restrictor plugging during off-line supercritical extraction. *Journal of Chromatography*, **609**, 321–32.
31. Chester, T.L., Innis, D.P. and Owens, G.D. (1985) Separation of sucrose polyesters by capillary supercritical-fluid chromatography/flame ionization detection with robot-pulled capillary restrictors. *Analytical Chemistry*, **57**, 2243–7.

32. Berger, T.A. (1989) Modeling linear and tapered restrictors in capillary supercritical fluid chromatography. *Analytical Chemistry*, **61**, 356–61.
33. Guthrie, E.J. and Schwartz, H.E. (1986) Integral pressure restrictor for capillary SFC. *Journal of Chromatographic Science*, **24**, 236–41.
34. Hawthorne, S.B. and Miller, D.J. (1987) Directly coupled SFE–gas chromatographic analysis of polycyclic aromatic hydrocarbons and polychlorinated biphenyls from environmental solids. *Journal of Chromatography*, **403**, 63–76.
35. Rose, K.A. and Schomburg, G. (1988) Instrumentation for SFC systems: different sampling and restriction designs. *Journal of High Resolution Chromatography*, **11**, 191–7.
36. Ramsey, E.D., Berry, A.J., Lawrence, S.D. *et al.* (1995) Ionophores: potential screening via supercritical-fluid chromatography combined with mass spectrometry and tandem mass spectrometry. *Rapid Communications Mass Spectrometry*, **9**, 712–16.
37. Cortes, H., Pfeiffer, C.D, Richter, B.E and Stevens, T.S. (1988) *US Patent*, 4,793,920.
38. Chang, H.-C.K and Taylor, L.T. (1990) The performance of electron-capture detection after capillary supercritical fluid chromatography. *Journal of Chromatographic Science*, **28**, 29–33.
39. Keobler, D.J. (1994) *US Patent*, 5,316,262.
40. Nickerson, M.A., Poole, J.S and Frank, L.G.R. (1991) *US Patent*, 5,009,778.
41. Allington, R.W., Walters, H.L., Jameson, D.G. and Tehrani, Y. (1997) *US Patent*, 5,614,089.
42. Jameson, D.G., Macomber R.J and Allington, R.W. (1993) *US Patent*, 5,268,103.
43. Bruce, M.L. and Keebler, D.J. (1995) *US Patent*, 5,379,790.
44. Burford, M.D., Hawthorne, S.B., Miller, D.J. and Braggins, T. (1992) Comparison of methods to prevent restrictor plugging during off-line SFE. *Journal of Chromatography*, **609**, 331–2.
45. Wenclawiak, B., Maio, G., Holst, H.V. and Darskus, R. (1994) Solvent trapping of some chlorinated hydrocarbons after supercritical fluid extraction from soil. *Analytical Chemistry*, **66**, 3581–6.
46. Calvey, E.M., Matusik, J.E., White, K.D. *et al.* (1994) Off-line supercritical fluid extraction of thiosulfinates from garlic and onion. *Journal of Agricultural Food Chemistry*, **42**, 1335–41.
47. Wright, B.W., Wright, C.W., Gale, R.W. and Smith, R.D. (1987) Analytical supercritical fluid extraction of adsorbent materials. *Analytical Chemistry*, **59**, 38–44.
48. Thompson, P.G., Taylor, L.T., Richter, B.E. *et al.* (1993) Trapping efficiencies of various collection solvents after supercritical fluid extraction. *Journal of High Resolution Chromatography*, **16**, 713–16.
49. Jameson, D.G. (1996) *US Patent*, 5,584,989.
50. Yang,Y., Hawthorne, S.B. and Miller, D.J. (1995) Comparison of sorbent and solvent trapping after supercritical fluid extraction of volatile petroleum hydrocarbons from soil. *Journal of Chromatography A*, **699**, 265–76.
51. Husers, N. and Kleibohmer, W. (1995) Studies on trapping efficiencies of various collection devices for off-line supercritical fluid extraction. *Journal of Chromatography A*, **697**, 107–14.
52. Furton, K.G. and Lin, Q. (1992) The dependence of sorbent analyte type on observed differences in supercritical fluid extraction efficiencies employing extraction vessels of different dimensions. *Chromatographia*, **34**, 185–7.
53. Mulcahey, L.J., Hendrick, J.L. and Taylor, L.T. (1991) Collection efficiency of various solid-phase traps for off-line SFE. *Analytical Chemistry*, **63**, 2225–32.
54. Mulcahey, L.J. and Taylor, L.T. (1992) Collection efficiency of solid surface and sorbent traps in supercritical fluid extraction with modified carbon dioxide. *Analytical Chemistry*, **64**, 2352–8.
55. Ashraf-Khorassani, M., Houck, R.K. and Levy, J.M. (1992) Cryogenically cooled adsorbent trap for off-line SFE. *Journal of Chromatographic Science*, **30**, 361–6.
56. Messer, D.C., Taylor, L.T., Weiser, W.E. *et al.* (1994) Quantitative analysis of a drug in an animal feed employing sample preparation by supercritical fluid extraction. *Pharmaceutical Research*, **11**, 1546.
57. Meyer, A. and Kleibohmer, W. (1993) Supercritical fluid extraction of polycyclic aromatic hydrocarbons from a marine sediment and analyte collection via liquid–solid trapping. *Journal of Chromatography A*, **657**, 327–35.

58. AOAC (1996) Official Method Am3-96, AOAC International, 1970 Chain Bridge Road, McLean, VA 22109, USA.
59. Miller, D.J., Hawthorne, S.B. and McNally, M.E.P. (1993) Solventless collection of analytes by rapid depressurization after static SFE. *Analytical Chemistry*, **65**, 1038–42.
60. Greibrokk, T. (1995) Applications of supercritical fluid extraction in multidimensional systems. *Journal of Chromatography A*, **703**, 523–36.
61. Chester, T.L., Pinkston, J.D. and Raynie, D.E. (1996) Supercritical fluid chromatography and extraction. *Analytical Chemistry*, **68**, 487R–514R.
62. Riekkola, M.-L., Manninen, P. and Hartonen, K. (1992) SFE, SFE/GC and SFE/SFC: instrumentation and applications, in *Journal of Chromatography Library Series, Volume 53. Hyphenated Techniques in Supercritical Fluid Chromatography and Extraction* (ed. K. Jinno), Elsevier, Amsterdam, pp. 275–304.
63. Wallace, J.C., Krieger, M.S. and Hites, R.A. (1992) Reduction of contamination levels in on-line supercritical fluid extraction systems. *Analytical Chemistry*, **64**, 2655–6.
64. Hawthorne, S.B., Krieger, M.S. and Miller, D.J. (1988) Analysis of flavour and fragrance compounds using supercritical fluid extraction coupled with gas chromatography. *Analytical Chemistry*, **60**, 472–7.
65. Hawthorne, S.B. (1992) Coupled supercritical fluid extraction–capillary gas chromatography (SFE–GC), in *Analysis with Supercritical Fluids: Extraction and Chromatography*, (ed. B. Wenclawiak), Springer, Berlin, pp. 61–73.
66. Dunham, D.C., Desmarais, M.S., Breid, T. and Hills, J.W. (1995) Carbon dioxide supercritical fluid extraction with on-line fluorescence detection. *Analytical Letters*, **28**(14), 2561–74.
67. Jordan, S.L. and Taylor, L.T. (1995) Extracting and quantifying a polyester fiber finish using on-line SFE/FTIR with intermediate trapping. *Textile Research Journal*, **65**, 230–5.
68. Lischeski, P.B. (1996) Supercritical fluid extraction coupled directly to infrared spectroscopy for the analysis of lipids in food. *Seminars in Food Analysis*, **1**, 85–100.
69. Liescheski, P.B. (1996) Supercritical fluid extraction coupled to infrared spectroscopy for iodine number analysis of edible oils. *Journal of Agricultural Food Chemistry*, **44**, 823–8.
70. Masuda, M., Koike, S., Handa, M. *et al.* (1993) Application of SFE and chromatography to assay fat-soluble vitamins in hydrophobic ointment. *Analytical Science*, **9**, 29–32.
71. Ashraf-Khorassani, M. and Levy, J.M. (1990) Quantitative analysis of polymer additives in low density polyethylene using supercritical fluid extraction/supercritical fluid chromatography. *Journal of High Resolution Chromatography*, **13**, 742–7.
72. Ashraf-Khorassani, M., Kumar, M.L., Koebler, D.L. and Williams, G.P. (1990) Evaluation of coupled supercritical fluid extraction–cryogenic collection–supercritical fluid chromatography (SFE–CC–SFE) for quantitative and qualitative analysis. *Journal of Chromatographic Science*, **28**, 599–604.
73. Hawthorne, S.B. (1993) Methodology for off-line supercritical fluid extraction, in *Supercritical Fluid Extraction and its Use in Chromatographic Sample Preparation* (ed. S.A. Westwood), Blackie Academic & Professional, London, pp. 39–64.

# 3 Off-line supercritical fluid extraction for solid matrices

J.M. BAYONA

## 3.1 Introduction

Off-line analytical supercritical fluid extraction (SFE) has been largely studied for the analysis of solid samples. The versatility of the technique has led to a large number of applications, a diverse range of analytes being reported (Chester *et al.*, 1994; Hawthorne, 1990; Taylor, 1995). Moreover, SFE of environmental matrices constitutes one of the largest groups of applications and has been previously reviewed (Barnabas *et al.*, 1994; Bøwadt and Hawthorne, 1995; Camel *et al.*, 1993; Janda *et al.*, 1993; McNally, 1995). In this chapter the discussion of analytical applications of SFE of solid samples will be updated, including general principals and current trends.

### *3.1.1 Sample preparation for supercritical fluid extraction*

Solid samples are subjected to several steps, depending on the sample characteristics prior to SFE, to ensure satisfactory extraction. Common procedures prior to sample determination are presented below. Sampling and sample preservation are beyond the scope of this chapter.

*Grinding*
Grinding is a common step prior to SFE for a variety of solid matrices, including both biotic and abiotic. SFE of additives from polymeric materials is frequently performed using ground polymer pellets since extraction kinetics are diffusion controlled.

Usually, samples with irregular shapes and a broad particle size distribution are ground prior to SFE to facilitate analyte mass transfer during the extraction process. In the case of biotic samples, grinding is usually combined with the drying step, discussed below. If volatile analytes are expected, it is of prime importance to grind the sample at low temperatures to avoid losses. It should be noted that some drying agents react exothermally with the moisture content of the sample.

*Sieving*
Solid samples may be heterogeneous in nature, containing a wide range of particle sizes whose organic or inorganic content may vary. Since extraction kinetics are frequently controlled by sample particle size it is very important to prepare the sample using a sieving step to obtain reproducible SFE results. In principle, small particle size is preferred to promote extraction kinetics but can lead to an increase in pressure drop along the extraction vessel. In addition, the passage of very fine particulate matter through frits can cause blockage of valves, restrictors, etc. Consequently it is advisable to use a filter membrane to protect the frits of the extraction vessel.

*Homogenization*
This pretreatment step is particularly important in SFE in order to obtain reproducible results, since the sample size is usually smaller than that used for conventional extraction techniques. Tumbling devices are commonly used for a relatively long period of time. In the case of labile analytes, precautions must be taken during sample homogenization; for example the use of an inert atmosphere is advisable.

*Drying*
Water is a common component in most real-world solid samples. Its presence may vary from trace levels to situations where it becomes the major component (e.g. in biotic samples, it constitutes *c.* 70% fresh weight). The low solubility of water in supercritical fluid carbon dioxide can cause difficulties in SFE of solid samples, particularly those containing large amounts of water, because of restrictor blockage caused by ice formation. Therefore, a drying step prior to SFE to ensure satisfactory recovery of analytes becomes mandatory when dealing with solid samples that contain considerable amounts of water. However, in some instances doping the sample with small amounts of water can produce a positive effect on the SFE efficiency for a variety of analytes (section 3.2.4). Drying the sample at elevated temperatures (typically 105°C) or even at room temperature can lead to losses of volatiles (e.g. *n*-tetradecane) or semivolatile compounds (e.g. octadecane) from the sample (Bøwadt and Hawthorne, 1995). Freeze drying is the common procedure used for sample preparation for conventional extraction techniques (e.g. Soxhlet, sonication) and it is also used in SFE. However, sample contamination in trace analysis or losses of volatile analytes during the lyophilization process are typical drawbacks of this sample preparation technique.

Several drying procedures have been evaluated to circumvent the freeze-drying step (Table 3.1). An extensive study concerning the characteristics of 21 drying agents has been published (Burford *et al.*, 1993). They found that when large percentages of water were present in the sample, losses of volatile species may occur owing to the possibility of exothermal reaction

**Table 3.1** Selected drying procedures used for solid sample preparation

| Analytes | Matrix | Drying agent | Reference |
|---|---|---|---|
| Polychlorinated biphenyls | fish tissue | basic $Al_2O_3$ | Lee *et al.*, 1995 |
| Carbamates | tissue (incurred; spiked) | Hydromatrix | Murugaverl *et al.*, 1993 |
| Alachlor, carbofuran, atrazine, benomyl, 2,4-dichlorophenoxyacetic acid | meat (spiked) | Hydromatrix | Souk and King, 1994 |
| Methamidophos | vegetables | magnesium sulphate | Valverde-Garcia and Fernández-Alba, 1995 |
| Pesticides (40) | fruit and vegetables | Hydromatrix–dry ice | Lehotay *et al.*, 1995 |

when the drying agent is mixed with the sample. Under these circumstances, volatilization losses can be minimized by first loading the drying agent into the outlet section of the extraction vessel. However, some polar analytes, such as phenols and anilines, can also be retained by some drying agents (Burford *et al.*, 1993). Hydromatrix blended with the sample in the presence of dry ice is one of the most effective methods, allowing the determination of volatile and semivolatile analytes (Lehotay *et al.*, 1995). Other authors have combined drying and cleanup steps with the use of activated adsorbents such as alumina for biotic matrices with a high lipid content (Lee *et al.*, 1995).

Figure 3.1 shows a flow chart which summarizes the stages involved in preparing a solid sample for SFE.

### *3.1.2* In situ *supercritical fluid derivatization extraction schemes*

*Covalent bond formation*

The extraction of polar analytes from abiotic matrices requires the addition of large amounts of modifiers to the supercritical fluid, which causes deterioration in extraction selectivity. *In situ* derivatization has been proposed to circumvent the lack of selectivity by using a procedure such that either polar functional groups of the analyte are converted to less polar derivatives or retentive sites associated with the matrix are derivatized. The successful implementation of either step can promote the selective extraction of analyte(s) using neat carbon dioxide at moderate temperatures and pressures. Supercritical fluid derivatization extraction (SFDE) is advantageous compared with room-temperature derivatization in terms of reaction completeness since the diffusion coefficients of supercritical fluids are higher than those of liquids. The derivatization reaction is usually performed in static SFE mode for a variable period of time (10–30 min) followed by dynamic SFE to extract the derivatized compounds (Table 3.2). A variety of derivatization reagents have been evaluated for the extraction of polar or ionic compounds from solid samples. Silylation reactions have been used to derivatize sterols from dried sewage (Nguyen *et al.*, 1995). They have also been successfully used to derivatize matrix active sites, which facilitates the extraction of several analytes [polycyclic aromatic hydrocarbons (PAHs), halogenated phenols, polychlorinated dibenzo-*p*-dioxins] from fly ash, sediment and urban dust (Hills and Hill, 1993; Hills *et al.*, 1994). Some authors have noticed the reversibility of the silylation reaction, particularly when the supercritical fluid contains moisture or when protic modifiers are added to the supercritical fluid. Therefore, alkylation reactions which produce more stable derivatives compared with trimethylsilyl ethers deserve greater interest. Lopez-Avila *et al.* (1993) have evaluated several variables that affect the derivatization yields for a variety of chlorophenoxy acid herbicides from soils. They concluded that a derivatization mixture of

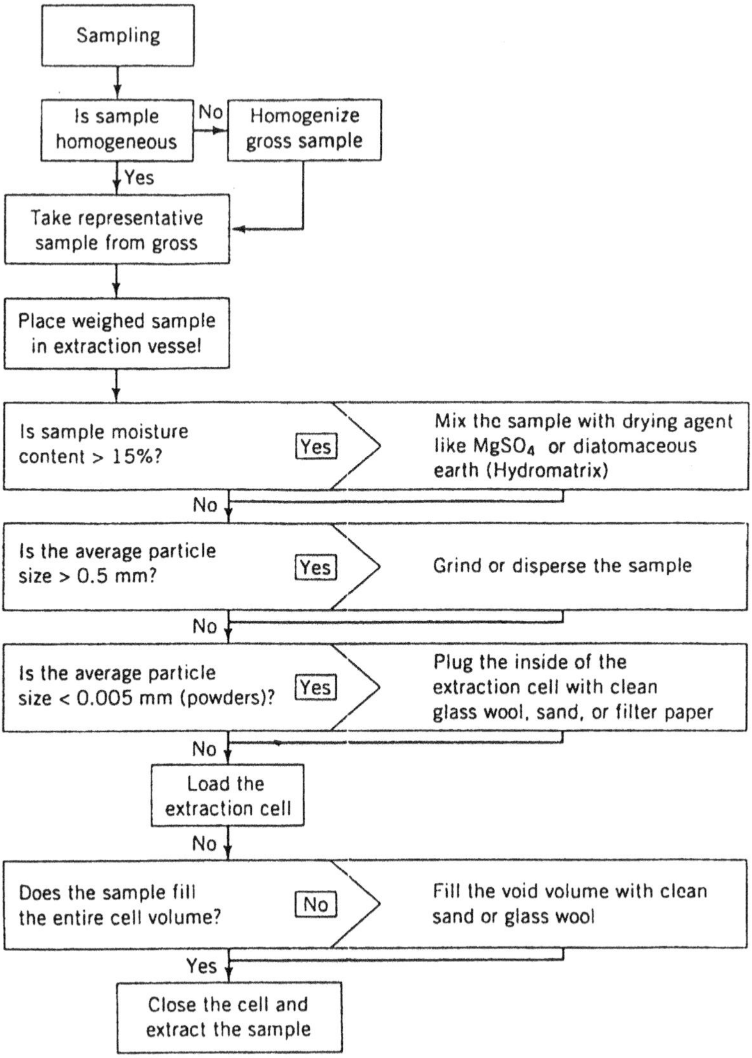


**Figure 3.1** Flowchart summarizing the stages involved for preparing a solid sample for supercritical fluid extraction. Reproduced with permission from Dionex Corporation and from Hewlett Packard.

tetrabutylammonium hydroxide–methyl iodide was most effective but that some analytes currently analysed by the US Environmental Protection Agency (EPA) SW-846 method (e.g. dalapon, dinoseb and 4-nitrophenol) do not undergo derivatization under these SFDE conditions. Hillmann and Bachmann (1994) have found the use of potassium carbonate necessary

**Table 3.2** Selected *in situ* derivatization reactions used in supercritical fluid extraction

| Analytes | Matrix | Reaction | Reagent | Extraction conditions | Recovery (%) | Reference |
|---|---|---|---|---|---|---|
| Sterols | sewage sludge | trimethylsilylation | BSTFA | $CO_2$, 36.5 MPa, 80°C | 60–80 | Nguyen *et al.*, 1995 |
| Halogenated benzenes and phenols | fly ash | trimethylsilylation | BSTFA | $CO_2$, 50 MPa, 100°C | 96–111 | Hills *et al.*, 1994 |
| Fatty acids | bacteria (*E. coli*) | methylation | TMPA | $CO_2$, 40 MPa, 80°C | variable | Hawthorne *et al.*, 1992a |
| Chlorophenoxy acids | soil | methylation | TBA–MI | $CO_2$, 40 MPa, 80°C | 46–93 | López-Avila *et al.*, 1993 |
| Chlorophenoxy acids | $C_{18}$ | alkylation (on-line) | PFBBr + $K_2CO_3$ | $CO_2$, 21 MPa, 50°C | quantitative | Hillmann and Bachmann, 1994 |
| Chlorinated phenolics | sediment | acetylation | AA + TEA | $CO_2$, 37 MPa, 110°C | quantitative | Lee *et al.*, 1992 |
| Organotin | sediment | hexylation | Grignard | $CO_2$, 35 MPa, 40°C | 63–90 | Cai *et al.*, 1994 |

Note: BSTFA = *N*,*O*-bis(trimethylsilyl)trifluoroacetamide; TMPA = trimethylphenylammonium hydroxide; TBA = tetrabutylammonium hydroxide; MI = methyl iodide; PFBBr = pentafluorobenzyl bromide; AA = acetic anhydride; TEA = triethylamine.

to obtain quantitative derivatization of chlorophenoxy acids with pentafluorobenzyl bromide (PFBBr). Acetylation using acetic anhydride has been optimized for the derivatization of phenol from soil (Llompart *et al.*, 1996a) by using multivariate techniques. Optimization results suggest that the extraction vessel temperature and the amount of derivatizing agent used are statistically significant to the overall extraction yield.

SFDE is useful in the speciation of organometallic species based on gas chromatography (GC) separation methods. SFDE with thioglycolic acid methylester in supercritical carbon dioxide has been applied to the speciation of organic and inorganic arsenic species (Wenclawiak and Krah, 1995). Cai *et al.* (1994) have performed the derivatization reaction for organotin determination in sediment samples in the extraction vessel at atmospheric conditions prior to SFE. The Grignard reagent used in the derivatization reaction of organotins is too reactive at supercritical conditions and excess reagent can cause damage to analytical instrumentation (e.g. valve and vessel seals). The *in situ* derivatization procedure is very useful in the determination of dialkyl, trialkyl and aryltin species from sediment, yielding comparable results with conventional solvent extraction techniques (Figure 3.2).

One requirement of SFDE is the need for excess derivatization reagent to ensure quantitative extraction of analytes. This is necessary to avoid poor analyte recoveries arising from the possibility of extraneous compounds or matrix active sites reacting with the derivatization reagent, causing its premature depletion. An example of by-product formation during *in situ* derivatization is shown in Figure 3.2, where alkylsulphides were obtained from the reaction between the Grignard reagent and the elemental sulphur contained in the sample. Therefore, the use of surrogates is mandatory for quantitative analysis involving SFDE.

*Ion-pair formation*

Direct SFE of ionic species is not feasible because of the very limited solubility of such compounds in supercritical fluids. An exception is the extraction of ditallowdimethyl ammonium chloride (a quarternary aliphatic amine, cationic surfactant) from sewage and sediments, which was accomplished with carbon dioxide (380 atm, 100°C) modified with large amounts of methanol (30%) (Fernández *et al.*, 1996). This could be attributable to these cationic compounds already occurring as ion pairs in sewage matrices; thus additional ion-pairing agents were not needed to solubilize them in carbon dioxide. In fact, ion-pair agents are needed only for their extraction in sediments where the occurrence of anionic ion pairs is limited.

Most SFE applications of ionic analytes require the use of ion-pair agents to neutralize the charged ions which otherwise exhibit very low solubility in supercritical carbon dioxide, with or without modifiers. To this end, tetrabutylammonium hydroxide forms ion pairs with secondary alkane

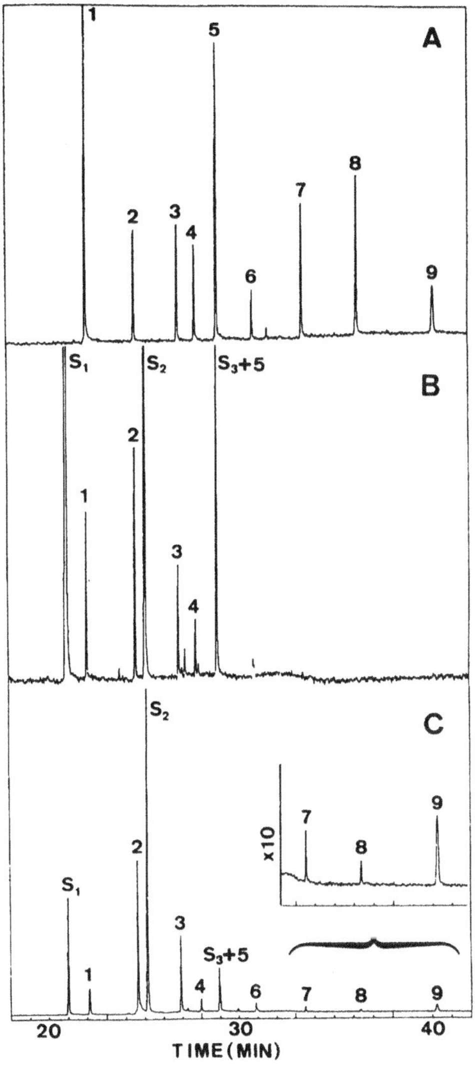


**Figure 3.2** Gas chromatography–flame photometric detection chromatograms obtained for Grignard derivatizations of: (A) the organotin standard solution; (B) the *in situ* derivatization supercritical fluid extraction (SFE) extract of the standard reference material PACS-1 (extracted with $CO_2$ at 40°C, 350 atm, 10 min static SFE then dynamic SFE with 10 ml $CO_2$); (C) the *in situ* derivatization SFE extract of Masnou marina sediment. Peak identification: 1 = $Bu_4Sn$; 2 = $Bu_3SnHex$; 3 = $Bu_2SnHex_2$; 4 = $Pe_3SnHex$; 5 = $BuSnHex_3$; 6 = $SnHex_4$; 7 = $PhSnHex_3$; 8 = $Ph_2SnHex_2$; 9 = $Ph_3SnHex$. $S_1$, $S_2$ and $S_3$ are hexylated sulphides. Certified reference sediment PACS-1 was obtained from the National Research Council of Canada. Reprinted with permission from Cai *et al.* (1994).

**Table 3.3** Selected complexing reagents used for the supercritical fluid extraction of metal ions

| Metal ions | Complexing agent | Extraction conditions | Reference |
|---|---|---|---|
| $Cu^{2+}$ | FDDC | $CO_2$, 35°C, 0.6 g/ml | Laintz *et al.*, 1992 |
| $Hg^{2+}$ | FDDC | $CO_2$/MeOH, 50°C, 150 atm | Wai *et al.*, 1993 |
| $Cu^{2+}$, $Co^{2+}$, $Cd^{2+}$, $Zn^{2+}$ | FDDC | $CO_2$/MeOH, 200 atm, 40°C | Liu *et al.*, 1993a |
| $La^{3+}$, $Eu^{3+}$, $Lu^{3+}$ | TBP + HFA | $CO_2$, 60°C, 150 atm | Lin and Wai, 1994 |
| $UO_2^{2+}$ | FOD | $CO_2$/MeOH | Lin *et al.*, 1993 |

Note: FDDC = bis(trifluoroethyl)dithiocarbamate; TBP = tributylphosphate; HFA = hexafluoroacetylacetone; FOD = 2,2-dimethyl-6,6,7,7,8,8,8-heptafluoro-3,5-octanedione.

sulphonates and linear alkylbenzene sulphonates in supercritical carbon dioxide (400 atm, 80°C, 5 min static) which can then be quantitatively extracted from sewage sludges by using neat carbon dioxide (Field *et al.*, 1992). An additional advantage of this ion-pair formation was that *in situ* sample derivatization occurred in the GC injection port, leading to the formation of butyl ester derivatives which were readily analysed. Similarly, a variety of polar analytes such as dichlorophenoxyacetic acids, phenols and phospholipids were extracted by using trimethylphenylammonium hydroxide as an ion-pair and methylating reagent (Hawthorne *et al.*, 1992a).

*Chelate formation*

This is an alternative approach to ion-pair formation and is useful for the extraction of metal ions and organometallic species from a variety of environmental samples (i.e. biotic and abiotic) (Table 3.3). In fact, the solubility of charged inorganic species in supercritical carbon dioxide or even in the presence of modifiers is almost negligible and charge neutralization becomes necessary to perform the extraction. Chelation reactions can be performed in supercritical carbon dioxide by adding the chelating agent to the fluid. In order to obtain an effective extraction it is of prime importance that the resulting chelate from the reaction of the ligand and the ion should be soluble in the fluid and form stable chelates.

Three of the most successful classes of ligands used for the extraction of metal ions are: fluorinated dithiocarbamates [bis(trifluoroethyl)dithiocarbamate (FDDC)], fluorinated β-diketones [$R^1COCH_2COR^2$; e.g. $R^1 = -C(CF_3)_3$, $R^2 = n\text{-}C_3F_7$] and tributylphosphate (TBP) in the presence of fluorinated β-diketones (Furton *et al.*, 1995; Lin *et al.*, 1993; Lin and Wai, 1994). The β-diketones react with metal ions to form neutral chelates through the enolate anions (Lin *et al.*, 1993). Fluorinated β-diketones react with metals to form complexes which are more soluble in supercritical fluids than are the non-fluorinated analogues. Solubilities of metal complexes range over eight orders of magnitude (Smart *et al.*, 1997). A general trend is for increasing complex solubility with increasing metal oxidation state; this property is associated with the increased number of coordinating ligands that result in a higher degree of interaction with the supercritical fluid (Smart *et al.*, 1997). The use of a variety of complexing agents have enabled the SFE of Cu(II), Hg(II), Cd(II), Co(II), Zn(II), La(III), Eu(III), Lu(III) and the uranyl ion ($UO_2^{2+}$) from a variety of matrices (Table 3.3). The quality of procedural blanks and the instrumentation itself can provide metal ions, giving rise to interferences. The situation can be partially improved by replacing stainless-steel tubing with poly(ether-ether-ketone) (PEEK) tubing (Liu *et al.*, 1993a). When the extraction is performed with neat carbon dioxide, TBP has shown a synergistic effect with the fluorinated ligands for the extraction of trivalent lanthanide ions from soil. The synergistic effect is effective only when fluorinated ligands are used. Methanol (5%) and water (10 µl) enhance the extraction efficiency of lanthanides and uranyl ions from cellulose. This can be accounted for by an increase in the metal-complex solubility and/or the modifier competing with the active sites of the matrix. Additionally, water in supercritical fluid carbon dioxide gives rise to the production of carbonic acid (section 3.2.4) (Toews *et al.*, 1995). This may lead to the displacement of the ionization equilibria of the chelating agents, promoting metal complexation. The use of ethanol-modified carbon dioxide with mixed ligands (fluorinated β-diketone and TBP) for the extraction of uranyl ions from clay has been demonstrated to provide higher recoveries in shorter extraction times compared with extraction with organic solvents (Furton *et al.*, 1995).

It should be noted that most of these preliminary investigations have been conducted with the use of spiked matrices where the analyte–matrix interactions are not well established. Further research is needed to validate the former SFE methodologies and to demonstrate their applicability to real-world samples.

Organometallic species, such as organotin and organolead, in soil and sediment samples can be extracted with carbon dioxide in the presence of ligand agents which enhance the extractability of the ionic organotins (Chau and Brown, 1995; Liu *et al.*, 1993b; López-Avila *et al.*, 1993). However, the extraction efficiency of monoalkyltins and monoaryltins exhibited a poor extraction efficiency even in spiked samples. Extraction of native organotin from sediment and biota can be improved by using acetic acid as a carbon dioxide modifier (Cai, personal communication; Fernández-Escobar and Bayona, 1997), since soluble acetates are presumably formed during the extraction, and acetic acid can act as a reactive modifier competing with the analytes for the active sites of the matrix. Organomercury compounds (e.g. methyl and dimethyl) are more soluble in carbon dioxide and can be extracted from spiked cellulose or as native methylmercury from sediment by using carbon dioxide modified with methanol (Wai *et al.*, 1993). Also, the extraction of methylmercury from sediment has been achieved by using neat carbon dioxide (Emteborg *et al.*, 1995). The solubility of native methylmercury in neat carbon dioxide is attributable to the fact that methylmercury is extracted as a sulphide produced during the SFE conditions.

### *3.1.3 In-line supercritical fluid extraction cleanup procedures*

SFE conditions can be selected to emphasize the differences in solubility which exist between interferences and target analytes present in complex matrices to achieve an effective cleanup or fractionation.

From the different steps in the SFE process – namely removal of compounds from the surface of the matrix, the solvation of analytes in the fluid and the mass transport of the solubilized molecules to the bulk of fluid to allow their removal from the vessel – only the solvation of analytes in the fluid can allow some extraction selectivity provided that large differences in the solubility parameter ($\delta$) exist between the analyte and interference. This requirement is rarely fulfilled in most SFE applications and a third phase (e.g. a solid sorbent) which provides different partitioning constants ($K$) for the analyte(s) and interference(s) is necessary to achieve a complete separation. The differences in $K$ values between analytes and adsorbent must be maximized to yield an effective cleanup which depends on the type of interactions (e.g. dispersive, ligand exchange, charge transfer, acid–base, dipolar or ionic).

*In-line chemical class fractionation*

The nature of the adsorbent and the characteristics of the supercritical fluid are crucial to achieve an effective separation between different chemical classes of analytes. Alexandrou *et al.* (1992) have nicely illustrated the potential of in-line fractionation of chlorinated benzenes (CBs) and polychlorinated biphenyls (PCBs) from polychlorinated dibenzo-*p*-dioxins (PCDDs), by using stepwise elution of Florisil® with carbon dioxide (204 atm, 15 min) followed by nitrous oxide (408 atm, 90 min). Approximately 75% of CBs and PCBs are eluted with carbon dioxide, whereas the PCDDs were quantitatively recovered in the nitrous oxide fraction. Alternatively, the elution of the PCDD fraction can be achieved by using high-temperature/high-pressure SFE with neat carbon dioxide (650 atm, 250°C), which is a more acceptable fluid than nitrous oxide and eliminates the requirement for fluid changeover during the extraction. This approach has been successfully applied to the extraction and cleanup of fly ash samples from municipal solid waste incinerators (Alexandrou *et al.*, 1992; Miao *et al.*, 1994).

Another way to obtain extraction selectivity when subcritical water is used is by modifying the extraction temperature, since a large change in the dielectric constant ($\epsilon$) is obtained (Figure 3.3) (Yang *et al.*, 1997). At low temperatures (50°C–150°C) only the highly polar compounds (phenols) or volatile compounds [benzene, toluene, ethylbenzene and xylenes (BTEX)] are extracted by liquid water. At higher temperatures (250°C–300°C) most of the hydrophobic analytes are extracted (PAHs and $n$-$C_{12-20}$) by liquid

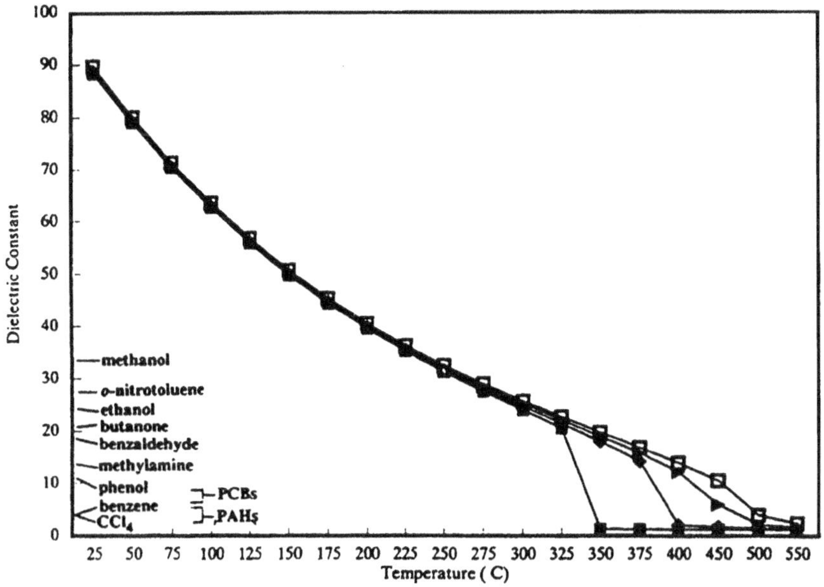


**Figure 3.3** Effect of temperature and pressure on the dielectric constant ($\epsilon$) of water compared with the dielectric constants of some representative organic compounds. ■ = 100 bar; ◆ = 200 bar; ▶ = 300 bar; □ = 400 bar. Reproduced with permission from Hawthorne *et al.* (1994).

water but the higher molecular weight alkanes (greater than $n$-$C_{20}$) required a superheated steam (5 atm, 250°C–300°C).

*In-line lipid removal*

For matrices that contain high levels of lipids, loading the exit end of the extraction vessel with activated alumina can provide an effective means of in-line sample cleanup. This approach was used for the determination of PCBs in fish tissue (Hale and Gaylor, 1995; Lee *et al.*, 1995) and chlorinated pesticides in fat (France *et al.*, 1991). A mixture of diol and $C_{18}$ or pure $C_{18}$ sorbents provided a sufficiently clean SFE fraction for the determination of carbamates in tissue or fat (Murugaverl and Voorhees, 1991; Murugaverl *et al.*, 1993). The main limitation of this approach is the need to employ neat carbon dioxide as an extractant agent at moderate pressures to preserve the cleanup efficiency. Lee *et al.* (1995) have reported co-extracted lipid levels, following in-line activated alumina SFE cleanup, varying from less than 0.1% to 2.2%, despite some samples containing over 30% m/m lipid. The lipid in the SFE extract was conveniently removed from the PCBs by a small Florisil® column.

An in-line cleanup procedure is particularly useful for the extraction of lipophilic xenobiotics from biota samples which contain large amounts of fat. The efficiency of in-line lipid removal depends upon the SFE conditions

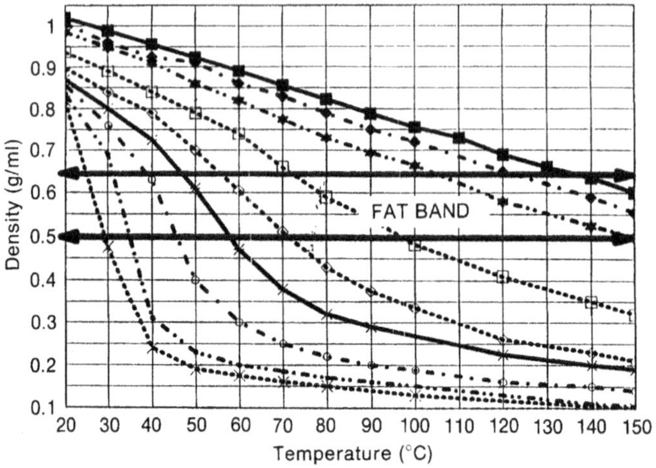


**Figure 3.4** The density of supercritical fluid carbon dioxide plotted as a function of temperature and constant pressure. ■ = 400 bar; ◆ = 350 bar; ★ = 300 bar; □ = 200 bar; ◇ = 150 bar; ∗ = 125 bar; ○ = 100 bar; • = 83 bar; × = 75 bar. Reproduced with permission from Hewlett Packard.

used for analyte recovery and the lipid content of the sample. By using the appropriate amount of cleanup medium, near quantitative retention of highly supercritical fluid soluble lipids can be achieved. Apart from providing an overall reduction in analysis time, these SFE cleanup procedures eliminate the need for organic solvents that are used for solid-phase extraction (SPE) fractionation of lipid-containing samples. Figure 3.4 serves as a useful guide to determine which SFE conditions are likely to result in high levels of extracted fats from appropriate matrices. The density of supercritical fluid carbon dioxide is shown plotted as a function of temperature and constant pressure, resulting in curved isobaric lines. A region exists called the 'fat band', the significance of which is that at any set of SFE conditions below the lower boundary no fats or lipids (monoglycerides, diglycerides or triglycerides) will be extracted. Above the upper 'fat band' boundary, sets of SFE conditions exist at which lipids will almost certainly be extracted. Within the band itself, the extractability of lipids is unpredictable, depending upon the precise nature of the matrix, extraction time and flow rate.

There are other techniques involving *in situ* enzymic reactions which produce less complex SFE fractions for analysis. These methods are described in section 4.6.

*In-line desulphurization*

Sediment and soil matrices contain large amounts of elemental sulphur which is soluble in supercritical carbon dioxide. This can give rise to two problems associated with such matrices using off-line SFE sample preparation: first, blockage of fixed restrictors arising from sulphur precipitation during fluid expansion; second, interferences for several GC detection systems such as electron capture detection (ECD) or flame photometric detection (FPD). Traditional methods for sulphur elimination involve treatment with activated copper. This approach has also been applied to SFE extracts (Tilio *et al.*, 1994) and can prove more effective to conventional solvent preparation procedures. Several authors have developed an in-line desulphurization method which involves the addition of activated copper to the extraction vessel (Bøwadt and Johansson, 1994; Reindl and Hofler, 1994), alternatively a copper 'scavenger column' can be placed after the extraction vessel (Pyle and Setty, 1991). Whereas the former method requires a short, static SFE period to allow reaction between copper and sulphur, with the latter approach extracted sulphur can be removed by using dynamic SFE without the need for a static period. The main advantage of in-line sulphur removal using SFE conditions is faster reaction kinetics compared with sample treatment at room temperature, which requires extended periods of time for quantitative removal. The high-efficiency of in-line sulphur removal has been demonstrated for sediments containing as high as 1.5% m/m of sulphur, allowing the determination of PCBs by GC/ECD without any noticeable interferences.

## 3.2 Experimental parameters of supercritical fluid extraction

SFE is influenced by a large variety of extraction parameters (Table 3.4). However, despite the large number of variables that can potentially affect SFE efficiency, optimization studies are usually carried out on a selected number of variables such as pressure, temperature and modifier. The following discussion will be focused on the most relevant extraction parameters.

### *3.2.1 Type of fluid*

Carbon dioxide is by far the most widely used fluid for SFE because of its low cost, non-toxicity, low reactivity, low critical pressure and temperature

**Table 3.4** Supercritical fluid extraction experimental parameters

| Variable | Units | Optimization range |
|---|---|---|
| Pressure | atm | 100–700 |
| Temperature | °C | 40–250 |
| Time | min | 10–80 |
| Flow rate | ml/min | 0.5–2.5 |
| Moisture content | % m/m | 0.1–2.0 |
| Sample size | g | 0.5–5 |
| Fluid type | | |
| Modifier | % vol/vol | 0–30 |
| Vessel volume | ml | 1.5–6 |

and non-flammability. The main drawback is its lack of polarity and, to a far lesser extent, its reactivity with primary amines. More polar fluids than carbon dioxide (see Figure 4.1) include nitrous oxide ($N_2O$), and in particular chlorodifluoromethane ($CHClF_2$) and trifluoromethane ($CHF_3$) (Camel *et al.*, 1993; Janda *et al.*, 1993). Although $CHClF_2$ has provided remarkably high SFE recoveries for PCBs, PAHs and phenols from a variety of environmental matrices (Hawthorne *et al.*, 1992b), its use has been abandoned because of the phase-out of chlorofluorocarbons (CFCs). Alternatively, $CHF_3$ has been evaluated for the extraction of polar analytes. Despite its large dipolar moment, the SFE efficiency for sulphonyl urea herbicides using $CHF_3$ is poorer than that obtained using $CHClF_2$ or methanol-modified carbon dioxide (Howard *et al.*, 1993). Hillmann and Bachmann (1995) found higher recoveries for a variety of agrochemicals (i.e. *s*-triazines, carbamates, organophosphorus pesticides and anilides) when supercritical $CHF_3$ was used compared with carbon dioxide. However, quantitative recoveries were not obtained.

Nitrous oxide has been mostly used for the extraction of aliphatic and aromatic amines (Ashraf-Khorassani *et al.*, 1990; Oostdyk *et al.*, 1993; Sauvage *et al.*, 1993) since carbon dioxide can react with primary amines. Nevertheless, the oxidant properties of this fluid may cause explosive oxidation reactions in matrices rich in organic matter (Raynie, 1993) and thus its application in SFE is of limited interest. Hence there are no reasonable supercritical fluids that can be considered as alternatives to carbon dioxide for the extraction of polar analytes.

Recently, subcritical water has been introduced because its polarity is strongly dependent on temperature (Figure 3.3). To this end, at very high temperatures (greater than 250°C) and moderate pressures (*c.* 50 bar) water behaves as an organic solvent, allowing the extraction of hydrophobic compounds (e.g. PAHs and PCBs) from environmental solids (Yang *et al.*, 1995a). The main drawbacks to the utilization of subcritical water is that its application is restricted to thermally stable analytes and that there is difficulty in handling extracts which are diluted with liquid water. Supercritical water is highly corrosive and of little practical interest to analytical SFE because of the high temperatures involved.

### 3.2.2 *Effect of density*

The density of a supercritical fluid depends on the extraction pressure and temperature. Although increasing supercritical fluid density is generally associated with higher solvating powers, the relationship is not linear. The contribution of temperature is rather unpredictable, varying according to the analyte(s) being extracted (Janda *et al.*, 1993). The analyte solubility in the supercritical fluid depends on the Hildebrand solubility parameter ($\delta$)

proposed by Giddings *et al.* (1968):

$$\delta = \frac{5}{8} p_c \frac{\rho}{\rho_l} \tag{3.1}$$

where $p_c$ is the critical pressure of the fluid, $\rho$ is the reduced density of the supercritical fluid and $\rho_l$ is the reduced density of the fluid in the liquid state. The closer the analyte and supercritical fluid solubility parameters, the higher the analyte solubility (King, 1989). Since the analyte solubility parameter is not always known, solubility can be obtained by means of theoretical calculations (e.g. the Peng–Robinson state equation, described in section 1.3). Alternatively, the solubility of analytes can be obtained experimentally from view-cell experiments and other experimental measurements (Smart *et al.*, 1997). Such considerations are useful when the target analytes represent a significant percentage of the bulk sample (e.g. extraction of fats from meat products).

The following discussion will be focused on the effect of pressure; the effect of temperature will be addressed in the next section. King (1989) proposed three basic parameters related to pressure which are useful for understanding the behaviour of major solutes or analytes in a supercritical fluid. These parameters – the pressure at which the analyte starts to become soluble, the pressure at which maximum solubility of the analyte is achieved and the pressure range for fractionation of the sample – enable predictions of analyte solubility and serve as a basis for establishing conditions in method development. The first of these parameters, the threshold or miscibility pressure, is defined as the pressure at which the solute starts to dissolve in the supercritical fluid. Its use depends on the sensitivity of the analytical technique used for determination, and this pressure can be used as a starting point for method development. The second parameter is the pressure at which the analyte achieves maximum solubility in the fluid. This corresponds to the fluid pressure to which the analyte Hildebrand solubility parameter approximates [equation (3.1)]. The third parameter is the fractionation pressure range, where the solubility of one analyte will vary between zero and the maximum solubility value. In this range, the relative solubilities between analytes can be controlled. Unfortunately, in very few SFE applications can differences in solubility be exploited for analyte fractionation. Exceptions include the isolation of components which possess functional groups that give rise to large differences in polarity or which possess large differences in relative molecular mass such as monomers or oligomers compared with polymers. In section 3.1.3 some strategies to improve extraction selectivity were addressed.

When the analyte(s) are present at trace levels, optimization of supercritical fluid solubility is not so important since lower pressures (hence lower $\delta$) are sufficient for extraction. For these cases, the extraction kinetics of analytes retained by matrix active sites can be controlled by extraction temperature rather than by pressure. In the following section, the impact of temperature on extraction is addressed.

### *3.2.3 Selection of supercritical fluid extraction temperature*

Temperature is a key parameter for the extraction of volatile and semivolatile organic compounds. An increase in temperature can lead to a decrease in the interaction between analyte and the matrix, promoting more efficient extraction. Chemisorbed analytes need an activation energy that is reduced at higher temperatures to bring about their release from the matrix (Figure 3.5). High-temperature SFE (HTSFE) has been applied to the extraction of a variety of hydrophobic thermally stable contaminants (e.g. PCBs and PAHs) from environmental matrices (Langenfeld *et al.*, 1993) and can be considered as an alternative to modifier addition to carbon dioxide. At conventional extraction temperatures (50°C), raising the pressure (e.g. 350–650 atm) has no effect on the extraction efficiency of PAHs from urban air particulate matter and PCBs from sediment. However, at higher temperatures (*c.* 200°C) the extraction efficiency increases, and in the case of PCBs they are extracted independently of pressure. These results indicate

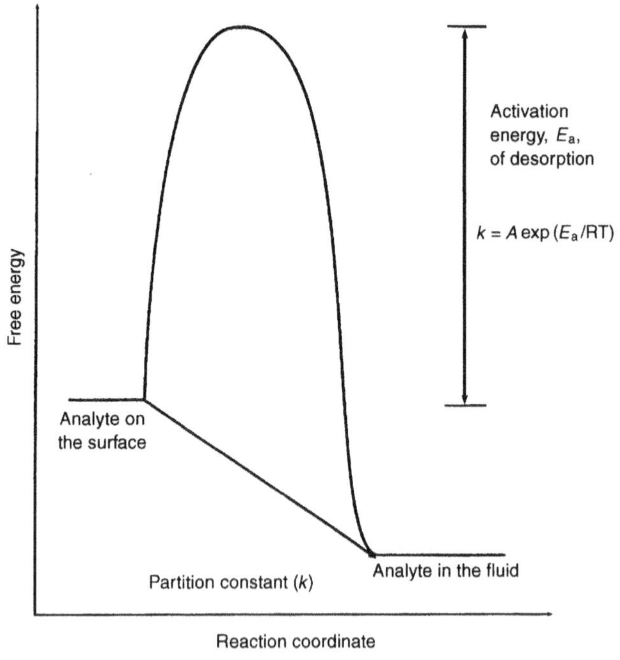


**Figure 3.5** An illustration of free energy and reaction coordinate, indicating the energy requirements for desorption of chemisorbed analytes in solid matrices. Reproduced with permission from Alexandrou *et al.* (1992).

that a reduction in the kinetic limitations associated with overcoming the energy barrier of the desorption step is needed when the analytes are strongly interacted with the matrix. HTSFE at extremely high temperatures (*c.* 350°C) and relatively high pressures (*c.* 340 atm) has been demonstrated to be very effective for the extraction of aliphatic and aromatic hydrocarbons (naphthalenes, phenanthrenes and their methyl and dimethyl derivatives) from shales (Furton *et al.*, 1994). By keeping the extraction pressure constant (340 atm), the extraction efficiency increases with temperature from 50°C to 350°C. At the highest temperature, the extraction efficiency of the aliphatic and aromatic hydrocarbons ranges from 108% to 196% higher relative to Soxhlet extraction. Indeed at 380°C the recoveries for several compounds are 700% higher than with Soxhlet, but small amounts of alkenes are detected in the extract, suggesting that pyrolysis of the organic matter may also occur at such high extraction temperatures.

Hawthorne and Miller (1994) have further evaluated the effect of temperature in SFE with neat carbon dioxide for a variety of analytes and matrices (e.g. soil and soot). They have investigated its effect on the extraction of PAHs, *N*-heterocyclics and *S*-heterocyclics, chlorinated phenols and pesticides and the results were compared with Soxhlet extraction. At lower temperatures (50°C) the extraction efficiency is lower than with Soxhlet extraction, but when the temperature is increased (200°C) results are similar to those achieved with Soxhlet extraction (18 h). A further temperature increase to 350°C did not provide increased recoveries (except for PAHs from soot) but did show evidence of causing thermal degradation for some analytes (e.g. aromatic amines) and possible production of low molecular weight PAHs. Interestingly, the increase in SFE efficiency cannot be solely explained by volatility considerations, since thermal desorption at these temperatures does not yield recoveries as high as in SFE. This finding can be attributed to the contribution of analyte supercritical fluid solvation. The same authors have found that extraction kinetics of PAHs differ greatly according to the matrix, confirming the existence of different analyte–matrix interactions.

The effect of modifiers using HTSFE has been evaluated for the extraction of PAHs from marine sediments, diesel soot and air particulate matter. These results were compared with those obtained under Soxhlet extraction (Yang *et al.*, 1995b). Extractions were performed at conventional (80°C) and high (200°C) temperatures and also with the modifiers methanol, toluene and diethylamine. The PAH recoveries increased with temperature, being independent of the matrix. In contrast, modifier effects were dependent on the matrix and modifier properties. Methanol was the poorest modifier for all matrices, often failing to provide improved recoveries compared with neat carbon dioxide. Both toluene and diethylamine yield increased PAH recoveries at both temperatures for the air particulate matter, but only diethylamine provided significantly enhanced recoveries from all three matrices. The enhancement in PAH recoveries using high temperatures and modifiers could be summed, indicating that temperature and modifier mechanistic interactions were different. Extractions at 200°C with diethylamine as modifier yielded the highest PAH recoveries, which agreed well with recoveries obtained by Soxhlet extraction (14–48 h).

Similarly, the extraction rates of PCDDs from Florisil® increased with temperature from 40°C to 200°C and were slower with a 5 ml vessel (12 cm long) than with a 0.5 ml vessel (6 cm long) (Langenfeld *et al.*, 1995). Increasing the temperature from 40°C to 120°C and then to 200°C enhanced the SFE rates of spiked [$^{13}$C]-PCDDs and native PCDDs from fly ash, as well as [$^{2}$H]-PAHs and native PAHs from marine sediment and railroad soil. In all cases, native analytes were extracted more slowly than were spiked analytes, suggesting additional analyte–matrix interactions for the native analytes.

In summary, HTSFE using neat carbon dioxide can be considered as an alternative to SFE using organic modifiers for the isolation of thermally stable analytes from environmental matrices. Unfortunately, this approach is limited by the specifications of commercially available SFE instrumentation. Current systems do not enable SFE operation above 150°C, and analyte trapping following high-temperature extraction could prove difficult for volatile species. Additionally, the high-pressure seals of most extraction vessels will not tolerate temperatures above 250°C. Despite these technical problems, HTSFE method development is relatively straightforward since fewer extraction variables need to be optimized. Further research is necessary to understand the extraction mechanisms of HTSFE and to predict the highest temperature that a sample can be heated to without causing significant matrix alteration. Developments in HTSFE will also provide further understanding of analyte–matrix interactions not obtainable by conventional solvent extraction techniques.

### *3.2.4 Selection of organic modifier*

Alternative terminology, frequently encountered in engineering literature, designating the incorporation of an organic solvent (or solvents) into a supercritical fluid, is the use of the terms 'co-solvent' and 'entrainer'. The term 'modifier' is now widely used in analytical SFE and has been adhered to throughout this chapter.

Two different experimental set-ups to deliver modifier to the supercritical fluid have been widely used:

- continuous addition throughout the SFE process, generally achieved by using an additional pump (slave pump) or, less commonly, by using a fluid supply cylinder that has been doped;
- static addition, whereby the modifier along with the sample is placed into the extraction vessel prior to extraction.

Using the latter method the modified supercritical fluid is generally allowed to equilibrate with the matrix prior to the onset of dynamic SFE such that analyte–modifier interactions are established to promote desorption. Although less flexible, static modifier addition can be very useful in situations where reactive modifiers can bring about pump and/or valve seal deterioration.

The selection of an organic modifier can represent a crucial aspect for SFE method development and depends on the physical and chemical properties of the analyte and the nature of interaction with the matrix. Despite a large variety of organic modifiers being used in SFE there are few predictive models concerning their selection. As a starting point, a commercial software package (Isco, Lincoln, NE, USA) which estimates the overall solvent polarity (strength) of commonly modified supercritical fluids can be used. The programme considers the contribution of individual Hildebrand solubility parameters for the components of modified supercritical fluids at varying compositions. Another approach to predict the extractability of analytes is based on the use of Henry's constant (Macnaughton *et al.*, 1995). This technique uses data describing the equilibrium between the pure modifier and the analyte that have been measured and subsequently used to predict ternary system solubility data.

Although several modifiers have been used with miscellaneous supercritical fluids, the present section will be focused on those which have been widely used with carbon dioxide. Several preliminary considerations should be kept in mind before selecting the modifier: first, its miscibility with the primary fluid at the concentration and extraction conditions to be used; second, the modifier reactivity with either the fluid or the analyte; last, the critical point of the mixture, particularly when large portions of modifiers are used. An excellent review provides binary phase diagrams for a variety of modifiers with carbon dioxide, being particularly useful in helping establish the optimization range (Page *et al.*, 1992). However, several authors have found that the extraction efficiency of strongly polar analytes such as phenolic and nitroaromatic contaminants is independent of the extraction conditions (subcritical or supercritical) and they have coined the term 'enhanced-fluidity liquid modifier–carbon dioxide' for the subcritical SFE conditions (Reighard and Olesik, 1996).

Two different mechanisms of modifier action in SFE have been postulated. A primary mechanism is the enhancement of the solubility of the analyte (or analytes) in the supercritical fluid which facilitates its (their) mass transport. The second mechanism involves modifier-mediated displacement of the analyte(s) from matrix active sites. The first mechanism is common for analytes with low functionalization (e.g. hydrocarbons), whereas the second applies to polar and functionalized molecules occurring at low concentrations in environmental samples.

Most of the literature dealing with the role of organic modifiers in SFE method development have involved the study of spiked matrices. In general, improved SFE efficiencies are obtained for these samples as the modified supercritical fluid is made more polar. However, when these procedures are applied to real-world samples, analyte recoveries can be very poor (Hawthorne *et al.*, 1994b). An increase in the amount of modifier relative to that used for spiked samples is required to obtain comparable analyte recoveries (Reindl and Hofler, 1994).

The difference in modifier behaviour towards spiked, weathered or native analytes reflects different levels of matrix interaction. These analyte–matrix interactions tend to be minimal with spiked samples, becoming established in real-world samples subjected to weathering processes (Fahmy *et al.*, 1993). For instance, it has been shown that polar pesticides such as *s*-triazines and pirimicarb, which are soluble in supercritical carbon dioxide, cannot be extracted from weathered samples using equivalent supercritical conditions (Alzaga *et al.*, 1995, 1996).

In order to select the most appropriate modifier it is of prime importance to know which step controls the extraction kinetics. Several aspects of method development have been identified for various types of sample (Hawthorne *et al.*, 1995). For samples whose extraction rates are effectively controlled by solubility or elution (situations which tend to involve high concentrations of analytes weakly bound to the matrix), the following statements apply.

- Faster flow rates will yield higher extraction rates. Hence, provided the analyte trapping is efficient, flow rates should be increased. Increasing the extraction period will also promote SFE recoveries.
- Smaller samples will extract faster than will larger samples (assuming the same flow rate).
- Static extraction will be much less effective than using dynamic extraction for the same duration.
- Efforts should be focused to increase the analyte solubility by increasing pressure, increasing or decreasing temperature (depending on the analyte vapour pressure) or adding modifier, which is selected to increase analyte solubility and/or decrease the equilibrium adsorption to matrix active sites.

For samples whose extraction rates are effectively controlled by desorption kinetics (situations which tend to involve low concentrations of analytes strongly bound to the matrix), the following statements apply.

- The SFE flow rate has little or no effect on the extraction rate. Increasing extraction time is less effective for such samples since the rate of extraction tends to decrease with time.
- The sample size has little or no effect on the extraction rate.
- Static extraction is often as effective as dynamic extraction performed for the same duration.

- Improving SFE efficiencies should be based on consideration of analyte–matrix interactions rather than analyte solubility. Effective approaches for improving rates of extraction for these samples include grinding (e.g. for polymers where extraction is controlled by the diffusion), addition of a modifier to disrupt analyte–matrix interactions (rather than to increase the analyte solubility) and raising the extraction temperature to increase the rate of desorption, provided the analyte is thermally stable.

The use of several reactive additives have been proposed to overcome strong analyte–matrix interactions encountered in analytical SFE (Alzaga *et al.*, 1995, 1996; Friedrich *et al.*, 1995). Usually, the reactive additives are added at low concentrations to the organic solvent used as modifier. Generally, the interactions between additive and analyte seem less important than those between additive and matrix. For PAHs, reactive additives are selected according to the nature of the sample matrix. Improved PAH SFE efficiencies were obtained by using triethylamine (TEA) for a carbonaceous fly ash sample, whereas trifluoroacetic acid (TFA) proved superior in promoting the extraction of PAHs from acidic matrices (Friedrich *et al.*, 1995). In another study, it was demonstrated that TEA and pyridine were the most efficient supercritical carbon dioxide modifiers for the extraction of the basic compound pirimicarb from weathered soil samples (Figure 3.6) (Alzaga *et al.*, 1995).

The extraction of atrazine and its metabolites (deethyl- and deisopropyl-) which are acidic ($pK_a = 1.7$) from weathered soil samples was achieved only with $10^{-2}$ M TFA in MeOH as modifier (Alzaga *et al.*, 1996). Similarly,

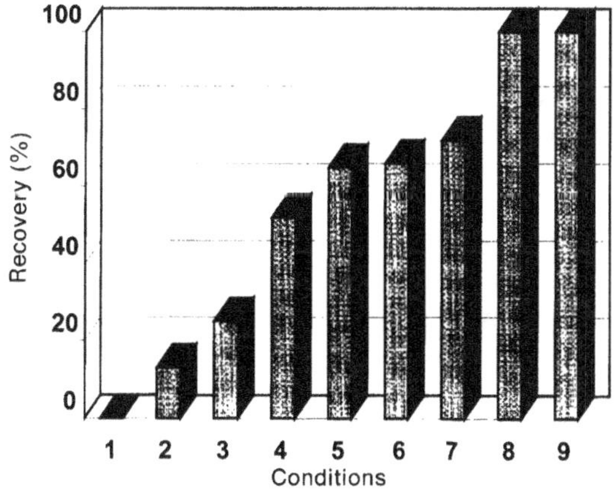


**Figure 3.6** The effect of modifiers on the extraction efficiency of native pirimicarb from soil: 1 = carbon dioxide at 100°C; 2 = chlorodifluoromethane at 150°C; 3 = carbon dioxide at 100°C modified with 5% methanol; 4 = nitrous oxide at 100°C modified with 5% methanol; 5 = nitrous oxide at 100°C modified with 5% pyridine; 6 = carbon dioxide at 50°C modified with 5% pyridine; 7 = chlorodifluoromethane at 100°C modified with 5% pyridine; 8 = carbon dioxide at 100°C modified with 5% pyridine; 9 = carbon dioxide at 100°C modified with 5% triethylamine. Pressure = 300 atm.

triazine and simazine have been extracted from 10 different types of soil samples, ranging from sand to clays, by using MeOH/TFA modified supercritical carbon dioxide. Other modifiers previously reported effective for the extraction of s-triazines from soils (e.g. acetone and methanol) were also evaluated during these studies but proved ineffective for the recovery of triazine and simazine (Alzaga, private communication).

This observation serves to illustrate the need for an understanding of the acid–base properties of analytes in order that appropriate modifiers be selected for SFE. Two main mechanisms have been postulated for the improvement of SFE efficiencies with use of modifier or reactive additives: that the modifier or reactive additive should compete with the analytes for retentive matrix active sites; that the analytes should be extracted as neutral species which are more soluble in supercritical fluids. Therefore, the acid–base properties of the modifier should be stronger than those of the analyte to keep it in neutral form. This hypothesis was tested in the extraction of the long-chain trialkylamines (TAMs), $CH_3NCl(C_{18}H_{37})_2$ and $CH_3NCl(C_{16}H_{33})_2$ from sediment. As with pirimicarb, the TAMs are basic (estimated $pK_a = 12.5$), hence SFE was performed with TEA-modified ($pK_a = 11.0$) supercritical carbon dioxide. However, this modifier was less effective than for pirimicarb, since the TAMs have two bulky alkyl groups, bringing about less interaction with the matrix (Alzaga *et al.*, 1998).

In a recent study the mechanisms of phenanthrene desorption from five subsurface soils by means of supercritical carbon dioxide were investigated. The technique involved measuring isotherms in the presence of fixed quantities of a polar modifier (methanol) and by calculating the desorption enthalpies from the temperature dependence of the isotherms (Young and Weber, 1997). Results indicate that the primary mechanism for polar modifier desorption enhancement is the competitive displacement of phenanthrene from polar sites associated with soil organic matter and mineral surfaces.

*Moisture content*

The effect of moisture content on SFE efficiency appears to be dependent upon the levels at which water is present. At low concentrations (*c.* 1%–2% m/m, which is below the solubility of water in supercritical fluid carbon dioxide) there is a tendency for water to promote SFE efficiencies, whereas at higher levels its effects are less predictable. It is thought that water at low levels promotes analyte recoveries as a result of competition with the matrix adsorption sites (Bøwadt and Hawthorne, 1995). At higher water levels, analyte distribution coefficients within a complex phase system can exert either positive or negative effects on SFE efficiency. Additionally, the presence of water in contact with supercritical carbon dioxide can bring about a dramatic reduction in pH (Toews *et al.*, 1995). In fact, over the temperature and pressure ranges of 25°C–75°C and 70–200 atm, respectively, the pH of water can vary from 2.8 to 2.95. This is accounted for by a reversible equilibrium involving the formation of carbonic acid:

$$CO_2 + H_2O \rightleftharpoons H_2CO_3 \rightleftharpoons H^+ + HCO_3^-$$

The pH increases with increasing temperature and decreases with increasing pressure. Therefore, the influence that pH exerts by changing these fundamental extraction parameters can be predicted for acidic or basic organic compounds and for metal ions extracted using complexing agents.

Polar modifiers such as water have an effect on matrix swelling, particularly for clays (e.g. montmorillonite) and plant materials (Fahmy *et al.*, 1993). For these matrices the swelling effect brought about by using water-saturated supercritical fluid carbon dioxide correlated with the extractability of several herbicides. In fact, when neat carbon dioxide was used no matrix swelling was observed and the extraction efficiency was extremely poor from the 'closed' matrices. A direct correlation between the extraction efficiencies and swelling from saturated matrices was obtained for the clay and from pea leaves as a function of pressure.

For the montmorillonite clay the correlation was achieved over the pressure range where the intracrystalline water of the clay should predominate. In the case of pea leaves, the swelling effects caused by six modifiers (including water-saturated carbon dioxide) were studied. It was demonstrated that modifier polarity correlated with swelling and herbicide extractability. Since these matrices swelled quickly, extractability was not a function of contact time. Kaolinite clay did not swell, and herbicide recoveries did not change with pressure.

## 3.3 Extract collection

Most SFE extract-collection techniques involve the use of solvents or adsorbents at subambient temperatures and are separately described in the following sections (also see sections 2.8 and 4.2.5). However, the possibility of alternative trapping techniques have also been highlighted (Bayona, 1993; McNally *et al.*, 1992). For example, Miller *et al.* (1993) have found reasonably high trapping efficiencies for PCBs, PAHs, gasoline and diesel fuels when the SFE extract was transferred into an empty vial via stainless tubing (178 mm) for a short period of time (3–30 s) following static SFE. The main advantage of this collection procedure is the elimination of the need for a restrictor, which can be prone to plugging, depending upon the nature of the sample. This collection method, however, cannot be combined with dynamic SFE procedures.

### *3.3.1 Extract trapping using solvents*

Owing to its simplicity and low cost, solvent collection is one of the most widely used techniques for off line SFE procedures. Usually, the heated restrictor is introduced directly into a vial containing an organic solvent at subambient temperatures. Langenfeld *et al.* (1992) have evaluated the effect of solvent collection variables for semivolatile pollutants, flavour and fragrance compounds. The effect of solvent polarity, temperature, volume and depth were investigated. Heating the solvent resulted in 20%–50% losses of the more volatile analytes, whereas higher than 90% trapping efficiency could be attained by controlling the solvent temperature at 5°C. In this study, restrictor plugging caused by ice formation, brought about by Joule–Thomson cooling as the supercritical fluid expands, precluded the use of lower solvent trapping temperatures. Dichloromethane at 5°C was found to be effective for the collection of most of the 66 compounds involved in this study. In general, excessively long extraction times and/or high supercritical fluid flow rates, with solvent collection, can result in the loss of volatile analytes as a result of purging.

Dual chamber trapping vials have been introduced for off-line SFE to allow restrictor heating at high temperatures to prevent plugging, whilst the collection solvent can be maintained at low temperatures (Porter *et al.*, 1992). Since the residence time within the restrictor heated zone is very short, this system prevents decomposition of thermally labile compounds. However, for matrices containing low volatility extractables restrictor plugging can still occur as a result of their deposition in the vicinity of supercritical fluid decompression. Thompson and Taylor (1994) have evaluated the trapping efficiency of this system using a 'polarity test mixture' (comprising acetophenone, *N*,*N*-dimethylaniline, naphthalene, decanoic acid, 2-naphthol and *n*-tetracosane) which they found to be dependent upon the supercritical fluid characteristics (neat compared with modified carbon dioxide). Nevertheless, hexane exhibited the best trapping efficiency for all the analytes and extracting conditions evaluated. The same authors have evaluated collection efficiencies using single solvents and binary mixtures for a 'polarity test mixture' of analytes using the dual-chamber collection vial system (Thompson *et al.*, 1993). Higher trapping efficiencies (greater than 90%) were obtained with the binary mixture of solvents. Wenclaviak *et al.* (1994) have modified the dual-chamber collection vial by introducing a frit at the end of the empty tube to improve the collection efficiency of volatile chlorinated hydrocarbons. This method allowed an efficient trapping (95%) of chlorobenzenes (dichloro to hexachloro) and hexachlorocyclohexane isomers in cold solvent. Another improvement of this collection system was achieved by introducing an adsorbent (i.e. silica) in the collection tube, which led to cleaner extracts than trapping into organic solvents (Meyer *et al.*, 1993).

In order to improve the partitioning of the analytes contained in the expanding supercritical fluid, a continuous solvent recirculation system has been developed by Vejrosta *et al.* (1994a). Solvent is continuously recirculated from the collection vial to the upflow of the restrictor expansion zone, this being performed within the interior of a fused silica tube. Analytes are partitioned into the solvent and transferred to the vial. This collection device has proven effective for the collection of the 3–5 aromatic ring PAHs and *s*-triazine herbicides using carbon dioxide as the fluid. The same authors (Vejrosta *et al.*, 1994b) have proposed a modification of the collection device for use with modifiers based on cryogenic trapping of the expanded fluid. Nebulization of the expanding supercritical/condensing modifier mixture, followed by analyte trapping into a moving liquid layer, is assumed. The collection device was successfully applied for the trapping of a urea herbicide using carbon dioxide modified with 10% methanol.

Another approach has been proposed for the minimization of collection losses of five groups of analytes (EPA PAHs, alkanes, nitro-PAHs, fatty acid esters and chlorinated hydrocarbons). This involves the use of a Dewar condenser connected to the top of the SFE collection vial (Figure 3.7) (Wenclaviak *et al.*, 1995). Collection losses were reduced to *c.* 2% m/m with the condenser maintained at −25°C, using ethyl acetate as the collecting

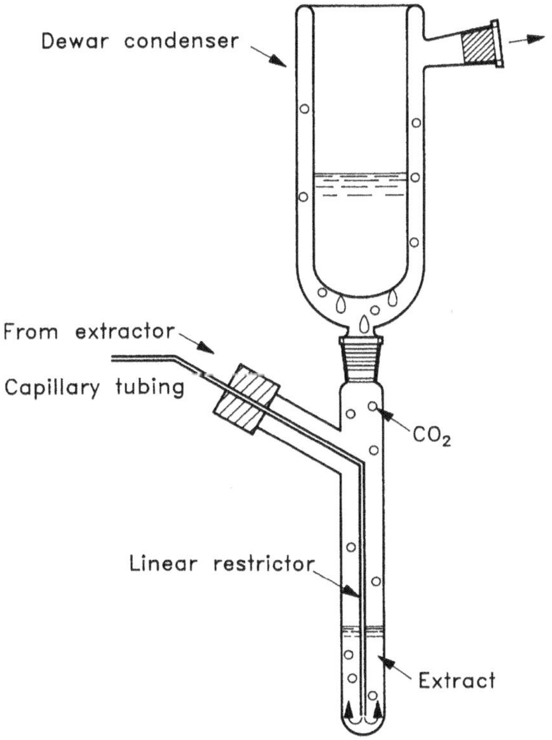


**Figure 3.7** Cryogenic trapping device for the off-line collection of volatile analytes in organic solvents. Adapted with permission from Wenclaviak *et al.* (1995).

solvent, whereas without the condenser losses ranged from 8% to 15% m/m, depending upon compound volatility.

### *3.3.2 Extract trapping using solid-phase sorbents*

Solid-phase sorbents were introduced to trap volatile analytes efficiently using off-line SFE procedures (Ashraf-Khorassani *et al.*, 1992; Mulcahey *et al.*, 1991). Solid-phase trapping is performed by depressurizing the supercritical fluid prior to the trap, and collecting the analytes from the gas phase onto sorbents such as silica gel and Florisil® or bonded phase packing materials, or onto cooled inert surfaces such as glass or stainless-steel beads. After trapping, the analytes are eluted with liquid solvents for subsequent analysis.

Unfortunately, solid phase trapping efficiency depends on a large number of variables, including collection temperature, type of fluid, adsorbent, fluid volume and eluting solvent, making method development difficult. However, solid-phase trapping has advantages compared with solvent collection. For example, under appropriate circumstances it is possible to bring about the fractionation of different extracted compound classes during solvent elution of the adsorbent. Additionally, solid-phase trapping is well suited to automation.

Solid-phase trapping is generally accomplished by two mechanisms: cryotrapping and adsorption. The predominance of one mechanism over the other depends on the physicochemical properties of the analytes (Mulcahey *et al.*, 1991). Trapping materials should be chosen to maximize the trapping efficiency, with appropriate chemical class selectivity. With low concentrations of modifiers and octadecylsilyl-modified silica (ODS) as adsorbent, trap temperatures below 20°C were necessary for the collection of volatile (boiling point less than 200°C) analytes. With less volatile analytes, trapping was near 100% regardless of the trap temperature. To achieve maximum recovery at moderate modifier concentrations (greater than 2% methanol), elevated trap temperatures were necessary, but at high modifier concentrations (8% of methanol) recoveries dropped to 50% for all the analytes (Mulcahey and Taylor, 1992). Stainless-steel was generally less satisfactory than ODS for volatile analytes (boiling point less than 200°C), but at low modifier concentrations (1% methanol) collection efficiency using stainless-steel beads improved (Mulcahey and Taylor, 1992). Although reduction of trap temperature usually improves the analyte recoveries, two problems can arise which can serve to limit the use of cryogenic trapping. The first of these is associated with moisture-containing samples which can give rise to ice formation within the trap, which causes immediate blockage and system overpressure. The second is when modifiers are used that condense within the trap, leading to uncontrollable elution of trapped analytes. These limitations can be overcome by raising the trap temperature, but this can lead to a low trapping efficiency of the more volatile analytes if the adsorbent is not properly chosen.

More recently, the use of a tandem system involving solid-phase adsorbent and solvent trapping has proven effective for the collection of high molecular weight glycosides using high percentages of modifier (Moore and Taylor, 1995). Optimum collection efficiency was obtained when the trap temperature was operated at high temperatures to boil away the modifier from the trap, with low supercritical fluid flow rates to maximize contact time with the trap. Therefore, sorbent selection is very important to obtain high trapping efficiencies of volatile analytes and it must be emphasized that when extraction conditions are modified (e.g. flow rate, modifier content) trapping conditions must be newly optimized. Yang *et al.* (1995c) have compared the efficiencies of sorbent and solvent trapping for the collection of diesel and gasoline hydrocarbons extracted from soil samples. Both sorbent (Porapak Q) and solvent (dichloromethane) trapping could effectively (>90%) collect BTEX hydrocarbons (benzene, toluene, ethylbenzene and xylenes). Sorbent trapping yielded quantitative collections (>88%) of *n*-alkanes as volatile as *n*-hexane, while the solvent trapping effectively collected *n*-alkanes as volatile as *n*-heptane. Quantitation of BTEX, total petroleum hydrocarbons (TPH) and individual components from contaminated soils obtained by the two SFE systems were in good agreement. Higher recoveries of these hydrocarbons were obtained by the SFE methods relative to Soxhlet extraction, suggesting greater losses being associated with the latter technique.

## 3.4 Mathematical models used for optimizing supercritical fluid extraction parameters

Although several kinetic models have been developed to fit SFE experimental data obtained using different extraction conditions (Bartle *et al.*, 1990; Dooley *et al.*, 1990; Langenfeld *et al.*, 1995; Pawliszyn, 1993; Young and Weber, 1997), the great complexity of real-world samples serves to limit the general applicability of these models. The limitations of these theoretical models include: difficulty in predicting the precise nature of analyte–matrix interactions which exist in the vast variety of different types of sample; sample heterogeneity; that specific SFE experimental parameters need to be optimized by method development. In the following discussion, an overview of the SFE models is presented and the optimization strategies for method development are discussed.

### *3.4.1 Supercritical fluid extraction kinetic models*

Packed-tube extractor geometry with a fluid flow is a very efficient means of accomplishing removal and collection of analytes from a matrix (Pawliszyn,

1993). When optimum conditions exist, only two void volumes of fluid are necessary to remove adsorbed analytes quantitatively, providing no retentive analyte–matrix interactions occur. Therefore, the objective of this approach is to determine the nature of the supercritical fluid (including the use of modifiers) to accomplish this goal.

Deconvolution of the experimental data that describe the extraction rate of spiked analytes from real sample matrices allowed an investigation of analyte release from the sample matrix to the fluid interface. Increasing the temperature resulted in faster analyte desorption for all analyte–matrix combinations and a three-rate-constant desorption model generally fitted the experimental data. Therefore, it would appear that both chromatographic elution (partitioning and bulk transport in the supercritical fluid) and desorption (matrix–fluid mass transport) have a significant impact on SFE rates of native analytes from real samples and that increasing the temperature may be a useful approach to improve SFE extraction efficiencies regardless of the rate-limiting step (Langenfeld *et al.*, 1995). In the presence of modifiers, a fixed-bed desorption equilibrium model indicated that contaminant solubilities in supercritical fluids are enhanced when soil samples are analysed (Dooley *et al.*, 1990).

### *3.4.2 Strategies for the optimization of supercritical fluid extraction variables*

The large number of experimental parameters associated with SFE makes method development very time consuming if the variables are optimized one at a time or by univariate approaches. Furthermore, variables can be interacted and their optimization by classical methods can lead to biased results. Therefore, multivariate techniques for variable optimization are more useful in SFE to save time in the optimization process. Although SFE involves several processes, namely desorption of analytes from the matrix, mass transport by the fluid and extract trapping, the optimization processes usually are focused on the first two processes. The last two variables can be studied by using spiked inert matrices (sand, glass beads) or even injecting the analytes into the fluid using an empty extraction vessel. In contrast, the optimization of the extraction process should be carried out using real-world samples or standard reference materials (SRM) with a certified value of the analytes of interest, when they are available, since matrix effects play an important role in most SFE applications. If the optimization of the extraction process is carried out on spiked samples it can lead to an overestimation of extraction recoveries since variables which are significant in real samples may not be adequately taken into account (van der Velde *et al.*, 1994a).

#### *Simplex optimization method*

The variable-size simplex method proposed by Morgan and Deming (1974) has been applied to SFE optimization (Ho and Tang, 1992). The variable-size simplex operation is a sequential statistical search which is a relatively straightforward method to locate the optimum combination of variables when there are multivariate, interactive factor effects. A geometry figure of the simplex is defined by a number of points ($n + 1$) equal to one more than the number of dimensions of the factor space ($n$). For example, in their study Ho and Tang optimized three fundamental variables relating to the extraction process, namely pressure, temperature and extraction time. Since three variables were selected the initial trials comprised four experiments at different points in factor space, geometrically represented by the vertices of a tetrahedron. The four experiments are arranged in order of decreasing response. Comparison of the coordinates of that point in factor space gave the worst response to the average of the other three, which gave a better response. The mathematical algorithm of this simplex method leads to a new set of coordinates for a proposed experiment. This experiment will give a response that falls within the ranking of previous results. The worst-case coordinates of the original experiment and the results are discarded from the ranking, the new experiment is substituted and the new set of four is re-ranked. The process is restarted with the new tetrahedron. This procedure is continued until all the coordinates of the group are near the region of the factor space which gives best results or until the responses approach plateau values. The simplex process is then stopped. Its combination with a factorial design (see the next subsection) helps to determine the initial conditions for SFE operation.

*Experimental design approach*

Experimental designs are one of the most useful approaches either to prescreen amongst a large number of variables (4–8 variables) or to determine conditions for a preselected number of them. The first approach, fractional factorial experimental design (e.g. $2^{(n-1)}$ experiments would be necessary using a half factorial design to study $n$ variables), has been used since the early days of SFE to evaluate the effect of temperature, pressure, moisture content, vessel volume, modifier content in the fluid, extraction time and the flow rate of fluid through the extraction vessel (Cai and Bayona, 1995; Dachs *et al.*, 1994; López-Avila *et al.*, 1990). This approach involves the following steps:

1. selection of the variables to be optimized;
2. definition of the optimization range;
3. building up of the matrix with the optimization variables;
4. performance of all the extractions at the preselected conditions in order to measure the response of each extraction variable;
5. analysis of variance (ANOVA) to find out the significantly statistical variables.

If a few of the variables are significant, the final optimization of the extraction variables is easily carried out by the simplex method (Ho and Tang,

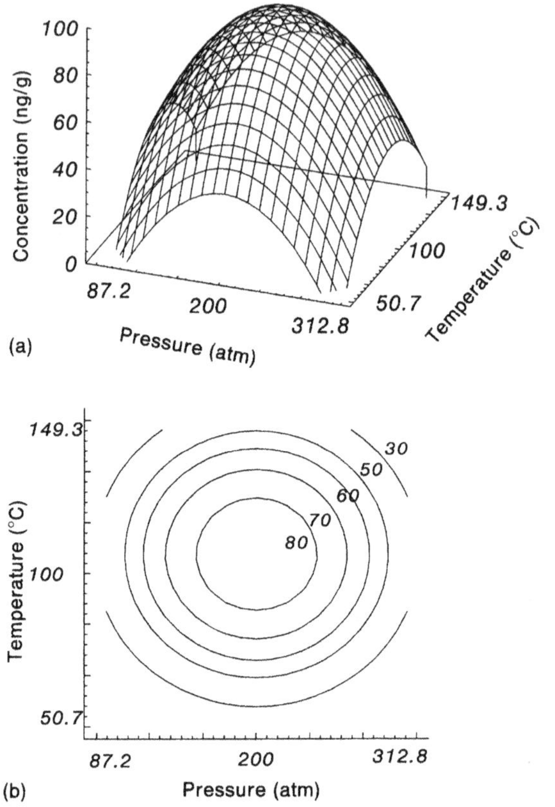


**Figure 3.8** (a) Surface response and (b) contour map (the concentration is shown beside each ring in ng/g) of the experimental design optimization of the extraction of native polychlorinated biphenyls from sediments. Reproduced with permission from Fernandez *et al.* (1996).

1992) or by the variable-at-a-time method. The response surface can be plotted or converted to a contour map, as shown in Figures 3.8(a) and 3.8(b), respectively. Polynomial models may be used empirically to approximate the true physical model over the experimental range. The most common models contain the linear interaction and quadratic terms, and recently a fourth, partial cubic, type has emerged which contains interaction terms of the third degree but does not contain cubic terms for a single variable (Bicking *et al.*, 1993; Fernández *et al.*, 1995, 1996; Fernández-Escobar and Bayona, 1997; Ho and Tang, 1992; Kane *et al.*, 1993; Llompart *et al.*, 1996a, 1996b). The significance of regression is tested and it should be above the 99% level of confidence; thus it is unlikely ($p < 0.01$) that the results are obtained by chance.

Usually, the experimental design approach has been used for the optimization of the SFE variables. However, very recently environmental variables such as the clay and organic matter content and the residence time of analytes within the matrix have also been considered in the optimization process [both van der Velde *et al.* (1994b) and Zhou *et al.* (1997) have considered these variables for the SFE of pesticides from soil]. Typical experimental designs used in SFE method development are two-level, involving two to four factors. However, in a few studies the experimental designs used allow direct optimization of a larger number of variables, but these techniques also require a large number of experiments. For example, the Plackett–Burman design allows 14 degrees of freedom and involves 24 randomized runs plus two centred points (26 experiments) which enable the optimization of up to nine variables (Llompart *et al.*, 1996a, 1996b). Although this design saves time (29 experiments would be necessary to optimize nine variables), 26 experiments is rather time consuming for most method developments.

Recently, Dachs *et al.* (1997) applied information theory combined with experimental design to minimize the qualitative differences between Soxhlet extraction and SFE (minimization of the information gain); the extraction efficiency is optimized by a hybrid experimental design. This work demonstrated that the optimized SFE conditions depended upon the experimental design procedure used; for example there can be variation between the values of optimized experimental conditions, determined using procedures based upon absolute analyte recoveries and those which are designed to provide comparable results between SFE and Soxhlet extraction. In fact, in addition to the concentration of contaminants, in environmental samples, their distribution is of primary importance since it provides information about their source. However, it is beyond the remit of this book to consider sampling chemometrics.

## References

Alexandrou, N., Lawrence, M. and Pawliszyn, J. (1992) Cleanup of complex mixtures using supercritical fluids and selective adsorbents. *Analytical Chemistry*, **64**, 301–11.

Alzaga, R., Bayona, J.M. and Barceló, D. (1995) Use of supercritical fluid extraction for pirimicarb determination in soil. *Journal of Agricultural Food Chemistry*, **43**, 395–400.

Alzaga, R., Bayona, J.M. and Barceló, D. (1996) Supercritical fluid extraction of atrazine and its metabolites from soil. *Journal of High Resolution Chromatography*, **19**, 23–6.

Alzaga, R., Maldonado, C. and Bayona, J.M. (1998) Intercomparison between SFE, ASE, sonication and Soxhlet for the extraction of trialkylamines in environmental matrices. *International Journal of Environmental Analytical Chemistry*, in press.

Ashraf-Khorassani, M., Houck, R.K. and Levy, J.M. (1992) Cryogenically cooled adsorbent trap for off-line supercritical fluid extraction. *Journal of Chromatographic Science*, **30**, 361–6.

Ashraf-Khorassani, M., Taylor, L.T. and Zimmerman, P. (1990) Nitrous oxide versus carbon dioxide for supercritical fluid extraction and chromatography of amines. *Analytical Chemistry*, **62**, 1177–80.

Barnabas, I.J., Dean, J.R. and Owen, S.P. (1994) Supercritical fluid extraction of analytes from environmental samples. A review. *Analyst*, **119**, 2381–94.

Bartle, K.D., Clifford, A.A., Hawthorne, S.B. *et al.* (1990) A model for dynamic extraction using a supercritical fluid. *Journal of Supercritical Fluids*, **3**, 143–9.

Bayona, J.M. (1993) The role of supercritical fluid extraction in the automation of sample preparation. *Laboratory Robotics and Automation*, **5**, 283–92.

Bicking, M.K.L., Hayes, T.G., Kiley, J.C. *et al.* (1993) An experimental design approach to the optimization of supercritical fluid extraction for the determination of oil and grease in soil. *Journal of Chromatographic Science*, **31**, 170–6.

Bøwadt, S. and Hawthorne, S.B. (1995) Supercritical fluid extraction in environmental analysis. *Journal of Chromatography A*, **703**, 549–71.

Bøwadt, S. and Johansson, B. (1994) Analysis of PCBs in sulfur-containing sediments by off-line supercritical fluid extraction and HRGC–ECD. *Analytical Chemistry*, **66**, 667–73.

Burford, M.D., Hawthorne, S.B. and Miller, D.J. (1993) Evaluation of drying agents for off-line supercritical fluid extraction. *Journal of Chromatography A*, **657**, 413–27.

Cai, Y. and Bayona, J.M. (1995) Simultaneous speciation of butyl-, phenyl-, and cyclohexyltin compounds in aqueous matrices using ethylation followed by solid-phase trace enrichment, SFE, and GC determination. *Journal of Chromatographic Science*, **33**, 89–97.

Cai, Y., Alzaga, R. and Bayona, J.M. (1994) In situ derivatization and supercritical fluid extraction for the simultaneous determination of butyltin and phenyltin compounds in sediment. *Analytical Chemistry*, **66**, 1161–7.

Camel, V., Tambuté, A. and Caude, M. (1993) Analytical-scale supercritical fluid extraction: a promising technique for the determination of pollutants in environmental matrices. *Journal of Chromatography A*, **642**, 263–81.

Chau, Y.K. and Brown, F.Y. (1995) Supercritical fluid extraction of butyltin compounds from sediment. *Analytica Chimica Acta*, **304**, 85–9.

Chester, T.L., Pinkston, J.D. and Raynie, D.E. (1994) Supercritical fluid chromatography and extraction. *Analytical Chemistry*, **66**, 106R–130R.

Dachs, J., Alzaga, R., Bayona, J.M. and Quevauviller, P. (1994) Development of a supercritical fluid extraction procedure for tributyltin determination in sediments. *Analytica Chimica Acta*, **286**, 319–27.

Dachs, J., Fernández, I. and Bayona, J.M. (1997) Combined experimental design and information theory for the optimisation of supercritical fluid extraction of organic priority pollutants from sediment. *Analytica Chimica Acta*, **351**, 377–85.

Dooley, K.M., Ghonasgi, D., Knopf, F. *et al.* (1990) Supercritical $CO_2$–cosolvent extraction of contaminated soils and sediments. *Environmental Progress*, **9**, 197–203.

Emteborg, H., Björklund, E., Ödman, F. *et al.* (1995) Determination of methyl mercury in sediments using supercritical fluid extraction and gas-chromatography coupled with microwave-induced plasma emission spectrometry. *Analyst*, **121**, 19–29.

Fahmy, T.M., Paulaitis, M.E., Johnson, D.M. *et al.* (1993) Modifier effects in the supercritical fluid extraction of solutes from clay, soil, and plant materials. *Analytical Chemistry*, **65**, 1462–9.

Fernández, P., Alder, A.C., Ster, M.J.F. *et al.* (1996) Determination of the quaternary ammonium surfactant ditallowdimethylammonium in digested sludges and marine sediments by supercritical fluid extraction and liquid chromatography with postcolumn ion-pair formation. *Analytical Chemistry*, **68**, 921–9.

Fernández, I., Dachs, J. and Bayona, J.M. (1995) Application of experimental design approach to the optimization of supercritical fluid extraction of polychlorinated biphenyls and polycyclic aromatic hydrocarbons. *Journal of Chromatography A*, **719**, 77–85.

Fernández-Escobar, I. and Bayona, J.M. (1997) Supercritical fluid extraction of priority organotin contaminants from biological matrices. *Analytica Chimica Acta*, **351**, 269–76.

Field, J.A., Miller, D.J., Field, T.M. *et al.* (1992) Quantitative determination of sulfonated aliphatic and aromatic surfactants in sewage sludge by ion-pair/supercritical fluid extraction and derivatization gas chromatography/mass spectrometry. *Analytical Chemistry*, **64**, 3161–7.

France, J.E., King, J.W. and Snyder, J.M. (1991) Supercritical fluid-based cleanup technique for the separation of organochlorine pesticides from fats. *Journal of Agricultural Food Chemistry*, **39**, 1871–4.

Friedrich, C., Cammann, K. and Kleiböhmer, W. (1995) Matrix independent supercritical fluid extraction of polycyclic aromatic hydrocarbons by employing binary modifiers. *Fresenius Journal of Analytical Chemistry*, **352**, 730–4.

Furton, K.G., Chen, L. and Jaffé, R. (1995) Rapid determination of uranium on solid matrices by synergistic in situ chelation supercritical fluid extraction and UV absorption spectroscopy. *Analytica Chimica Acta*, **304**, 203–8.

Furton, K.G., Huang, C.-W., Jaffé, R. *et al.* (1994) High-temperature supercritical fluid extraction of hydrocarbons from geological samples and comparison to Soxhlet extraction. *Journal of High Resolution Chromatography*, **17**, 679–81.

Giddings, J.C., Myers, M.N., McLaren, L. and Keller, R.A. (1968) High-pressure gas chromatography of non-volatile species. Compressed gas is used to cause migration of intractable solutes. *Science*, **162**, 67–73.

Hale, R.C. and Gaylor, M.O. (1995) Determination of PCBs in fish tissues using supercritical fluid extraction. *Environmental Science and Technology*, **29**, 1043–7.

Hawthorne, S.B. (1990) Analytical-scale supercritical fluid extraction. *Analytical Chemistry*, **62**, 633A–642A.

Hawthorne, S.B. and Miller, D.J. (1994) Direct comparison of soxhlet and low- and high-temperature supercritical $CO_2$ extraction efficiencies of organics from environmental solids. *Analytical Chemistry*, **66**, 4005–12.

Hawthorne, S.B., Galy, A.B., Scmitt, V.O. *et al.* (1995) Effect of SFE flow rate on extraction rates: classifying sample extraction behavior. *Analytical Chemistry*, **67**, 2723–32.

Hawthorne, S.B., Langenfeld, J.J., Miller, D.J. and Burford, M.D. (1992b) Comparison of supercritical $CHClF_2$, $N_2O$ and $CO_2$ for the extraction of polychlorinated biphenyls and polycyclic aromatic hydrocarbons. *Analytical Chemistry*, **64**, 1614–22.

Hawthorne, S.B., Miller, D.J., Nivens, D.E and White, D.C. (1992a) Supercritical fluid extraction of polar analytes using in situ chemical derivatization. *Analytical Chemistry*, **64**, 405–12.

Hawthorne, S.B., Yang, Y. and Miller, D.J. (1994) Extraction of pollutants from environmental solids. *Analytical Chemistry*, **66**, 2912–20.

Hillmann, R. and Bachmann, K. (1994) On-line supercritical fluid derivatization and extraction–capillary gas chromatography of polar compounds. *Journal of High Resolution Chromatography*, **17**, 350–2.

Hillmann, R. and Bachmann, K. (1995) Extraction of pesticides using supercritical trifluoromethane and carbon dioxide. *Journal of Chromatography A*, **695**, 149–54.

Hills, J.W. and Hill, H.H. (1993) Carbon dioxide supercritical fluid extraction with a reactive solvent modifier for the determination of polycyclic aromatic hydrocarbons. *Journal of Chromatographic Science*, **31**, 6–12.

Hills, J.W., Hill, H.H. Jr, Hansen, D.R. *et al.* (1994) Carbon dioxide supercritical fluid extraction of incinerator fly ash with a reactive solvent modifier. *Journal of Chromatography A*, **679**, 319–28.

Ho, J.S. and Tang, P.H. (1992) Optimization of supercritical fluid extraction of environmental polutants from a liquid–solid extraction. *Journal of Chromatographic Science*, **30**, 344–50.

Howard, A.L., Yoo, W.J., Taylor, L.T. *et al.* (1993) Supercritical fluid extraction of environmental analytes using trifluoromethane. *Journal of Chromatographic Science*, **31**, 401–8.

Janda, V., Bartle, K.D. and Clifford, A.A. (1993) Supercritical fluid extraction in environmental analysis. *Journal of Chromatography A*, **642**, 283–99.

Kane, M., Dean, J.R., Hitchen, S.M. *et al.* (1993) Experimental design approach for supercritical fluid extraction. *Analyst*, **271**, 83–90.

King, J.W. (1989) Fundamentals and applications of supercritical extraction in chromatographic sciences. *Journal of Chromatographic Science*, **27**, 355–64.

Laintz, K.E., Wai, C.M., Yonker, C.R. *et al.* (1992) Extraction of metal ions from liquid and solid materials by supercritical carbon dioxide. *Analytical Chemistry*, **64**, 2875–8.

Langenfeld, J.J., Burford, M.D., Hawthorne, S.B. and Miller, D.J. (1992) Effects of collection solvent parameters and extraction cell geometry on supercritical fluid extraction efficiencies. *Journal of Chromatography*, **594**, 297–307.

Langenfeld, J.J., Hawthorne, S.B., Miller, D.J. *et al.* (1993) Effects of temperature and pressure on supercritical fluid extraction efficiencies of polycyclic aromatic hydrocarbons and polychlorinated biphenyls. *Analytical Chemistry*, **65**, 338–44.

Langenfeld, J.J., Hawthorne, S.B., Miller, D.J. *et al.* (1995) Kinetic study of supercritical fluid extraction of organic contaminants from heterogenous environmental samples with carbon dioxide and elevated temperatures. *Analytical Chemistry*, **67**, 1727–36.

Lee, H.-B., Peart, T.E., Nimi, A.J. and Knipe, C.R. (1995) Rapid supercritical carbon dioxide extraction method for determination of polychlorinated biphenyls in fish. *Journal of AOAC International*, **78**(2), 437–43.

Lee, H.-B., Peart, T.E. and You, R.L. (1992) *In situ* extraction and derivatization of pentachlorophenol and related compounds from soils using a supercritical fluid extraction system. *Journal of Chromatography*, **605**, 109–13.

Lehotay, S.J., Aharonson, N., Pfeil, E. and Ibrahim, M.A. (1995) Development of a sample preparation technique for supercritical fluid extraction for multiresidue analysis of pesticides in produce. *Journal of AOAC International*, **78**, 831–40.

Lin, Y.L. and Wai, C.M. (1994) Supercritical fluid extraction of lanthanides with fluorinated β-diketones and tributyl phosphate. *Analytical Chemistry*, **66**, 1971–5.

Lin, Y., Brauer, R.D., Laintz, K.E. *et al.* (1993) Supercritical fluid extraction of lanthanides and actinides from solid materials with a fluorinated β-diketone. *Analytical Chemistry*, **65**, 2549–51.

Liu, Y., López-Avila, V., Alcaraz, M. *et al.* (1993a) Determination of metals in solid samples by complexation–supercritical fluid extraction and gas chromatography–atomic emission detection. *Journal of Chromatographic Scence*, **31**, 310–16.

Liu, Y., López-Avila, V., Alcaraz, M. and Beckert, W. F. (1993b) Determination of organotin compounds in environmental samples by supercritical fluid extraction and gas chromatography with an atomic emission detector. *Journal of High Resolution Chromatography*, **16**, 106–12.

Llompart, M.P., Lorenzo, R.A. and Cela, R. (1996a) Optimization of supercritical fluid extraction of phenol and cresols in soil samples. *Journal of Chromatography A*, **723**, 123–34.

Llompart, M.P., Lorenzo, R.A. and Cela, R. (1996b) Multivariate optimization of supercritical fluid derivatization and extraction of phenol in soil samples. *Journal of Chromatographic Science*, **34**, 43–51.

López-Avila, V., Dodhiwala, N.S. and Beckert, W.F. (1990) Supercritical fluid extraction and its application to environmental analysis. *Journal of Chromatgraphic Science*, **28**, 469–76.

López-Avila, V., Dodhiwala, N.S. and Beckert, W.F. (1993) Developments in the supercritical fluid extraction of chlorophenoxy acid herbicides from soil samples. *Journal of Agricultural Food Chemistry*, **41**, 2038–43.

McNally, M.E.P. (1995) Advances in environmental SFE. *Analytical Chemistry*, **67**, 308A–315A.

McNally, M.E.P., Deardorff, C.M. and Fahmy, T.M. (1992) Supercritical extraction: new directions and understandings, in *Supercritical Fluid Extraction in Supercritical Fluid Technology* (eds F.V. Bright and M.E.P. McNally), American Chemical Society, Washington, DC, pp. 144–64.

Macnaughton, S.J., Alessi, P., Cortesi, A. *et al.* (1995) Predictive and experimental methods for the choice of cosolvent in the supercritical fluid extraction of pesticides, in *Innovations in Supercritical Fluids* (eds K.W. Hutchenson and N.R. Foster), American Chemical Society, Washington, DC, pp. 126–39.

Meyer, A., Kleibhöhmer, W. and Cammann, K. (1993) SFE of PAHs from soils with a high carbon content and analyte collection via combined liquid/solid trapping. *Journal of High Resolution Chromatography*, **16**, 491–4.

Miao, Z., Zhang, Z. and Pawliszyn, J. (1994) Supercritical fluid extraction and clean-up with temperature fractionation: application to determination of polychlorinated dibenzo-*p*-dioxins. *Journal of Microcolumn Separations*, **6**, 459–65.

Miller, D.J., Hawthorne, S.B. and McNally, M.E.P. (1993) Solventless collection of analytes by rapid depressurization after static supercritical fluid extraction. *Analytical Chemistry*, **65**, 1038–42.

Moore, W.N. and Taylor, L.T. (1995) Solid phase trapping mechanisms involved in the supercritical fluid extraction of *Digitalis* glycosides with modified carbon dioxide. *Analytical Chemistry*, **67**, 2030–16.

Morgan, S.L. and Deming, S.N. (1974) Simplex optimization of analytical chemical methods. *Analytical Chemistry*, **46**, 1179–81.

Mulcahey, L.J. and Taylor, L.T. (1992) Collection efficiency of solid surface and sorbent traps in supercritical fluid extraction with modified carbon dioxide. *Analytical Chemistry*, **64**, 2352–8.

Mulcahey, L.J., Hedrick, J.L. and Taylor, L.T. (1991) Collection efficiency of various solid-phase traps for off-line supercritical fluid extraction. *Analytical Chemistry*, **63**, 2225–32.

Murugaverl, B. and Voorhees, K.J. (1991) On-line supercritical fluid extraction/chromatography system for trace analysis of pesticides in soybean oil and rendered fats. *Journal of Microcolumn Separations*, **3**, 11–16.

Murugaverl, B., Gharaibeh, A. and Voorhees, K.J. (1993) Mixed adsorbent for cleanup during supercritical fluid extraction of three pesticides in tissues. *Journal of Chromatography A*, **657**, 223–6.

Nguyen, D.-K., Bruchet, A. and Arpino, P. (1995) Determination of sterols in sewage sludge by combined in situ trimethylsilylation/supercritical fluid extraction and GC/MS. *Environmental Science and Technology*, **29**, 1686–90.

Oostdyk, T.S., Grob, R.L., Snyder, G.J.L. and McNally, M.E. (1993) Study of sonication and supercritical fluid extraction of primary amines. *Analytical Chemistry*, **65**, 596–600.

Page, S.H., Sumpter, S.R. and Lee, M.L. (1992) Fluid phase equilibria in supercritical fluid chromatography with carbon. *Journal of Microcolumn Separations*, **4**, 91–122.

Pawliszyn, J. (1993) Kinetic model of supercritical fluid extraction. *Journal of Chromatographic Science*, **31**, 31–7.

Porter, N.L., Rynaski, E.R., Campbell, E.R. *et al.* (1992) Studies of linear restrictors and analyte collection via solvent trapping after supercritical fluid extraction. *Journal of Chromatography*, **30**, 367–73.

Pyle, S.M. and Setty, M.M. (1991) Supercritical fluid extraction of high sulfur soils, with use of a copper scavenger. *Talanta*, **38**, 1125–8.

Raynie, D. (1993) Warning concerning the use of nitrous oxide in supercritical fluid extractions. *Analytical Chemistry*, **65**, 3127–8.

Reighard, T.S. and Olesik, S.V. (1996) Comparison of the extraction of phenolic and nitroaromatic pollutants using supercritical and enhanced-fluidity liquid methanol–$CO_2$ mixtures. *Journal of Chromatography A*, **737**, 233–42.

Reindl, S. and Hofler, F. (1994) Optimization of the parameters in supercritical fluid extraction of polynuclear aromatic hydrocarbons from soil samples. *Analytical Chemistry*, **66**, 1808–16.

Sauvage, E., Rocca, J.-L. and Toussaint, G. (1993) The use of nitrous oxide for supercritical fluid extraction of pharmaceutical compounds from animal feed. *Journal of High Resolution Chromatography*, **16**, 234–8.

Smart, N.G., Carlesson, T., Kast, T. *et al.* (1997) Solubility of chelating agents and metal-containing compounds in supercritical fluid carbon dioxide. *Talanta*, **44**, 137–50.

Souk, K. and King, J.W. (1994) Supercritical fluid extraction and enzyme immunoassay for pesticide detection in meat products. *Journal of Agricultural Food Chemistry*, **42**, 1469–74.

Taylor, L.T. (1995) Strategies for analytical scale SFE. *Analytical Chemistry*, **67**, 364A–370A.

Thompson, P.G. and Taylor, L.T. (1994) Liquid trapping after supercritical fluid extraction with modified carbon dioxide. *Journal of High Resolution Chromatography*, **17**, 759–64.

Thompson, P.G., Taylor, L.T., Richter, B.E. *et al.* (1993) Trapping efficiencies of various collection solvents after supercritical fluid extraction. *Journal of High Resolution Chromatography*, **16**, 713–16.

Tilio, R., Kapila, S., Nam, K.S. *et al.* (1994) Reduction/elimination of sulfur interference in organochlorine residue determination by supercritical fluid extraction. *Journal of Chromatography A*, **662**, 191–7.

Toews, K.L., Shroll, R.M., Wai, C.M. *et al.* (1995) pH defining equilibrium between water and supercritical $CO_2$. Influence of SFE of organics and metal chelates. *Analytical Chemistry*, **67**, 4040–3.

Valverde-García, A. and Fernández-Alba, R. (1995) Extraction of methamidophos residues from vegetables with supercritical fluid carbon dioxide. *Journal of AOAC International*, **78**(3), 867–73.

van der Velde, E.G., Dietvorst, M., Swart, C.P. *et al.* (1994a) Optimization of supercritical fluid extraction of organochlorine pesticides from real soil samples. *Journal of Chromatography A*, **683**, 167–74.

van der Velde, E.G., Ramlal, M.R., van Beuzekom, A.C. *et al.* (1994b) Effects of parameters on supercritical fluid extraction of triazines from soil by use of multiple linear regression. *Journal of Chromatography A*, **683**, 125–39.

Vejrosta, J., Ansorgová, A., Planeta, J. *et al.* (1994a) Solute trapping in off-line supercritical fluid extraction using controlled modifier condensation. *Journal of Chromatography A*, **683**, 407–10.

Vejrosta, J., Planeta, J., Mikesová, M. *et al.* (1994b) Solute collection after off-line supercritical fluid extraction into a moving liquid layer. *Journal of Chromatography A*, **685**, 113–19.

Wai, C.M., Lin, Y., Brauer, R. *et al.* (1993) Supercritical fluid extraction of organic and inorganic mercury from solid materials. *Talanta*, **40**, 1325–30.

Wenclaviak, B.W., Heemken, O.P., Sterzenbacch, D. *et al.* (1995) Device for efficient solvent collection of environmentally relevant compounds in off-line SFE. *Analytical Chemistry*, **67**, 4577–80.

Wenclaviak, B.W., Maio, G., van Hoist, C. *et al.* (1994) Solvent trapping of some chlorinated hydrocarbons after supercritical fluid extraction from soil. *Analytical Chemistry*, **66**, 3581–6.

Wenclawiak, B.W. and Krah, M. (1995) Reactive supercritical fluid extraction and chromatography of arsenic species. *Fresenius Journal of Analytical Chemistry*, **351**, 134–8.

Yang, Y., Bøwadt, S., Hawthorne, S.B. and Miller, D.J. (1995a) Subcritical water extraction of polychlorinated biphenyls from soil and sediment. *Analytical Chemistry*, **67**, 4571–6.

Yang, Y., Gharaibeh, A.S., Hawthorne and Miller, D.J. (1995b) Combined temperature/modifier effect on supercritical $CO_2$ extraction efficiencies of polycyclic aromatic hydrocarbons from environmental samples. *Analytical Chemistry*, **67**, 641–56.

Yang, Y., Hawthorne, S.B. and Miller, D.J. (1995c) Comparison of sorbent and solvent trapping after supercritical fluid extraction of volatile petroleum hydrocarbons from soil. *Journal of Chromatography A*, **699**, 265–76.

Yang, Y., Hawthorne, S.B. and Miller, D.J. (1997) Class-selective extraction of polar, moderately polar, and nonpolar organics from hydrocarbon wastes using subcritical water. *Environmental Science and Technology*, **31**, 430–7.

Young, T.M. and Weber, W.J. (1997) A distributed reactivity model for sorption by soils and sediments. 7. Enthalpy and polarity effects on desorption under supercritical fluid conditions. *Environmental Science and Technology*, **31**, 1692–6.

Zhou, M., Trubey, R.K., Keil, Z.O. and Sparks, D.L. (1997) Study of the effects of environmental variables and supercritical fluid extraction parameters on the extractability of pesticide residues from soils using a multivariate optimization scheme. *Environmental Science and Technology*, **31**, 1934–9.

# 4 Supercritical fluid extraction strategies of liquid-based matrices

E.D. RAMSEY, B. MINTY AND R. BABECKI

## 4.1 Introduction

When one considers the many different types of environmental sample submitted for analysis of trace-level organic pollutants, aqueous samples appear to be amongst the least complex. The same applies for the analysis of drugs and their metabolites in biological matrices. In spite of this view, and considering supercritical fluid extraction (SFE) has been available since about 1982, it is interesting to note that the vast majority of reported SFE methods involve the analysis of solid matrices. This is surprising considering that as early as 1983, Ehntholt *et al.* [1] reported the potential of SFE for the isolation of organic compounds in water at parts-per-billion level using supercritical fluid carbon dioxide.

Perhaps the principal reasons for the relatively slow development of liquid SFE procedures are that:

- since current commercial SFE instruments have been developed primarily for the analysis of solid samples, then studies involving direct analysis of liquids require the fabrication of safe SFE vessels suitable for the containment of liquid samples;
- there is a need for the manufacture of reliable restrictors, resilient to blockages caused by ice formation resulting from extracted water freezing as a result of Joule–Thomson cooling;
- the solubilities of many organic compounds in aqueous solution are sufficiently high to preclude their useful recovery using SFE for subsequent analytical procedures.

The difficulties which co-extracted water can impose on SFE combined on-line with gas chromatography (GC) and capillary supercritical fluid chromatography (cSFC) are described in Chapters 5 and 6, respectively.

Insofar as all current sample preparation procedures have limitations which prevent their universal application, the objective of this chapter is to describe how liquid SFE techniques can be successfully applied to a range of analytical problems. In doing so it is hoped that some of the features and benefits that can make liquid SFE an attractive option, in appropriate circumstances, will become apparent.

## 4.2 General considerations

An attempt to develop a successful liquid SFE procedure will largely depend on four basic considerations. First, the solubility of the analytes to be extracted within the supercritical fluid, usually $CO_2$, need to be established. Bartle *et al.* [2] have published detailed tables listing the solubilities of compounds in SF-$CO_2$ that had been reported in the literature to the end of 1989. The solubility data of 261 compounds in near-critical liquid $CO_2$ (900 psi, 25°C) have been reported by Francis [3] on a functional group basis. Taylor [4] has reviewed supercritical fluid (SF) solubility data for various classes of naturally occurring compounds. Luque de Castro *et al.* [5] have provided a comprehensive review concerning the theoretical aspects of the solvating characteristics of supercritical fluids; factors considered include solubility and chemical structure, temperature, pressure and the influence and selection of modifiers. The solvating potential of various supercritical fluids compared with conventional organic solvents is shown in Figure 4.1. Although this serves as a useful guide, it should be remembered that particularly in the case of direct aqueous SFE a knowledge of analyte SF distribution coefficients would be ideally required to design an extraction protocol. The distribution coefficients of a complex mixture of phenols and chlorinated hydrocarbons between water and SF-$CO_2$ have been determined by Roop and Akgerman [6] and Sengupta *et al.* [7], respectively. In our experience at the University of Glamorgan [8] compounds which are sparingly soluble in water, and which are also directly amenable to GC, can be successfully extracted from aqueous samples. This view is also held by Hedrick *et al.* [9] who have conducted a detailed study of the optimisation of extraction parameters for the analysis of phenols using direct aqueous SFE.

Second, one must select either a direct or an indirect liquid SFE procedure. Direct liquid SFE involves the extraction of the liquid in the absence of a support medium, whereas indirect liquid SFE involves loading the liquid onto a support such as solid phase extraction (SPE) cartridges or discs. Direct liquid SFE has the benefit of simplicity, but custom-built extraction vessels of fixed volume are required. Indirect liquid SFE involves more stages but can offer several advantages; for example selection of appropriate sorbent can provide improved analyte recovery. Additionally, indirect liquid SFE can be performed using commercial SFE systems since it is possible to place support media such as SPE discs and cartridges into extraction vessels which have been designed for solid samples.

Third, the volume of liquid sample that needs to be extracted should be considered if a choice of direct or indirect liquid SFE is available. Koester

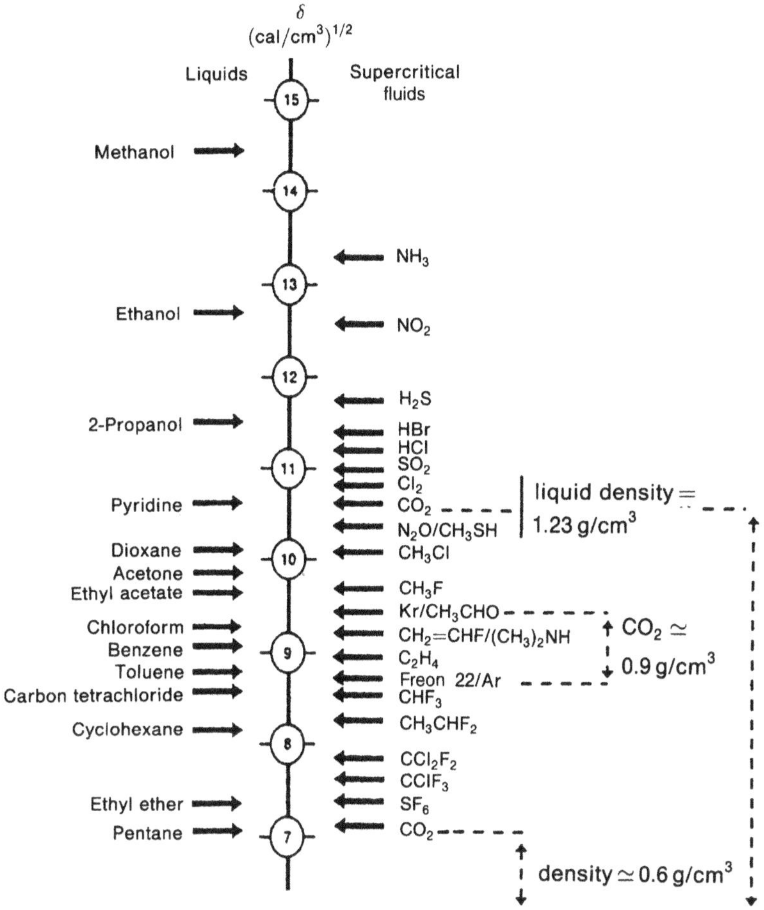


**Figure 4.1** Solubility parameters, δ, of various liquids (1 atm) and supercritical fluids. Reproduced with permission from ref. 5.

and Clement [10] have published a critical review concerning the analysis of organic pollutants in drinking water. Often large-volume water samples (typically 100 ml to 1 l) need to be extracted to recover sufficient amounts of compounds to enable their detection by established analytical procedures. Typical SPE cartridge extraction times, assuming that no particulates are present, are quoted by Koester and Clement as being 1–2 h/l using a 1 g cartridge. Faster filtration times are possible using SPE discs but problems still arise when high levels of particulates are present within the sample. Direct liquid SFE involving the extraction of large-volume samples offers the potential for rapid extraction even in the presence of particulates.

Fourth, selection of dynamic, static or a combination of both modes of SFE must be made. Dynamic SFE allows the continuous flow of fresh SF

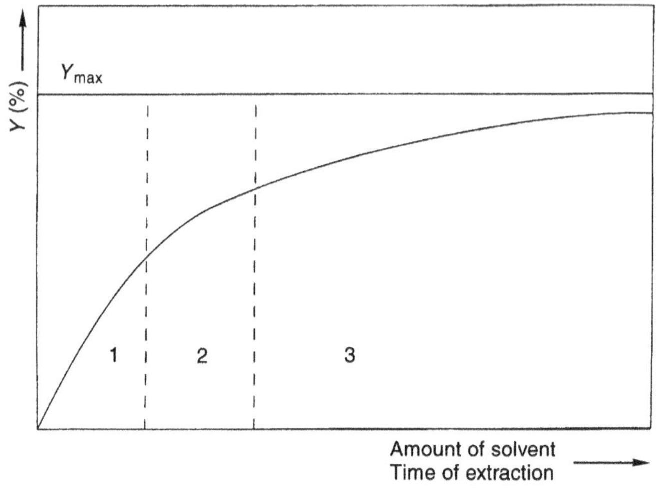


**Figure 4.2** Typical supercritical fluid extraction profile: $Y$ = percentage yield; $Y_{max}$ = maximum possible percentage yield; region 1 = rapid extraction of solvated analyte; region 2 = disruption of established analyte–matrix interactions; region 3 = diffusion-controlled extraction. Reproduced with permission from ref. 11.

through the sample, whereas a fixed volume of SF is equilibrated with the sample in the static mode. Most liquid SFE methods involve the use of dynamic extraction at some stage to improve extraction efficiency. Even analytes which possess poor SF–matrix distribution coefficients have the potential to be quantitatively extracted given sufficient time and flow of SF with appropriate extraction conditions. The theoretical extraction profile of an analyte using dynamic SFE is shown in Figure 4.2 [11], a detailed account concerning the theory of extraction processes can be found in Chapter 1. In region 1, the analyte is solvated by the SF and is rapidly extracted. The rate of extraction at this point is related to the SF solubility of the analyte. In region 2, disruption of analyte–matrix association starts to control the extraction rate. Finally, in region 3, the extraction process becomes diffusion controlled. An initial static SFE stage is often used to allow sufficient time for methods involving *in situ* derivatisation procedures which promote analyte SF solubility and for the use of an organically modified SF, which requires the estabishment of analyte–modifier interactions prior to extraction.

### *4.2.1 Vessels for direct liquid supercritical fluid extraction*

Hedrick and Taylor [12] have described a direct aqueous SFE vessel whose dimensions were length 10 cm, inner diameter (i.d.) 1 cm, volume 8 ml supplied by the Suprex Corp (Pittsburgh, PA, USA). Figure 4.3 shows details

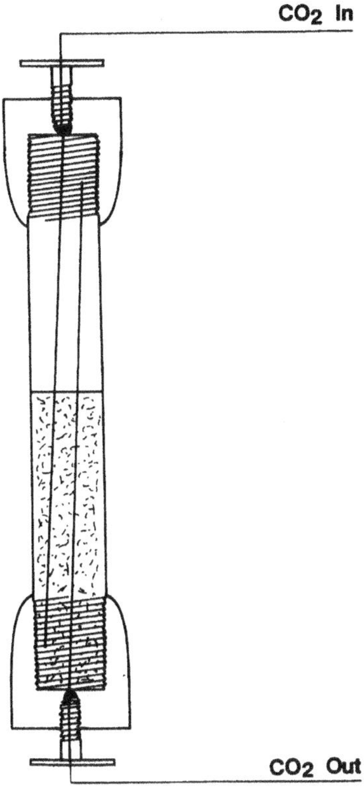


**Figure 4.3** Small-volume, direct aqueous supercritical fluid extraction vessel. Reproduced with permission from ref. 12.

of the vessel that was originally used for the analysis of diisopropyl methylphosphonate and triprolidine. The same vessel was also used by Hedrick and Taylor [13] for the extraction of phosphonates, drugs and phenols from aqueous samples. A 6.94 ml capacity SFE vessel, of similar design and dimensions to that shown in Figure 4.3, supplied by Keystone Scientific (Bellefonte, PA, USA), was modified by Hedrick *et al.* for the analysis of phenols [9] and nitrogenous bases [14] from aqueous solution. This second direct aqueous SFE vessel was equipped with finger-tight fittings and Kel-F ferrules. These type of fittings were reported to provide a more reliable seal which could be broken and resealed many times compared with standard stainless-steel compression fittings. The inclusion of screens at either end of the vessel prevents the transport of particulates through the 0.01 inch i.d. inlet and $\frac{1}{16}$ inch outer diameter (o.d.) outlet stainless-steel supply lines. This type of SFE vessel has also been used by Combs *et al.* [15] for the analysis of phenols in water and by Mulcahey and Taylor [16] for the extraction of sulphamethoxazole and trimethoprim from a liquid drug formulation. The extraction of metals species from aqueous media using this versatile small-volume direct SFE vessel have also been reported. Examples include the isolation of Th and $UO_2^{2+}$ ions by Lin *et al.* [17, 18] and the recovery of Zn(II), Cd(II), Pb(II) [19] and Cu(II) ions [20] in a series of studies performed by Wang *et al.*

Ong *et al.* [21] have described the construction of a direct liquid SFE vessel, using commercially available compression fittings, the design of which resulted in a large number of high-pressure stainless-steel seals. The dimensions of the extraction vessel were 60 mm by 4 mm i.d. With this system Ong *et al.* [21, 22] have demonstrated the feasibility of directly extracting cholesterol from horse blood serum. A 0.85 ml Keystone Scientific liquid extraction vessel was used by Liu and Wehmeyer [23] as a sample preparation technique for the direct isolation of the drugs flavone and ketorolac from dog plasma.

Brewer and Kruus [24] have constructed direct aqueous SFE vessels equipped with a removable lid which can be sealed using a Teflon® gasket. These vessels could accommodate maximum sample sizes of 25 ml and 75 ml and were used for the analysis of pentachlorophenol. Barnabas *et al.* [25] have used direct aqueous SFE for the analysis of several organochlorine pesticides (OCPs) using an extraction vessel which accommodated a 45 ml aqueous sample. Solid phase extraction of the OCPs loaded onto Celite® was used to help optimise the direct aqueous SFE procedure. Kane *et al.* [26] used the same vessel in a later study for the extraction of surfactants from water. The high-pressure seal of the screwtop vessel is made by Teflon®-coated threads and a poly(ether-ether-ketone) (PEEK) O-ring. A 10 μm porous stainless-steel frit attached to the SF-$CO_2$ head-space draw-off tube helped prevent water carryover to the analyte collection vial.

A custom-built 14.9 ml vessel was used by Laintz *et al.* [27] to study the extraction of Cu(II) ions from aqueous solution. The vessel was equipped with quartz windows and mixing was performed with a magnetically driven Teflon®-coated stir bar. The section of the vessel containing the quartz windows was housed in an ultraviolet/visible (UV/VIS) spectrometer in order that the extraction process could be dynamically monitored. A 30 ml extraction vessel was used by Sengupta *et al.* [7] in their studies to determine the distribution coefficients of chlorinated hydrocarbons between SF-$CO_2$ and water. The direct aqueous SFE vessel was equipped with a window to allow the visual observation of the phase separation process. Using a custom-built 150 ml extraction vessel combined on-line with Fourier transform infra-red spectroscopy (FTIR), Minty *et al.* [28] demonstrated the potential of the technique for the determination of oil in industrial process waters prior to their discharge. Roop and Akgerman [6] used a 300 ml high-pressure autoclave fitted with a stirrer to determine the distribution coefficient of phenols between SF-$CO_2$ and water. The autoclave was also fitted with a valve system which permitted the sampling of the aqueous phase during the extraction.

Ehntholt *et al.* [1] used a 600 ml vessel which permitted the extraction of 400 ml water samples that had been spiked with 23 organic compounds. This procedure was subsequently scaled up to 10 l [29] although no design details of either vessel were reported. Shimoda *et al.* [30] used a direct aqueous SFE vessel of 120 ml capacity for the extraction of volatile organic compounds from water and the technique was successfully applied for the deodorising of a Japanese soy sauce. A custom-built direct aqueous SFE vessel of 300 ml capacity connected to a variety of high-pressure valve configurations has been used for the analysis of testosterone [31], phenols [32] and drugs in water [33]. Details of the last two applications can be found in sections 9.6 and 9.7, respectively. An unusual direct aqueous SFE approach for the analysis of phenols in water and urine samples has been described by Thiebaut *et al.* [34]. A segmented flow system involving separate introduction of aqueous sample and SF-$CO_2$ to an extraction coil (dimensions 1.3 m by 0.5 mm i.d.) was used. After extraction, the phases were separated by means of a device which used a hydrophobic membrane (see section 7.4.2).

In order to promote the extraction of compounds from liquids using direct SFE it is extremely important to ensure that the SF makes good interfacial contact and is efficiently dispersed with the liquid matrix. The actual kinetics of the extraction process itself are largely determined by the diffusivity of the analyte within the liquid and SF phases and the phase exchange efficiency.

In their early studies, Ehntholt *et al.* [1] placed a plug of silanised glass wool within the extraction vessel to promote the dispersion of SF-$CO_2$ as fine bubbles in the aqueous sample. Brewer and Kruus [24] have used a disc with small holes (termed the 'drubbler') to break SF-$CO_2$ into small 'drubbles' (droplets/bubbles) to increase the SF–water interface. With the 'drubbler' the recovery of pentachlorophenol from water was 38 $\pm$ 2% compared with 23 $\pm$ 1% without this SF-$CO_2$ inlet tube attachment. The effect of SF-$CO_2$ flow rate on pentachlorophenol extraction efficiences was also studied. As flow rate increases, contact time decreases, lowering extraction efficiency. However, increased turbulence associated with higher SF flow will tend to promote mixing, thereby increasing extraction efficiency. Brewer and Kruus [24] reported a twofold increase in pentachlorophenol extraction efficiency when the flow of liquid $CO_2$ supplied to the extraction vessel was increased from 0.3 ml/min to 0.75 ml/min (20 ml of liquid $CO_2$ delivered in each case with other extraction parameters held constant). In order to promote mixing, Barnabas *et al.* [25] attached a conventional high-pressure liquid chromatography (HPLC) solvent filter to the terminus of the SF-$CO_2$ inlet. In spite of this, it was suggested that poor diffusion of SF-$CO_2$ through the aqueous sample may have been responsible for the *c.* 20% lower recoveries of the OCPs (lindane, aldrin and dieldrin) compared with an indirect SPE–SFE procedure used in the same study. This view was substantiated by the fact that varying the SF-$CO_2$ flow rate had no apparent effect on direct aqueous OCP extraction efficiencies, indicating that extraction was diffusion controlled. Using the same vessel, Kane *et al.* [26] demonstrated improved surfactant extraction efficiencies when the $CO_2$ inlet tube was fitted with an HPLC solvent filter (19% increase compared with the open $CO_2$ delivery tube). Shimoda *et al.* [30] have also conducted a comparative study involving the extraction of volatile organics from water with different SF-$CO_2$ delivery arrangements. It was demonstrated that when the SF-$CO_2$ inlet tube was fitted with a 10 µm porous stainless-steel nozzle, extraction efficiencies significantly increased (resulting in more than 95% recovery of volatiles consisting of 6–12 carbon atoms). This is attributed to the formation of $CO_2$ microbubbles.

Janda *et al.* [35] have used an aluminium coil placed within the extraction vessel to increase the retention of SF-$CO_2$. This is an important consideration in the event that slow diffusion and/or slow phase exchange result in analyte extraction efficiencies being independent of SF flow rate.

In their studies concerned with the optimisation of various extraction parameters including temperature and pressure, Hedrick *et al.* [9] reported reduced SFE efficiencies for phenols from aqueous solution at 50°C, as the pressure of SF-$CO_2$ was increased from 250 atm to 300 atm. This trend was continued as the pressure was further increased from 300 atm to 350 atm. Several factors were suggested to account for this surprising result, including: that an increase in pressure may result in lower $CO_2$ diffusion leading to increased $CO_2$ bubble size and a reduction in numbers that percolate through the sample; that the SF–water phase boundary becomes more defined at increased $CO_2$ density, leading to reduced analyte phase exchange efficiency; and that phenols may react with SF-$CO_2$ at high pressure to form a more water soluble, less extractable product. This clearly serves as a warning not to make the assumption that increasing SF density will necessarily afford a means of improving analyte extraction efficiencies when liquids are involved.

### *4.2.2 Vessels for indirect liquid supercritical fluid extraction*

Since indirect liquid SFE involves initial extraction or adsorption of the liquid matrix onto support media prior to SFE, then custom-built SFE vessels suitable for liquid containment (see above) are generally not required. A further advantage of this approach stems from the use of SPE cartridges and discs, that is, the volume of liquid extracted can be varied to achieve the necessary level of analyte enrichment to enable the use of specific analytical procedures.

Modifications to commercial extraction vessels have been made for indirect SFE studies. For example the incorporation of a stainless-steel cartridge for housing SPE cartridges (cut to length after SFE) or discs within a standard SFE vessel has been described by Tang *et al.* [36] for the extraction of polycyclic aromatic hydrocarbons (PAHs) from water, and an SPE method involving the insertion of a cylindrical aluminium foil derivatisation reactor within an SFE vessel has been developed by Alzaga and Bayona [37] for the recovery of organotins from aqueous samples. HPLC guard columns equipped with 2 μm stainless-steel frits and finger-tight compression fittings have been used by Furton and Rein [38] for studies involving the quantitative effect of SFE vessel geometry on the SFE efficiences of PAHs loaded onto octadecylsilyl-bonded silica ($C_{18}$) sorbent. The SFE efficiences were increased by more than a factor of two by decreasing the SFE vessel diameter:length ratio from 1:20 to 1:1 for the largest PAHs studied. In a subsequent study, Furton and Lin [39] demonstrated that the recoveries of polychlorinated biphenyls (PCBs) loaded onto $C_{18}$ and phenyl sorbents were significantly influenced by the dimensions of HPLC guard columns, used as SFE vessels. Furton *et al.* [40] have also reported that vessel orientation and void volume can affect recoveries of analytes loaded onto $C_{18}$ sorbent. The internal diameter of a 7 ml Hewlett Packard SFE vessel ('thimble') permits the tight fitting of an SPE cartridge (flange removed from standard 3 ml, 500 mg sorbent size) within the extraction chamber, as shown in Figure 4.4. This arrangement was exploited by Ahmad *et al.* [41] to develop indirect liquid SFE procedures for the determination of pesticide residues in vegetable oils (see section 8.4).

Many of the small-volume direct liquid SFE vessels are also suitable for indirect SFE studies. For example, Ong *et al.* [42] have used $C_{18}$ sorbent placed within a vessel originally designed for direct liquid SFE [21, 22] for indirect SFE combined with capillary supercritical fluid chromatography (cSFC) for the determination of PAHs in water. With this technique aqueous samples, which had been previously concentrated by rotary evaporation, were introduced into the vessel by an HPLC injection valve.

### 4.2.3 *Liquid supercritical fluid extraction vessel safety considerations*

Throughout the preceding sections concerned with the description of vessels used for direct and indirect liquid SFE, several references have been made to the use of custom-built extraction vessels, especially for direct liquid SFE studies.

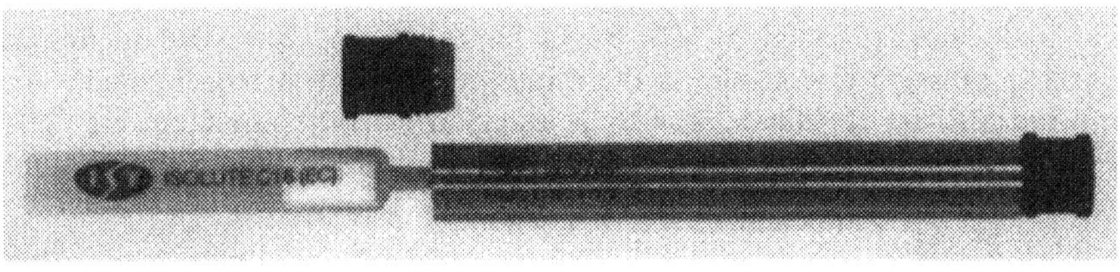

**Figure 4.4** A solid phase extraction cartridge ready for insertion into a 7 ml Hewlett Packard supercritical fluid extraction 'thimble'.

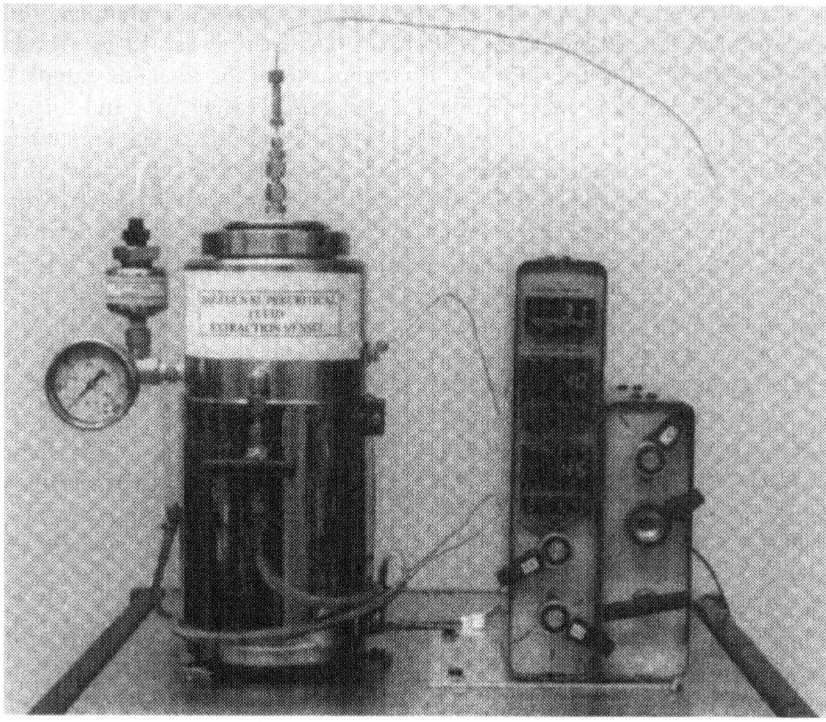

**Figure 4.5** Custom-built one-litre capacity supercritical fluid extraction vessel (4000 psi upper pressure rating) with high-pressure multivalve control system.

Figure 4.5 shows a custom-built direct aqueous SFE vessel of 1 l capacity whose inclusion is used to describe several important safety features to any readers who may be considering fabricating their own system. The vessel has been manufactured to rigorous safety standards which require the vessel chamber be milled from a solid, seamless stainless-steel bar. The vessel is equipped with a removable lid assembly whose housing contains an atmospheric exhaust port which is engaged before the lid can be completely removed. This arrangement prevents accidental explosive release of the lid should the operator attempt to loosen it whilst the vessel is still under pressure. Since the upper working pressure limit for the vessel is 4000 psi it was hydraulically tested to 6000 psi (a factor of 1.5 above the maximum working pressure is considered the minimum vessel test requirement). An appropriate test certificate was then issued from the manufacturers. The serial number and upper pressure limit are stamped on the main body of the vessel. The vessel is also equipped with a 4100 psi rupture disc assembly to prevent vessel overpressure should, for example, the liquid $CO_2$ pump develop a fault preventing its pressure control. The minimum wall thickness of the SFE vessel (circular cross-section) can be calculated [43] from the

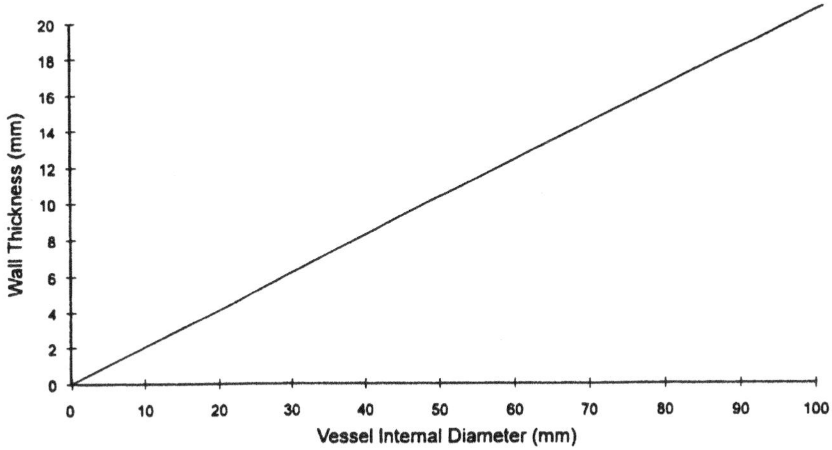


**Figure 4.6** Relationship between the internal diameter and minimum wall thickness of a stainless-steel cylindrical supercritical fluid extraction vessel for 350 kgf/cm$^2$ (343 bar, 4980 psi) pressure operation. Reproduced with permission from ref. 43.

following equation:

$$t(\mathrm{mm}) = \frac{PD_\mathrm{o}}{200\sigma_\mathrm{a}\eta + 0.8P} \tag{4.1}$$

where $P$ is the maximum operating pressure in kgf/cm$^2$, $D_\mathrm{o}$ is the outer diameter of the tubing, $\sigma_\mathrm{a}$ is the allowable tensile stress of the material in kgf/mm$^2$ and $\eta$ is the welding efficiency ($\eta = 1$ for seamless construction – a mandatory requirement for liquids).

Figure 4.6 shows the relationship between the vessel internal diameter and minimum wall thickness (stainless-steel construction). In the event of the vessel walls becoming pitted, the vessel would be milled to remove the defect(s). The vessel would then be brought back into service, only after successfully completing a new pressure test with the test certificate issued.

### *4.2.4 Selection of support media for indirect liquid supercritical fluid extraction*

The majority of reported indirect SFE procedures involve the use of either SPE cartridges or discs for the preconcentration of analytes. The manufacturers of these products can supply comprehensive application details [44, 45] designed to aid the selection of appropriate SPE sorbents for specific studies. SPE discs which are usually composed of silica-based sorbents enmeshed in a network of poly(tetrafluoroethene) (PTFE) fibrils, producing a strong porous membrane, offer several advantages over SPE cartridges.

- They offer shorter sample processing times because of the higher flow rates permitted by the large cross-sectional area of the discs. This is an important consideration where large-volume liquid samples have to be analysed, for example trace-level analysis of organic pollutants in water may involve the analysis of 500–1000 ml aqueous samples [10]. Typical SPE disc flow rates range from 20 to 50 ml/min compared with generally less than 10 ml/min for SPE cartridges.

- They have improved analyte mass transfer characteristics. The porosity of the membranes is 0.5–1.5 μm, which provides multiple flow-paths, resulting in close contact between the liquid and sorbent phases.
- The large cross-sectional area of the discs and multiple flow-paths help reduce plugging problems caused by the presence of particulates within the sample.
- There is a reduction in the number of problems associated with channelling of liquid samples through the sorbent phase. This is because of the improved mechanical stability of the PTFE–sorbent matrix.
- They prevent the extraction of interfering compounds from polypropene housing, typically used for SPE cartridge manufacture.

Most $C_8$ and $C_{18}$ SPE disc procedures involve an 'activation' stage where the disc is first subjected to a solvent rinse procedure (often involving methanol) prior to application of the aqueous sample. The purpose of this is to 'wet' the disc surface, opening up the alkyl chain ligands and increasing their availability to extract analytes. Tang *et al.* [36, 46] also recommend the addition of 5 ml methanol per litre of aqueous sample prior to SPE disc application. This prevents the SPE alkyl chains from 'matting', which can result in low analyte recoveries and sample filtration rates. Ho and Tang [47] have applied experimental design techniques (see section 3.4.2) for the optimisation of SFE of an SPE cartridge, used for the isolation of 29 environmental pollutants (including PAHs and OCPs) from water. A statistical $2^3$ factorial design experiment was initially performed to study three SFE variables: pressure, temperature and extraction time. Following the factorial design, a variable size simplex optimisation was used to demonstrate that the SFE conditions had been optimised to obtain the best overall recoveries of the compounds studied.

Ezzel and Ritcher [48] have conducted a comparative study involving the use of various SPE cartridges (loaded with: $C_{18}$, Florisil®, Tenax-GC and a polystyrene divinyl benzene resin, XAD-4 sorbents) and a $C_8$ SPE disc for the SFE of five OCPs and phthalate esters from water. Of the SPE cartridges the $C_{18}$ provided the highest mean pesticide SPE recovery [90%, with relative standard deviations (RSDs) 1%–4%, $n = 3$] and SFE efficiency (84%, with RSDs 1%–3%, $n = 3$). However, the $C_8$ SPE disc was found to provide the most efficient mean SFE pesticide recoveries (88%, with RSDs 3%–6%, $n = 3$) using pure SF-$CO_2$. The use of methanol-modified SF-$CO_2$ did not result in a high GC/electron capture detection (ECD) background due to the extraction of impurities from the SPE disc. Since the SPE cartridges could not be directly loaded into the extraction vessel used in these studies, transfer of the solid phase material was reported as being tedious and a potential source of sample loss.

Environmental water samples are often acidified (pH adjusted to 2) prior to SPE application. This procedure has been used by Davi *et al.* [49] to prevent plugging of the SPE disc pores. Acidification promotes the dissolution of particles of inorganic salts which can persist following sample prefiltration. Since some compounds are acid labile, including several triazine and organophosphate pesticides (OPPs), then due consideration should be given prior to sample acidification. Kane *et al.* [26] have conducted experiments to determine whether $C_{18}$ SPE discs could be reused for the isolation of alcohol phenol ethoxylate (APE) surfactants from water following SFE. The discs were cleaned using methanol. The recovery of APE ($n = 3$) decreased from 87%±7.65% to 60%±13.2% when the discs were reused.

Sandra *et al.* [50, 51] have described the use of an automated SPE–SFC system equipped with a photodiode-array detector (DAD). The system was used to evaluate the recoveries of 30 pesticides in the on-line mode, using $C_{18}$ SPE cartridges. The most important stage in the SPE–SFC procedure was drying the SPE cartridge (15 min) prior to SFE. Residual water resulted in a two-phase desorption process, leading to peak broadening and splitting. Detection limits set by the European Community were achieved (further details of these studies can be found in section 7.5).

Supports other than SPE cartridges and discs have been successfully used for indirect liquid SPE procedures. In their early studies Pawliszyn and Alexandrou [52] used an HPLC guard column loaded with Tenax-GC sorbent for the analysis of polychlorinated benzenes in water at the 1 ppb level. Zegers *et al.* [53] have used a $C_{18}$ pre-column to isolate three OPPs from spiked river water, analyses being performed on-line with packed capillary SFC and thermionic detection (TID). As reported by Sandra *et al.* [50, 51], it was necessary to dry the sample loaded pre-column carefully prior to the onset of SFE (further details of this study can be found in section 7.5).

Bengtsson *et al.* [54] have described a procedure involving the use of loose $C_{18}$ sorbent held on a glass-fibre filter disc placed within the filtration apparatus for the rapid extraction of pesticides from river water prior to SFE. The method does not require sample prefiltration or acidification to prevent plugging. The sample was filtered at a rate of 200 ml/min and losses due to analyte breakthrough could be minimised by increasing the amount of sorbent used. Despite these advantages this study demonstrated that SFE efficiencies for the pesticides were higher using a $C_{18}$ SPE disc procedure prior to extraction.

Edder *et al.* [55] have evaluated eight different silica-based HPLC stationary phases for the SFE of morphinic alkaloids from urine and other aqueous solutions. Each phase was slurry packed into a 75 mm by 3.5 mm i.d. vessel. The breakthrough volumes of individual sorbents were determined by passing an aqueous solution of morphine through the packed vessel, whilst a diode-array UV/VIS detector monitored the level of morphine eluted. The $C_{18}$, phenyl and Bond Elut Certify (BEC) silicas and PRP1 resin were found to possess significantly higher morphine breakthrough volumes than did the cyanopropyl, diol, aminopropyl silicas and pure silica. It was demonstrated that good SFE recoveries of the morphinic alkaloids could be obtained from urine loaded onto either $C_{18}$ or BEC silicas. However, the SFE selectivity of the BEC silica resulted in a cleaner GC/FID chromatogram since lower levels of endogenous compounds were co-extracted.

Murugaverl and Voorhees [56] have developed an SFE–cSFC procedure for the detection of pesticides in soybean oil. The vegetable oil was first mixed with $C_{18}$ sorbent and then loaded onto the top of an extraction column which had been dry packed (tap and fill method) with various sorbents, including silica, $C_{18}$, $C_2$, cyanopropyl and aminopropyl. A photodiode-array detector, equipped with a high-pressure cell, was situated between the extraction column and the restrictor. Investigations were conducted to determine which sorbents retained the lipids, allowing the pesticides to be trapped onto the cSFC column. For each sorbent, using SF-$CO_2$, the pesticides eluted before the lipids. The lipids were strongly retained on the polar media, aminopropyl and silica sorbents. The retention time differences between the pesticides and lipids fell within reasonable practical limits using the $C_{18}$ and cyanopropyl sorbents. The $C_{18}$ sorbent was selected for final SFE–cSFC method development. In order to prevent plugging of $C_{18}$ sorbent pores, Johansen *et al.* [57] treated milk samples with aqueous acetonitrile (1 : 1) to precipitate the proteins prior to loading (further details of this indirect liquid SFE study, involving the analysis of PCBs in human milk, blood serum and crab hepatopancreas can be found in section 8.4). Ahmad *et al.* [41] have also published details of indirect SFE procedures for the analysis of pesticide residues in vegetable oils.

Slack *et al.* [58] have used a phenyl sorbent contained within an SFE vessel for the analysis of explosives in water. With this arrangement water samples were aspirated through the support media prior to SFE. The SFE recoveries of explosives from spiked river water were found to be lower than from reagent water. By linking two extraction vessels (each loaded with phenyl sorbent) in series it was demonstrated that the presence of additional organic compounds in river water reduced the number of active sites available for analyte adsorption rather than giving losses due to analyte–organic compound association. Johnson *et al.* [59] have also reported that the presence of dissolved organic compounds interfering with solid phase sorbents can lead to low analyte recoveries from aqueous samples.

Koski *et al.* [60] have used XAD-2 for the analysis of prostaglandins by SFE–cSFC. Purification of XAD-2 by repeated SFE prior to sample application is recommended by Yunker *et al.* [61], since this sorbent can release many aromatic compounds after coming into contact with water.

Croft *et al.* [62] have developed a simultaneous indirect SFE and methylation procedure (section 4.4) for the determination of chlorophenoxyacetic acids in water and urine samples. Hydromatrix, Celite® and Extrelut (dry packed into an SFE vessel) were evaluated as support media. Extrelut was selected since it provided the cleanest GC/ECD chromatogram for quantitation studies.

Mulcahey and Taylor [16] have loaded an aqueous-based sample of Septra Infusion drug formulation onto filter paper for the recovery of sulphamethoxazole and trimethoprim by SFE. However, Celite® was reported to be a more convenient support for this matrix in terms of sample handling. Scalia *et al.* [63] have also used filter paper as a means of supporting various cosmetric creams and lotions for the SFE of vitamin A palmitate and vitamin E acetate. The extraction vessel void volume was filled with chromatographic-grade sand prior to extraction. Improved recoveries of the vitamins were achieved when the cosmetic products were dispersed within sand or Celite® supports.

### *4.2.5 Restrictors and analyte traps for direct and indirect liquid supercritical fluid extraction*

The solubility of water in SF-$CO_2$ (normally ranging from 0.3% to 3% [64], depending on SF density) can lead to several problems: the Joule–Thomson cooling effect at the point at which the SF expands may result in restrictor plugging due to ice formation; and water carryover into collection traps containing immiscible organic solvents can result in poor recoveries as a result of the analyte redissolving back into the aqueous phase. These problems are particularly relevant to direct aqueous SFE procedures. Also, indirect aqueous SFE techniques are prone to such problems in situations where thorough drying of SPE cartridges, discs and sorbents is prevented because of the potential loss of volatile analytes.

In their original direct aqueous SFE studies, Ehntholt *et al.* [1] selected 23 organic compounds as representatives of contaminants usually encountered in aqueous systems. These compounds were divided into 10 classes based upon final analytical methods for detection following extraction. The trapping system consisted of three sequential U-tubes maintained at −76°C (note that $CO_2$ freezes at −78°C). With this 'open trap' system, losses ranging from 69% to 80% of volatile compounds (including 1-chlorododecane, 2,2′,5,5′-tetrachlorobiphenyl, biphenyl, bis(2-ethylhexyl)phthalate and 2,4-dichlorobiphenyl) were reported, following mass balance calculations based upon levels of analytes trapped and remaining in the extracted aqueous sample. Trap breakthrough was also reported for three semivolatile phenols, since carryover from the first U-tube was detected.

Inspection of Table 4.1 reveals that the majority of reported liquid SFE procedures involve the use of fixed i.d. fused-silica capillaries. Although

**Table 4.1** Applications of analytical SFE for liquid samples

| Analyte | Sample extraction mode | Volume of sample | Restrictor | Ref. |
|---|---|---|---|---|
| 23 organic compounds: acids, bases and neutrals | Direct | 400 ml | Pressure reduction valve | 1 |
| Phenols | Direct | 3 ml | Fused-silica capillary used for SFC interfacing or SPE collection | 9 |
| DIMP, triprolidine | Direct | 3 ml | Fused-silica capillary used for SFC interfacing | 12 |
| Triprolidine, DIMP, phenol, pseudoephedrine | Direct | 3 ml | Fused-silica capillary used for SFC interfacing or SPE collection | 13 |
| Succiononitrile, 2,5-lutidine, picolinic acid, caffeine, nicotine, pseudoephedrine, triprolidine, sulphamethazine | Direct | 3 ml | Fused-silica capillary used for SFC interfacing or SPE collection | 14 |
| Phenols | Direct | 5 ml | Suprex variable restrictor | 15 |
| Sulphamethoxazole, trimethoprim | Direct and indirect | Direct: 5 ml Indirect: 100 μl | Tapered fused-silica capillary restrictor used for direct SFE. Indirect SFE performed using an HP 7680 A instrument with variable restrictor | 16 |
| Cholesterol | Direct | Not quoted. Vessel volume *c.* 0.75 ml | Tapered fused-silica capillary | 21 |
| Cholesterol | Direct | *c.* 0.38 ml | No details | 22 |
| Neutral (flavone) and acidic (ketorolac) drugs | Direct | 0.4 ml | Linear fused-silica capillary | 23 |
| Pentachlorophenol | Direct | 25 ml or 75 ml | Pressure reduction valve | 24 |
| OCPs | Direct and indirect | Direct: 45 ml Indirect: 200 ml | Jasco variable restrictor | 25 |
| Non-ionic surfactants (alcohol phenyl ethoxylates) | Direct and indirect | Direct: 45 ml Indirect: 45 ml | Jasco variable restrictor | 26 |
| Hydrocarbons | Direct | 50 ml | Gilson variable restrictor | 28 |
| Volatile alcohols, aldehydes and esters | Direct | Not quoted. Vessel volume 120 ml | Pressure reduction valve | 30 |
| Testosterone | Direct | 230 ml | Gilson variable restrictor | 31 |

**Table 4.1** (*cont.*)

| Analyte | Sample extraction mode | Volume of sample | Restrictor | Ref. |
|---|---|---|---|---|
| Phenols | Direct | 250 ml | On-line SFE–SFC: HP SFC G1205 A variable restrictor | 32 |
| Estrone, hexestrol, zeranol | Direct | 250 ml | On-line SFE–LC: Tescom variable restrictor | 33 |
| Phenols | Direct | Variable, segmented flow extractor | Packed fused-silica capillary | 34 |
| PAHs, PCBs, OCPs and phthalate esters | Indirect | 500–1000 ml | Linear fused-silica capillary | 36 |
| Dibutyltin and tributyltin compounds | Indirect | 500–1000 ml | Linear fused-silica capillary | 37 |
| OCPs, OPPs | Indirect | 0.5 ml | HP 7680 A instrument with variable restrictor | 8, 41 |
| PAHs | Indirect | 1 μl (sample first concentrated using rotary evaporation) | Tapered fused-silica capillary | 42 |
| Phenols | Indirect | 500–1000 ml | Linear fused-silica capillary | 46 |
| OCPs, OPPs and phthalate esters | Indirect | SPE cartridges 10 ml; SPE discs 500 ml | Off-line studies used a Dionex SFE-703 heated restrictor. On-line cSFC studies performed with a Dionex SFE-1A instrument | 48 |
| Acidic pesticides, phenyl urea pesticides, triazine pesticides, caffeine, theophylline, theobromine | Indirect | 10–1000 ml | Off-line studies: HP 7680 T instrument equipped with a variable restrictor. On-line studies: details of the interface between a Gilson 233 XL and an HP SFC G1205 A system not provided | 50 |
| OCPs, triazine pesticides | Indirect | 10 ml | On-line studies: details of the interface between a Gilson 233 XL and an HP SFC G1205 A system not provided | 51 |
| Polychlorinated benzenes | Indirect | 10 ml | Linear fused-silica capillary | 52 |
| OPPs | Indirect | 47 ml | On-line SFE–cSFC: linear fused-silica capillary | 53 |
| OCPs, OPPs | Indirect | 1000 ml | Isco linear stainless-steel heated restrictor | 54 |

**Table 4.1** (*cont.*)

| Analyte | Sample extraction mode | Volume of sample | Restrictor | Ref. |
|---|---|---|---|---|
| Morphine, codeine, cocaine, papaverine, ethylmorphine, monoacetyl-morphine | Indirect | 2–20 ml | Variably compressed PEEK tubing | 55 |
| OCPs, carbamate pesticides | Indirect | *c.* 6 μl | On-line SFE–cSFC: linear fused-silica capillary | 56 |
| PCBs | Indirect | 5 ml | On-line SFE–LC: linear fused-silica capillary | 57 |
| Explosives: 2,3-dinitrotoluene, 2,4-dinitrotoluene, trinitrotoluene | Indirect | 500 ml | Linear fused-silica capillary | 58 |
| Prostaglandins | Indirect | 100 μl | On-line SFE–cSFC: linear fused-silica capillary | 60 |
| Chlorophenoxy acids: 2,4-D, 2,4,5-T | Indirect | 3 ml | Linear fused-silica capillary | 62 |
| Vitamin A acetate, vitamin E acetate | Indirect | *c.* 150 μl | Dionex SFE-703 system used | 63 |
| Butyl, phenyl-, and cyclohexyltin compounds | Indirect | 150 ml | Linear fused-silica capillary | 65 |
| β-Blockers: oxprenolol, metoprolol, propranolol | Indirect | 15 ml | Linear fused-silica capillary | 67 |
| Morphine, methadone, cocaine, codeine, papaverine | Indirect | 20 ml | Variably compressed PEEK tubing | 68 |
| OCPs, OPPs | Indirect | 200 ml | Heated capillary stainless-steel tubing | 69 |
| OCPs, herbicides | Indirect | 200 ml | Jasco variable restrictor | 70 |
| Sulphometuron methyl, chlorsulphuron | Indirect | 100–1000 ml | HP 7680 A with variable restrictor | 71 |
| Mebeverine alcohol | Indirect | 1.1 ml | Linear fused-silica capillary | 73 |

Note: CSFC = capillary supercritical fluid chromatography; DIMP = diisopropyl methylphosphonate; HP = Hewlett Packard; LC = liquid chromatography; OCP = organochlorine pesticide; OPP = organophosphate pesticide; PAH = polycyclic aromatic hydrocarbon; PCB = polychlorinated biphenyl; PEEK = poly(ether-ether-ketone); SFC = supercritical fluid chromatography; SFE = supercritical fluid extraction; SPE = solid phase extraction; 2,4-D = 2,4-dichlorophenoxyacetic acid; 2,4,5-T = 2,4,5-trichlorophenoxyacetic acid.

these restrictors are inexpensive they are particularly vulnerable to blockage by ice formation and the deposition of extracted analytes and/or particulate matter. The use of organically modified SFs for indirect aqueous SFE has been reported, in some instances, to prevent restrictor plugging. Tang and Ho [46] obtained increased phenol recoveries from SPE discs using acetone-modified SF-$CO_2$. The presence of the acetone also prevented the fused-silica restrictor from freezing as a result of residues of water extracted from the disc which had been partially dried by vacuum. In a similar study involving the use of methanol-modified SF-$CO_2$ for the extraction of PAHs from an SPE disc, Tang *et al.* [36] arranged an air flow outside the collection vial to prevent the tip of the fused-silica restrictor from freezing. Slack *et al.* [58] used a hot air gun to prevent the fused-silica restrictor from plugging, despite having vacuum dried the phenyl sorbent used for extracting explosives from water for 15 min prior to SFE. The incorporation of a stainless-steel collection chamber (heated to 100°C) between the extraction vessel and fused-silica restrictor eliminated freezing problems, allowing Ong *et al.* [21, 42] to perform a series of direct and indirect aqueous SFE studies. A heated stainless-steel restrictor was used by Bengtsson *et al.* [54] which prevented freezing, although a two-phase system formed in the collection vial containing cyclohexane as pesticide-trapping solvent. To prevent phase separation, a methanol/acetone mixture was used for analyte trapping.

In order to prevent the restrictor freezing as a result of residual water remaining in vacuum dried SPE discs, Cai and Bayona [65] packed anhydrous magnesium sulphate in the outlet section of the SFE vessel. A detailed study has been conducted by Burford *et al.* [66] to evaluate the performance characteristics of 21 common drying agents for SFE applications. Five (anhydrous and monohydrated magnesium sulphate, molecular sieves 3A and 5A and Hydromatrix) prevented restrictor plugging by retaining the majority of the water but not the target analytes. The enthalpy of hydration of some drying agents (e.g. magnesium sulphate) was sufficient to cause the loss of volatile components.

Several reports involving indirect liquid SFE studies have described restrictor problems associated with *in situ* derivatisation procedures. Croft *et al.* [62] have reported blockage of fused-silica restrictors during methylation of chlorophenoxyacetic acids with methyl iodide. This was attributed to the deposition of silica, dissolved from Extrelut support media, when 0.1 M sodium hydroxide was used to provide basic conditions necessary for sample derivatisation. Substitution with 0.2 M sodium hydrogen carbonate buffer (pH 10) eliminated this problem. Efforts to improve the yields of acetylated β-blockers by increasing the quantity of acetic anhydride available during static SFE have been investigated by Hartonen and Riekkola [67]. This resulted in serious blockage of the fused-silica restrictor which also became brittle and prone to breakage during the dynamic SFE stage.

In order to eliminate blockage problems associated with fused-silica or crimped stainless-steel restrictors, Edder *et al.* [55, 68] constructed a restrictor made from $\frac{1}{16}$ inch o.d. PEEK tubing (see Figure 2.12). The back pressure of the PEEK restrictor (stable to 150°C) is regulated by varying its compression between two metal plates by adjustment of a locking screw. Since PEEK will regain its original form, loosening of the screw permits a higher $CO_2$ flow rate, which serves to unplug blockages.

Although higher SF density generally affords a means of improving SFE efficiency for analytes from solid matrices (including dried indirect liquid SFE support media), losses of volatile analytes may occur as a result of increased aerosol formation as the SF expands to a gas which rapidly passes through a collection solvent. In order to prevent such losses, Barnabas *et al.* [25, 69, 70] have linked a dry SPE cartridge to the $CO_2$ exhaust tube leading from the collection vial. Any escaping analytes can then be recovered by back flushing the SPE cartridge with a small volume of collection solvent. Hedrick and Taylor [13] have described the use of an SPE cartridge holder directly connected to a fused-silica restrictor, serving as an analyte trap. A thermocouple was installed in the cartridge void volume. By selecting a fused-silica restrictor of appropriate i.d. it was possible to perform cryogenic trapping (e.g. −30°C to −35°C using a 50 μm i.d. restrictor with a liquid $CO_2$ flow rate of 2 ml/min) to improve analyte recovery.

Production of particulates, formed during direct aqueous SFE, can also result in restrictor blockage. Liu and Wehmeyer [23] suggest that plasma proteins, denatured during SFE, were responsible for restrictor plugging during a drug study. The use of an antifoam agent eliminated this problem. Mulcahey and Taylor [16] prevented restrictor plugging by using an in-line filter to prevent transport of precipitated sulphamethazole, formed during SFE of a Septra infusion.

Many of the above problems can be circumvented by the use of a variable orifice restrictor as shown in Figure 4.7. The diameter of the orifice can be varied by changing the position of a valve needle within its conic seating. A pressure transducer is used to regulate the position of the solenoid-driven needle assembly. If a blockage starts to occur, the orifice is widened to blow the offending matter free. Sensing the sudden pressure drop, the restrictor orifice is again narrowed to maintain the original SFE target pressure. The Hewlet Packard model 7680A SFE system uses a variable orifice restrictor and a sorbent trap, the temperature of which can be lowered to perform low-temperature analyte collection. At the end of the extraction period the trap can be automatically rinsed with the fractions being collected into vials mounted on a carousel. A diagram of the HP SFE system is shown in Figure 4.8. With this system Ahmad *et al.* [41] developed a fully automated procedure for determining pesticide residues in vegetable oils, which were first loaded onto SPE cartridges before being placed in the SFE vessel (see Figure 4.4; further details of this application can be found in section 8.4).

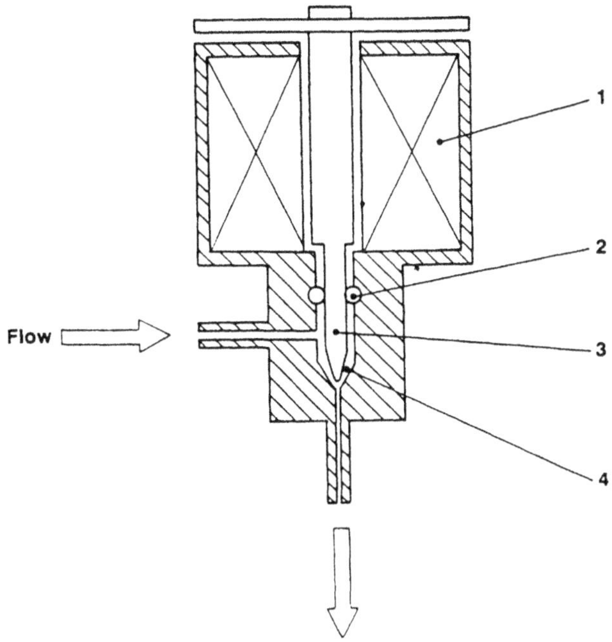


**Figure 4.7** Cross-sectional view of a variable orifice supercritical fluid extraction restrictor: 1 = needle drive solenoid; 2 = needle seal; 3 = valve needle; 4 = valve seal. Redrawn with permission from ref. 88.

Howard and Taylor [71] have conducted a detailed series of investigations, with the HP 7680A SFE system, aimed at optimising solid phase trapping material, trap temperature and rinse solvent. Also, HPLC columns have been connected to several different manufacturers' variable orifice restrictors as a means of trapping analytes following direct aqueous SFE [31–33]. Using liquid $CO_2$ flow rates ranging from 4 to 8 ml/min no restrictor blockage problems arose with these systems. The principal drawback of variable restrictors is cost. Such capital outlay, however, seems a good investment for liquid SFE in order to minimise restrictor blockage problems and improve method reproducibility. Further details of restrictors can be found in section 2.7.

## 4.3 Procedures involving pH control and use of additives to improve supercritical fluid extraction efficiencies of analytes from aqueous samples

The pH of $CO_2$-saturated water at moderate $CO_2$ pressures is approximately 3–3.5 [15]. Ehntholt *et al.* [1] reported poor recoveries of basic compounds using direct aqueous SFE. This was attributed to the production of carbonic acid, resulting in reduced SF-$CO_2$ solubility of the protonated bases. With

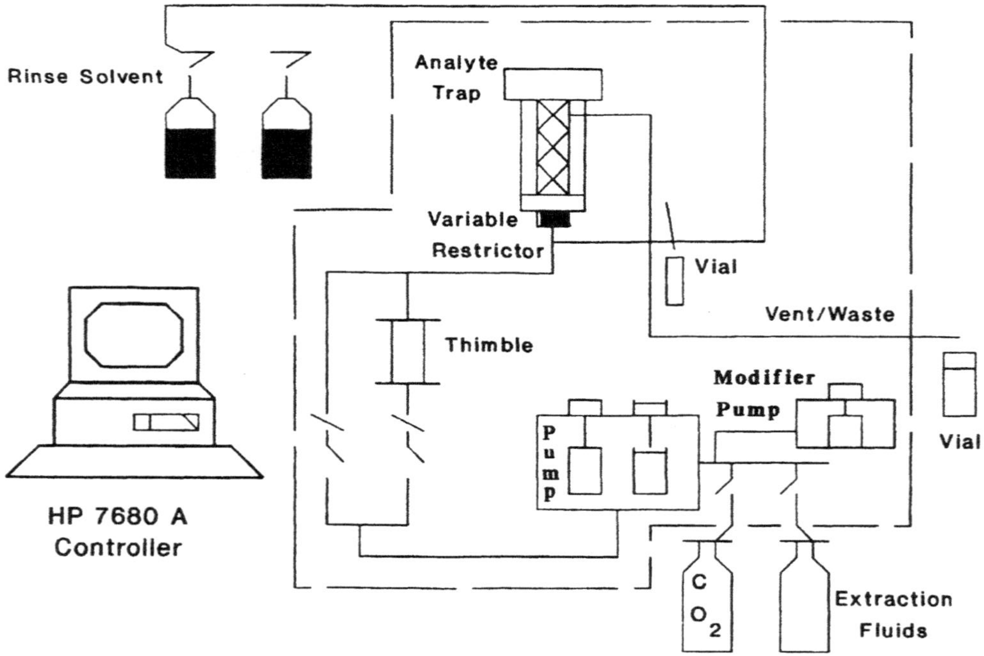


**Figure 4.8** A diagram of the HP 7680 A supercritical fluid extraction system, including modifier pump.

the SFE conditions used ($CO_2$ at 2500 psi, temperature 45°C), recoveries of quinoline and caffeine were 3.7% and 0%, respectively.

Hedrick and Taylor [14] have studied the direct SFE of nitrogenous bases from water, the bases selected having different hydrocarbon moieties. Caffeine was recovered using SF-$CO_2$ (at 250 atm and 50°C). Although pure nicotine is more soluble in SF-$CO_2$ than is pure caffeine, it was not extracted from water using the previous conditions. It was suggested that the greater basicity and water solubility of nicotine may be responsible for this behaviour. None of the small nitrogen-containing compounds involved in this study [i.e. succinonitrile (neutral), 2,5-lutidine (basic) and picolinic acid (amphoteric)] were successfully extracted from aqueous solution. Since the larger organic bases (triproline, pseudoephedrine and sulphamethazine) could be extracted, Hedrick and Taylor concluded that the extractability of an organic base may have more to do with the size of the SF soluble organic moiety rather than the $pK_a$ value.

The hydrochloride salts of triproline and pseudoephedrine have also been subjected to direct aqueous SFE. In order to liberate the SF-$CO_2$-soluble free bases, Hedrick and Taylor [14] used tetramethylammonium hydroxide (TMAOH) to neutralise their acidic salt solutions prior to extraction.

Combs *et al.* [15] have investigated the effect of pH on the direct SFE recoveries of phenol and 2,4,6-trichlorophenol (TCP) from water. These studies involved the use of a series of pH 3, 5 and 8 buffered and non-buffered solutions of the two phenols. The extraction efficiency for TCP was consistently greater than that for phenol throughout the pH range for both sets of solutions. Since TCP ($pK_a = 6.0$) is more acidic than phenol ($pK_a = 9.9$) it was concluded that the high solubility of phenol (9.3 g per 100 ml water at pH 7) relative to TCP (less than 2 g per 100 g water at pH 7) was more significant than the extent of dissociation for determining SF solubility. As anticipated, both phenols showed highest recoveries at low pH, with TCP showing the greatest percentage improvement. The high level of carbonic acid produced during SFE resulted in buffer capacities being exceeded. The use of indirect aqueous SFE can eliminate many of the problems associated with pH regulation and poor analyte recoveries due to high water solubility, especially when sample-loaded support media can be completely dried prior to extraction.

The underlying theory of increasing the ionic strength to promote SFE efficiencies of organic compounds from aqueous solution has been described by Barnabas *et al.* [25]. Within pure water a well-defined hydrogen bonding scheme exists which will tend to be distorted by the dissolution of a hydrophobic organic compound, such a disturbance being energetically unfavourable. Addition of an inorganic salt, however, results in the establishment of strong bonds between the ionic species and water molecules which more than compensate for the reduction of hydrogen bonding. Consequently, the addition of a salt can promote the aggregation of hydrophobic molecules which become more amenable to extraction at the SF–water interface. This process will continue to occur until the concentration of hydrophobic species falls below the 'aggregation point'.

Hedrick and Taylor [12] demonstrated that the addition of sodium chloride to an aqueous solution of diisopropyl methylphosphonate (DIMP) reduced the time required to establish equilibrium distribution between water and SF-$CO_2$ from 90 min to 5 min. Slack *et al.* [58] added 20 ml of saturated sodium sulphate per litre of water prior to sample preconcentration using a phenyl sorbent. However, no details of improved extraction efficiency of explosives from water were reported. Hedrick and Taylor [14] failed to alter the direct aqueous SFE distribution coefficients of succinonitrile, 2,5-lutidine and picolinic acid by saturating the solution with sodium chloride and/or sodium hydroxide. Neither of these three compounds were extracted.

Barnabas *et al.* [25] reported no significant increase in direct aqueous SFE recoveries of lindane, dieldrin and aldrin (each present at the 220 ppb level) following the addition of 8 g of sodium chloride to 45 ml water sample. Presumably, at this level, the OCPs were below their 'aggregation point'. Furthermore, it was reported that the addition of salt led to increased water carryover into the collection solvent during these direct studies. The phenomenon of analyte aggregation has also been used by Kane *et al.* [26] to explain the influence of critical micelle concentration on direct aqueous SFE recoveries of surfactants.

Roop and Akgerman [6] have investigated the use of modifiers for aqueous systems and have developed a method for qualitatively predicting the effect of adding a small amount of co-solvent. A 50% increase in the distribution of a complex mixture of phenols into SF-$CO_2$ from water using benzene as a modifier was reported, with the addition of methanol as predicted with the developed model having no effect.

Several reports have described the use of ion-pair reagents to improve SFE efficiency. Croft *et al.* [62] have evaluated the use of methyltrioctylammonium chloride (MTOA) and tetrahexylammonium hydrogen sulphate (THA) as phase-transfer catalysts for the derivatisation of chlorophenoxyacetic acids. Aqueous herbicide solutions and spiked urine samples (pH 10 following adjustment with sodium hydrogen carbonate buffer) containing THA were first loaded onto Extrelut sorbent. THA was used as a phase-transfer catalyst since MTOA has low water solubility. Static SFE was used whilst the herbicides underwent derivatisation in the SF phase with methyl iodide. It was demonstrated that an initial concentration of 13 mM THA was sufficient for efficient sample derivatisation. The THA which was carried over into the collection solvent was removed by Bio-beads SM7 prior to final GC/ECD analysis. Tena *et al.* [72] have conducted studies aimed at enhancing the yield of sulphonamides adsorbed onto various supports. Of the five sulphonamide drugs studied only the SFE efficiency of sodium sulphaquinoxaline could be significantly improved by the addition of a buffer (pH 4.76). However, the addition of 1 ml ion-pair reagent (0.1 M trimethylphenylammonium hydroxide in methanol) to spiked diatomaceous earth prior to SFE resulted in improved extraction efficiencies, between 12–15 fold compared with the addition of 1 ml pure methanol. It is interesting to note that Hedrick and Taylor [14] used TMAOH to bring about neutralisation of the hydrochloric salts of triproline and pseudoephedrine prior to direct aqueous SFE. A potential benefit of this 'neutralising agent' lies in its surfactant character. Any reduction of water–SF interfacial surface tension would tend to promote the phase exchange efficiency of the free bases.

The use of organic bases to block highly retentive sorbent active sites, resulting in improved indirect aqueous SFE efficiencies, has been reported. Liu *et al.* [73] added either dimethylhexylamine (DMHA) or triethylamine (TEA) to plasma samples to block residual $C_{18}$ sorbent silanol groups during the SPE stage. Edder *et al.* [55] added TEA to methanol-modified SF-$CO_2$ to block residual silanol and cation exchange sites present in $C_{18}$ and BEC sorbents, respectively, thereby promoting the SFE recoveries of morphinic alkaloids from urine.

### 4.4 Aqueous sample derivatisation procedures

A standard approach for improving the solubility of polar analytes in non-polar SF-$CO_2$ involves the use of organic modifiers. Sample derivatisation in which polar compounds are converted into less polar derivatives offers an alternate means for improving SFE efficiency and extending the range of compounds amenable to SFE (also, see the subsection on Covalent bond formation in section 3.1.2, which describes the derivatisation of analytes within solid matrices).

Tang and Ho [46] have described an indirect SFE procedure which involves the acetylation of 11 target phenols, preconcentrated from 500 ml aqueous samples (pH 10, adjusted with sodium hydroxide) using an anion exchange SPE disc. The loaded disc was acetone washed before being placed into an SFE vessel, to which acetone and acetic anhydride were also added. On-disc acetylation was performed using static SFE for 5 min after which dynamic SFE recovered the acetylated phenols for GC analysis. An alternative approach involving *in situ* aqueous acetylation of the phenols prior to SPE was also investigated. The procedure involved the addition of sodium hydrogen carbonate and acetic anhydride to 1 l aqueous phenol samples. At the end of the derivatisation period the water sample was applied to an SPE disc which was subsequently washed (pH 7 reagent water) and dried before being subjected to direct SFE. Styrene–divinylbenzene discs were found to provide improved acetylated phenol recoveries compared with $C_{18}$ discs. Improved phenol detection limits using the *in situ* aqueous acetylation procedure were demonstrated. This approach eliminated the breakthrough problem associated with SPE trapping of underivatised phenols.

Alzaga and Bayona [37] have used Grignard alkylation to derivatise organotins concentrated from synthetic and real-world seawater. The procedure involved loading $C_{18}$ SPE discs with 500–1000 ml acidified samples. The loaded discs were carefully dried, cut into small pieces and then placed into an aluminium reaction cell contained within an SFE vessel. A tetrahydrofuran (THF) solution of ethylmagnesium chloride was added to the SFE vessel which was kept open for 15 min following sample derivatisation (2 min) to allow the removal of excess THF. The vessel was then sealed and subjected to both static and dynamic SFE. The SFE conditions used had been developed following investigations to establish that losses of relatively volatile organotin derivatives would not occur during the collection stage. These preliminary studies involved GC with flame photometric detection (FPD) analysis of derivatised organotin standards spiked into collection solvent through which $CO_2$ from a blank SFE was vented.

Since Grignard alkylation involves the use of organic solvent and can only be performed in completely dry media, Cai and Bayona [65] have developed an *in situ* aqueous alkylation derivatisation SFE procedure for organotins in seawater. Sodium tetraethylborate is added to the aqueous sample (pH 8, adjusted with sodium acetate) and thoroughly agitated for 6 min. Preconcentration of the ethylated organotin derivatives was then performed using $C_{18}$ SPE discs. Addition of 30 µl hexane to the aqueous reaction mixture was found to improve analyte recoveries. Analysis of nine derivatised organotins remaining in spiked solution, using a conventional liquid–liquid extraction (LLE) method, confirmed excellent SPE recoveries. After sample loading the disc was dried and placed in an SFE vessel (containing anhydrous sodium sulphate to prevent extracted water causing restrictor freezing) and subjected to dynamic SFE. Cai and Bayona used a simplex algorithm to optimise pH, sodium tetraethylborate concentration and reaction time for sample derivatisation. The SFE method was developed using factorial design in which high and low values were applied to pressure, temperature and volume of SF-$CO_2$ used for extraction. The *in situ* derivatisation procedure enabled the detection of organotins in river and seawater at the low parts per thousand level using GC/FPD.

Croft *et al.* [62] have developed a rapid derivatisation procedure involving the methylation of chlorophenoxyacetic acids and simultaneous SFE from aqueous solution. Urine and water samples (pH 10, adjusted with sodium hydrogen carbonate) spiked with 2,4-dichlorophenoxyacetic acid (2,4-D) and 2,4,5-trichlorophenoxyacetic acid (2,4,5-T), mixed with a solution of THA (section 4.3), were loaded onto loose Extrelut sorbent. The sorbent was then placed into an SFE vessel to which an aliquot of methyl iodide was added. Sample derivatisation was performed using static SFE for

10 min, after which dynamic SFE was used to recover the methyl ester derivatives. The $CO_2$ from the collection vial was exhausted through charcoal to prevent release of excess methyl iodide. Extracts were analysed by GC/ECD. A series of studies were performed which conclusively demonstrated that methylation with THA and methyl iodide takes place within the SFE vessel rather than the GC injection port. Since underivatised 2,4-D and 2,4,5-T could not be extracted using pure SF-$CO_2$, the use of methanol as an SF modifier resulted in reduced methyl ester recoveries owing to extraction of free acids prior to derivatisation. A dynamic SFE method in which a methanolic solution of methyl iodide was continuously supplied to the extraction vessel was found to provide less reproducible results than did the static procedure.

An indirect aqueous SFE procedure involving the acetylation of three β-blockers (propranolol, metoprolol and oxprenolol) has been described by Hartonen and Riekkola [67]. Urine samples (pH 10, adjusted with borax buffer and sodium hydroxide) obtained from volunteers administered with the drugs (samples collected 0–4 h after intake to minimise collection of metabolites) were loaded onto $C_{18}$ SPE discs. The discs were then vacuum dried and transferred to an SFE vessel to which acetic anhydride and pyridine were also added. On-disc acetylation was performed using static SFE for 15 min, after which time the acetylated derivatives were recovered using dynamic SFE. Analysis was performed using GC–MS. It was demonstrated that a certain volume of pyridine was necessary to promote β-blocker SF solubilities for acetylation, beyond which the SFE conditions became sufficiently modified to prevent derivatisation.

## 4.5 Supercritical fluid extraction of metal ions from aqueous media

Although SFE of organic compounds has been widely studied there is a paucity of information regarding the extraction of metal ions from aqueous solution. The use of SF-$CO_2$ for direct extraction is inefficient because of weak Van der Waals interaction and the inability to neutralise the charge on the metal. Complexation with a suitable ligand (or ligands) provides a viable answer to this problem, leading to enhanced solubility of the metal species in SF-$CO_2$. The complexing agent should possess high thermodynamic stability constants at acidic pH, high SF solubility and fast complexation kinetics. Table 4.2 provides a summary of the applications involving SFE of metals from aqueous media.

Passage of SF-$CO_2$ through a saturation vessel (loaded with the complexing agent) en route to the SFE vessel provides a general means of introducing the complexation agent to the aqueous metal sample.

Although diethyldithiocarbamate (DDC) is widely used as a chelating agent, Laintz *et al.* [74] have reported that it produces metal complexes which have low solubility in SF-$CO_2$. This limitation can be overcome by

**Table 4.2** Applications of direct aqueous supercritical fluid extraction of metal ions

| Metal ion | Complexing agent[a] | Extraction conditions | Restrictor | Ref. |
|---|---|---|---|---|
| $UO_2^{2+}$, $Th^{4+}$ | TTA, TBP | $CO_2^b$, 150 atm, 60°C | Linear fused-silica capillary | 17 |
| $UO_2^{2+}$, $Th^{4+}$ | TBP, TBPO, TTA, TOPO, TPPO | $CO_2$, 200 atm, 60°C | Isco linear stainless-steel capillary | 18 |
| $Zn^{2+}$, $Cd^{2+}$, $Pb^{2+}$ | DBDTC, DEDTC, PDTC | $CO_2$, 24.05 MPa, 50°C | Linear fused-silica capillary | 19 |
| $Cu^{2+}$, $Zn^{2+}$ | DBDTC | $CO_2$, 24.06 MPa, 50°C[c] | Linear fused-silica capillary | 20 |
| $Cu^{2+}$ | FDDC | $CO_2$, 79.3 bar, 35°C[c] | None. Vessel equipped with inspection cell to monitor extraction kinetics using ultraviolet/visible spectrometer | 27 |
| $Sm^{3+}$, $Eu^{3+}$, $Gd^{3+}$, $Dy^{3+}$, $La^{3+}$, $Ce^{3+}$, $Yb^{3+}$, $Lu^{3+}$ | TTA, TBP | $CO_2$, 350 atm, 60°C | Linear fused-silica capillary | 75 |
| $As^{3+}$, $Cd^{2+}$, $Cu^{2+}$, $Mn^{2+}$, $Pb^{2+}$, $Se^{2+}$, $Zn^{2+}$ | DBDTC | $CO_2$, 24.05 MPa, 50°C[c] | Linear fused-silica capillary | 19 |
| $Cu^{2+}$, $Zn^{2+}$, $Cd^{2+}$, $Mn^{2+}$, $Fe^{2+}$ | DBDTC | $CO_2$, 24.1 MPa, 50°C[c] | Linear fused-silica capillary | 76 |
| $HCrO_4^-$, $SeO_4^{2-}$, $SeO_3^{2-}$ | THAB, THPAB, TBA | | | |

Complexing agents: TTA = thenoyltrifluoroacetylacetone; TBP = tributylphosphate; TBPO = tri-*n*-butylphosphine oxide; TOPO = tri-*n*-octylphosphine oxide; TPPO = triphenylphosphine oxide; DBDTC = dibutyldithiocarbamate; DEDTC = diethyldithiocarbamate; PDTC = pyrollidinedithiocarbamate; FDDC = bis(trifluoroethyl)dithiocarbamate; THAB = tetrahexylammonium bromide; THPAB = tetraheptylammonium bromide; TBA = tetrabutylammonium cation.
[a] Some applications also involve the use of mixtures of the quoted complexing agents.
[b] 5% methanol-modified SF-$CO_2$ was also evaluated.
[c] Typical extraction conditions quoted.

the use of a fluorinated analogue, bis(trifluoroethyl)dithiocarbamate (FDDC) whose metal complexes exhibit solubilities several orders of magnitude higher in SF-$CO_2$ (approximately $10^{-3}$ M at 100 atm and 50°C). Laintz *et al.* [27] have used LiFDDC to study the extraction kinetics of Cu(II) ions from water. The direct aqueous SFE vessel was equipped with a high-pressure quartz cell which permitted the use of a UV/VIS spectrometer to monitor the extraction of $Cu(FDDC)_2$ by measuring the charge-transfer absorbance at 416 nm in SF-$CO_2$. Quantitative extraction of Cu(II) was readily achieved by this method (SF-$CO_2$ at 79.3 bar and 35°C).

The direct SFE of $UO_2^{2+}$ and Th(IV) ions from aqueous samples, using thenoyltrifluoroacetylacetone (TTA) and tributyl phosphate (TBP) has been studied by Lin *et al.* [17]. In the absence of the ligand species the recovery of uranyl and Th(IV) ions was less than 2%, even with the use of 5% methanol-modified SF-$CO_2$. Using SF-$CO_2$ modified with TTA, the recoveries of uranyl and Th(IV) ions were 38% and 70%, respectively. Using TBP alone poor SFE recoveries were obtained for both metal ions (5%–6%), but by utilisation of SF-$CO_2$ containing a 1:1 mixture of TBP and TTA the extraction efficiencies for uranyl and Th(IV) ions improved to 70% and 87%, respectively. Using 5% methanol-modified SF-$CO_2$ with TBP and TTA the recoveries of uranyl and Th(IV) ions were improved yet further, to 85% and 90%, respectively. The amount of water transferred out of the extraction vessel using methanol-modified SF-$CO_2$ was estimated as small, since no phase separation was observed in the chloroform collection solvent. The feasibility of the mixed-ligand SFE technique was tested on samples of uranium mine water. The levels of uranyl ions present in these samples was determined using neutron activation analysis (NAA) prior to SFE. Using neat SF-$CO_2$ with a 1:1 mixture of TBP and TTA (at 150 atm and 60°C) the recoveries of uranyl ions from two different mine waters was 81% ± 4% and 78% ± 5% (for each sample, $n = 3$) following 40 min extraction.

Lin *et al.* [18] have also demonstrated quantitative direct SFE of uranyl and Th(IV) ions from nitric acid solutions using organophosphorous ligands. Uranyl ions can be extracted from nitric acid solutions using organophosphorous reagents (L) by formation of the $[UO_2(NO_3)_2(L)_2]$ complex, which is soluble in organic solvents. By increasing the concentration of nitric acid and/or nitrate salts it is possible to increase production of the complex by shifting the equilibrium to suppress the effect of increased nitrate ion concentrations. This effect was also confirmed using SF-$CO_2$ modified with TBP, since the SFE recoveries of uranyl ions were 91%, 52% and 12% in 6 M, 1 M and 0.1 M nitric acid, respectively (200 atm and 60°C). Higher recoveries were possible for both uranyl and Th(IV) ions using tri-*n*-butylphosphine oxide (TBPO) under equivalent SFE conditions; for example for uranyl ions, 99%, 99% and 97% in 6 M, 1 M and 0.1 M nitric acid, respectively. It was suggested that stronger complexation with TBPO relative to TBP was a result of the inductive effect of the alkyl groups, promoting the electron density of the TBPO phosphoryl oxygen relative to the electron-withdrawing effect of the butoxyl oxygens in TBP.

A procedure for the extraction of trivalent lanthanide ions from 6 M nitric acid to 3 M lithium nitrate solution, using TBP-modified SF-$CO_2$, has been described by Laintz and Tachikawa [75]. A syringe pump was first filled with TBP, followed by the addition of liquid $CO_2$ to create 10%, 20% and 30% TBP-modified liquid $CO_2$. The pump was ramped several times to ensure thorough mixing. A solution of selected lathanides was placed in a commercial direct liquid SFE vessel (Jasco, Tokyo, Japan) of similar design to that shown in Figure 4.3. A further series of extractions, aimed at studying the synergistic effect of TTA upon the TBP-modified SF-$CO_2$, were also performed. A quantity of TTA was placed in the ligand saturation vessel, sufficient to provide, for all the lanthanides present in the aqueous sample, slightly more than the threefold molar excess required for 1:3 lanthanide–TTA complexation. A period of 15 min static SFE followed by 30 min dynamic SFE (at 350 atm and 60°C) was used, with ethanol as collection solvent. The highest recoveries of lanthanides were obtained using 30% TBP-modified SF-$CO_2$ with TTA: near quantitative for Sm(III), Eu(III), Gd(III) and Dy(III) and ranging from 56% to 83% for La(III), Ce(III), Yb(III) and Lu(III). The data obtained indicate that this process is more selective for the mid-period lanthanides.

Wang and Marshall [19] have measured the solubility of a series of tetraalkylammonium dialkyldithiocarbamate ligands in SF-$CO_2$ to investigate their suitability for direct SFE of Zn(II), Cd(II) and Pb(II) from aqueous solution. Using a by-pass valve (direct aqueous SFE vessel isolated) it was possible to collect individual complexing agents entrained in SF-$CO_2$ by subjecting the ligand saturating vessel to 30 min static SFE. The quantities of ligands extracted were determined off-line using UV/VIS spectroscopy. It was determined that the SF-solubility of the ligands was largely determined by the chain length of the alkyl substituent attached to the carbamate nitrogen and to a lesser extent the alkyl substituent(s) of the ammonium counterion. Following a period of static SFE it was possible to extract more than 73% and more than 94% for each of the ions Zn(II), Cd(II) or Pb(II) within 5 min and 15 min dynamic SFE, respectively, using SF-$CO_2$ modified with tetrabutylammonium dibutyldithiocarbamate (DBDTC). Addition of nitric acid to the aqueous sample resulted in more than 95% recovery of each of the metal ions after 10 min using SF-$CO_2$ and DBDTC.

Wang *et al.* [20, 76, 77] have published several reports dealing with the coupling of direct aqueous SFE with atomic absorption spectrometry (AAS) for studying complexometric extraction kinetics of metal ions from aqueous media. The AAS nebulising assembly consists of a hollow needle which is used to position a heated fused-silica capillary restrictor within the spray chamber. The flame–AAS interface can accommodate SF-$CO_2$ flow rates equivalent to 2.2 l/min decompressed gas. Subnanogram to low picogram detection limits were reported for a range of metal ions, including Cu(II), Zn(II) and Cd(II) using DBDTC as a chelating agent.

## 4.6 Supercritical fluid extraction of analytes from enzymic reactions

The first report demonstrating the feasibility of SFs as solvent media for performing enzymic reactions was published in 1986 by Nakamura *et al.* [78]. This study involved the interesterification of triglycerides using lipase. Since then several reports have confirmed that the relatively low critical pressure and temperature of SF-$CO_2$ provides potential for the use of enzymes with thermally labile biopolymers as substrates. Nakamura [79] has reviewed such applications. By performing enzymic reactions in non-aqueous media one is offered the following advantages:

- stabilisation of enzymes – increased stability of enzymes in non-aqueous media relative to aqueous media has often been reported; for example Zaks and Klibanov [80] have determined that the half-life of porcine pancreatic lipase in organic media at 100°C is approximately 10 h compared with instantaneous denaturation in 0.1 M phosphate buffer (pH 8) at the same temperature;
- dissolution of hydrophobic substrates.

Supercritical fluids offer additional advantages:

- adjustment of temperature and pressure to selectively recover products from reactants;
- high diffusivity and low surface tension, promoting improved reaction rates as a result of higher rates of substrate mass transfer for heterogeneous chemical or biochemical catalysis.

The majority of published reports dealing with the use of SFs as alternative solvents for enzyme catalysis [79] have involved the use of immobilised enzymes in feasibility studies for batch-scale processes. The stability of enzymes in SFs are dependent on pressure, temperature, organic modifier and water content. Taniguchi *et al.* [81] have studied the stability of nine commercial enzyme preparations in SF-$CO_2$ containing ethanol (3%–6%) and water (0.1%) at 35°C, 200 atm for 1 h. With these conditions the activities of the enzymes were largely unaffected. Marty *et al.* [82] have conducted detailed investigations to determine the reaction kinetics of the esterification of *cis*-9-octadecanoic acid with ethanol, catalysed by an immobilised lipase. The study compared two solvents: SF-$CO_2$ and *n*-hexane. It was demonstrated that a 10% reduction in lipase activity in SF-$CO_2$ (13 MPa and 40°C) occurred after 6 days. The presence of water had a dramatic influence on enzyme activity, for example less than 30% residual activity after 30 μl water was added to 50 mg immobilised enzyme (incubated in SF-$CO_2$ for 1 day at 13 MPa and 40°C). It was suggested that denaturing of the enzyme may be due to hydrolysis. The optimum water content adsorbed onto the polymeric support for lipase activity was 10%. The presence of water was necessary to allow the enzyme to form the correct conformation. Using *n*-hexane, esterification of *cis*-9-octadecanoic acid proceeded more rapidly relative to the SF-$CO_2$ solvent systems. This was attributed to the lowering of pH brought about by carbonic acid production, affecting the enzyme when SF-$CO_2$ was used as a solvent.

King *et al.* [83] have used scanning electron microscopy to study morphological changes of several synthetic resin adsorbents treated with SF-$CO_2$.

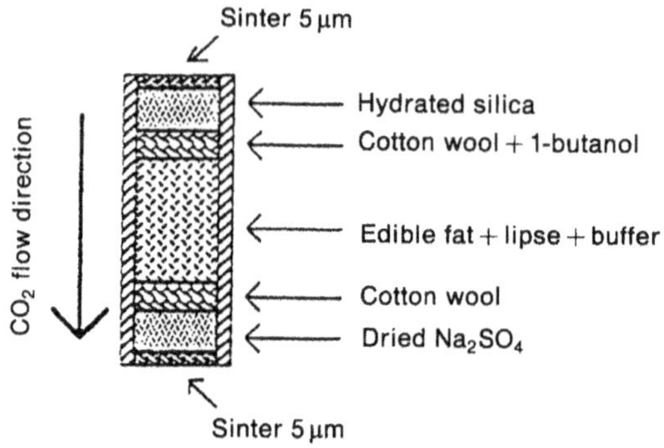


**Figure 4.9** Supercritical fluid extraction vessel loading arrangement of edible fat with lipase for interesterification reaction. Reproduced with permission from ref. 85.

It was recommended that careful selection of polymeric support material should be considered to minimise problems associated with change in breakthrough volume, swelling and plasticisation.

An on-line SFE–cSFC method involving the use of an immobilised lipase for interesterification of triglycerides in edible fat to produce fatty acid propyl and butyl esters has been developed by Berg *et al.* [84]. Figure 4.9 shows the manner in which the 0.5 ml SFE vessel was loaded. Approximately 50 mg of edible fat:immobilised lipase mixture (3:1 by mass, respectively) was loaded into the vessel. Hydrated silica placed above the substrate served to supply the necessary level of water to promote enzyme catalysis. In order to minimise reduction of enzyme activity arising from carbonic acid production, pH 7 phosphate buffer was added to the substrate mixture. Dried sodium sulphate placed at the outlet end of the vessel was added to prevent water dissolved in SF-$CO_2$ causing any restrictor blockage as a result of ice formation. Approximately 1 μl of butanol was then added prior to 15 min static SFE and 15 min dynamic SFE using SF-$CO_2$ (150 bar and 50°C). With this extraction protocol, extracts were collected off-line using a solvent trap with cyclohexane. Using on-line SFE–cSFC short dynamic SFE periods were used to prevent column overloading. A 1.8 m × 100 μm i.d. deactivated fused-silica capillary serving as a retention gap or a thin coated precolumn [0.1 μm film of poly(dimethylsiloxane)] connected to the cSFC column via a venting valve arrangement was used for collecting the extract in the on-line mode. Co-extracted butyl alcohol was removed by a venting procedure prior to cSFC analysis. The procedure provided high yields of fatty acid esters, and the large sample size meant that the enzyme-catalysed reaction could be studied over a long period of time.

In a subsequent study Berg *et al.* [85] used the on-line SFE–cSFC procedure for the quantification of cholesterol in spray-dried egg yolk. Figure 4.10(a) shows the SFE–cSFC chromatogram obtained for dried egg yolk not subjected to interesterification. Although cholesterol can be detected it

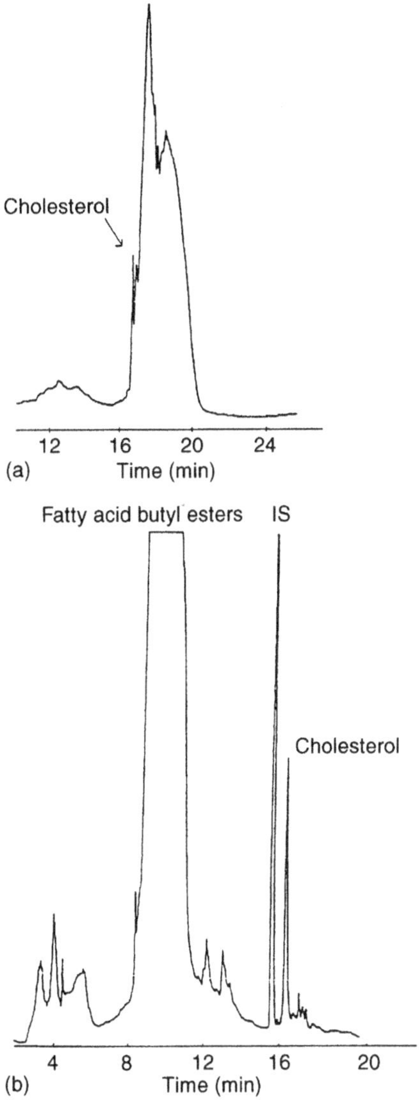


**Figure 4.10** On-line supercritical fluid extraction–capillary supercritical fluid chromatography analysis of spray-dried egg yolk. Approximately 1 mg of a mixture of egg yolk/Hydromatrix (1.02% m/m) was weighed into the extraction vessel: (a) without lipase; (b) with the addition of immobilised lipase (1 mg) to the sample. IS = internal standard. Reproduced with permission from ref. 86.

is not sufficiently resolved from an interfering 'packet' of glycerides to be quantified. Using an immobilised lipase interesterification procedure the triglyceride butyl ester derivatives are sufficiently resolved [Figure 4.10(b)] to enable quantification of the extracted cholesterol. By adding soybean oil (loaded onto Hydromatrix, 1.02% oil by mass) into the SFE vessel with the enzyme–substrate mixture it was demonstrated that a threefold improvement in cholesterol recovery was obtained. Without soybean oil the recovery of cholesterol was approximately 20%.

Ramsey *et al.* [31] have demonstrated the feasibility of direct aqueous SFE for the recovery of testosterone liberated following the enzymic hydrolysis of testosterone-β-D-glucuronide. Since the vast majority of testosterone is excreted in urine, in the form of glucuronic acid and sulphate conjugates which are too polar to be analysed by gas chromatography–mass spectroscopy (GC–MS) dope-testing methods involve incubation of urine samples with glucuronidase. The liberated testosterone is then isolated using LLE prior to analysis. Using a direct aqueous SFE vessel of 300 ml capacity, a 230 ml aqueous sample spiked with testosterone-β-D-glucuronide was incubated with HP-2 β-glucuronidase. Following a period of 15 min, required to reach SF-$CO_2$ target pressure (24.1 MPa at 55°C), the enzymic hydrolysis mixture was subjected to 120 min dynamic SFE. A variable orifice restrictor was used and the extract was depressurised onto a 250 mm × 4.6 mm i.d. $C_{18}$ HPLC column serving as an analyte trap. At the end of the extraction period the column was rinsed with ethyl acetate and quantitative GC–MS was performed. Dynamic SFE efficiency was determined by heating the contents of the vessel at 100°C to denature the enzyme, followed by an LLE procedure to recover the remaining testosterone following SFE. Results of quantitative GC–MS analyses indicated that 70% testosterone-β-D-glucuronide had been hydrolysed after 135 min enzymic digestion, resulting in 88% of the liberated testosterone being trapped after 120 min dynamic SFE.

## 4.7 Inverse supercritical fluid extraction

This technique, first described by Messer and Taylor [86], involves extracting the sample matrix away from the analyte of interest. Five parameters have been identified for the successful implementation of this procedure.

1. The analyte should be totally insoluble in the SF, for example many water-soluble drugs which possess polar functional groups are not soluble in organic-solvent-modified SFs.
2. Conversely, the matrix should be highly soluble in the SF, for example mineral oil.
3. Following SFE an extremely efficient method should be employed to dissolve the analyte remaining in the extraction vessel.
4. The analyte concentration should be greater than 2% within the matrix.
5. The final analytical procedure should provide low detection limits for the analyte.

Perhaps not suprisingly in view of the above criteria, pharmaceutical creams in which a polar drug is entrained within a non-polar matrix represent ideal candidates for application of this technique. In their original studies Messer and Taylor [86] developed a sample preparation procedure for the analysis of acyclovir, the water-soluble, active ingredient in the hydrocarbon-based Zovirax® Ointment. Approximately 100 mg of ointment (5% acyclovir) was loaded into an empty polypropylene SPE cartridge placed within a stainless-steel SFE vessel. Dynamic SFE was performed using 2% methanol-modified SF-$CO_2$ (500 atm and 50°C) for 20 min. The restrictor could be manually opened to prevent blockage occurring as the large amount of hydrocarbon-based matrix was extracted. At the end of the extraction period the SPE cartridge was sonicated with 0.01 M NaOH to dissolve the isolated acyclovir whose recovery was determined using HPLC. Using inverse SFE, the recoveries of acyclovir fom the 'matrix-stripped' ointment were 99% ± 5.3% ($n = 3$).

Moore and Taylor [87] have developed an inverse SFE procedure for the isolation of the polar drug, polymyxin B sulphate (a class of about eight compounds whose individual molecular masses are approximately 1200 amu) from two pharmaceutical preparations. Polymyxin B sulphate is present in semi-solid Neosporin® Ointment and semi-liquid Neosporin® Cream at 0.8% m/m and 0.17% m/m levels, respectively. Exhaustive SFE using 5% methanol-modified SF-$CO_2$ was found to be effective in removing both sets of hydrocarbon-based matrices. The effectiveness of inverse SFE was compared with an established SPE procedure. The SPE method involves loading 1 g of either Neosporin® preparation onto a silica cartridge, which is then eluted with hexane and ethyl acetate, isolation of polmyxin B sulphate being achieved in approximately 2 h. Sample sizes ranging from 200 to 800 mg of either Neosporin® preparation were loaded into an empty polypropylene SPE cartridge placed within an SFE vessel. A restrictor operated at 300°C was used in these studies to avoid blockage problems. Following SFE the isolated drug was dissolved by sonicating the SPE cartridge in 0.1 M hydrochloric acid:methanol:Tween 80 (75:25:0.1 vol/vol) prior to HPLC analysis. Initial inverse SFE recoveries of the drug were poor (relative to the SPE method) using 5% methanol-modified SF-$CO_2$ (500 atm, with temperatures ranging from 65°C to 40°C). This problem was overcome by loading the cream between two layers of silica gel whose presence was designed to promote selective trapping of the drug during SF decompression. The final recovery of the drug was 137% ± 1.9% ($n = 6$) relative to the SPE procedure, using 5% methanol-modified SF-$CO_2$ (450 atm at 60°C) with dynamic SFE for 45 min. The high inverse SFE recoveries indicated that this procedure was more effective than was the SPE method.

Finally, providing the analytes are sufficiently SF soluble, it is possible selectively to isolate such compounds from hydrocarbon-based cream preparations using conventional SFE. Such an example involves the SFE recovery of vitamin A palmitate and vitamin E acetate from cosmetic creams and lotions loaded onto Celite® using SF-$CO_2$ (250 atm and 40°C). Using this SFE procedure, Scalia *et al.* [63] were able to demonstrate recoveries greater than 91.6% for the vitamin esters compared with a conventional liquid extraction procedure.

## 4.8 Selected liquid supercritical fluid extraction applications

Throughout the preceding sections a diverse range of applications of liquid SFE procedures have been described, and these are summarised in Tables 4.1 and 4.2. The purpose of this section is to describe in further detail selected applications of liquid SFE to illustrate the flexibility of these techniques.

Liu *et al.* [73] developed an indirect aqueous SFE procedure using $C_{18}$ SPE cartridges for the ultratrace analysis of a drug metabolite, mebeverine alcohol (MEBOH), in dog plasma. Addition of an organic base, DMHA or TEA to the plasma sample prior to SPE resulted in high recoveries of the metabolite using 5% methanol-modified SF-$CO_2$ (350 atm, 40°C, extraction time 10 min). This was attributed to the blockage of highly retentive silanol groups: recovery of MEBOH was less than 1% in the absence of the base with the previous SFE conditions. The SFE recoveries of MEBOH using DMHA and TEA were 95% ± 5% ($n = 3$) and 85% ± 1% ($n = 3$), respectively, at the 250 ng/ml plasma level. The higher metabolite recoveries using DMHA were attributed to losses of more volatile TEA during the $C_{18}$ sorbent drying stage prior to SFE. Mass balance studies were performed using [$^{14}$C]-MEBOH. Analysis of the $C_{18}$ sorbent following SFE confirmed that incomplete extraction rather than SPE breakthrough was responsible for the small losses of [$^{14}$C]-MEBOH. A conventional $C_{18}$ SPE procedure provided greatest [$^{14}$C]-MEBOH recoveries (100.6% ± 1.5%, $n = 3$) as a result of the elutropic strength of water/methanol.

A direct liquid SFE procedure involving sample preparation of drugs in dog plasma has also been evaluated by Liu and Wehmeyer [23]. The absolute recovery, selectivity, accuracy and precision of this technique was compared with a conventional LLE procedure. Water, plasma and water/plasma mixtures spiked with either [$^{14}$C]-flavone (neutral) or [$^{14}$C]-ketorolac (acidic) were used to study experimental extraction parameters, including time and pressure. Addition of a food-grade antifoaming agent proved effective in preventing denatured plasma proteins, formed during direct SFE, causing blockage of a fused-silica capillary restrictor. Using SF-$CO_2$ (30 MPa and 60°C) the recovery of [$^{14}$C]-flavone from water (spiked at the 250 ng/ml level) was quantitative and complete after 30 min. However, the recovery of [$^{14}$C]-flavone from plasma, spiked at the same level, was only 60% after 50 min extraction. The binding of the drug to non-extracted plasma proteins was suggested as a possible cause for reduced recoveries. Dilution of the plasma sample with water resulted in 85% recovery after 20 min SFE. The average absolute SFE (30 MPa, 60°C, extraction time 20 min) recovery of [$^{14}$C]-flavone from diluted plasma, spiked at the 25 ng/ml level, was 98% ± 5.2% ($n = 3$), compared with LLE efficiencies of 92% ± 5% ($n = 3$) and 80% ± 7.1% ($n = 3$) using hexane and dichloromethane, respectively. At the 25 ng/ml level the SFE recovery for [$^{14}$C]-ketorolac from diluted dog plasma was 80% ± 11.8% ($n = 3$); LLE recoveries for this drug were not reported. HPLC analysis (UV detection at 300 nm) of the SFE and LLE diluted plasma extracts indicated fewer endogenous components were extracted using the LLE procedures. Despite this, direct SFE provided suitably interference-free samples for quantitative analysis of the drugs by HPLC.

Howard and Taylor [71] have demonstrated the use of an indirect aqueous SFE procedure for the quantitative recovery of sulphonyl urea herbicides. An HP 7680 A SFE system (Figure 4.8) was used to investigate experimental extraction parameters, including solid phase trapping material, trap rinse solvent, trap temperature, influence of spiking solvent and minimum extraction time. Celite®, spiked with an acetonitrile solution of sulphometuron methyl and chlorsulphuron, was used for SFE method development. Two trapping materials were evaluated for analyte collection following SFE: 40 µm $C_{18}$ particles and 100 µm stainless-steel beads. Initial experiments were conducted using 2% methanol-modified $CO_2$ to determine the optimum trap temperature. Although setting the trap temperature higher than the boiling point of the organic modifier ensures its removal at the decompression stage, highest herbicide recoveries were obtained using the stainless-steel trap at 15°C. Dissolution of the analytes in a thin film of methanol on the surface of the stainless-steel beads improved their recovery by preventing losses associated with the high flow of decompressed $CO_2$ gas vented to waste. Stainless-steel beads were selected as an analyte trapping medium to avoid problems associated with the interaction of the relatively polar herbicides with residual silanol groups of the $C_{18}$ sorbent. Acetonitrile was used as the rinse solvent for the stainless-steel trap since all extracted herbicides could be recovered within the first 1 ml rinse. The SFE acetonitrile solution was collected in a vial containing sulphachloropyridazine as the internal standard for subsequent HPLC analysis. The minimum time required to obtain quantitative SFE recoveries of both herbicides from spiked Celite® was determined to be less than 5 min. Following these studies, $C_{18}$ and $C_8$ SPE discs were evaluated for use with the SFE method. The discs were loaded by applying either 100 ml or 1 l water samples spiked with each herbicide at the 500 ppb and 50 ppb level, respectively. Mass balance calculations involving HPLC analysis of the SFE acetonitrile extracts and aqueous raffeinates indicated that the highest herbicide losses due to incomplete SFE occurred with the $C_{18}$ SPE discs. In order to improve the final analyte recoveries, the sample-loaded $C_8$ SPE discs were washed with pH 9 aqueous solution prior to SFE. This effectively prevented the establishment of hydrogen bonding between the relatively polar herbicides and residual silanol groups of the SPE $C_8$ particles. Final indirect aqueous SFE recoveries obtained for sulphometuron methyl and chlorsulphuron were, respectively, 97.2% ± 7.8% and 93.6% ± 5.5% ($n = 5$) for 1 l water samples spiked with each herbicide at the 50 ppb level.

Tang *et al.* [36] have evaluated the use of $C_{18}$ SPE cartridges and discs in conjunction with SFE for the recovery of 36 organic compounds, including selected PAHs, PCBs, OCPs and phthalate esters from aqueous samples. The selected compounds are those analysed using EPA Method 525. The relative effects of controllable SFE parameters were studied and optimised for the recovery of analytes adsorbed onto SPE media. These studies were conducted using $C_{18}$ SPE discs spiked with 14 PAHs, 400 μl methanol being added to each disc prior to SFE. A methanol solution of four deuterated PAHs, acting as internal standards, was used for SFE analyte collection prior to quantitative GC–MS analysis. Initially, three SFE parameters were studied: volume of SF-$CO_2$ required for quantitative analyte recoveries; temperature; and pressure. Conditions of 300 atm at 60°C and 30 ml SF-$CO_2$ provided maximum PAH recoveries. Following these studies it was demonstrated that no significant difference in analyte recoveries were obtained using dynamic or static–dynamic modes of SFE. It was further demonstrated that the level of residual water remaining on SPE discs did not affect analyte recovery and the methanol used in SFE was required to prevent restrictor blocking rather than enhancing analyte recovery. In EPA Method 525, 1 l of aqueous sample is passed through an SPE cartridge containing 1 g $C_{18}$ sorbent prior to elution with dichloromethane. Tang *et al.* reported that 1–2 h were required to pass 500 ml spiked water sample (free from particulates, 8 ppb each compound) through an SPE cartridge containing 500 mg $C_{18}$ sorbent, compared with 20 min to pass 1 l spiked water sample (2 ppb each compound) through a 47 mm diameter $C_{18}$ SPE disc. Furthermore, the SFE efficiency using the disc was superior to that using the cartridge, for the compounds studied. The mean recoveries of 34 specific compounds, using the indirect aqueous SFE procedures, were compared with those published for EPA Method 525. The results were: SPE disc–SFE procedure, 95% ± 21% ($n = 7$); SPE cartridge–SFE procedure, 84% ± 22% ($n = 4$); US Environmental Protection Agency (EPA) Method 525, 100% ± 15%. It was concluded that fewer experimental errors should be associated with the SFE of real-world samples, since a reduction in the number of sampling handling stages could be anticipated.

In a subsequent study, Tang and Ho [46] evaluated the use of several different SPE disc types for the recovery of 11 phenols from water using indirect SFE. These studies also involved two different derivatisation strategies to promote SFE recoveries and final GC–MS performance. One approach involved on-disc acetylation of the phenols prior to SFE, whereas the other procedure involved *in situ* aqueous acetylation of the phenols prior to SPE (details of both derivatisation procedures are given in section 4.4). The dynamic SFE conditions used throughout were SF-$CO_2$ at 400 atm, 60°C and 10 min extraction time. The addition of 1.5 ml acetone to each disc prior to extraction promoted the recovery of all phenols and prevented the restrictor from plugging. The extraction efficiencies of 47 mm diameter $C_{18}$, 47 mm diameter styrene–divinylbenzene (SDB) and 90 mm diameter SDB discs were investigated for the recovery of the phenols from a 500 ml spiked water sample (20 ppb each phenol). Poor phenol SFE recoveries were obtained using the $C_{18}$ SPE disc, only five phenols being recovered at levels greater than 60%, with phenol itself providing only 4% recovery. The 47 mm diameter SDB disc provided quantitative recovery for most of the compounds, except phenol, for which the recovery was 16%. The 90 mm SDB disc provided quantitative recovery for all phenols. However, when 1 l spiked water samples (10 ppb each phenol) were applied to 90 mm SDB discs, the SFE phenol recovery fell to 37%. This was attributed to its high water solubility giving rise to breakthrough. An anion exchange SPE disc was also used to concentrate phenols from spiked aqueous samples, with an on-disc acetylation procedure being used to promote SF phenol solubility. This procedure proved effective for the recovery of phenol (*c.* 90% at 50 ppb) from 250 ml water samples but again fell rapidly (*c.* 25% at 50 ppb) as the sample volume was increased to 1 l. In order to minimise SPE disc breakthrough, an *in situ* aqueous acetylation procedure was developed to convert phenols into their less-water-soluble acetate derivatives. With this approach, the SFE recovery of phenol was 102% ± 8.1% ($n = 7$) for 1 l spiked water samples (1 ppb level) applied to a 47 mm diameter SDB disc.

Eight different commercially available sorbents have been evaluated by Edder *et al.* [55] (section 4.2.4) for an indirect SFE procedure involving the analysis of morphine alkaloids in urine. An inexpensive custom-built restrictor was used in these studies (section 4.2.5), whose design prevented blockage problems due to ice formation. It was demonstrated that two silica-based sorbents, $C_{18}$ and Bond Elut Certify (BEC), provided the necessary level of analyte preconcentration for a drug-testing procedure. Aqueous solutions of morphine, applied to both sorbents, were used for SFE method development. Morphine was selected as the test compound since it had the lowest SF-$CO_2$ solubility of the basic alkaloids studied. Owing to its strong interactions with $C_{18}$ residual silanol groups, morphine could be quantitatively extracted from this sorbent using 8.5% methanol: 1.5% TEA (vol/vol) modified SF-$CO_2$ (25 MPa, 40°C, extraction time 25 min). The presence of trace levels of water (used to rinse the $C_{18}$ sorbent following sample application) during SFE was also required since morphine recoveries fell to 19% in its absence. The use of 14% methanol: 6% TEA (vol/vol) modified SF-$CO_2$ (using previous extraction conditions) failed to recover morphine from the BEC sorbent. This was because of the establishment of ionic forces of attraction between the basic alkaloids and the sorbent cation exchange sites. Using 6% methanol:6% TEA:3% water (vol/vol) modified SF-$CO_2$, quantitative recovery of morphine from BEC sorbent was possible using an extraction time of 20 min. Spiked urine samples were then extracted using both indirect SFE procedures. Analysis of the SFE extracts was performed using gas chromatography/flame ionisation detection (GC/FID). As shown in Figure 4.11(a), the $C_{18}$ SFE procedure resulted in the co-extraction of endogenous compounds from urine, whose presence resulted in a high chromatographic background making drug quantitation difficult. However, the BEC indirect SFE procedure proved more selective, providing a cleaner chromatogram, as shown in Figure 4.11(b). The BEC indirect SFE procedure provided high drug recoveries from spiked urine: morphine, 99% $\pm$ 4.4% ($n = 6$); codeine, 102% $\pm$ 1% ($n = 6$); ethylmorphine, 93% $\pm$ 2.5% ($n = 6$). In this instance, careful selection of the sorbent with appropriate SFE conditions resulted in high analyte recoveries and clean-up stage combined in one step.

Hedrick and Taylor [13] have described an on-line direct aqueous SFE–SFC system for the determination of the distribution coefficients of phenols between water and SF-$CO_2$. The system uses the same multivalve configuration that they originally developed for the direct SFE analysis of phosphonates [12] and nitrogenous bases [14] and from water. The valve configurations are shown in Figure 4.12. SF-$CO_2$ is passed through the spiked water sample using a recirculating pump. A 1 m length of deactivated fused silica was used to interface the direct aqueous SFE vessel to a 250 mm $\times$ 1 mm i.d. cyanopropyl column used for SFC. Since a 20 μl injection loop was used, each injection representing only 0.3% of the SF head-space (taking into account the void volume of the column and ancillary tubing), replicate injections could be made at specific time intervals following the onset of SFE. Once the steady-state concentration of phenols in the SF phase had been established, remaining concentrations in the spiked aqueous phase were calculated by mass balance. Hence individual phenol distribution coefficients were determined. With this technique SFE efficiencies through a wide range of extraction conditions could be rapidly investigated.

The different solubilities of OCPs and OPPs in SF-$CO_2$ has been exploited by Barnabas *et al.* [69] to develop a class-selective indirect aqueous SFE method. The OCPs and OPPs, loaded (200 ml aqueous sample) onto a $C_{18}$ SPE disc, were extracted using first neat SF-$CO_2$ and then methanol-modified SF-$CO_2$. Studies were performed at different temperatures and pressures. Hexane was used as the collection solvent, with the $CO_2$ from the vial being exhausted through a $C_{18}$ SPE cartridge. With this system any analytes that had escaped solvent collection could be recovered from the cartridge by using a small volume of hexane. The OCPs and OPPs were fractionated using a two-stage SFE procedure. Using neat SF-$CO_2$ (13.5 MPa and 50°C) the OCPs were recovered, along with trace levels of OPPs, as shown

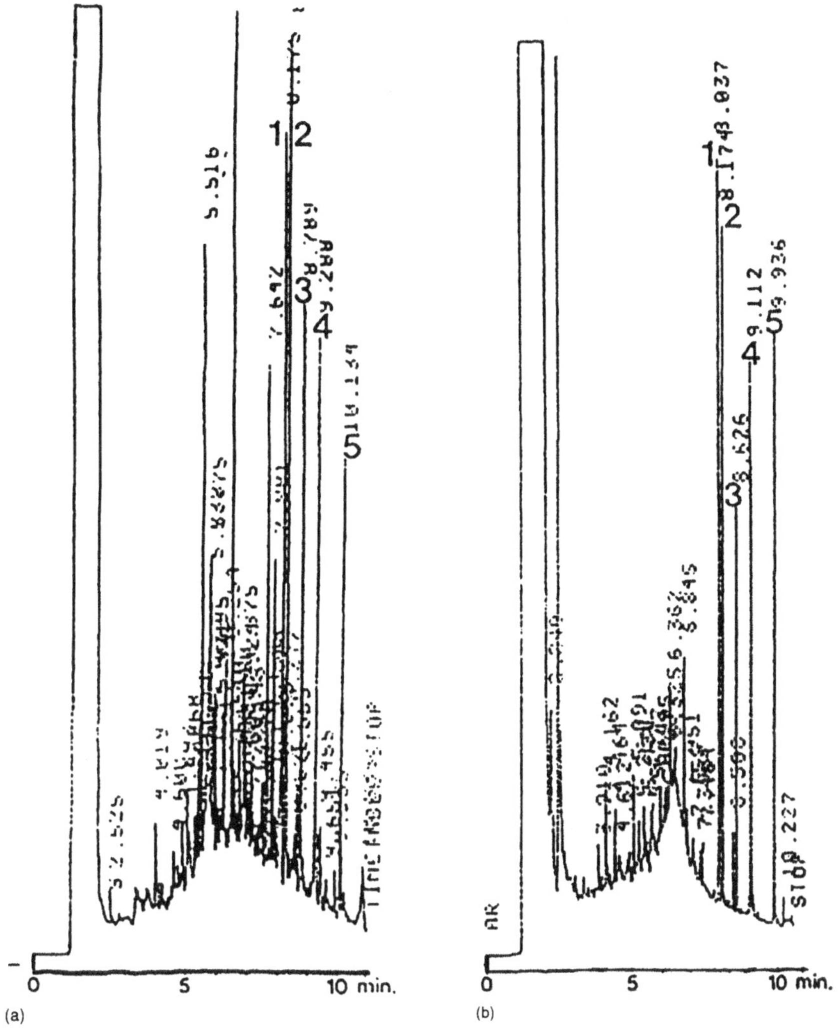


**Figure 4.11** Gas chromatograms of morphinic alkaloids spiked in urine obtained by indirect aqueous supercritical fluid extraction after loading onto: (a) $C_{18}$ bonded silica; (b) Bond Elut Certify support. Peaks: 1 = codeine; 2 = ethylmorphine; 3 = 6-monoacetylmorphine; 4 = morphine; 5 = nalorphine (internal standard). Reproduced with permission from ref. 55.

in Figure 4.13. The OPPs were then recovered using methanol-modified SF-$CO_2$ (35 MPa and 50°C) along with residual traces of OCPs, as shown in Figure 4.14. With this technique the OCP recoveries were: aldrin, 98.1% ± 7.6% ($n = 5$); dieldrin, 72% ± 5.7% ($n = 5$); lindane, 91.4% ± 7.5% ($n = 5$). The OPP recoveries were: dichlorovos, 95.6% ± 6.6% ($n = 5$); diazinon, 93% ± 6.1% ($n = 5$), malathion, 84.6% ± 2.1% ($n = 5$).

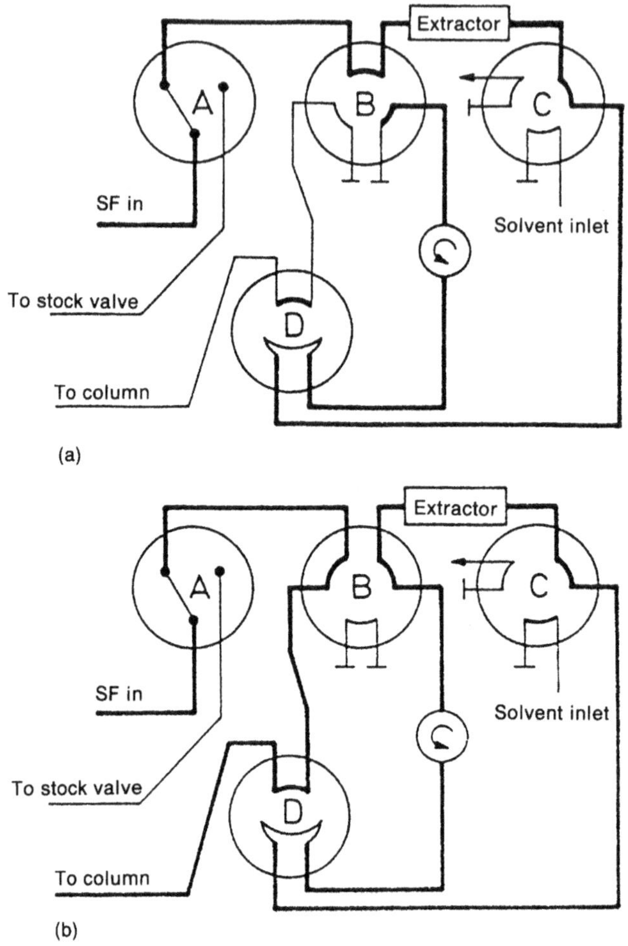


**Figure 4.12** (a) Pressurisation scheme for an equilibrium extractor; (b) equilibration and injection scheme for an equilibrium extractor. Injection onto a column is made by rotating high-pressure valve D. The bold dark lines represent the flow paths. A–D = high-pressure valves; SF = supercritical fluid. Reproduced with permission from ref. 13.

In a subsequent study, Barnabas *et al.* [70] developed another class-selective procedure, involving indirect aqueous SFE, for the analysis of three OCPs by GC–MS, and six herbicides by HPLC (for further details, see section 8.3).

## 4.9 Conclusions

The applications described within this chapter serve to demonstrate, within the limits of modified SF-$CO_2$ solvating strength, the potential of both direct and indirect aqueous SFE sample preparation procedures. In order

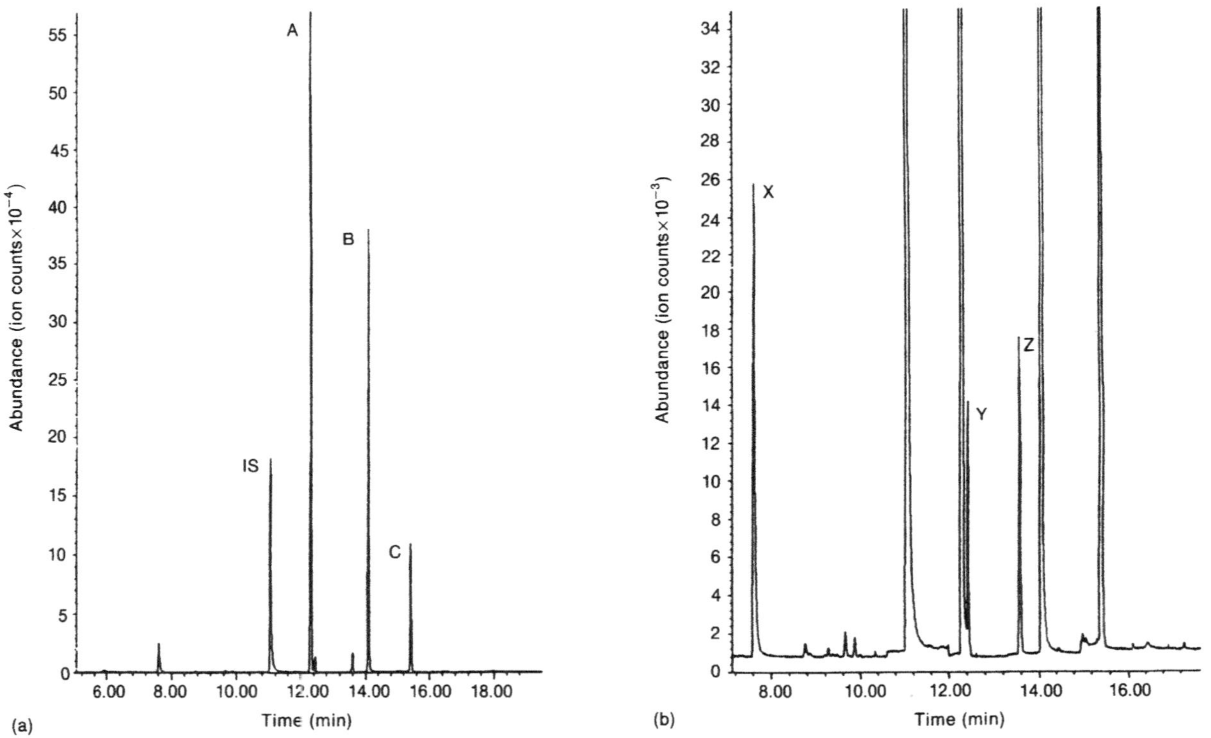


**Figure 4.13** (a) Gas chromatography–mass spectrometry (GC–MS) single-ion monitoring of organochlorine pesticides after indirect aqueous supercritical fluid extraction (SFE) with supercritical fluid carbon dioxide (SF-$CO_2$) only. Peaks: IS = internal standard; A = lindane; B = aldrin; C = dieldrin. (b) GC–MS single-ion monitoring of organophosphorus pesticides after indirect aqueous SFE with SF-$CO_2$ only. Peaks: X = dichlorvos; Y = diazinon; Z = malathion. Reproduced with permission from ref. 69.

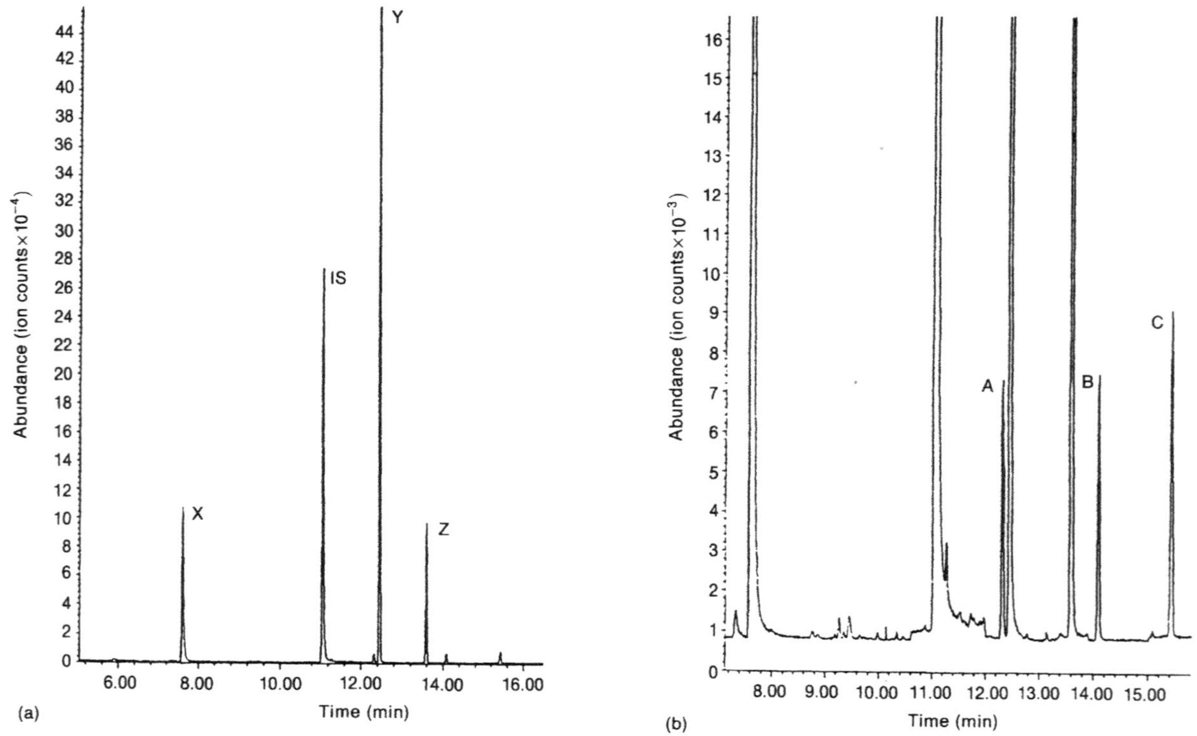


**Figure 4.14** (a) Gas chromatography–mass spectrometry (GC–MS) single-ion monitoring of organophosphorus pesticides after extraction with methanol-modified supercritical carbon dioxide (SF-$CO_2$). Peaks: IS = internal standard; X = dichlorvos; Y = diazinon; Z = malathion. (b) GC–MS single-ion monitoring of organochlorine pesticides after exhaustive extraction with methanol modified SF-$CO_2$. Peaks: A = lindane; B = aldrin; C = dieldrin. Reproduced with permission from ref. 69.

to obtain reproducible results with either technique it is essential to use reliable restrictors, resistant to blockages caused by the freezing of low levels of SF-$CO_2$ dissolved water. The inclusion of drying agents between the sample being extracted and the restrictor module can afford some degree of protection, although care must be exercised in selecting the drying agent, to avoid retention of extracted analytes.

The use of chelating agents provides an effective means of charge suppression for the extraction of metal ions, whereas the use of enzymes offers great potential to allow the SFE recovery of drugs from solution by hydrolysing their polar metabolites. Sample derivatisation and ion-pair formation provide alternative means by which relatively polar compounds can be extracted using direct and indirect aqueous SFE.

Direct aqueous SFE has the advantage of simplicity, with minimum handling stages, and can be successfully used with samples that contain a high level of particulates. In appropriate circumstances the addition of salt to the aqueous sample can improve SFE efficiency for some analytes. Indirect aqueous SFE is considered to be more flexible, since varying volumes of liquid samples can be readily accommodated using the wide range of SPE products available. Careful selection of the support matrix can also provide an effective means of analyte trapping or extraction and cleanup in one stage. A further advantage of indirect SFE, especially when the support media can be completely dried prior to extraction, stems from the fact that many of the problems associated with pH regulation and poor analyte recoveries arising from high water solubility can be avoided; for example, although nicotine is readily soluble in SF-$CO_2$, it is extremely difficult to extract this basic compound from aqueous solution by direct aqueous SFE.

In summary, the full potential of both direct and indirect aqueous SFE has yet to be established; clearly the flexibility of both approaches warrants further investigation.

## References


1. Ehntholt, D.J., Thrun, K. and Eppig, C. (1983) The concentration of model organic compounds present in water at parts-per-billion levels using supercritical fluid carbon dioxide. *International Journal of Environmental Analytical Chemistry*, **13**, 219–35.
2. Bartle, K.D., Clifford, A.A., Jafar, S.A. and Shilstone, G.F. (1991) Solubilities of solids and liquids of low volatility in supercritical carbon dioxide. *Journal of Physical and Chemical Reference Data*, **20**, 728.
3. Francis, A.W. (1954) Ternary systems of liquid carbon dioxide. *Journal of Physical Chemistry*, **58**, 1099.
4. Taylor, L.T. (1996) *Supercritical Fluid Extraction*, John Wiley, New York, pp. 28–41.
5. Luque de Castro, M.D., Valcarcel, M. and Tena, M.T. (1994) *Analytical Supercritical Fluid Extraction*, Springer, Berlin, pp. 88–123.
6. Roop, R.K. and Akgerman, A. (1990) Distribution of a complex phenolic mixture between water and supercritical fluid carbon dioxide. *Journal of Chemical Engineering Data*, **35**, 257–60.

7. Sengupta, S., Gupta, S., Dooley, K.M. and Knopf, F.C. (1994) Measuring and modeling of the extraction of chlorinated hydrocarbons from water with supercritical carbon dioxide. *Journal of Supercritical Fluids*, **7**, 201–9.
8. Ahmad, M. (1993) *Pesticide Residue Analysis Using Supercritical Fluid Extraction*, PhD thesis, School of Applied Sciences, University of Glamorgan, Pontypridd.
9. Hedrick, J.L., Mulcahey, L.L. and Taylor, L.T. (1992) Supercritical fluid extraction of phenols from water, in *Supercritical Fluid Technology: Theoretical and Applied Approaches to Analytical Chemistry* (eds F.V. Bright and M.E.P. McNally), ACS Symposium Series 488, American Chemical Society, Washington, DC, pp. 206–20.
10. Koester, C.J. and Clement, R.E. (1993) Analysis of drinking water for trace organics. *Critical Reviews in Analytical Chemistry*, **24**(4), 263–316.
11. Stahl, E., Quirin, K.-W. and Gerard, D. (1988) *Dense Gases for Extraction and Refining*, Springer, Berlin, pp. 5–7.
12. Hedrick, J.L. and Taylor, L.T. (1989) Quantitative supercritical fluid extraction/supercritical fluid chromatography of a phosphonate from aqueous media. *Analytical Chemistry*, **61**, 1986–8.
13. Hedrick, J.L. and Taylor, L.T. (1990) Supercritical fluid extraction strategies of aqueous based matrices. *Journal of High Resolution Chromatography*, **13**, 312–16.
14. Hedrick, J.L. and Taylor, L.T. (1992) Direct supercritical fluid extraction of nitrogenous bases from aqueous solution. *Journal of High Resolution Chromatography*, **15**, 151–4.
15. Combs, M.T., Khorassani, M.A. and Taylor, L.T. (1996) pH effects on the direct supercritical fluid extraction of phenols from aqueous matrices. *The Journal of Supercritical Fluids*, **9**, 122–7.
16. Mulcahey, L.J. and Taylor, L.Y. (1992) Supercritical fluid extraction of active components in a drug formulation. *Analytical Chemistry*, **64**, 981–4.
17. Lin, Y., Wai, C.M., Jean, F.M. and Brauer, R.D. (1994) Supercritical fluid extraction of thorium and uranium ions from solid and liquid materials with fluorinated β-diketones and tributyl phosphate. *Environmental Science and Technology*, **28**, 1190–3.
18. Lin, Y., Smart, N.G. and Wai, C.M. (1995) Supercritical fluid extraction of uranium and thorium from nitric acid solutions with organophosphorous reagents. *Environmental Science and Technology*, **29**, 2706–8.
19. Wang, J. and Marshall, W.D. (1994) Recovery of metals from aqueous media by extraction with supercritical carbon dioxide. *Analytical Chemistry*, **66**, 1658–63.
20. Wang, J., Marshall, W.D. and Gamble, D.S. (1996) Influence of selected natural complexants on the mobilization and purging of copper from aqueous media into supercritical carbon dioxide. *Analyst*, **121**, 289–95.
21. Ong, C.P., Ong, H.M., Li, S.F.Y. and Lee, H.K. (1990) The extraction of cholesterol from solid and liquid matrices using supercritical $CO_2$. *Journal of Microcolumn Separations*, **2**, 69–73.
22. Ong, C.P., Li, S.F.Y. and Lee, H.K. (1989) A liquid sample extraction chamber for supercritical fluid extraction. *Bulletin of Singapore National Institute of Chemistry*, **17**, 51–3.
23. Liu, H. and Wehmeyer, K.R. (1994) Supercritical fluid extraction as a sample preparation technique for the direct isolation of drugs from plasma prior to analysis. *Journal of Chromatography B*, **657**, 206–13.
24. Brewer, S.E. and Kruus, P. (1993) Direct supercritical fluid extraction from water. *Journal of Environmental Science and Health*, **B28**(6), 671–85.
25. Barnabas, I.J., Dean, J.R., Hitchen, S.M. and Owen, S.P. (1994) Supercritical fluid extraction of organochlorine pesticides from an aqueous matrix. *Journal of Chromatography A*, **665**, 307–15.
26. Kane, M., Dean, J.R., Hitchen, S.M. *et al.* (1995) Extraction of surfactants from aqueous media by supercritical fluid extraction. *Analyst*, **120**, 355–9.
27. Laintz, K.E., Wai, C.M., Yonker, C.R. and Smith, R.D. (1992) Extraction of metal ions from liquid and solid materials by supercritical carbon dioxide. *Analytical Chemistry*, **64**, 2875–8.
28. Minty, B., Ramsey, E.D. and Lewis, R. (1996) Hydrocarbons in water: analysis using on-line aqueous supercritical fluid extraction–Fourier transform infrared spectroscopy. *Analytical Communications*, **33**, 203–4.

29. Ehntholt, D.J., Eppig, C. and Thrun, K.E. (1987) Isolation of organic compounds in water at low concentrations using supercritical fluid carbon dioxide, in *Organic Pollutants in Water. Sampling Analysis and Toxicity Testing* (ed. J.M. Comstock), ACS Symposium Series 214, American Chemical Society, Washington, DC, pp. 483–501.
30. Shimoda, M., Ishikawa, H., Kawano, T. and Osajima, Y. (1994) Extraction of volatile compounds from aqueous solution using micro bubble, gaseous, supercritical and liquid carbon dioxide. *Journal of Food Science*, **59**(1), 231–3.
31. Ramsey, E.D., Minty, B. and Rees, A.T. (1996) Dynamic aqueous supercritical fluid extraction of the enzymic hydrolysis of testosterone-β-D-glucuronide. Analysis of liberated testosterone by gas chromatography–mass spectrometry. *Analytical Communications*, **34**, 307–9.
32. Ramsey, E.D., Minty, B., McCullagh, M.A. *et al.* (1997) Analysis of phenols in water at the ppb level using direct supercritical fluid extraction of aqueous samples combined on-line with supercritical fluid chromatography–mass spectrometry. *Analytical Communications*, **34**, 3–6.
33. Ramsey, E.D., Minty, B. and Rees, A.T. (1997) Drugs in water: analysis at the part-per-billion level using direct supercritical fluid extraction of aqueous samples coupled on-line with ultraviolet–visible diode-array liquid chromatography–mass spectrometry. *Analytical Communications*, **34**, 261–4.
34. Thiebaut, J.-P., Chervet, R.W., Vannoort, G.J. *et al.* (1989) Supercritical fluid extraction of aqueous samples and on-line coupling to supercritical fluid chromatography. *Journal of Chromatography*, **477**, 151–9.
35. Janda, V., Mikesova, M. and Vejrosta, J. (1996) Direct supercritical fluid extraction of water-based matrices. *Journal of Chromatography A*, **733**, 35–40.
36. Tang, P.H., Ho, J.S. and Eichelberger, J.W. (1993) Determination of organic pollutants in reagent water by liquid–solid extraction followed by supercritical fluid elution. *Journal of AOAC International*, **76**, 72–82.
37. Alzaga, R. and Bayona, J.M. (1993) Supercritical fluid extraction of tributyltin and its degradation products from seawater via liquid–solid phase extraction. *Journal of Chromatography A*, **655**, 51–6.
38. Furton, K.G. and Rein, J. (1991) Effect of microextractor cell geometry on supercritical fluid extraction recoveries and correlations with supercritical fluid chromatographic data. *Analytica Chimica Acta*, **248**, 263–70.
39. Furton, K.G. and Lin, Q. (1993) Variation in the supercritical fluid extraction of polychlorinated biphenyls as a function of sorbent type, extraction cell dimensions, and fluid flow rate. *Journal of Chromatographic Science*, **31**, 201–6.
40. Furton, K.G., Jolly, E. and Rein, J. (1993) Variables affecting the supercritical fluid extraction of analytes from octadecylsilane solid-phase sorbents. *Journal of Chromatography*, **629**, 3–9.
41. Ahmad, M., Berry, A.J. and Ramsey, E.D. (1993) Analysis of pesticide residues in vegetable oils and seed crops using combined SFE/LC, in *The 2nd European Symposium on Analytical Supercritical Chromatography and Extraction* (eds P. Sandra and K. Markides), Huethig, Heidelberg, pp. 282–3.
42. Ong, C.P., Lee, H.K. and Li, S.F.Y. (1991) Direct coupling of supercritical fluid extraction–supercritical fluid chromatography for the determination of selected polycyclic aromatic hydrocarbons in aqueous environmental samples. *Environmental Monitoring and Assessment*, **19**, 63–71.
43. Saito, M. and Yamauchi, Y. (1994) Instrumentation, in *Fractionation by Packed-column SFC and SFE* (eds M. Saito, Y. Yamauchi and T. Okuyama), VCH Publishers, New York, pp. 107–10.
44. International Sorbent Technology Ltd. (1997) Catalogue of solid phase extraction products and services, IST House, Duffryn Industrial Estate, Hengoed, Mid Glamorgan CF82 7RJ, UK.
45. Varian (1997) Sample preparation products, 24201 Frampton Avenue, Harbor City, CA 90710, USA.
46. Tang, P.T. and Ho, J.S. (1994) Liquid–solid extraction followed by supercritical fluid elution and gas chromatography of phenols from water. *Journal of High Resolution Chromatography*, **17**, 509–17.

47. Ho, J.S. and Tang, P.H. (1992) Optimization of supercritical fluid extraction of environmental pollutants from a liquid–solid extraction cartridge. *Journal of Chromatographic Science*, **30**, 344–50.
48. Ezzell, J.L. and Richter, B.E. (1992) Supercritical fluid extraction of pesticides and phthalate esters following solid phase extraction from water. *Journal of Microcolumn Separations*, **4**, 319–23.
49. Davi, L.D., Baldi, M., Penazzi, L. and Liboni, M. (1992) Evaluation of the membrane approach to solid-phase extractions of pesticide residues in drinking water. *Pesticide Science*, **35**, 63–7.
50. Sandra, P., Kot, A., Medvedovici, A. and David, F. (1995) Selected applications of the use of supercritical fluids in coupled systems. *Journal of Chromatography A*, **703**, 467–78.
51. Sandra, P., Medvedovici, A., Kot, A. *et al.* (1996) SPE–SFC–DAD: a new hyphenated system for monitoring organic micropollutants in aqueous samples. *LC–GC International*, **9**(9), 540–54.
52. Pawliszyn, J.B. and Alexandrou, N. (1989) Indirect supercritical fluid extraction of organics from water matrix samples. *Water Pollution Research Journal of Canada*, **24**, 207–14.
53. Zegers, B.N., de Geus, H.-J., Wildenburg, H.J. *et al.* (1994) Large-volume injection in packed-capillary supercritical fluid chromatography. *Journal of Chromatography A*, **677**, 141–50.
54. Bengtsson, S., Berglof, T., Granat, S. and Jonsall, G. (1994) Solid-phase extraction of pesticides from surface water using discs, bulk sorbents and supercritical fluid extraction (SFE). *Pesticide Science*, **41**, 55–60.
55. Edder, P., Veuthey, J.L., Kohler, M. *et al.* (1994) Sub-critical fluid extraction of morphine alkaloids in urine and other liquid matrices after adsorption on solid supports. *Chromatographia*, **38**, 35–42.
56. Murugaverl, B. and Voorhees, K.J. (1991) On-line supercritical fluid extraction/chromatography system for trace analysis of pesticides in soybean oil and rendered fat. *Journal of Microcolumn Separations*, **3**, 11–16.
57. Johansen, H.R., Becher, G. and Griebrokk, T. (1994) Determination of planar PCBs by combining on-line SFE–HPLC and GC–ECD or GC/MS. *Analytical Chemistry*, **66**, 4068–73.
58. Slack, G.C., McNair, H.M., Hawthorne, S.B. and Miller, D.J. (1993) Coupled solid phase extraction–supercritical fluid extraction on-line gas chromatography of explosives from water. *Journal of High Resolution Chromatography*, **16**, 473–8.
59. Johnson, W.E., Fendinger, N.J. and Plimmer, J.R. (1991) Solid-phase extraction of pesticides from water: possible interferences from dissolved organic material. *Analytical Chemistry*, **63**, 1510–13.
60. Koski, I., Jansson, B.A., Markides, K.E. and Milton, L.L. (1991) Analysis of prostaglandins in aqueous solution by supercritical fluid extraction and chromatography. *Journal of Pharmaceutical and Biomedical Analysis*, **9**(4), 281–90.
61. Yunker, M.B., McLaughlin, F.A., Macdonald, R.W. *et al.* (1989) Measurement of natural trace dissolved hydrocarbons by in situ column extraction: an intercomparison of two adsorption resins. *Analytical Chemistry*, **61**, 1333–43.
62. Croft, M.Y., Murby, J.E. and Wells, R.J. (1994) Simultaneous extraction and methylation of chlorophenoxyacetic acids from aqueous solution using supercritical carbon dioxide as a phase transfer solvent. *Analytical Chemistry*, **66**, 4459–65.
63. Scalia, S., Renda, A., Ruberto, G. *et al.* (1995) Assay of vitamin A palmitate and vitamin E acetate in cosmetic creams and lotions by supercritical fluid extraction and HPLC. *Journal of Biomedical and Pharmaceutical Analysis*, **13**(3), 273–7.
64. Paulitis, M.E., Penninger, R.D., Gray, R.D. and Davidson, K.P. (eds) (1983) *Chemical Engineering at Supercritical Fluid Conditions*, Ann Arbor Science, Ann Arbor, MI, USA.
65. Cai, Y. and Bayona, J.M. (1995) Simultaneous speciation of butyl-, phenyl-, and cyclohexyltin compounds in aqueous matrices using ethylation followed by solid-phase trace enrichment, SFE, and GC determination. *Journal of Chromatographic Science*, **33**, 89–97.
66. Burford, M.D., Hawthorne, S.B. and Miller, D.J. (1993) Evaluation of drying agents for off-line supercritical fluid extraction. *Journal of Chromatography A*, **657**, 413–27.
67. Hartonen, K. and Riekkola, M.-J. (1996) Detection of β-blockers in urine by solid-phase extraction and gas chromatography–mass spectrometry. *Journal of Chromatography B*, **676**, 45–52.

68. Edder, P., Haerdi, W., Veuthey, J.L. and Staub, C. (1992) Quantitative capillary supercritical chromatography and supercritical fluid extraction of basic drugs of abuse. *Chimia*, **46**, 141–8.
69. Barnabas, I.J., Dean, J.R., Hitchen, S.M. and Owen, S.P. (1994) Selective extraction of organochlorine and organophosphorus pesticides using a combined solid phase extraction–supercritical fluid extraction approach. *Analytica Chimica Acta*, **291**, 261–7.
70. Barnabas, I.J., Dean, J.R., Hitchen, S.M. and Owen, S.P. (1994) Selective supercritical fluid extraction of organochlorine pesticides and herbicides from aqueous samples. *Journal of Chromatographic Science*, **32**, 547–51.
71. Howard, A.L. and Taylor, L.T. (1992) Quantitative supercritical fluid extraction of sulfonyl urea herbicides from aqueous matrices via solid phase extraction disks. *Journal of Chromatographic Science*, **30**, 374–82.
72. Tena, M.T., Luque de Castro, M.D. and Valcarcel, M. (1995) Improved supercritical fluid extraction of sulphonamides. *Chromatographia*, **40**, 197–203.
73. Liu, H., Cooper, L.M., Raynie, D.E. *et al.* (1992) Combined supercritical fluid extraction/solid-phase extraction with octadecylsilane cartridges as a sample preparation technique for the ultratrace analysis of a drug metabolite in plasma. *Analytical Chemistry*, **64**, 802–6.
74. Laintz, K.E., Yu, J.J. and Wai, C.M. (1992) Separation of metal ions with sodium bis(trifluoroethyl)dithiocarbamate chelation and supercritical fluid chromatography. *Analytical Chemistry*, **64**, 311–15.
75. Laintz, K.E. and Tachikawa, E. (1994) Extraction of lanthides from acidic solution using tributyl phosphate modified supercritical carbon dioxide. *Analytical Chemistry*, **66**, 2190–3.
76. Wang, J. and Marshall, W.D. (1994) Metal speciation by supercritical fluid extraction with on-line detection by atomic absorption spectometry. *Analytical Chemistry*, **66**, 3900–7.
77. Wang, J. and Marshall, W.D. (1996) Anion mobilization from aqueous media by ion associate extraction into supercritical carbon dioxide with on-line detection by flame atomic absorption spectrometry. *Analyst*, **121**, 817–23.
78. Nakamura, K., Chi, Y.M., Yamada, Y. and Yano, T. (1986) Lipase activity and stability in supercritical carbon dioxide. *Chemical Engineering Communications*, **45**, 207–10.
79. Nakamura, K. (1990) Biochemical reactions in supercritical fluids. *TIBTECH*, **8**, 288–92.
80. Zaks, A. and Klibanov, A.M. (1984) Enzymatic catalysis in organic media at 100°C. *Science*, **224**, 1249–51.
81. Taniguchi, M., Kamihara, M. and Kobayashi, T. (1987) Effect of treatment with supercritical carbon dioxide on enzymatic activity. *Agricultural and Biological Chemistry*, **51**, 593–4.
82. Marty, A., Chulalaksananukul, W., Willemot, R.M. and Condoret, J.S. (1992) Kinetics of lipase-catalyzed esterification in supercritical $CO_2$. *Biotechnology and Bioengineering*, **39**, 273–80.
83. King, J.W., Eissler, R.L. and Friedrich, J.P. (1988) in *Supercritical Fluid Extraction and Chromatography* (eds B.A. Charpentier and M.R. Sevenants), ACS Symposium Series 366, American Chemical Society, Washington, DC.
84. Berg, B.E., Hansen, E.M., Gjorven, S. and Greibrokk, T. (1993) On-line enzymatic reaction, extraction, and chromatography of fatty acids and triglycerides with supercritical carbon dioxide. *Journal of High Resolution Chromatography*, **16**, 358–63.
85. Berg, B.E., Lund, H.S. and Greibrokk, T. (1997) Separation and quantification of components of edible fat utilizing open tubular columns in SFC. Sample introduction by direct injection and SFE coupled on-line SFC. *Chromatographia*, **44**, 399–404.
86. Messer, D.C. and Taylor, L.T. (1994) Inverse analytical supercritical fluid extraction of Zovirax ointment 5%. *Analytical Chemistry*, **66**, 1591–2.
87. Moore, W.N. and Taylor, L.T. (1994) Analytical inverse supercritical fluid extraction of polar pharmaceutical compounds from cream and ointment matrices. *Journal of Pharmaceutical and Biomedical Analysis*, **12**, 1227–32.
88. Saito, M., Hondo, T. and Yamauchi, Y. (1988) Fractionation by coupled micro-supercritical fluid extraction and supercritical fluid chromatography, in *Supercritical Fluid Chromatography* (ed. R.M. Smith), Royal Society of Chemistry, London, p. 209.

# 5 Supercritical fluid extraction coupled on-line with gas chromatography

M.D. BURFORD

This chapter is an expanded version of the chapter 'Directly coupled (on-line) SFE–GC: instrumentation and application', by M.D. Burford, K.D. Bartle and S.B. Hawthorne, published in *Advances in Chromatography, Volume 37*, eds P.R. Brown and E. Grushka, Marcel Dekker Inc., 1997.

## 5.1 Introduction

The saying 'time is money' is becoming ever more common in the laboratory environment. There has always been a need to control analytical costs in terms of solvent usage and disposal, consumable expenditure and administration overheads. However, there is currently an even greater emphasis on the time an analysis takes, and this has led to the development of automated commercial extraction methods using solvent, microwave-assisted and supercritical fluid extraction (SFE) systems. To automate the analytical process fully the extraction and analysis procedures need to be combined to produce a coupled on-line technique that minimises the sample handling and preparation steps which contribute to analyte loss, degradation and/or contamination. This direct transfer of extracted analytes into the analyte separation and detection system produces a more sensitive analysis and increases the laboratory sample throughput. For example, the extraction and chromatographic analysis of real-world samples with low (ppm–ppb) analyte concentrations can be achieved within 1 h using an on-line approach [1–3].

Conventional extraction methods such as Soxhlet and sonication techniques do not readily lend themselves as convenient means of direct sample introduction for the most popular method of analysis, namely chromatographic separation. These liquid solvent extraction methods generally lack selectivity, producing extracts that often contain species which interfere with the chromatographic analysis of the target analytes and so require a clean-up step prior to the analysis. This sample preparation step makes the coupling of the extraction and analysis procedures complicated. Furthermore, the typical injection volumes of the chromatographic techniques such as capillary supercritical fluid chromatography (cSFC, $<0.1\ \mu l$), capillary gas chromatography (GC, *c.* $1\ \mu l$), or high-performance liquid chromatography (HPLC, *c.* 10 μl) are incompatible with the large solvent volumes used in the extraction process. Even the relatively large injection volumes (e.g. 100 μl), which are now possible with capillary GC [4–6], still only represent a small fraction of the original solvent extract. So an additional preconcentration step is required prior to the chromatographic analysis and this can result in the co-evaporation of the more volatile compounds [7].

The volume of organic solvent is a particular problem for trace analysis where the concentration of the analytes in the extract has to be significantly enhanced, since the solvent volume cannot be accurately reduced to a volume of less than *c.* 100 μl. Therefore to obtain adequate sensitivity in the analysis of analytes in the low ppm–ppb concentration range the sample size (and corresponding collection time for samples such as air particulates) has to be increased. However, commercial on-line extraction/analysis systems such as head-space–GC, purge and trap–GC, thermal desorption–GC, and pyrolysis–GC are suitable for trace analysis, since they quantitatively transfer the entire extract to the analytical system without introducing large volumes of liquid into the chromatographic column. While these on-line techniques are valuable extraction methods capable of recovering volatile analytes from a wide range of samples and are relatively easy procedures to combine, since the extraction fluid is in the same phase as the chromatographic mobile phase (i.e. a gas), they are limited by the requirement of high analyte volatility. For example, head-space and purge and trap capillary GC techniques are routinely used to analyse petroleum-contaminated samples containing very volatile gasoline range organics (e.g. $C_4$–$C_{12}$ boiling point range). However, when the sample contains only moderately volatile kerosene and diesel fuel organics (e.g. $C_9$–$C_{25}$ boiling point range) the fuels are poorly detected [8, 9]. Thus for the moderate to low volatility analytes conventional liquid solvent extraction methods are required; however, as stated above these methods often yield low recoveries of volatile analytes and are not easily combined with chromatographic systems.

Ideally, what is required is an extraction technique which combines the capability of a liquid solvent extraction method (e.g. Soxhlet) to recover quantitatively a broad range of analytes, with quantitative gas phase introduction of the extracted analytes into the chromatographic system. Such a technique, that can successfully bridge the gap between the extraction and analysis steps, is supercritical fluid extraction. Supercritical fluids have solvent strengths approaching that of a liquid so that a wide range of analytes can be extracted and yet at ambient conditions the fluid becomes a gas and is therefore amenable to a range of chromatographic techniques. Under SFE conditions (e.g. carbon dioxide at 400 atm and 60°C) the supercritical fluid has an order of magnitude lower viscosity ($10^{-4}$ versus $10^{-3}$ N s/m$^2$) and higher solute diffusivity ($10^{-4}$ versus $10^{-5}$ cm$^2$/s) than liquid solvents, so that quantitative analyte recoveries can be achieved within 30 min compared with conventional solvent extraction procedures which may require several hours or days. This rapid mass transfer of analytes into the supercritical fluid enables the extract to be transferred efficiently to the chromatographic column, thus minimising the potential for band broadening and poor chromatographic separation.

The solvent strength of the supercritical fluid can also be varied with temperature, pressure and/or polarity (e.g. by the addition of a modifier or more polar supercritical fluid) so that a cleaner more selective extract can be obtained with fewer chromatographically interfering components. Using these various extraction conditions supercritical fluids have been used to extract quantitatively a broad range of analytes, including polycyclic aromatic hydrocarbons (PAHs) [10–12], polychlorinated biphenyls (PCBs) [2], pesticides [13, 14], hydrocarbons [15, 16], ionic surfactants [17], metals [18, 19], polymer additives [20], essential oils [21, 22], fatty acids [23, 24], vitamins [25], soft resins [26] and caffeine [27]. Very pure liquified gases are commercially available as SFE/SFC grade solvents [99.999% pure (Air Products plc, Weston Road, Crewe CW1 1DF, UK)] and this has enabled fluids such as carbon dioxide to be directly linked to several detection systems such as flame ionisation (FID) [28], electron capture (ECD) [29], ultraviolet/visible (UV/VIS) absorption [30], mass spectrometry (MS) [31], Fourier transform infra-red spectroscopy (FTIR) [32] and atomic emission spectrometry (AES) [33] detectors. The majority of these detectors have a very low response to carbon dioxide and any interference is usually a result of contamination problems in the pump or the associated plumbing of the system. The high-grade fluids can be directly depressurised into a chromatographic column without any detrimental effects. For example SFE/SFC grade fluids do not strip the column stationary phase, cause column contamination problems or decrease the chromatographic resolution. The latter two problems are most often observed when low-grade fluids containing traces of hydrocarbons and/or water are used [29].

SFE can be coupled to a wide range of chromatographic techniques including GC [1–3], SFC [34–37], HPLC [38, 39], thin-layer chromatography (TLC) [40] and gel permeation chromatography (GPC) [41]. The choice of chromatographic system depends on the nature of the sample under investigation, but generally the analytes extracted using supercritical carbon dioxide are non-polar to moderately polar in nature, with sufficient volatility to be analysed by GC. Consequently combined SFE–GC is the most popular on-line technique. This chapter will discuss the practical requirements needed to assemble and operate an SFE–GC system, with particular attention to the modifications required to obtain both qualitative and quantitative analyses. It will be demonstrated that the coupling of the extraction and analysis procedures requires a detailed understanding both of the extraction and of the chromatographic parameters.

## 5.2 Techniques for coupling supercritical fluid extraction with gas chromatography

To couple SFE successfully with GC a series of procedures or steps must be undertaken to enable the analytes of interest to be removed quantitatively from the sample matrix and efficiently transferred and focused in the chromatographic column. Figure 5.1 details the steps involved. In the initial step the extraction fluid is pumped into a heated extraction vessel (alternatively termed the extraction cell) containing the sample of interest. Conditions inside the extraction vessel are above the critical temperature and pressure of the extraction fluid (i.e. for $CO_2$, above 31°C and 73 atm) so that the fluid is supercritical. As the supercritical fluid comes into contact with the sample matrix and penetrates the material, the soluble analytes are partitioned into the bulk of the fluid which is then swept out of the extraction vessel and through a flow restrictor. The restrictor is used to maintain the pressure in the extraction system. Once the supercritical fluid has passed through the SFE system it is depressurised at the head of the chromatographic column, where the analytes are deposited in the stationary phase and the gaseous extraction fluid is vented through the injection port. At the end of the extraction step the flow of supercritical fluid through the

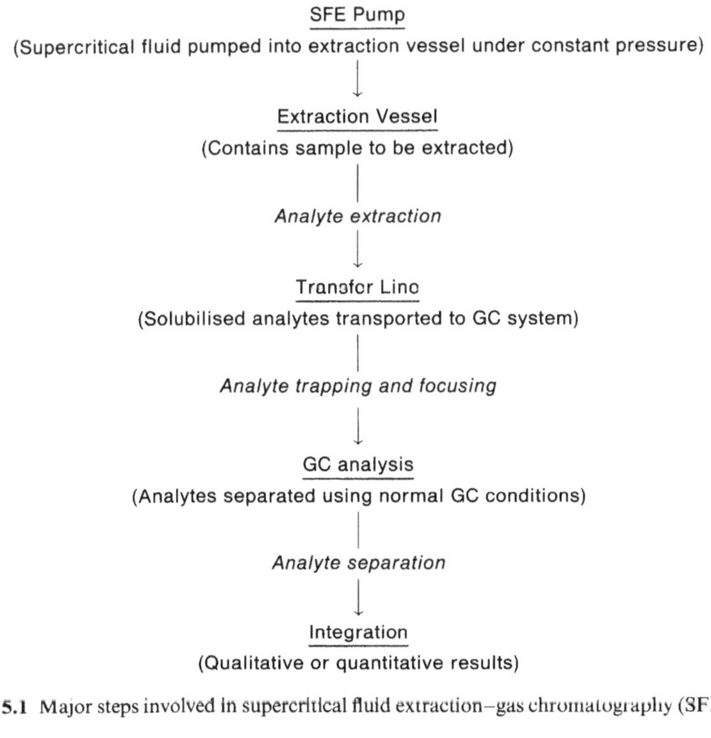


**Figure 5.1** Major steps involved in supercritical fluid extraction–gas chromatography (SFE–GC).

sample is halted and the gaseous extraction fluid is allowed to dissipate through the column prior to conventional GC analysis being performed.

To obtain quantitative SFE recoveries, the extraction conditions are usually optimised in terms of the extraction temperature, pressure and supercritical fluid (e.g. non-polar $CO_2$, polar freons or organically modified $CO_2$/methanol) [42–47]. However, to achieve optimum recoveries the extraction and subsequent focusing of the analytes on the column requires *c.* 30–60 min, compared with the several seconds needed to introduce a liquid solvent extract into the column using a conventional manual injection technique. Furthermore, the depressurisation of the supercritical fluid generates a high gas flow rate at the head of the GC column which can result in poor focusing of the analytes and severe chromatographic band broadening. For example, an SFE–GC analysis which involves the use of 1 ml/min liquid $CO_2$ results in the production of *c.* 550 ml/min $CO_2$ gas, which then has to be separated and removed from the analytes at the SFE–GC interface. Thus the fundamental problem with coupling SFE to GC is how to collect and focus the extracted analytes efficiently over a long period of time, taking into account the high gas flow rates generated from the depressurisation of the supercritical fluid.

## 5.3 External trapping of analytes

Several methods have been employed to overcome the problems encountered in retaining the extracted analyte for chromatographic analysis. These various approaches can be loosely divided into two main groups, as indicated in Table 5.1 [2, 3, 48–50, 55–61] and shown schematically in Figure 5.2. The extracted analytes obtained from the SFE system can either be collected on an analyte accumulation device external to the GC using an 'analyte trap' or be collected internally in the GC using the GC column itself. With an external accumulation device the supercritical fluid extract is depressurised inside a cold trap (e.g. cryogenically cooled stainless steel or silica tubing) [48, 49] or a sorbent trap (e.g. Tenax) [50], with analytes selectively retained in the trap while the gaseous extraction fluid is vented out of the purge line at the base of the device (Figure 5.2). In order to prevent plugging of the restrictor during the extraction (as depressurisation of the supercritical fluid within the restrictor typically causes the analytes to precipitate and accumulate) the restrictor is either heated with the tip situated just inside the cold trap [48], or the restrictor and the trap are heated, as is possible when using a sorbent resin [50]. For example, a Tenax-GC trap has been used at 50°C to simultaneously maintain the extraction flow rate and analyte collection during the SFE–GC analysis of pesticides [50]. On completion of the extraction, the analytes are thermally desorbed from the trap and transferred to the GC column by closing the trap purge vent, using a switching

**Table 5.1** Supercritical fluid extraction–gas chromatography (SFE–GC): analyte collection methods

| Location of analyte collection relative to GC column | Collection method | Path taken by supercritical fluid (SF) | Heated depressurisation zone | Reference |
|---|---|---|---|---|
| External | Cooled sample loop on switching valve | SF vented out of sample loop | Yes | 49 |
| External | Cold trap | SF vented out of purge vent on trap | Yes | 48 |
| External | Sorbent trap (Tenax-GC) | SF vented out of purge vent on trap | Yes | 50 |
| Internal | Direct injection on head of GC column | SF vented through GC column | Yes<br>No | 55, 56<br>57 |
| Internal | Cooled on column injection port | SF vented through the GC column and injection port | No | 2, 3 |
| Internal | Split/splitless injection port | SF vented through GC column and split vent | Yes | 58–60 |
| Internal or external | Programmed temperature vaporiser injection port | SF vented through GC column and split vent | No | 61 |

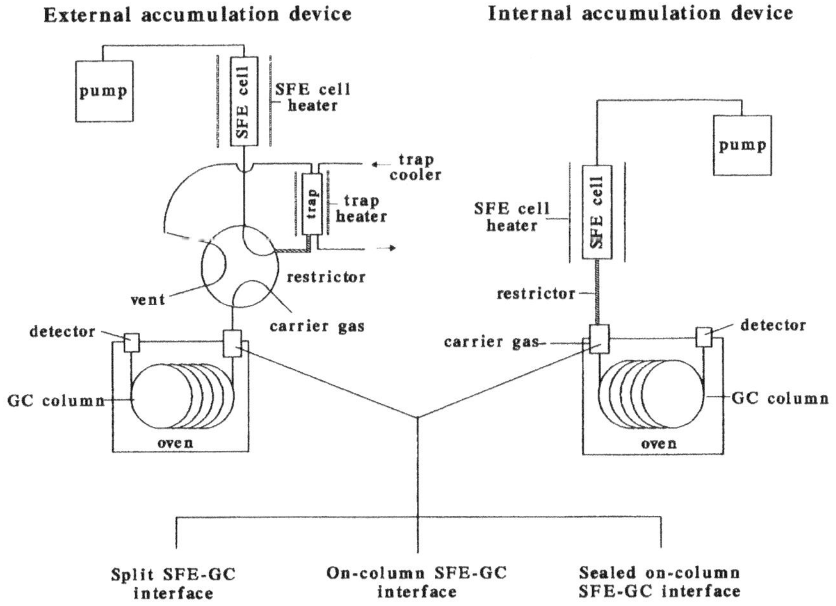


**Figure 5.2** Schematic of internal and external accumulation devices used in supercritical fluid extraction–gas chromatography (SFE–GC).

valve to divert the carrier gas through the rapidly heated trap into the column. Alternatively, the analytes can be recovered by extracting the trap with a supercritical fluid so that, in effect, an SFE–SFE–GC analysis is undertaken [50].

The advantage of collecting the analyte in an external trap is that during the extraction step the depressurised gas is not vented through the GC column so that potentially larger supercritical fluid extraction volumes can be used than are possible when using a direct SFE–GC interface (however, the successful use of external trapping will partially depend on the volatility of the analytes involved). For example, the sorbent Tenax-GC can successfully retain extracted analytes over a relatively long period (e.g. 30 min) during the SFE step, after which the analytes can be quickly transferred to the GC (e.g. in less than 10 min) by extracting the resin with a supercritical fluid [50]. Thus the extraction fluid volume and the transfer time required to transport the analytes into the GC column can be significantly reduced and this enables the analytes to be more efficiently focused in the column. Alternatively, if the analytes are recovered from the trap using the GC carrier gas, so that none of the supercritical fluid is introduced into the column during SFE–GC, the potential chromatographic and detection problems from the supercritical fluid can be avoided. This is particularly important when an organically modified supercritical fluid is used, because if the extraction fluid is introduced into the column the modifier will generate a 'solvent' peak in the resulting chromatogram which could co-elute with the target analytes. The use of an external trap such as Tenax-GC enables modified supercritical fluids such as SF-$CO_2$/methanol to be used without significant detection of the modifier [50], as the sorbent can selectively retain the semivolatile hydrocarbons during the extraction (as they have very large retention volumes [51]), but does not retain polar compounds such as methanol which are purged from the trap and diverted away from the GC column.

The disadvantage of external analyte traps is that they can have a limited collection efficiency compared with the wide range of analytes that are retained when depressurising the extract directly into the chromatographic column. For example, volatile and semivolatile analytes which can be retained in a fused-silica tube cold trap operated at conventional 10–60 ml/min gas flow rates [52] are poorly retained when operated under SFE–GC conditions where gas flow rates of 300–600 ml/min are routinely used [48]. The cold trap also has a restricted SFE–GC working temperature range (typically −10°C to −50°C), since if too high a temperature is chosen volatile analytes are lost, and if too low a temperature is used carbon dioxide or extracted water will freeze (at below −78°C and 0°C, repectively) and plug the trap.

These problems are not as severe for sorbent resins, which can retain a wider range of semivolatile analytes at ambient conditions [51]. Water has little impact on the affinity of the resin for organic compounds [53], but the sorbent will not retain the analytes indefinitely during an extraction and this may lead to problems when extracting difficult samples which require long extraction times. The sorbent resins are also unable to retain hazardous organic solvents efficiently such as alcohols, ketones, freons and chlorinated solvents which can be successfully analysed when depressurised directly into the GC column [54]. External traps are therefore easier to optimise in terms of the SFE conditions, such as extraction fluid volume and use of a modifier, but are somewhat limited by the range of possible analytes.

## 5.4 Internal accumulation of analytes

A simpler, more popular, method of coupling SFE with GC systems which avoids switching valves or external traps involves collecting the extract within the chromatographic system. By using the chromatographic column as the accumulation device improved quantitative and reproducible recoveries are obtained with little or no modifications required to the GC system. The same GC injection port can be used for conventional liquid solvent injection or SFE–GC without conversion. A schematic of the process is shown in Figure 5.2. This 'internal accumulation method' which directly couples the SFE effluent to the GC column can be undertaken using 'on-column' SFE–GC, where either the SFE effluent is depressurised directly into the chromatographic column, so that all the extracted analytes are deposited in the column stationary phase [2, 50, 62], or the analytes are collected in a length of uncoated fused-silica tubing which serves as a retention gap at the head of the column [55, 56]. Alternatively, 'split' SFE–GC can be performed, where the SFE effluent is depressurised in a conventional split/splitless injection port operated in the split mode. With this approach only a fraction of the extract is deposited in the column stationary phase (or retention gap), with the majority of the extract being vented through the injection port split vent [54, 58, 60, 63]. Table 5.1 summarises these two approaches, and Figure 5.3 illustrates common SFE–GC interface configurations.

Both the on-column and spilt SFE–GC approaches usually require the GC column to be cooled during the extraction so that the analytes (particularly volatile analytes such as *n*-pentane) can be as efficiently retained and focused in the column stationary phase [54]. Alternatively, the injection port can be cooled using a programmed temperature vaporiser (PTV) injection system so that the analytes are retained on the cold injection port liner during the extraction [61]. The chromatographic peak shapes and peak area reproducibilities obtained from these SFE–GC techniques are comparable to those generated from the conventional on-column or split injection of a liquid solvent extract. Relative standard deviations of replicate SFE–GC analyses have been reported to be in the region of 2%–10% [3, 54, 60]. The appropriate use of the SFE–GC techniques to transfer part (split) or all (on-column)

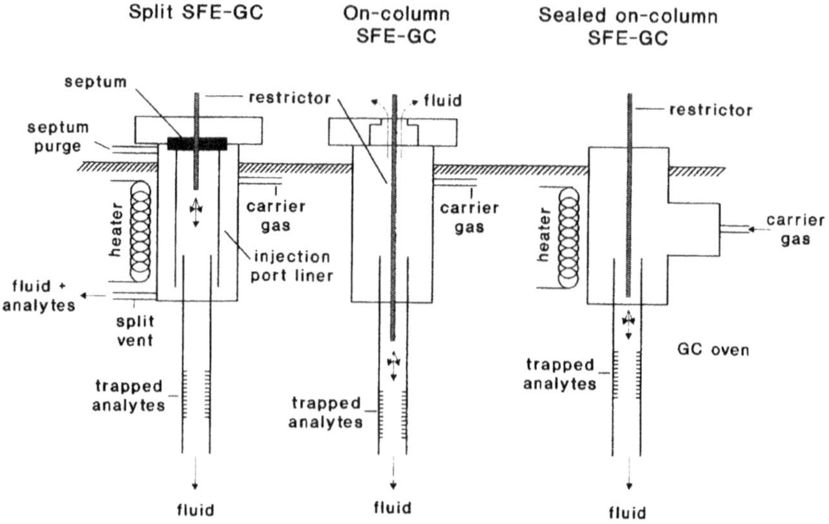


**Figure 5.3** Common supercritical fluid extraction–gas chromatography (SFE–GC) interfaces.

of the supercritical extract to the GC system will depend on the nature of the sample (e.g. water content), analyte concentration (e.g. ppb or ppm) and extraction conditions (e.g. flow rate, use of modifier, etc.). When only a small quantity of sample is available, such as a forensic or air particulate sample, the method of choice is on-column SFE–GC since all of the extracted analytes can be transferred to the GC column. In this way the maximum sensitivity and absolute quantitation of the extract can be obtained, with low parts-per-billion (ng/g) detection limits for milligram samples [2]. However, by depositing all the extract in the column stationary phase any extracted involatile components are also transferred to the GC system. Accumulation of such compounds can lead to poor chromatographic peak shapes and reduce the working life of the column. The use of a retention gap at the head of the column may reduce such column contamination problems [55, 56], but when a sample contains a high percentage of extractable components not amenable to GC the most practical approach is to use the alternative split SFE–GC technique.

In split SFE–GC only a fraction of the extract is transferred into the GC column and the non-volatile matrix components are retained in the injection port liner rather than the column stationary phase, so a 'cleaner' chromatogram is obtained. A potential disadvantage of split SFE–GC is that the extract is divided in the injection port; a fraction of the extract enters the column while the remainder is flushed out the split vent, so that this technique has a factor of 10–100 (depending on the split ratio) higher detection limit than on-column SFE–GC, though in practice this is rarely a drawback.

The composition of the sample matrix may also influence the choice of coupling technique. A cool on-column SFE–GC system is ideal for samples which contain thermally labile components since the restrictor and injection port are not heated [2, 3]. However, samples which contain a high percentage of extractable components, for example elemental sulphur [64] or lipids and waxes [65], are not suitable for this technique, since these analytes have a tendency to precipitate and plug in the unheated restrictor. In these instances split SFE–GC [58–60] or heated on-column SFE–GC [55, 56] are the more appropriate methods of analysis as the injection port and/or restrictor can be heated to *c.* 250°C–300°C and at these temperatures restrictor plugging is completely eliminated.

The presence of water in the sample may also create problems in both split and on-column SFE–GC. Most environmental samples contain at least trace quantities of water which when extracted can freeze and plug in the cooled GC column, since during the SFE step the column is usually kept below 0°C to trap the analytes efficiently. Heating the restrictor will ensure that a constant extraction flow rate is maintained with a wet sample matrix. However, this will not stop the extracted water freezing and plugging the top of the column. If the column becomes plugged during the extraction the further transfer of analytes from the extraction vessel to the GC column is prevented and only qualitative rather than quantitative results are obtained. To ensure that the extracted analytes are continuously transferred to the cooled GC column throughout the entire extraction, the sample can be placed on a bed of drying agent so that during the SFE step the water is selectively retained in the extraction vessel and only the analytes are deposited in the column [54]. Alternatively, the column can be maintained above 0°C [60] but here the chromatographic trapping and focusing of the volatile analytes (e.g. $C_4$ to $C_8$ *n*-alkanes) is greatly reduced [54]. Problems have also been encountered when analysing animal tissue such as crab meat containing a high percentage of extractable fat which, if deposited on the column, can cause gross contamination of the GC system. By using a basic alumina sorbent the fat can be selectively retained during the SFE step so that only the analytes of interest, namely PCBs, are deposited in the column stationary phase [65].

The maximum extraction flow rate that can be accommodated by the GC system will depend on whether all (on-column) or part (split) of the extract is transferred into the chromatographic column. If all the gaseous extraction fluid is vented through the column (e.g. in sealed on-column SFE–GC which involves a gas-tight seal between the restrictor and column; see Figure 5.3), then the maximum practical flow rate will depend on the inner diameter (i.d.) of the GC column. For example, when a narrow bore (0.25 mm i.d.) column is used, low gaseous extraction fluid flow rates of *c.* 5–10 ml/min are employed, as the small chromatographic void volume of the system limits the amount of fluid which can be vented through the column [56]. If higher flow rates are attempted with the narrow bore column a back pressure builds up in the GC system and will lead to poor chromatographic peak shapes and possible failure of the GC fittings, which are not designed to withstand high pressures. By increasing the i.d. of the column to 0.32 mm, higher gaseous flow rates of 30–80 ml/min can be used to focus the analytes efficiently [55]. With a conventional on-column injection port (Figure 5.3) gaseous extraction fluid flow rates of up to *c.* 350 ml/min can be used because there is no gas-tight seal between the restrictor and the injection port, so the effluent can vent back through the injection port and to a lesser extent through the column [3]. However, the highest flow rates are achieved with a split SFE–GC interface as the majority of the depressurised extraction fluid is vented through the split vent (split ratio typically 100:1) so gaseous effluent flows of *c.* 550 ml/min can be routinely used [54, 60].

The flow rate limitations of the various SFE–GC techniques also limit the maximum possible sample size, since the extraction flow rate can determine the time required to obtain quantitative recoveries. The optimum extraction flow rate will depend on the kinetics and mechanism of the extraction, the void volume of the SFE system and the time allocated to the SFE step (typically 30 min). For many environmental samples the fluid flow rate has a minimal effect on the analyte extraction rate, as the rate limiting step in SFE is the slow kinetics of the partitioning process between the sample matrix and the extraction fluid, and the diffusion of the analytes to the sample matrix surface [66, 67] (a detailed account of these processes can be found in Chapter 1). However, the fluid flow rate does affect the time taken to expel the void volume from the extraction system and this can have a significant effect on the efficiency of the SFE–GC system. For example, typical void volumes in a packed extraction vessel are *c.* 20%–80%, so several vessel volumes of supercritical fluid would be required for a quantitative extraction. If the extraction is to be completed in 30 min, then a sealed on-column supercritical fluid flow rate produced from the delivery of 60 μl/min liquid $CO_2$ (which results in the production of *c.* 33 ml/min depressurised gas) involves the use of an SFE volume generated from only 1.8 ml liquid $CO_2$, for extraction and transfer of the analytes into the GC system. This small extraction volume limits the sample size of sealed on-column SFE–GC to *c.* 20–50 mg, and requires the use of micro-extraction vessels with small internal volumes (typically less than 0.5 ml). Extraction vessels with a larger volume can be used by periodically sampling the vessel with a rotary valve so that a 'heart cut' of the extract is transferred into the column [68]. This approach can also be used to obtain class selective extracts such as the selective determination of alkane, alkene and aromatic components in gasoline [69], although quantitation is difficult.

Conventional on-column injection ports enable higher supercritical fluid flow rates, equivalent to the delivery of *c.* 0.6 ml/min liquid $CO_2$ (resulting in the production of *c.* 330 ml/min depressurised gas) to be used so that larger sample sizes of up to 1 g can be investigated. The most versatile system is the split SFE–GC technique which can accommodate supercritical fluid flow rates produced from the delivery of up to *c.* 2 ml/min liquid $CO_2$ (resulting in the production of *c.* 1100 ml/min depressurised gas) [2, 60], enabling the analysis of sample sizes up to 15 g [60]. However, the analyte concentration in the sample matrix must also be taken into account, for if the sample is highly contaminated then only a small sample size can be used. For example, only 50 mg of a gasoline-contaminated charcoal filter containing 17% m/m extractable hydrocarbons could be analysed to avoid gross overloading of the column stationary phase, even though a high (100:1) split ratio was used during split SFE–GC analysis [70]. All the coupled SFE–GC techniques therefore use much smaller sample sizes (typically 0.1–1 g) than can be accommodated by conventional solvent extraction methods (typically 25–100 g). Consequently, for the on-line methods sample homogeneity is of even greater importance (section 3.1.1).

## 5.5 Construction of supercritical fluid extraction–gas chromatography instrumentation

When constructing an SFE–GC system it is best to keep the apparatus as simple as possible so that contamination and void volume problems can be minimised. The main components of the system are illustrated in Figure 5.4. It consists of: a reciprocating or syringe pump capable of operating at pressures from 100 to 400 bar and delivering a constant flow rate from μl/min to ml/min; packless on/off valves to avoid contamination by lubricants and extractable species from components such as O-rings; high-purity supercritical fluids (e.g. SFE/SFC grade $CO_2$), since previous studies have shown that impurities in the extraction fluid can cause artifact peaks in the SFE–GC generated chromatograms [48, 57, 71]; transfer tubing constructed from narrow bore 316 grade stainless steel [0.0625″ outer diameter (o.d.) by 0.01″ inner diameter (i.d.)] fitted with zero dead volume metal-to-metal connections to reduce the void volume in the system (Teflon® tape should be avoided because of potential contamination problems); a small 316 grade stainless steel extraction vessel (e.g. 0.1–1.0 ml) designed to withstand working pressures of typically 400 atm; fine frits (e.g. 2 μm mesh size) placed at either end of the extraction vessel to retain the sample inside the vessel during the extraction; silanised glass beads (100 μm o.d.) added to either end of the sample to reduce vessel dead volume and prevent outlet frit blockage (beads can also be mixed with the sample matrix to avoid sample compaction and the formation of an impervious plug [72]); finger-tight fittings to connect the vessel to the extraction system, as these fittings are much less prone to leaking despite extensive usage; a thermostatically

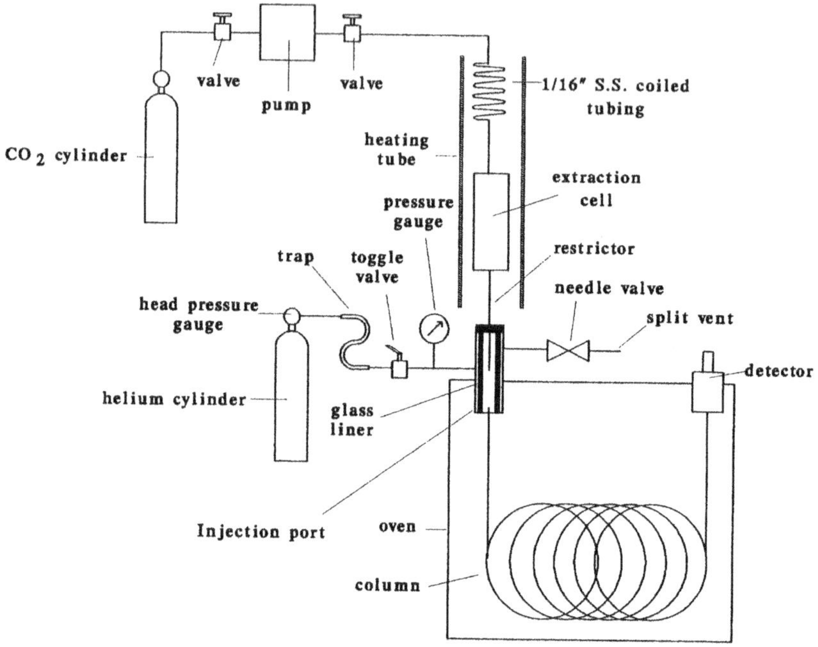


**Figure 5.4** Schematic of general supercritical fluid extraction–gas chromatography (SFE–GC) equipment.

controlled oven or tube heater (constructed from a steel pipe, a thermocouple, heating tape and a temperature control unit) to heat the extraction vessel; and a narrow-bore fused-silica restrictor (e.g. 15–30 μm i.d.) [2, 3] or crimped stainless tube (e.g. 0.01″ i.d.) [61, 69] to control the flow rate and, indirectly, the pressure inside the extraction system.

The internal diameter of the fused-silica restrictor controls the rate at which the supercritical fluid can flow through the extraction vessel: the larger the diameter the higher the flow. For example, typical extraction conditions involving the use of SF-$CO_2$ at 400°C and 40°C and a fused-silica restrictor (10 cm × 20 μm i.d.), result in a flow rate of *c.* 0.4 ml/min (measured as liquid $CO_2$ at the pump), whereas a restrictor with a 30 μm i.d. provides a flow rate of *c.* 1 ml/min of liquid $CO_2$. In general, larger internal diameter restrictors yield an initially faster extraction as the void volume of the extraction system is more rapidly transferred to the GC system, although the amount of analyte extracted decreases with time as other rate-limiting steps such as diffusion and analyte–matrix interactions become more significant (Chapter 1). Consequently, there is a compromise between optimising the extraction flow rate to extract and transfer the analytes to the column efficiently, and the flow rate which can be accommodated in the GC system to provide chromatograms with well resolved peaks.

Once assembled, the SFE system can quickly be connected to the GC apparatus. The easiest and most reliable approach is to use split SFE–GC as only a few minor modifications (depending on the gas chromatograph) need to be made to the GC instrumentation. The principal steps involve the insertion of the extraction vessel restrictor into the split injection port so that the end of the restrictor is about 3 cm above the GC column inlet. In contrast, if a commercial SFE system is used, a heated transfer line is usually required to achieve the coupling because of the greater separation of the SFE and GC systems. However, if a home-made system is used, the extraction vessel and heater can be mounted directly above the injection port (as shown in Figure 5.4) so that over 90% of the restrictor can be inserted into the heated injection port liner, and no additional transfer line is required. This SFE–GC coupling can be laboriously achieved by loosening the injection port septum nut and carefully pushing the restrictor through the septum and then retightening the septum nut. Unfortunately, to undertake this operation the injection port must be cooled and reheated each time a restrictor is changed so a delay of up to 60 min can be involved. A more convenient alternative is to use a septumless injector (e.g. Scientific Glass Engineering model SLI-M) which can be installed on a conventional injection port and can be used for both GC and SFE–GC analysis. With this system the restrictor is passed through a cylindrical seal which can be opened and closed by simply pressing a lever, so that the coupling of the SFE and GC systems can be achieved within seconds.

With the restrictor installed inside the injection port liner the split ratio is adjusted using the needle valve on the split vent (Figure 5.4). It is important to note that the restrictor must form a gas-tight seal with the injection port in order to maintain a constant split ratio throughout the extraction. Problems can arise if a mass flow controller (controlling the carrier gas flow rate) and a back pressure regulator (controlling the column head pressure) are part of the GC instrumentation [54]. With this type of GC system the split ratio is controlled by the amount of carrier gas supplied by the mass flow controller, and the back pressure regulator allows the split flow to exit while still maintaining a constant column head pressure. Under SFE conditions the high flow rates of expanded gas in the injection port can cause the head-pressure regulator to open or close during the SFE step (in an attempt to maintain its set point pressure), thus varying the split ratio. Although this does not affect the qualitative peak shapes, a change in split ratio during the SFE step will yield poor quantitative results [54]. In an attempt to maintain the split ratio the mass flow controller and back pressure regulator have been isolated and replaced with a simple needle valve installed on the split vent and a manual head pressure regulator situated on the carrier gas inlet line (Figure 5.4) [54]. This simplified system, used by many GC manufacturers, is easy to maintain and clean, and results in good peak area reproducibility.

One additional modification required regardless of the GC system used is a shut off or toggle valve located prior to the injection port (Figure 5.4). This valve is closed during the SFE step to prevent the SFE effluent back flushing into the carrier gas line and contaminating the carrier gas. This can be a problem in SFE–GC as the internal volume of the vaporising chamber (i.e. the glass liner) inside the injection port is relatively small (*c.* 0.5 ml) compared with the gaseous flow of the supercritical fluid (*c.* 275 ml/min of gaseous $CO_2$ at a typical $CO_2$ liquid flow rate of 0.5 ml/min). The pressure inside the injection port during the extraction can therefore be higher than the carrier gas head pressure. A glass liner with a larger void volume can alleviate the problem but a shut off valve is still recommended.

With the apparatus assembled the following SFE–GC procedure is recommended:

1. cool the GC oven to the appropriate cryogenic trapping temperature (e.g. 25°C to −50°C) using liquid nitrogen, liquid carbon dioxide or ambient air;
2. place the extraction vessel into the heater above the injection port; shut off the GC carrier gas and insert the restrictor into the injection port; the position of the restrictor should correspond to the syringe needle length appropriate for the injection port;
3. pressurise the heated extraction vessel and extract the sample for 10–30 min; a fraction of the gaseous extract entering the injection port goes into the GC column for focusing in the stationary phase; the remainder of the extract is flushed out the split vent;
4. at the end of the extraction the restrictor is withdrawn from the injection port and the extraction fluid allowed to dissipate from the GC system (this usually takes *c.* 1 min);
5. switch on the carrier gas and commence the GC analysis, rapidly heating the column to the initial starting temperature, after which the GC analysis is conducted in the usual manner.

## 5.6 Optimisation of supercritical fluid extraction–gas chromatography

### *5.6.1 Extraction flow rate*

The ability of SFE–GC to yield good peak shapes depends on the capability of the coupling method to collect and focus the analytes efficiently during the extraction step. This is dependent on several factors: the extraction flow rate, the column trapping temperature and the type of column used. Previous studies have shown that successful SFE–GC is dependent on the extraction flow rate and extraction time [2, 3, 60]. High extraction flow rates may be desirable as the potential sample size can be increased (e.g. to several grams) and the extraction time decreased [2, 60], though additional factors such as the kinetics and mechanism of the extraction may also effect the extraction rate [46]. The impact of the extraction flow rate on the chromatography of the analysis can be investigated by comparing conventional split injections of a text mix, containing analytes with a broad range of volatilities [e.g. $C_4$ to $C_{20}$ *n*-alkanes and benzene, toluene, ethylbenzene and xylene (BTEX)] with the split SFE–GC analysis of the same test mix spiked onto an SFE compatible matrix (e.g. a sorbent resin such as Tenax-TA). Figure 5.5 shows the effect of the SFE flow rate on the peak shapes obtained by split SFE–GC using a thick film column (30 m × 0.32 mm i.d., 5 μm film thickness), cryogenically cooled to −50°C for analyte trapping [54]. At extraction flow rates of 0.6 ml/min (measured as liquid $CO_2$ at the pump, which corresponds to a gaseous flow rate of *c.* 330 ml/min) good chromatographic peak shapes are obtained which are indistinguishable from those in a conventional split GC injection, indicating that SFE–GC does not introduce any significant splitter discrimination. However, differences in the absolute peak intensities do occur, with peak areas decreasing with increasing extraction flow rate (Figure 5.5). This is related to changes in the split ratio discussed later.

The symmetry or shape of the peaks obtained by SFE–GC at moderate extraction flow rates (using 0.2–0.6 ml/min liquid $CO_2$) compare favourably with those obtained by split injection. As shown in Table 5.2, the peak width at half-height for both the very volatile (*n*-butane) and semivolatile (*n*-eicosane) analytes analysed by SFE–GC are essentially identical to those obtained in a conventional solvent injection of the test mix. The slight band broadening of the early eluting peaks is a common occurrence and is to be expected with the coupled technique as these are from the most volatile analytes, which are the hardest to focus chromatographically. The reproducibility of the SFE–GC and GC techniques are comparable, both methods having low relative standard deviations (RSDs), being less than 5% (Table 5.2). This is a particularly striking value when it is considered that the RSDs of the SFE–GC method include all the possible errors associated with the spiking procedure, the extraction of the analytes, the chromatographic focusing of the extract, the GC separation and the chromatographic peak integration. Conversely, conventional solvent injection involves only the latter two variables and variations in the injection volume.

At high extraction flow rates (e.g. 1.0 ml/min liquid $CO_2$ equivalent to *c.* 550 ml/min gaseous $CO_2$) SFE–GC can produce poor chromatographic peak shapes, with broadening of early eluting peaks and fronting of later eluting peaks (Figure 5.5). One might initially conclude that the poor focusing of the analytes is due to the high gaseous extraction fluid flow entering the injection port during the SFE step, but this is not the case as higher injection port flow rates can be tolerated under normal GC conditions (Table 5.3). It is more likely that poor peak shapes are associated with the high volumetric

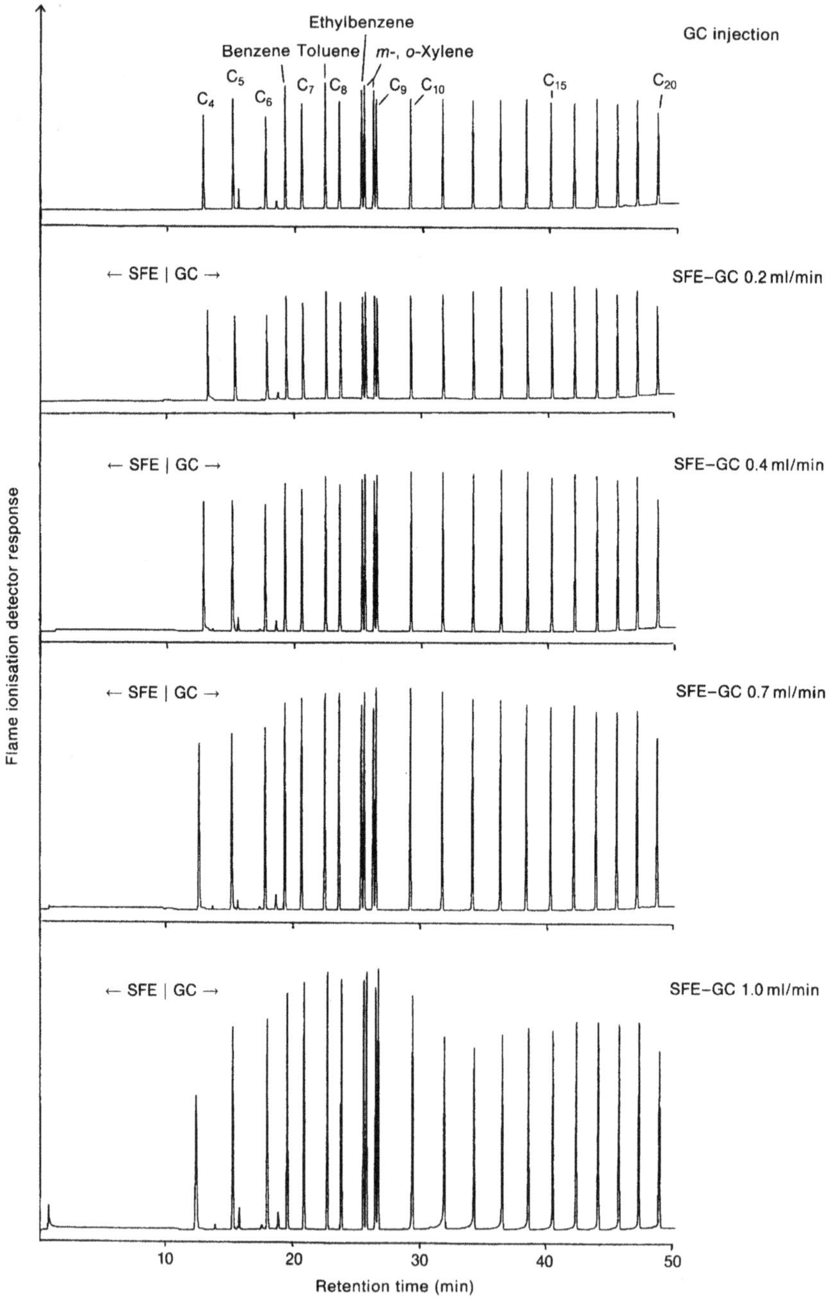

Ethylbenzene
Benzene
Toluene
m-, o-Xylene
GC injection
$C_4$
$C_5$
$C_6$
$C_7$
$C_8$
$C_9$
$C_{10}$
$C_{15}$
$C_{20}$
← SFE | GC →
SFE–GC 0.2 ml/min
SFE–GC 0.4 ml/min
SFE–GC 0.7 ml/min
SFE–GC 1.0 ml/min
Flame ionisation detector response
10
20
30
40
50
Retention time (min)

**Table 5.2** Comparison of the half-height peak widths for a volatile analyte test mix obtained by split supercritical fluid extraction–gas chromatography (SFE–GC) and conventional GC solvent injection

| | Peak width at half-height (s)[a] | |
|---|---|---|
| Analyte | Split injection | SFE–GC |
| *n*-Butane ($C_4$) | 0.96 (3) | 1.18 (3) |
| *n*-Hexane ($C_6$) | 1.03 (2) | 1.09 (4) |
| Benzene | 0.97 (2) | 1.06 (3) |
| Toluene | 0.97 (2) | 1.04 (3) |
| Ethylbenzene | 0.97 (2) | 1.01 (2) |
| *m*-Xylene | 0.97 (2) | 0.99 (1) |
| *n*-Decane ($C_{10}$) | 0.98 (1) | 0.97 (3) |
| *n*-Pentadecane ($C_{15}$) | 0.95 (3) | 0.95 (4) |
| *n*-Eicosane ($C_{20}$) | 1.09 (2) | 1.12 (5) |

[a] Values in parentheses are the percentage relative standard deviations of triplicate 10 min extractions or triplicate manual solvent injections [54].

flow rate through the chromatographic column. For example, peak fronting occurs when the gaseous $CO_2$ flow rate through a 30 m × 0.32 mm i.d. column exceeds *c.* 9 ml/min during the extraction (Table 5.3, Figure 5.5). By increasing the split ratio so that more of the $CO_2$ can be vented through the split vent and less through the column, peak broadening and peak fronting can be eliminated even though the SFE flow rate and hence the total gas flow through the injection port remains the same [54]. The best approach for SFE–GC is to use the highest SFE flow rate which will result in good extraction efficiencies in a reasonable time and yet be compatible with the GC system to maintain good peak shapes.

For split SFE–GC the maximum practical SFE flow rate depends on the split ratio, that is, higher split ratios (and thus lower column flows) allow higher SFE flow rates. However, a point is reached where the split ratio required to give the appropriate gaseous extraction fluid column flow rate will be so large that splitter discrimination may become a problem. Alternatively, if a lower extraction flow rate is used to ensure that the analytes are efficiently retained on the column, the extraction time required to obtain quantitative recoveries may increase to a point where the more volatile analytes will start to elute through the column. This elution of analytes during

**Figure 5.5** Effect of supercritical fluid extraction (SFE) extraction flow rate on the peak shapes of benzene, toluene, ethylbenzene and xylene (BTEX) and $C_4$–$C_{20}$ *n*-alkanes obtained by split SFE–gas chromatography (GC). The hydrocarbon mixture was either injected onto the capillary column (top chromatogram) or was spiked onto Tenax-TA and extracted on-line for 10 min with supercritical fluid (SF) $CO_2$ (400 atm, 60°C) at various extraction fluid flow rates. The corresponding split ratios are given in Table 5.3. The 30 m × 0.32 mm inner diameter (5 μm film) DB-1 capillary column was kept at −50°C during the injection or extraction step. After each extraction or 10 min after the injection, the GC oven was heated at 50°C/min to 40°C and then at 8°C/min to 300°C. Reproduced with permission from ref. 54.

**Table 5.3** Split ratio measured under gas chromatography (GC) and split supercritical fluid extraction–GC (SFE–GC) conditions. Source: ref. 54

| Analysis | Restrictor inner diameter (μm) | SFE flow rate (ml/min)[a] | Column head pressure (psi)[b] | Column volumetric flow (ml/min)[c] | Split vent volumetric flow (ml/min) | Split ratio (column/split) |
|---|---|---|---|---|---|---|
| GC | – | – | 15 | 5.1 | 789 | 1:155 |
| Split SFE–GC | 15 | 0.18 | 1 | 0.8 | 86 | 1:107 |
| Split SFE–GC | 22 | 0.38 | 4 | 2.8 | 267 | 1:95 |
| Split SFE–GC | 25 | 0.58 | 9 | 5.0 | 400 | 1:80 |
| Split SFE–GC | 30 | 0.96 | 16 | 9.2 | 577 | 1:63 |

[a] Flow rate measured as liquid $CO_2$ at the pump.
[b] Flow rate measured, by means of a bubble flow meter, as volume of gas.
[c] SFE–GC conditions: 400 atm, 60°C $CO_2$ depressurised into a thick phase (30 m × 0.32 mm i.d., 5 μm) column at −50°C.

the extraction step can be seen by comparing Figure 5.5 with Figure 5.6, where the extraction time is increased from 10 min to 30 min, respectively. So a compromise must be reached between using the shortest practical dynamic extraction time to facilitate the column focusing of the analytes and the extraction time required to ensure that quantitative recoveries are achieved. This in turn will be related to the highest practical split ratio which can be used before splitter discrimination becomes the limiting factor.

Although the use of the appropriate split ratio will ensure that the analytes are adequately focused at the top of the column, it can be seen that as the extraction flow rate is increased the chromatogram starts to 'spread out' and the retention times of the volatile analytes decrease and the analyte peak widths increase (Figure 5.5). This suggests that at the high extraction flow rates the analytes are less focused in the analytical column. A short (less than 1 m) retention gap of deactivated fused silica can be used as a method of obtaining good peak shapes as it enables the analytes to be refocused in the chromatographic column at the start of the GC analysis [39].

For high extraction flow rates (greater than 1 ml/min) several minor modifications need to be made to the GC apparatus. For example, even though most of the SFE effluent (typically 95%–99%) is vented through the split vent, the gas flow through the column may still be sufficient to extinguish a flame ionisation detection (FID) flame. Therefore, the hydrogen flow may need to be increased to sustain the detector signal. The mass spectrometer is also sensitive to the extraction flow rate, as high flow rates in the column can cause excessively high pressures (e.g. greater than $10^{-4}$ torr) within the ion source. However, this can be avoided in both on-column and split SFE–GC by either using a longer and narrow GC column (e.g. a 60 m × 0.25 mm i.d. column instead of a 30 m × 0.32 mm i.d. column), or by placing a 1 m section of smaller diameter (e.g. 0.15 mm i.d.) transfer line between the column and the ion source [2]. For split SFE–GC the use of a narrower column or transfer line results in an increased split ratio and thus poorer detection limits. It is therefore more convenient to adjust the split ratio until an acceptable mass spectrometer ion source pressure is obtained.

### *5.6.2 Column trapping temperature*

The cryogenic trapping temperature employed during the extraction step will also affect the ability of SFE–GC to focus the analytes efficiently, particularly the volatile components. Figure 5.6 shows the effect of the cryogenic trapping temperature on the chromatographic peak shape of a test mix (representing the major components in gasoline and diesel fuel) analysed by split SFE–GC using identical extraction conditions (using 0.6 ml/min liquid $CO_2$) and a 30 m × 0.32 mm i.d., 5 μm film thickness column [54]. To retain efficiently the volatile analytes in the $C_4$ to $C_9$ *n*-alkane boiling point

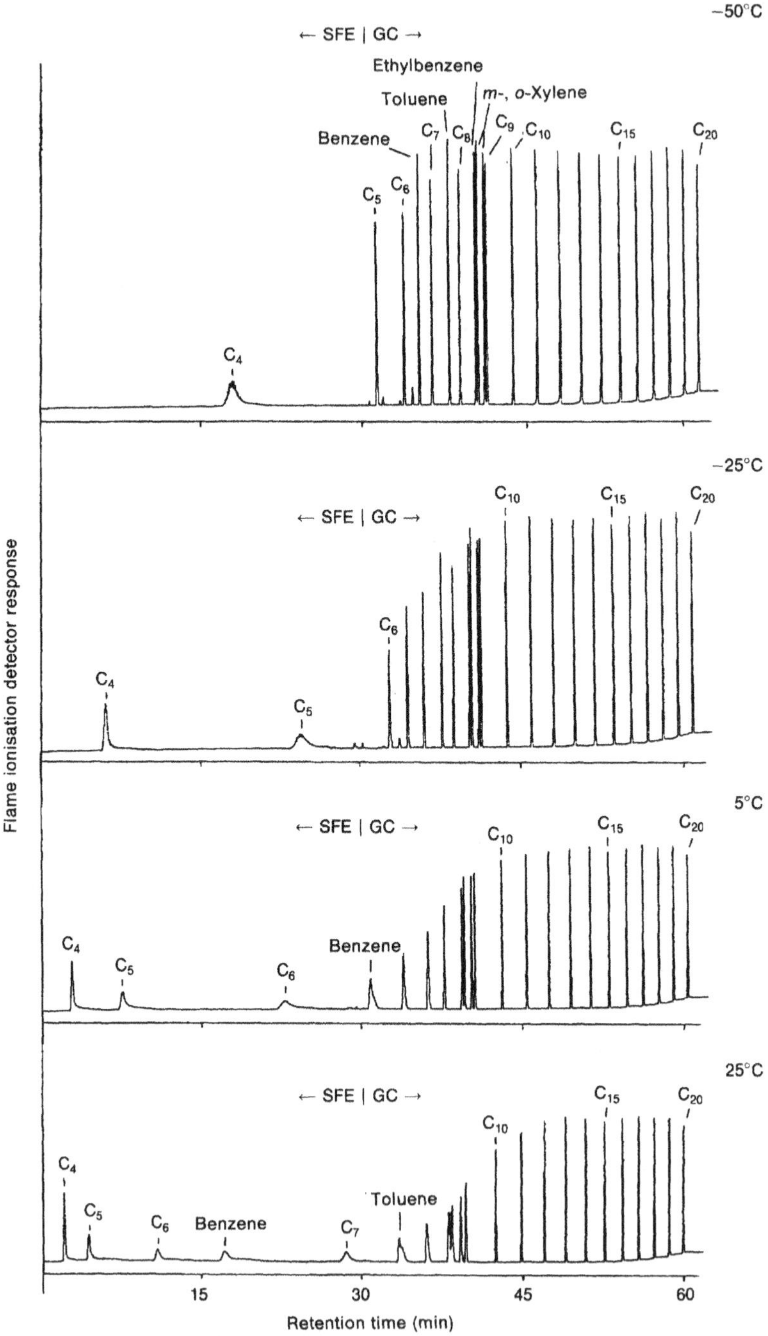
−50°C
← SFE | GC →
Ethylbenzene
Toluene
m-, o-Xylene
Benzene
C7
C8
C9
C10
C15
C20
C5
C6
C4
−25°C
← SFE | GC →
C10
C15
C20
C6
C4
C5
5°C
← SFE | GC →
C10
C15
C20
C4
C5
C6
Benzene
25°C
← SFE | GC →
C15
C20
C10
C4
C5
C6
Benzene
C7
Toluene
Flame ionisation detector response
15
30
45
60
Retention time (min)

range the column was cryogenically cooled to *c.* −50°C. However, even at this very low trapping temperature the most volatile analyte, *n*-butane, is not indefinitely retained in the column stationary phase, as the analyte will pass through the column when the extraction time is increased from 10 min (Figure 5.5) to 30 min (Figure 5.6). The rest of the analytes (e.g. $C_5$ to $C_{20}$ *n*-alkanes) are efficiently focused over at least a 60 min extraction time (Figure 5.7). It is not possible to increase the trapping efficiency of the column by using a lower trapping temperature (e.g. less than −50°C) as the column stationary phase is a solid below −60°C [73, 74], so at this temperature the column will become more characteristic of a retention gap as the analytes can no longer partition into the stationary phase. Furthermore, at −78°C $CO_2$ freezes, which would plug the column.

When the cryogenic trapping temperature is raised from −25°C to 25°C the trapping efficiency of the column stationary phase correspondingly decreases and an increasing number of analytes passes through the column during the extraction step (Figure 5.6). The choice of cryogenic trapping temperature will depend on the analyte composition. For example, a sample with gasoline range organics (defined as compounds in the $C_6$ to $C_{10}$ boiling point range [8]) will require a column trapping temperature of −50°C to −25°C, whereas a kerosene-contaminated sample ($C_9$ to $C_{16}$ boiling point range) requires a less rigorous procedure with a trapping temperature in the region of 5°C. It is interesting to note that with no cryogenic cooling during the extraction, and the column maintained at room temperature (25°C), analytes as volatile as *n*-decane can still be efficiently retained. Peak shapes of the retained analytes are generally broader for more volatile species (e.g. toluene to *n*-nonane), but less volatile analytes (e.g. $C_{10}$ to $C_{20}$ boiling point range) have very good chromatographic peak shapes, so that the 25°C trapping temperature is suitable for the retention of analytes in the diesel range ($C_{10}$ to $C_{25}$).

Using column trapping temperatures above 0°C may be advantageous, as wet samples can be analysed without the problem of extracted water freezing and plugging the column. However, if the wet sample contains volatile analytes then quantitative analysis of the whole extract is no longer possible as the volatile fraction will have eluted through the column before the GC analysis has commenced. An alternative procedure, as mentioned earlier, is to place the sample on a bed of drying agent inside the extraction vessel so that the water is selectively retained by the reagent inside the vessel while

**Figure 5.6** Effect of cryogenic trapping temperature on the retention of benzene, toluene, ethylbenzene and xylene (BTEX) and $C_4$–$C_{20}$ *n*-alkanes on a thick-film (5 μm) 30 m × 0.32 mm inner diameter DB-1 capillary column during split supercritical fluid extraction–gas chromatography (SFE–GC) analysis. The hydrocarbon mixture was extracted from Tenax-TA using supercritical fluid (SF) $CO_2$ (400 atm, 60°C) at 0.6 ml/min (liquid $CO_2$ delivery) for 30 min. After each extraction the GC oven was heated at 50°C/min to 40°C and then at 8°C/min to 300°C. Reproduced with permission from ref. 54.

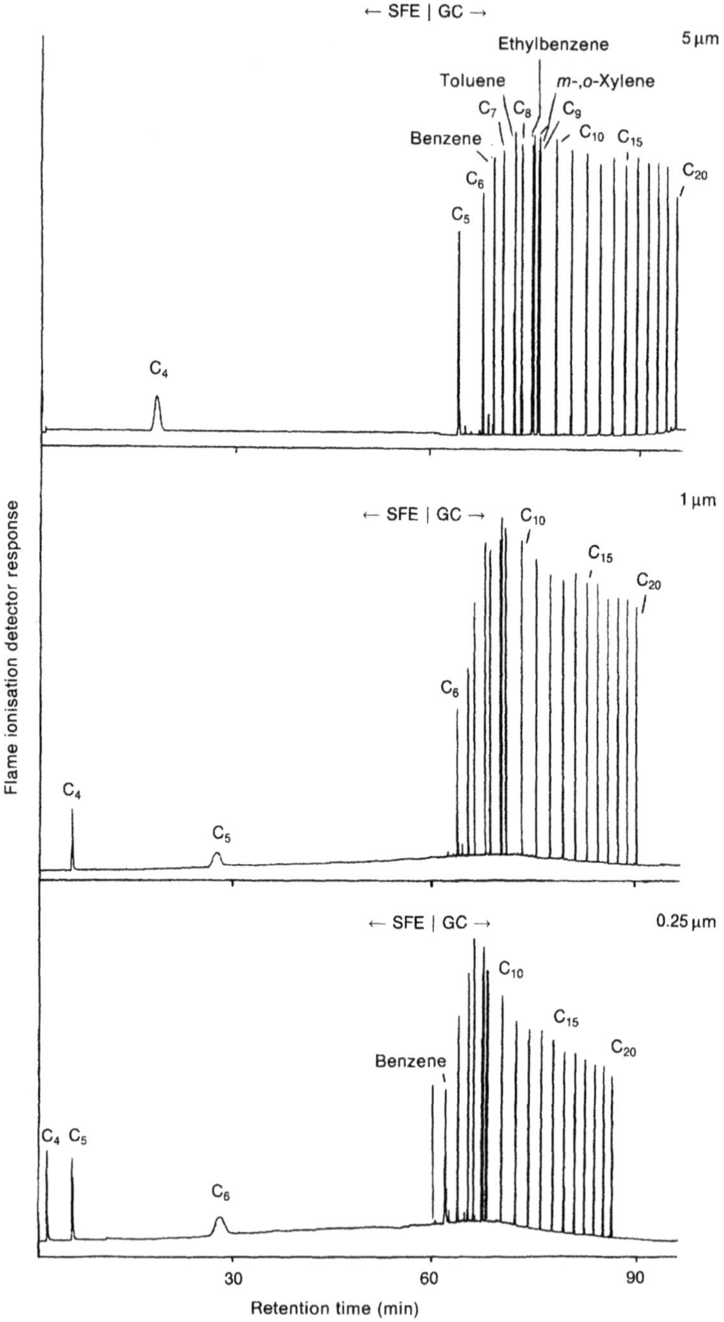
← SFE | GC →
Ethylbenzene
5 μm
Toluene
m-,o-Xylene
C7
C8
C9
Benzene
C10
C15
C6
C20
C5
C4
← SFE | GC →
1 μm
C10
C15
C20
C6
C4
C5
← SFE | GC →
0.25 μm
C10
C15
C20
Benzene
C4
C5
C6
Flame ionisation detector response
30
60
90
Retention time (min)

the analytes of interest are transferred to the GC column. The most popular drying agents in SFE with a high water capacity but low affinity for the analytes of interest are: Hydromatrix (diatomaceous earth) [75, 76]; anhydrous magnesium sulphate [15, 16]; and molecular sieves 3A and 5A [72]. The use of molecular sieve 3A has enabled very wet (25% m/m water) gasoline-contaminated samples to be successfully analysed by split SFE–GC at a low cryogenic trapping temperature of −50°C, the problems of frozen water in the GC system having been eliminated [70]. To ensure that the drying agent is efficiently retaining the water it is suggested that the column gas flow rate is monitored during the entire extraction. This is easily achieved by switching off the detector and placing a gas flow meter at the end of the column, and if FID is used this simply means inserting an appropriate adapter in the cooled detector and measuring the flow directly.

Monitoring the column flow during the SFE step is a very important and yet often underestimated procedure. Poor SFE–GC quantitation and reproducibility are often blamed on inefficient column analyte trapping or irreproducible split ratios when in fact the problem is with frozen water or extracted analytes slowly building up at the head of the column, so that during the extraction the column flow rate decreases. For example, at very low cryogenic trapping temperatures (e.g. −50°C) a diesel and/or motor oil extract has a 'waxy' consistency which causes the column to become plugged on commencing the extraction. However, a continuous column flow is achieved if a less severe cryogenic trapping temperature of −25°C is used [70]. Similar plugging problems have been encountered when analysing plant material containing cuticle waxes, and these problems have been overcome by increasing the column trapping temperature from −50°C to −30°C [62]. It is often preferable to start an SFE–GC investigation using a −25°C cryogenic trapping temperature as this experimental condition allows a wide range of analytes to be correctly focused and retained while still ensuring a continuous gaseous extraction flow rate through the column.

### *5.6.3 Column stationary phase thickness*

Another important experimental parameter is the capillary column stationary phase film thickness, for as the film thickness increases the ability of the column to retain and focus the analytes during the extraction also

**Figure 5.7** Effect of capillary column stationary phase thickness on the chromatographic peak shape of benzene, toluene, ethylbenzene and xylene (BTEX) and *n*-alkanes during split supercritical fluid extraction–gas chromatography (SFE–GC). The hydrocarbon mixture was extracted from Tenax-TA using supercritical fluid $CO_2$ (400 atm, 60°C) at 0.6 ml/min (liquid $CO_2$ delivery) for 60 min. The extracted analytes were cryogenically focused onto: 5 µm film DB-1 capillary column [30 m × 0.32 mm inner diameter (i.d.)]; 1 µm film DB-5 capillary column (30 m × 0.32 mm i.d.); 0.25 µm film DB-5 capillary column (20 m × 0.25 mm i.d.). After each extraction the GC oven was heated at 50°C/min to 40°C and then at 8°C/min to 300°C. Reproduced with permission from ref. 54.

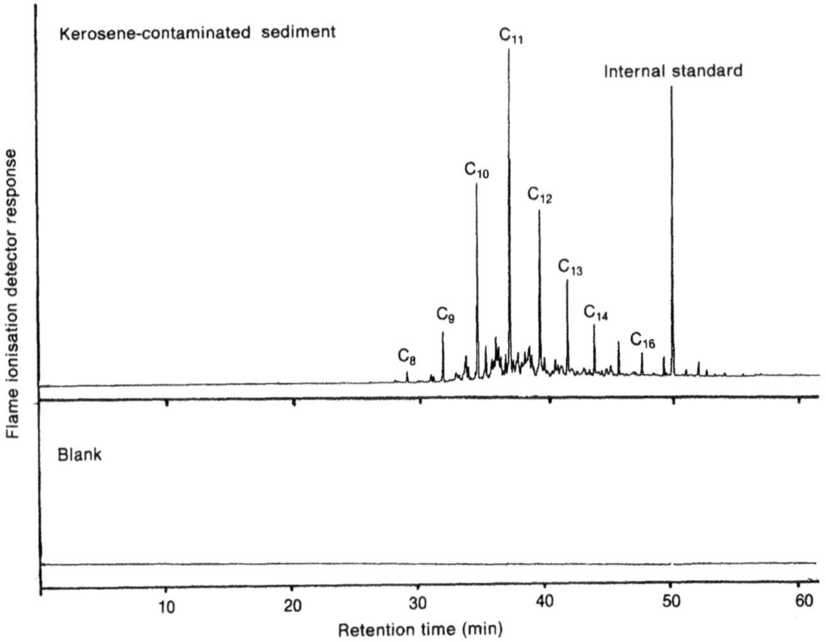
Kerosene-contaminated sediment
$C_{11}$
Internal standard
$C_{10}$
$C_{12}$
$C_{13}$
$C_9$
$C_{14}$
$C_8$
$C_{16}$
Flame ionisation detector response
Blank
10
20
30
40
50
60
Retention time (min)

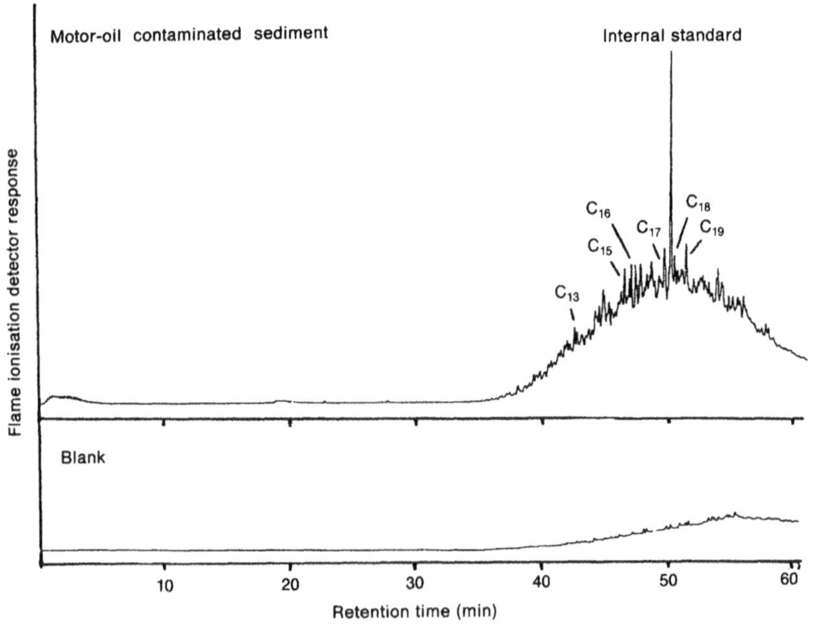
Motor-oil contaminated sediment
Internal standard
$C_{16}$
$C_{18}$
$C_{17}$
$C_{19}$
$C_{15}$
$C_{13}$
Flame ionisation detector response
Blank
10
20
30
40
50
60
Retention time (min)

increases. For example, Figure 5.7 shows the split SFE–GC analysis of a volatile analyte test mix using thick (5 μm), medium thickness (1 μm) and thin (0.25 μm) film columns. The results demonstrate that the column with the thickest stationary phase is the most efficient in retaining the analytes, since the column of 5 μm film thickness is able to give good peak shapes for analytes as volatile as *n*-pentane, even after 60 min of SFE [54]. When chromatographic columns are used with thinner stationary phase films the trapping efficiency of SFE–GC decreases and more analytes pass through the column during the SFE step (Figure 5.7). The advantage of using a thinner stationary phase film is that a wider boiling point range of components can be resolved. For example, the 5 μm thick-film column can resolve analytes up to the $C_{25}$ boiling point range in a typical 30 min GC run, but the thin-film 0.25 μm column can resolve analytes up to *c.* $C_{40}$ in the same analysis time. It follows that the choice of stationary phase will depend on the requirements of efficient trapping of the volatile analytes during the extraction step and the ability to elute the semivolatile analytes during the GC analysis.

The importance of selecting the appropriate column stationary phase film thickness can be demonstrated with the analysis of fuel-contaminated samples. A thick-film (5 μm) column was used in split SFE–GC to investigate both a kerosene-contaminated and motor-oil-contaminated sediment (Figure 5.8). For both samples quantitative fuel recoveries were possible, as the hydrocarbon recoveries from a 15 min SFE–GC extraction were comparable to those obtained by sonicating the samples in methylene chloride [70]. By performing a second GC analysis without an extraction or injection (i.e. a blank run) it can be seen that the kerosene extract had completely eluted from the thick-film column. However, this was not the case for the motor oil extract, as the less volatile analytes accumulated on the column to produce a rising baseline during subsequent GC analysis (Figure 5.8). If several motor oil samples were investigated using the thick-film column then poor quantitative results would have quickly resulted. The degradation of the column performance can be rectified by trimming *c.* 15 cm off the injection end of the column. A more practical alternative is to use a thin-film (0.25 μm) column which can efficiently elute all the components in the sample.

**Figure 5.8** Memory effect from low and high molecular weight petroleum hydrocarbons during split supercritical fluid extraction–gas chromatography (SFE–GC) analysis. The samples (1 g) and a piece of filter paper spiked with the internal standard (octahydroanthracene) were placed on a bed of drying agent (molecular sieve 3A) inside an extraction vessel and extracted for 15 min with supercritical fluid $CO_2$ (400 atm, 60°C), at 0.6 ml/min (liquid $CO_2$ delivery). The extracted analytes were trapped onto a thick-film (5 μm) 30 m × 0.32 mm inner diameter DB-1 capillary column at −25°C. After the extraction, the GC oven was heated to 50°C/min to 40°C then at 8°C/min to 300°C. To assess column contamination the GC temperature programme was repeated without making a further injection. Reproduced with permission from ref. 70.

## 5.7 Quantitative supercritical fluid extraction–gas chromatography

After optimisation of the chromatographic conditions, a quantitative assessment of the SFE–GC technique can be undertaken. The ability of the on-line method to enable a quantitative and reproducible transfer of the analytes from the SFE system to the GC column has been investigated using both spiked and real-world samples [1, 2, 50, 54, 63]. Also, studies involving comparisons of SFE–GC recoveries with conventional liquid solvent extraction have been reported [1, 3, 10, 11, 60]. Ideally, a calibration method is required in which the calibration standard is introduced into the GC system in the same way as the real sample, that is, through the SFE restrictor. This would enable the SFE–GC collection parameters to be investigated independent of the SFE efficiencies. The calibration standard should consist of a matrix spiked with the analytes of interest and should be easy to prepare, convenient to handle and be stable. Previous off-line SFE collection efficiency studies have used an inert matrix such as silanised glass beads [67, 77], but this unretentive matrix is unsuitable for use with volatile analytes because of analyte losses occurring during the spiking process through volatilisation [70]. Conversely, sorbent resins such as Tenax-TA have proved to be ideal matrices for a calibration standard, since volatile and semivolatile analytes have large gas retention volumes on the resin at ambient conditions [51], yet analytes can be quantitatively recovered by SFE [10, 50, 54]. As shown in Table 5.4, quantitative SFE–GC recoveries for a wide range of analytes (e.g. $C_4$ to $C_{20}$ *n*-alkanes) can be achieved from spiked Tenax-TA using a 10 min SFE period. In addition, the sorbent also has a good shelf-life as

**Table 5.4** Supercritical fluid extraction–gas chromatography recovery of a volatile analyte test mix 5 min (fresh) and 24 h (aged) after spiking onto several matrices. Source: ref. 54

| Analyte | Recovery (%)[a] | | | |
|---|---|---|---|---|
| | Glass beads | | Tenax-TA | |
| | Fresh | Aged | Fresh | Aged |
| *n*-Butane ($C_4$) | 61 (20) | ND | 109 (6) | 94 (5) |
| *n*-Pentane ($C_5$) | 69 (18) | ND | 105 (4) | 109 (4) |
| *n*-Hexane ($C_6$) | 91 (9) | ND | 103 (3) | 107 (3) |
| Benzene | 93 (8) | ND | 97 (3) | 100 (4) |
| Toluene | 97 (8) | ND | 97 (3) | 96 (6) |
| Ethylbenzene | 101 (4) | 5 (59) | 96 (1) | 95 (5) |
| *m*-Xylene | 101 (3) | 7 (65) | 96 (1) | 95 (4) |
| *o*-Xylene | 101 (3) | 8 (69) | 98 (3) | 95 (4) |
| *n*-Decane ($C_{10}$) | 100 (0) | 52 (53) | 100 (0) | 100 (0) |
| *n*-Pentadecane ($C_{15}$) | 100 (5) | 94 (7) | 98 (2) | 102 (9) |
| *n*-Eicosane ($C_{20}$) | 100 (2) | 97 (7) | 99 (5) | 100 (11) |

[a] The value in parenthesis is the percentage relative standard deviation of triplicate 10 min extractions using supercritical fluid $CO_2$ (400 atm, 60°C).
ND = not detected.

the analytes are retained by the Tenax-TA for at least 24 h at room temperature and pressure (Table 5.4). This enables the preparation of several calibration standards which can be stored for use throughout the working day.

Depending on the method of sample introduction, either an internal or external calibration method can be used to quantify the analyte recoveries achieved by SFE–GC. For sealed on-column SFE–GC both calibration methods are possible since a gas-tight seal exists between the column and the syringe needle or SFE restrictor, and this enables all the analytes to be transferred into the GC column. Sealed on-column SFE–GC peak areas can therefore be compared to those obtained with either a conventional GC injection of the external calibration standard, or can be compared to the internal standard which is added to the sample matrix and co-extracted with the sample analytes for GC analysis. However, the use of an external calibration method is often not a suitable method for split SFE–GC as the on-line split ratio can significantly differ from that of conventional GC. For example, as shown in Table 5.3, a split ratio of *c.* 155:1 during GC analysis decreases to *c.* 65:1 under split SFE–GC conditions, even though the needle valve (which controls the split) has not been adjusted and the pressure inside the injection port is similar [54]. It should be stressed that the split ratio during the extraction or GC injection is constant as good quantitative reproducible raw peak areas are achievable by both methods. Furthermore, no splitter discrimination between the high and low molecular weight analytes is observed as a similar chromatographic profile is obtained by both split SFE–GC and conventional GC analysis (Figure 5.5). So to quantify the split SFE–GC analysis an internal calibration method is recommended as any variation in the split ratio between the techniques will not affect the analyte response factors (e.g. area ratio of analyte to internal standard) [70]. By using an internal standard split SFE–GC extracts can be quantified based on the conventional GC injection of the calibration standards as demonstrated in Table 5.5. If an internal standard is not to be used in split SFE–GC then the calibration standards should be introduced into the GC under the same conditions as the extracted analytes (i.e. through the SFE restrictor).

It is thought that the difference between the GC and SFE–GC split ratios is due to significant differences in the gaseous flow profiles generated in the injection port during the analysis [54, 63]. In split GC analysis a relatively small volume of solvent is introduced into the injection port in a short period of time. Conversely, in split SFE–GC a relatively large volume of solvent ($CO_2$) is continually introduced into the injector port at near supersonic velocities [78]. During the SFE step the highest linear velocities in the injection port will be in the centre of the restrictor exit, and flows will be slower towards the wall of the injection port liner (Figure 5.9). As the column inlet is directly 'in line' with the restrictor outlet, it seems likely that a greater proportion of the $CO_2$ is vented into the column than if a more laminar flow

**Table 5.5** Comparison of peak area reproducibility for split supercritical fluid extraction–gas chromatography (SFE–GC) versus a conventional injection of a volatile analyte test mix. Source: ref. 54

| Analyte | Average GC area counts[a] | Average SFE–GC area counts[b] | Average GC int. stand. value[c] | Average SFE–GC int. stand. value[c] | Recovery based on raw area (%) | Recovery based on int. stand. (%) |
|---|---|---|---|---|---|---|
| *n*-Butane | 988 088 (4) | 1 643 680 (5) | 0.88 (4) | 0.94 (5) | 166 | 107 |
| *n*-Hexane | 1 066 732 (4) | 1 706 508 (3) | 0.95 (1) | 0.97 (3) | 160 | 102 |
| Benzene | 1 383 139 (4) | 2 060 167 (1) | 1.23 (1) | 1.18 (2) | 149 | 96 |
| Toluene | 1 330 885 (5) | 1 992 884 (2) | 1.18 (1) | 1.14 (2) | 150 | 97 |
| Ethylbenzene | 1 239 959 (4) | 1 856 394 (3) | 1.10 (1) | 1.06 (1) | 150 | 96 |
| *m*-Xylene | 1 269 779 (4) | 1 902 309 (3) | 1.13 (1) | 1.08 (1) | 150 | 96 |
| *n*-Decane (IS) | 1 124 422 (4) | 1 753 399 (3) | 0.00 (0) | 0.00 (0) | – | – |
| *n*-Pentadecane | 970 689 (1) | 1 430 895 (5) | 0.86 (3) | 0.82 (2) | 147 | 95 |
| *n*-Eicosane | 1 036 660 (2) | 1 587 179 (1) | 0.92 (3) | 0.91 (4) | 153 | 99 |

[a] Split ratio during GC analysis *c.* 155:1.
[b] Split ratio during split SFE–GC analysis *c.* 80:1.
[c] IS value = internal standardisation value (ratio of peak area to internal standard area).
Note: IS = internal standard; int. stand. = internal standardisation; values in parentheses are the percentage relative standard deviations of the triplicate manual solvent injection of a volatile analyte test mix, or triplicate 10 min extractions of Tenax-TA spiked with a volatile analyte test mix.

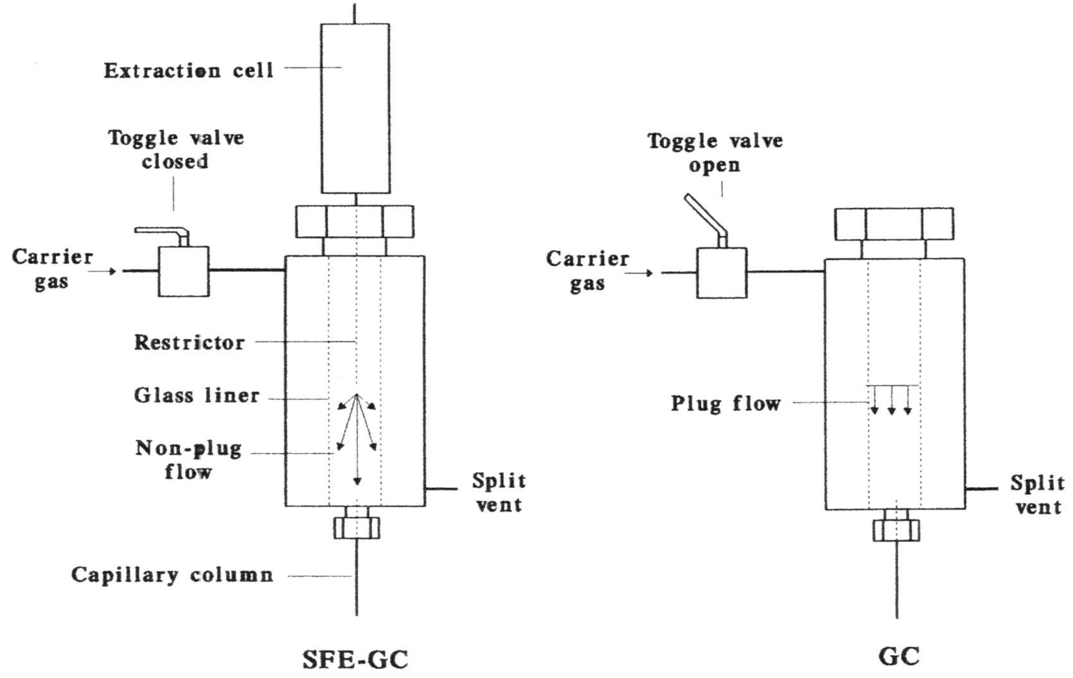


**Figure 5.9** Flow profile during split supercritical fluid extraction–gas chromatography (SFE–GC) and GC analysis.

had been generated. Thus as the SFE flow rate is increased a greater proportion of the $CO_2$ enters the column, resulting in a decrease in the split ratio (Table 5.3). Furthermore, if a 'packed' injection port liner is used (e.g. containing glass wool or a porous glass frit) to disperse the expanding stream of $CO_2$ exiting the restrictor, a split ratio comparable to that of conventional GC analysis can be achieved and raw peak areas can be obtained in split SFE–GC similar to those in GC [1, 60].

## 5.8 Optimisation of extraction conditions for supercritical fluid extraction–gas chromatography

For quantitative recoveries in SFE–GC the supercritical fluid must be able to solvate the analyte of interest and overcome the interactions between analyte and sample matrix so that a favourable partitioning of the analyte into the supercritical fluid can occur. With these considerations in mind the most popular supercritical fluid, namely $CO_2$, which is a non-polar fluid, has proved to be particularly suited for the extraction of hydrocarbons, as these analytes have a relatively high solubility in $CO_2$. Both selective and quantitative hydrocarbon recoveries are possible by varying the temperature and pressure of the supercritical fluid [79]. The selectivity of $CO_2$ has been used in SFE–GC to fractionate PAHs. As shown in Figure 5.10, progressively higher molecular weight material was extracted as the density of the extraction fluid was increased. Low extraction pressures (e.g. SF-$CO_2$ at 80 atm, 50°C) recovered two-ring and three-ring PAHs, and higher extraction pressures (e.g. SF-$CO_2$ at 200 atm, 50°C) recovered predominately four-ring and larger aromatics [56]. Although some overlap of the components occurred in the different fractions the results indicate that SFE–GC is a feasible method for undertaking a class fractionation analysis.

Using more extreme extraction conditions (typically SF-$CO_2$ at 400 atm and 50°C) quantitative SFE–GC of a wide range of analytes has been carried out for various sample matrices, including fuel-contaminated sediments [54, 60], PCBs in biological samples [65], PAHs from Tenax-GC [10], essential oils in plant material [62], N-heterocyclic compounds from cigarette smoke particulates [80] and phenolics from wood smoke particulates [3]. However,

**Figure 5.10** Capillary gas chromatography (GC) chromatograms of polycyclic aromatic fractions obtained by on-column supercritical fluid extraction–gas chromatography (SFE–GC). Polycyclic aromatic compounds were extracted from glass beads at 50°C with a 0.005 ml/min (liquid $CO_2$ delivery) at several extraction pressures. The extracts were trapped at 30°C onto a 15 m × 0.25 mm inner diameter (i.d.) (0.25 μm film) SE-54 capillary column fitted with a 30 cm × 0.53 mm i.d. deactivated fused-silica retention gap. After the extraction the GC oven was maintained at 30°C for 2 min, then the temperature was programmed at 4°C/min to 265°C. Compounds A and B are marked in each fraction to facilitate comparison. Reproduced with permission from ref. 56.

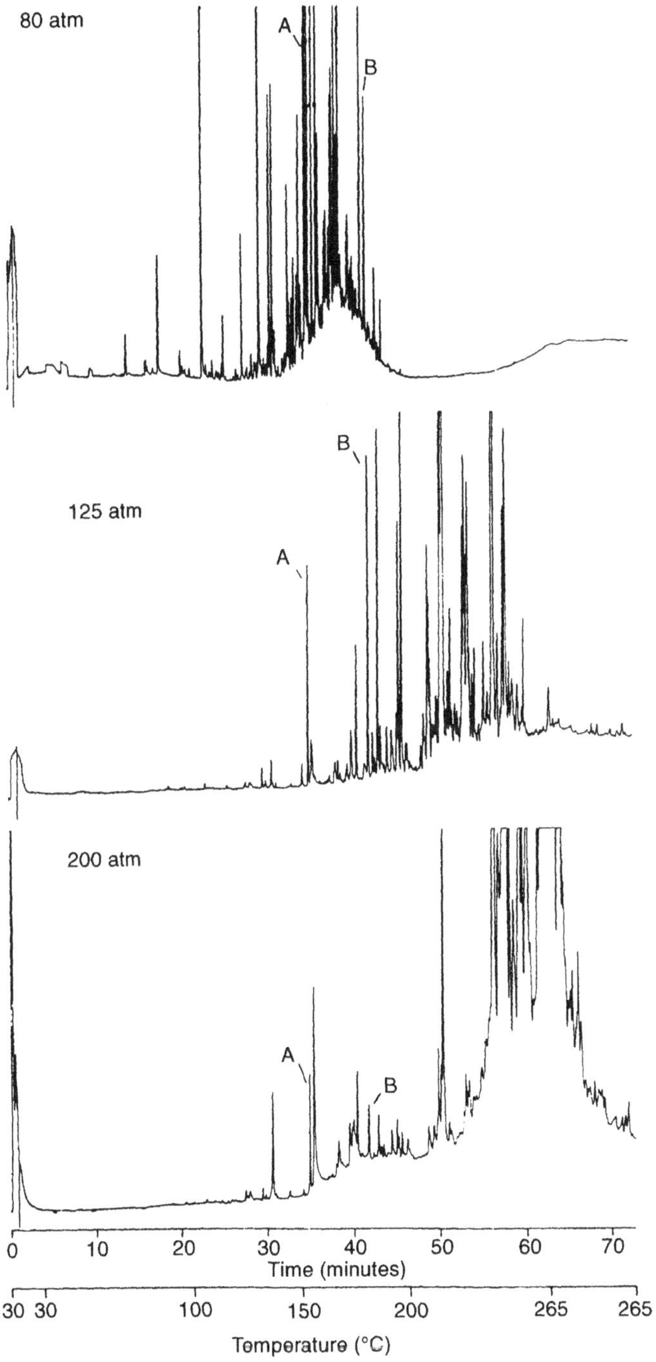
80 atm
A
B
125 atm
B
A
200 atm
A
B
0 10 20 30 40 50 60 70
Time (minutes)
30 30 100 150 200 265 265
Temperature (°C)

several of these 'quantitative recoveries' were determined by using a second sequential SFE extraction performed under conditions identical to the first to demonstrate that no further analytes were present in the sample matrix. Although this approach may be valid for some samples, it is often more reliable to re-extract the sample using more rigorous extraction conditions or to compare the recoveries with those from a conventional solvent extraction method [46]. SFE–GC analysis of certified reference material (quantitative recoveries based on an exhaustive Soxhlet extraction) has proved an ideal means of assessing the quantitative capability of the technique. Two certified reference materials from the National Institute of Standards and Technology (NIST), namely a marine sediment and an urban air particulate sample, have been analysed by both split and/or on-column SFE–GC. The results are given in Table 5.6. Using $CO_2$ at a high pressure (375 atm) and moderate temperature (50°C), quantitative SFE–GC recoveries of both low and high molecular weight PAHs were achieved from a 50 mg urban air particulate sample. What is particularly interesting is that the SFE–GC analysis could be completed within an hour compared with several days for the certified method. However, poor PAH recoveries were obtained when a larger urban air particulate sample size was used (e.g. 320 mg [46]) and when the PAHs were present in a more retentive or 'less accessible' sample matrix, such as a marine sediment [60]. Poor PAH recoveries were obtained from the sediment with $CO_2$ and quantitative recoveries were only achievable when a more polar fluid such as nitrous oxide ($N_2O$) was used (Table 5.6). The difference in PAH recoveries is related to the sample matrix and demonstrates the importance of the analyte–matrix interactions. The results also highlight the versatility of the coupled technique in that other fluids besides $CO_2$ can be used. Nitrous oxide is ideally suited for the on-line method, as it is a gas at ambient conditions and does not give an appreciable response in the FID. However, caution must be exercised as the fluid is an oxidant and should not be used with high organic content samples because of a possible explosion hazard. Other supercritical fluids such as sulphur hexafluoride ($SF_6$) have also been successfully used with split SFE–GC to obtain selective *n*-alkane extractions from a gasoline-contaminated soil [81].

Several analytes and sample matrices are not quantitatively extracted by pure SF-$CO_2$ and it has often proved necessary to use a small amount (typically 10% vol/vol) of organic modifier to achieve quantitative recoveries. The addition of the modifier enhances the solvation strength of the SF-$CO_2$ so that the analyte solubility and/or extraction fluid–matrix interaction are increased. The modifier is normally introduced into the SFE–GC system by using a dual pump delivery system, a premixed cylinder or by adding the modifier to the sample inside the extraction vessel. Depending on the modifier concentration a retention gap or thick-film capillary column is usually required to enable the solvent to be eluted from the column [81]. The use of a modifier may cause problems since organic solvents are

**Table 5.6** Supercritical fluid extraction–gas chromatography (SFE–GC) analysis of certified reference material

| Polycyclic aromatic hydrocarbon | Urban air particulates (NIST 1649) concentration (μg/g) | | | Marine sediment (NIST 1941) concentration (μg/g) | | |
|---|---|---|---|---|---|---|
| | Certified value[a] | $CO_2$[b] | $N_2O$[c] | Certified value[d] | $CO_2$[e] | $N_2O$[f] |
| Fluoranthene | 7.1±0.5 | 7.2±0.9 | 7.3±1.0 | 1.22±0.24 | 1.37±0.26 | 1.45±0.14 |
| Benz[*a*]anthracene | 2.6±0.3 | 2.6±0.8 | 2.6±0.8 | 0.55±0.08 | 0.38±0.01 | 0.60±0.08 |
| Indeno[1,2,3-*cd*]pyrene | 3.3±0.5 | 3.4±0.6 | 3.0±0.5 | 0.57±0.04 | 0.12±0.03 | 0.56±0.08 |
| Benzo[*ghi*]perylene | 4.5±1.1 | 3.9±1.0 | 3.6±0.9 | 0.52±0.08 | 0.15±0.05 | 0.56±0.09 |

[a] NIST-certified values based on 48 h Soxhlet extraction of 1 g samples using methanol or methylene chloride as the extraction solvent.
[b] Split SFE–GC analysis of a 50 g sample using SF-$CO_2$ (375 atm, 50°C), 30 min extraction [1].
[c] On-column SFE–GC analysis of a 2 mg sample using SF-$N_2O$ (350 atm, 45°C), 20 min extraction [2].
[d] NIST-certified values based on 16 h Soxhlet extraction of 7–25 g samples using both hexane and methylene chloride as the extraction solvents.
[e] Split SFE–GC analysis of a 30 mg sample using SF-$CO_2$ (400 atm, 50°C), 10 min extraction [60].
[f] Split SFE–GC analysis of a 30 mg sample using SF-$N_2O$ (400 atm, 50°C), 10 min extraction [60].
Note: NIST = National Institute of Standards and Technology; SF = supercritical fluid.

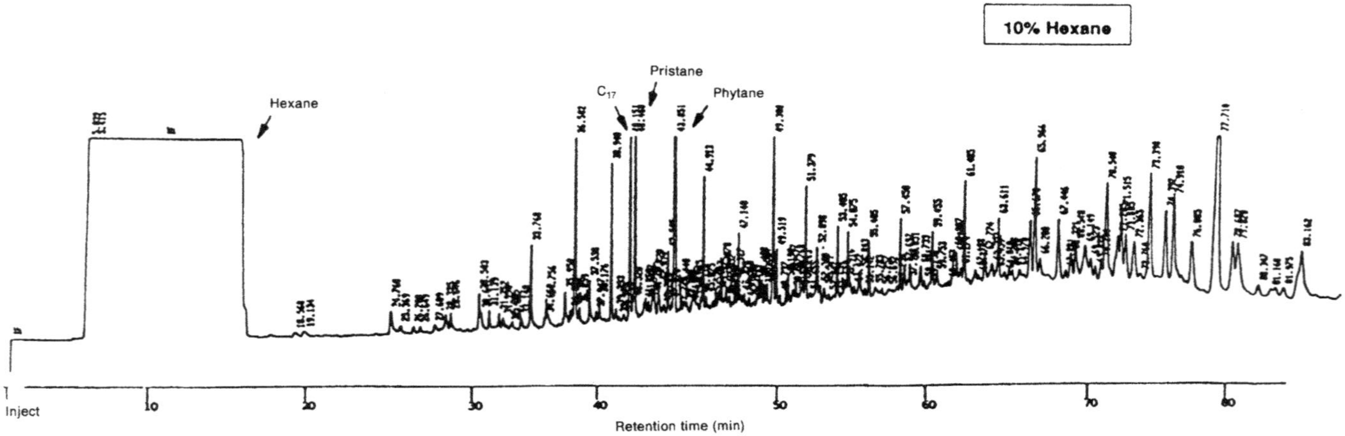


**Figure 5.11** Characterisation of shale rock using split supercritical fluid extraction–gas chromatography (SFE–GC) with modifier. The sample was initially extracted statically for 30 min with supercritical fluid carbon dioxide (SF-$CO_2$) (350 atm, 60°C) and 50 μl of hexane in the extraction vessel. The sample was then dynamically extracted for 10 min using pure SF-$CO_2$. The extract was trapped onto a 50 m × 0.20 mm inner diameter PONA column with a temperature of 100°C (10 min) to 330°C at 5°C/min with flame ionisation detection. Reproduced with permission from ref. 82.

generally not volatile under ambient conditions and appear as a large solvent peak in the chromatogram. Depending on the extract, the modifier may be eluted as a discrete peak away from the analytes of interest (Figure 5.11) and a range of solvents including methanol, hexane, carbon disulphide [82], ethanol [61], benzene, propylene carbonate [81], acetone, chloroform, methylene chloride, toluene [83] and formic acid [1] have all been successfully used in SFE–GC. However, careful selection of the modifier may be required as analyte discrimination has been reported when using a range of organic solvents [84]. To minimise the amount of modifier required the solvent is usually added to the sample inside the extraction vessel and then an initial static extraction is carried out, followed by a short dynamic extraction [81, 82]. With this approach, a small (50 μl) volume of modifier was successfully used to enhance the extraction efficiency of aromatic analytes from a sludge/fly ash sample matrix [81] (Table 5.7). Modifiers such as carbon disulphide [82] and formic acid [1] are also of particular interest to on-line SFE–GC techniques as these solvents have a very low response in FID and therefore have a minimum impact on the chromatography part of the method.

The effects of a modifier on the analyte chromatographic peak shape will depend on the amount of organic solvent transferred to the column during the extraction step. In split SFE–GC only a small fraction of the modifier is generally transferred to the column, as the majority is vented via the split line. For example, use of 10% organically modified $CO_2$ at a typical SFE–GC extraction flow rate of 0.5 ml/min (measured as liquid $CO_2$ at the pump head) for 20 min with a split ratio of between 300:1 and 100:1 results in only *c.* 3–10 μl of the organic solvent being transferred to the column. The effect of the modifier on the peak shape can be even further limited if the analytes are focused on the GC column above the boiling point of the modifier, so that the majority of the organic solvent is unretained. This approach has proved suitable for the on-line extraction of PAHs and polymeric material but is limited in its applicability by the boiling point of the modifier and the volatility of the analytes [84]. Alternatively, the

**Table 5.7** Use of modifiers in supercritical fluid extraction–gas chromatography (SFE–GC) for the extraction of aromatic analytes. Source: ref. 81

| Analyte | Recovery (%) | | | |
|---|---|---|---|---|
| | $CO_2$ | $CO_2$/10% propylene carbonate | $CO_2$/10% benzene | $CO_2$/10% methanol |
| Ethylbenzene | 74 | 96 | 95 | 80 |
| Cumene | 72 | 96 | 96 | 79 |
| 2-Chloronaphthalene | 93 | 93 | 92 | 82 |
| 1,2,4-Trimethylbenzene | 71 | 96 | 96 | 84 |

Note: SFE method: 225 mg of sludge/fly ash, supercritical fluid $CO_2$ (375 atm, 50°C), 10 min static extraction, 7 min dynamic extraction, 50 μl modifier in a 500 μl extraction vessel.

modifier can be retained on the GC column as a liquid and this has been shown to aid analyte focusing on thin-film columns (e.g. 0.18 μm film thickness) as the modifier acts as a solvent trap. This effect is less pronounced in thick-film columns as the retention power of these columns is already relatively strong [84].

When a programmed temperature vaporiser (PTV) injector is used to retain the extracted analytes the presence of the modifier can have a very significant effect on the chromatography [84]. In this instance the total extract is collected in a packed PTV liner and the organically modified supercritical fluid is vented through the split line. With pure supercritical fluids the PTV injector is cooled during the extraction so that the analytes are efficiently retained in the liner. However, when an organically modified supercritical fluid is used with the PTV operated below the boiling point of the solvent, the extracted analytes are poorly retained, as the modifier has liquified and washed the analytes from the injector [84]. To overcome this problem the PTV can be operated above the boiling point of the organic solvent, but this approach will be limited by the volatility of the analytes. A more robust approach may be the use of a more retentive matrix in the PTV such as sorbent resins, which can selectively retain the analytes relative to the organic solvent. However, as discussed earlier, these resins have a limited analyte range and will not retain the analyte indefinitely throughout the extraction.

To increase the range of analytes which can be investigated by SFE–GC, on-line derivatization of involatile components has been incorporated into the method of the coupled technique (section 3.1.2) [85, 86]. For example, a supercritical fluid reactor (SFR) consisting of a packed bed of alumina pretreated with methanol has been used *in situ* to transesterify extracted triglycerides from oilseeds [85]. The SFR is placed in line between the extraction vessel and GC system so that during the extraction the triglycerides are transesterified to their fatty acid methyl esters as they pass through the SFR under supercritical fluid conditions. This allows the resulting methyl esters to be collected and analysed by GC [85]. Alternatively, a derivatisation reagent such as pentafluorobenzylbromide (PFBBr) with potassium carbonate can be added directly to the extraction vessel, so that during the extraction polar compounds such as phenoxycarboxylic acids can be alkylated to produce esters which can be analysed by GC [86]. In-line derivatisation and reaction processes may prove to be a popular means of pretreating and analysing a wide range of samples.

A recent development has been the use of high-temperature (e.g. 150°C) supercritical fluids such as $CO_2$ [87, 88] or $CO_2$ modified with organic solvents [89] to enhance the recovery of environmental pollutants such as PAHs from real-world samples. This approach, as yet, has not been extensively applied to SFE–GC, but it is envisaged that the high-temperature extraction fluid will be compatible with the current SFE–GC techniques.

The increased recoveries and faster rate of extraction achieved by this approach make the use of high-temperature fluids in SFE–GC an attractive area of research for further investigation (section 3.2.3).

## 5.9 Supercritical fluid extraction–gas chromatography applications

### *5.9.1 Environmental samples*

SFE–GC has been used to analyse a wide variety of analytes and sample matrices ranging from natural products such as flavour and fragrance compounds in plant material to organic pollutants in soil, sediment, air particulate matter, animal tissue and even water. These diverse applications are summarised in Table 5.8 (and Table 5.9) and demonstrate the ability of SFE–GC to yield good chromatographic peak shapes and quantitative results. The most popular application for SFE–GC is the analysis of environmental samples, as these samples are generally contaminated in the parts-per-billion range and so require the high sensitivity that can be achieved by an on-line approach. A typical application is the analysis of PAH and PCB analytes from spiked [2, 48, 65, 90], certified [2, 57, 60] and/or real-world environmental samples [10, 60, 90]. It was demonstrated early on that these types of environmental samples can be qualitatively and quantitatively analysed using SFE–GC with flame ionisation detection (FID), mass spectrometry (MS) and electron capture detection (ECD) [2]. The simplest approach is on-column SFE–GC/FID, which was initially used to recover automobile exhaust organics collected on Tenax-GC [10]. The Tenax trap was directly coupled to a 40 × 0.25 mm i.d. (0.25 μm film thickness) DB-5 column, and the organics were recovered by extraction with SF-$CO_2$ (200 atm, 45°C) for 15 min with the GC column cryogenically cooled to 0°C. The viability of this direct approach in which the total extract is transferred to the column was proven, since the on-column SFE–GC determination of PAH concentrations in river sediment generated RSDs ranging from 2% to 21%, similar to the 3%–8% for a comparable conventional liquid injection of the extract [2].

The optimum SFE–GC/FID conditions for environmental samples have been evaluated in terms of extraction flow rate and column trapping temperature [2, 54, 80]. As a general rule, restrictors as large as 20 μm (for 0.25 mm i.d. GC columns) or 25 μm (for 0.32 mm i.d. GC columns) yield good chromatographic peak shapes [2, 80]. The trapping temperature required depends on the volatility of the analyte species [2, 62], but trapping temperatures of −30°C to 5°C are sufficient to yield good chromatographic peak shapes for most analytes; for example for species that are less volatile than *n*-heptane [54, 80].

SFE–GC/FID has proved to be particularly suited for the analysis of environmental samples contaminated with petroleum hydrocarbons, as the

**Table 5.8** Environmental applications of supercritical fluid extraction–gas chromatography

| Sample matrix | Analyte | Extraction fluid | Reference |
|---|---|---|---|
| Fuel-contaminated sediment | Petroleum hydrocarbons | $CO_2$ | 58, 60, 81 |
| | | $SF_6$ | 81 |
| | | $CO_2$/5% modifier (MeOH, toluene, $CH_2Cl_2CHCl_3$, acetone, hexane) | 83 |
| Shale rock | Petroleum hydrocarbons | $SF_6$ | 98 |
| | | $CO_2$/10% modifier (MeOH, $CS_2$, hexane) | 82 |
| Heavy hydrocarbon wax | Hydrocarbons | $CO_2$/10% $SF_6$ | 82 |
| Solid waste | Hydrocarbons | $CO_2$/0.3% formic acid | 1 |
| Dry sludge | Hydrocarbons | $CO_2$/20% $SF_6$ | 98 |
| Sludge/fly ash | Aromatics | $CO_2$/10% modifier (MeOH, benzene, propylene carbonate) | 81 |
| Biosludge | PAHs | $CO_2$ | 10 |
| Lamp black | PAHs | $N_2O$ | 2 |
| Marine sediment | PAHs | $CO_2$ and $N_2O$ | 60 |
| Marine sediment | PCBs | $CO_2$ | 90 |
| River sediment | PAHs, alkanes | $CO_2$ | 60 |
| River sediment | PCBs | $CO_2$/5% modifier (acetonitrile, $CS_2$, propylene carbonate) | 82 |
| Sediment | PCBs | $CO_2$/2% MeOH | 57 |
| Sediment | Purgable halocarbons | $CO_2$ | 28 |
| Urban air particulates | PAHs | $CO_2$ | 1 |
| | | $N_2O$ | 2 |
| Wood smoke particulates | Guaiacol and syringol derivatives | $CO_2$ | 1 |
| Cigarette smoke particulates | Phenols, nicotine, N-heterocycles | $CO_2$ | 55, 80 |
| | | $N_2O$ | 2 |
| Polyurethane foam (PUF), diesel exhaust, coal gasification water | Alkanes, alkyl benzenes, PAHs, phenols, N-heterocycles | $CO_2$ | 60, 80 |
| Spiked pesticides, herbicides, hexachlorobenzene, PCBs | Pesticides, herbicides, HCB, PCBs | $CO_2$ | 48, 50 |
| Herbicides on $C_{18}$ | Herbicides | $CO_2$/derivatisation | 86 |
| Soil | Pesticides | $CO_2$ | 95 |
| Organic solvents on Tenax-TA | Hazardous organic solvents | $CO_2$ | 70 |
| Soil | Explosives | $CO_2$/15% acetone | 94 |
| Water | Explosives | $CO_2$ | 93 |
| Water | Water contaminants | $CO_2$ | 91 |

Note: PAH = polycyclic aromatic hydrocarbon; PCB = polychlorinated biphenyl; HCB = hexachlorobenzene.

technique can analyse a wide range of gasoline, diesel and motor oil hydrocarbons. Typical SFE–GC/FID conditions for these samples involves the use of a thick film (30 m × 0.32 mm i.d., 5 μm film thickness) DB-1 column operated at a cryogenic trapping temperature of −25°C in the split SFE–GC mode, with a short, 15 min, extraction step with SF-$CO_2$ (400 atm, 60°C) [70]. A similar boiling point range of volatile and semivolatile hazardous organic solvents have been quantitatively extracted from Tenax-TA using the same SFE–GC/FID system [54].

SFE–GC is also compatible with more selective detectors such as electron capture detection (ECD) which are conventionally used for PCB analysis in the environment. Using an SFE–GC/ECD system, PCBs have been quantitatively recovered from a National Water Research Institute certified sediment sample with SF-$CO_2$ modified with 2% methanol [57]. The extract was trapped in an external accumulation device consisting of a short (2 m × 0.33 mm i.d., 1.2 μm film thickness) SE-54 coated silica tube cooled to 5°C. Using such a device the kinetics of the extraction can be evaluated by periodically analysing the extraction effluent. More polar supercritical fluids such as nitrous oxide have been used with on-column SFE–GC/ECD to obtain quantitative PCB recoveries from river sediment, but as the supercritical fluid has a relatively high ECD response, the column has to be flushed with the carrier gas prior to GC analysis [2]. The purity of the supercritical fluid is also important as the sensitivity of the ECD detector can vary with the quality of the fluid, so an on-line filtration system using high-purity $CO_2$ is recommended for SFE–GC/ECD analysis [28, 90].

To ensure that the sensitivity of the ECD is not reduced when analysing wet samples, a drying agent such as Hydromatrix has been used with matrices such as marine sediment (50% m/m water) or plant material (broccoli) contaminated with PCBs and pesticides, respectively [90]. For trace analysis of PCBs in animal tissue, a selective alumina trap is required between the extraction vessel and GC column to ensure that no co-extracted fat is deposited inside the GC column [91]. This selective retention of the fat during the analysis ensures good chromatography and enables the SFE–GC/ECD technique to analyse PCBs quantitatively in cod and crab tissue with a detection limit in the 10–15 pg range. Several biological matrices, such as human milk and blood, have required a more extensive sample clean up prior to SFE–GC/ECD analysis [92]. These materials were preconcentrated off-line by passing the samples through an octadecylsilyl-bonded silica ($C_{18}$) sorbent which is subsequently extracted on-line by SFE. Basic alumina is added to the $C_{18}$/sample matrix to retain lipids and any remaining water, but an additional column packed with alumina is recommended between the extraction vessel and GC column, as the activity of the alumina in the extraction vessel is occasionally reduced by the residual water in the $C_{18}$/sample matrix material. With this SFE–GC/ECD procedure *c.* 90% of spiked Arochlor (0.5 ng) was recovered from human milk, and *c.* 60% of the spike was recovered from serum. The detection limit for the method is *c.* 25 ng/l which is comparable to the regular GC–ECD method, but the SFE–GC method only requires a fraction of the original sample as the whole extract is transferred to the GC column.

This use of sorbents as a solid phase extraction (SPE) preconcentration method has also been used to retain explosives from water [93]. The average SPE recoveries of spiked nitrotoluenes with a phenyl sorbent are 52% to 95% from well and surface water at the low parts-per-billion and parts-per-trillion levels. The preconcentrated sorbents are quantitatively analysed using SFE–GC/ECD with SF-$CO_2$ and SF-$CO_2$ modified with toluene. The presence of toluene increases the rate of extraction but does not improve analyte recovery compared with unmodified SF-$CO_2$.There is also some evidence to suggest that for the on-line analysis of these explosives the use of an inverted cup liner in the split/splitless injector gives better chromatography than a straight tube liner [93].

Besides ECD several other selective detection systems have proved suitable for use in SFE–GC analysis [94, 95]. A chemiluminescence thermal energy analyser (TEA) which is specific for nitro and nitroso compounds has been coupled to SFE–GC to analyse explosives in soil samples [94]. Using this approach relatively volatile and thermally stable nitro compounds can be rapidly (*c.* 10 min) analysed, but thermally labile compounds such as the nitrate ester are shown to decompose in the thermal desorption modulator used to focus the extracted analytes prior to GC analysis. For the extraction of the higher molecular weight nitro compounds such as 1,3,5-trinitrotoluene and 1-nitropyrene, the SF-$CO_2$ has to be modified with 15% vol/vol acetone, but the presence of an organic solvent does not adversely affect the performance of the system. The detection limit for the on-line technique is 2.6 ppb for 2,4-dinitrotoluene [94].

An element selective radiofrequency plasma detector (RPD) has been used in SFE–GC for the analysis of chlorine-containing and sulphur-containing pesticides present in soil [95]. The RPD has a selectivity of 3500 (Cl/C) for chlorine, and 2600 (S/C) for sulphur, which enables the on-line system to have a detection limit of 24.8 pg/s for chlorine, and 9.2 pg/s for sulphur. It was noted that the performance of the RPD was altered when using $CO_2$ as the effluent gas through the GC column; however, by using a higher plasma level and a supporting helium flow the quenching of analyte signal by $CO_2$ was overcome. Both the SFE–GC/TEA and SFE–GC/RPD systems use a high-speed thermally modulated SFE–GC interface where the supercritical fluid extract is introduced into the modulator and concentrated on the stationary phase. The extract is then released as a sharp injection pulse to the analytical column by electrical resistance thermal desorption. This system enables simultaneous monitoring of extraction and separation processes by generating a number of consecutive high-speed gas chromatograms while the extraction proceeds [96]. With this capability of real-time monitoring of the extraction process, quantitative results can be extrapolated before the extraction is complete [97] and if this feature is combined with short trapping times the reliability of analyte quantitation is greatly improved.

When positive confirmation of the identities of contaminants extracted from environmental samples is required, SFE–GC–MS can be used (section 9.4). Unlike the previous detectors, the mass spectrometer is not suitable for use with wide-bore chromatographic columns, since an unacceptably high pressure (*c.* $10^{-3}$ torr) would be generated within the mass spectrometer ion source during the extraction step. In order to reduce the $CO_2$ flow into the mass spectrometer, a 60 m narrow-bore DB-5 column (0.25 mm i.d. ×0.25 μm film thickness) is recommended [2]. The use of a narrow-bore column results in an acceptable ion source pressure during the extraction stage (e.g. $5 \times 10^{-5}$ torr). After the extraction is complete the column is flushed with carrier gas for 3 min to remove residual $CO_2$, and the chromatographic separation is performed in the usual manner. An NBS standard reference material, urban dust (SRM 1649), was used to demonstrate the ability of on-column SFE–GC–MS to achieve rapidly (e.g. less than 1 h) quantitative results which are in agreement with the certified values that are based upon a 48 h Soxhlet extraction [2]. On-column SFE–GC–MS has successfully analysed diesel exhaust pollutants collected on a polyurethane foam sorbent plug, and cigarette smoke particulates collected on a quartz fibre filter [80]. Both sample matrices can be quantitatively extracted within 10 min using SF-$CO_2$ (300 atm, 45°C).

For problematic environmental samples which contain a significant quantity of water, sulphur, test analyte or extractable involatile matrix components split SFE–GC has proved to be the most suitable approach. For example, the NIST standard reference material 1941 (PAH-contaminated marine sediment) is not amenable to on-column SFE–GC analysis as the restrictor plugs with elemental sulphur (2% m/m in the sample) during the extraction step. However, a continuous extraction flow rate can be achieved with split SFE–GC if the injection port is heated to 300°C [60]. Furthermore, split SFE–GC is suitable for wet samples as the heated restrictor eliminates the problem of frozen water forming a plug at the restrictor outlet. Using the heated injection port and maintaining the GC column at 5°C, a wet (20% m/m water) fuel-contaminated sediment and a polyurethane foam sorbent resin soaked in coal gasification wastewater have been quantitatively extracted by split SFE–GC [60]. However, maintaining the column temperature above the freezing point of water during the extraction step limits the application of this SFE–GC technique to *n*-heptane and higher boiling point species. An alternative approach is to use a bed of drying agent with the sample matrix, so that wet (25% m/m water) fuel-contaminated sediments can be analysed with the column operated at −25°C [70]. Using a lower trapping temperature enables better chromatography to be achieved and more volatile analytes can be resolved (Figure 5.6).

Split SFE–GC and to some extent external collection devices can extend the range of environmental samples amenable to the on-line technique, since, analogous to conventional split injections compared with conventional on-column injections of liquid solvents, the use of a split injection port or external accumulation device helps prevent involatile matrix components from being transferred to the GC column. This ability selectively to retain the involatile analytes in the injection port has proved to be very advantageous when analysing relatively high molecular weight material such as motor-oil-contaminated soil [70], shale rock [90] and source rock [28].

Another major concern when working with coupled SFE–GC techniques is that the sample size is generally small (less than 100 mg) [1–3], particularly for on-column SFE–GC. Although this is clearly an advantage when the collection of the sample is difficult (e.g. air particulates), some analyses may require the use of larger sample sizes to ensure homogeneity. Split SFE–GC has been used successfully in the routine analysis of 1 g sample sizes of petroleum-contaminated soil and sediment samples [60, 70], and even 15 g samples of river sediment have been investigated [60]. However, for the large, 15 g, samples a 30 μm i.d. restrictor was required to provide a suitable extraction flow rate over a 30 min time period, and even with this arrangement only the *n*-alkanes were quantitatively recovered from the sediment, as a second sequential extraction recovered significant concentrations of PAHs. Thus, relatively large sample sizes are possible in SFE–GC, but the test analytes will need to be at a relatively low concentration, so the GC column is not overloaded, and the sample matrix must be amenable to SFE, so that quantitative recoveries can be achieved in a realistic period of time.

To enhance the analyte recoveries from complex environmental samples, several more polar supercritical fluids and organically modified supercritical fluids have been investigated. Sulphur hexafluoride has been used in SFE–GC selectively to extract *n*-alkanes from an alkane/PAH test mix spiked onto alumina [1]. However, sulphur hexafluoride generally seems a weaker extraction fluid than $CO_2$, as poor hydrocarbon recoveries were obtained from dried sludge and shale rock when using the $SF_6$ fluid [98]. Formic acid has proved to be an ideal modifier for SFE–GC/FID analysis, as the solvent has a minimal response in FID. Using SF-$CO_2$ modified with 0.3% vol/vol formic acid, a wide range of *n*-alkanes can be extracted from solid hydrocarbon waste without the problem of a solvent peak co-eluting with the analytes of interest [1]. Several more conventional organic solvents, namely methanol, toluene and methylene chloride, have been evaluated as modifiers for the on-line extraction of crude-oil-contaminated soil [83]. The modifiers are added directly to the extraction vessel (e.g. 50 μl for a 1 ml vessel), and a static equilibration step is required prior to the dynamic extraction of the sample matrix to ensure adequate matrix/modified supercritical fluid interaction. Using this approach methanol proved to be the most suitable modifier giving an extraction efficiency of 94% compared with the 74% which was achieved when pure SF-$CO_2$ was used [83]. To extend further the range of environmental analytes that are applicable to SFE–GC, on-line derivatisation methods have been developed. For example, involatile phenoxycarboxylic acids can be derivatised to their volatile esters using a static SFE (SF-$CO_2$, 207 atm, 50°C for 15 min) derivatisation procedure with PFBBr and potassium carbonate to alkylate the acids. The derivatised herbicides were then analysed using on-line GC/ECD [86].

### *5.9.2 Plant and plant-derived samples*

In addition to environmental samples other matrices such as food, spices and fragrances are amenable to SFE–GC (Table 5.9). The most common application is the extraction of essential oils from herbs and spices, as these analytes can be rapidly and quantitatively extracted with pure SF-$CO_2$ [3, 49, 62, 80, 99]. Generally, more problems have been associated with the focusing of the analytes on the GC column than with the extraction process. Several cryogenic trapping temperatures were investigated for the on-column SFE–GC/FID analysis of eucalyptus leaves using a typical 0.5 ml/min (liquid $CO_2$ delivery) extraction flow rate (25 μm i.d. restrictor) [3]. As expected, the two most volatile analytes ($\alpha$-pinene and 1,8-cineole) required a low (−30°C) column trapping temperature to yield narrow peaks, whereas the less volatile analytes such as the oxygenated sesquiterpenes were efficiently trapped at both high (30°C) and low (−30°C) temperatures. Similar findings were found for the on-column analysis of rosemary, thyme and cinnamon [62]. If too low a trapping temperature was used (e.g. −50°C) co-extracted lipids and waxes formed a frozen plug at the head of the column which prevented quantitative studies. A similar phenomenon has been observed with the on-line analysis of brewing hops which require the use of split SFE–GC, so that the non-volatile resins are retained in the injection port liner to allow analysis of the more volatile (i.e. separable by GC) components [60]. An alternative SFE–GC interface is the use of a temperature-programmable on-column injector, consisting of a packed glass liner connected directly to the GC column. With this system a seal exists between the column's outer polyimide coating and the inner surface of the tapered glass liner, so the total SFE extract is transferred to the glass liner and the non-volatile components are retained on the glass wool packing [99]. An additional feature is a valve that was fitted between the extraction vessel and restrictor, which allows short single SFE extract 'injections' to be made into the GC column. In this way the extraction rate of essential oils from cloves could be monitored with time and it was shown that while some compounds such as eugenol were present throughout the extraction other higher molecular weight compounds such as the sesquiterpenes were not extracted until later on in the extraction. Positive identification of these components was achieved using SFE–GC–MS [99].

A simpler means of focusing the analytes is to use an external cold trap/thermal desorption unit consisting of a stainless steel tube (7.9 mm o.d. × 0.5 mm i.d. × 105 cm) placed in a bath of dry ice/ethanol (−58°C), which on commencing the GC analysis can be removed from the bath and rapidly heated [49]. This device was used for the on-column SFE–GC analysis of *Thyme vulgaris* L. plant material and is seen as a means of extending the analytical life of the GC column, as the majority of non-volatile components are retained in the trap. Alternatively, it has been suggested that by regularly trimming the GC column, the column lifetime can be similar to that used for conventional injections (e.g. several months) [3]. Certain samples, such as cheeses that contain a high concentration of extractable semivolatile components, will usually always generate chromatograms with a rising baseline, but the identification and quantification of the free fatty acids and other flavour components is still achievable [100]. However, some sample matrices which contain high concentrations of extractable involatile components (e.g. fats in tissue samples) are not amenable to on-line SFE–GC regardless of the steps taken or the system used. For example, attempts to determine the flavour components in beef jerky using SFE–GC have proved to be impractical [80].

As the majority of plant matrices contain a significant amount of water (typically 70% m/m) and the GC column typically has to be cooled to below 0°C to enable the analytes to be efficiently focused, steps frequently need to be taken to avoid water freezing at the head of the column. As stated above for the environmental samples, a drying agent can be evaluated and a column packed with molecular sieve sorbent placed between the extraction vessel and GC column has been successfully used in the analysis of neat orange juice [81]. Alternatively, if the sample size is sufficiently small (e.g. 2 mg) so that a quantitative recovery can be achieved in a short period of time (e.g. 10 min) at a relative low extraction flow (e.g. using a 15 μm i.d. restrictor), the volatile plant components from fresh pine needles, lemon, lime and orange peels can be efficiently focused on column at 5°C [3].

Generally, for the analysis of flavour and fragrance components from plant material (*c.* 1 mg sample size) a quantitative recovery can be achieved within 10 min using SF-$CO_2$ (300 atm, 45°C), the quantitation being based on the fact that a second sequential extraction of the plant material does not recover any significant quantities of the test analytes [62]. Using these extraction conditions a range of samples has been successfully analysed, including: coriander, mint leaves, parsley, sage, rosemary, thyme, spruce needles, orange, lemon and lime peel, chewing gum, cedar wood and cinnamon [3, 62]. Some of the samples, such as thyme, can be quantitatively extracted at relatively low pressures (80 atm), but generally a pressure in the region of 300 atm is recommended. The extraction flow can also affect the recovery of the analytes from these matrices, because if too low an extraction flow rate is used (e.g. using a 15 μm i.d. restrictor that gives 0.05 ml/min liquid $CO_2$ measured at the pump) poor analytical recoveries are obtained, even though good chromatographic peak shapes are achievable. Conversely, if too high a flow rate is used (e.g. 40 μm i.d. restrictor that gives more than 1 ml/min liquid $CO_2$ measured at the pump) quantitative recoveries are achievable, but poor chromatography with severe band broadening will occur [62]. The optimum flow rate was achieved with 20, 25 or 30 μm i.d. restrictors to give liquid $CO_2$ flow rates in the region of 0.2–0.6 ml/min that generated both quantitative recoveries and good chromatography, using a 30 m × 0.32 mm i.d. (1 μm film thickness) DB-5 capillary column at −30°C. A positive identification of the components in the majority of the sample matrices was achieved using SFE–GC–MS [3, 62].

Using the optimised analyte extraction and focusing conditions of SF-$CO_2$ (300 atm, 45°C, for 10 min) with a 25 μm i.d. restrictor into a 30 m × 0.32 mm i.d. (1 μm film thickness) DB-5 capillary column at −30°C, the on-column SFE–GC analysis of 1 mg of basil and eucalyptus leaves yielded peak widths essentially the same as those obtained using an on-column injection of the methylene chloride extract [3]. The small, 1 mg, sample size is often desirable for the analysis of herbs and spices as the flavour and fragrance components in these matrices can be present in the order of 1 mg/g [62], which means that the upper limit of sample size to avoid severely overloaded chromatography peaks is *c*. 1 mg of the bulk sample. With such a small sample there will always be some concern that the sample is not representative of the bulk matrix. However, the raw integrated peak areas of individual compounds from the basil sample generated RSDs ranging from 6% to 17%, which was comparable to those obtained by replicate on-column injections of the liquid solvent extract [3]. This demonstrates that the SFE–GC technique is capable of good reproducibility even with the small sample sizes studied.

The versatility of SFE–GC for the analysis of plant materials can be further expanded by the addition of an on-line derivatisation method which is performed under supercritical fluid conditions. An integrated extraction and derivatisation technique has been applied to the determination of fatty acid composition in soybean, evening primrose and peanut seeds [85]. Triglycerides were extracted from the seed samples using SF-$CO_2$ (200 atm, 60°C) and transesterified to their methyl esters, *in situ*, with alumina pretreated with methanol. The fatty acid methyl ester distribution obtained by SFE–GC was comparable to the literature values and demonstrated that the transesterification of triglycerides with methanol is promoted under supercritical fluid conditions in the presence of alumina. As methanol is used in the derivatisation step a broad solvent peak is observed at the beginning of the chromatogram. However, by using a retention gap (7 m × 0.53 mm i.d. deactivated fused-silica capillary) at the

**Table 5.9** Applications of supercritical fluid extraction–gas chromatography for plant tissues and food

| Sample matrix | Analyte | Extraction fluid | Reference |
|---|---|---|---|
| Fir needles | Essential oils | $CO_2$ | 55 |
| Thyme (*Thymus vulgaris* L.) | Essential oils | $CO_2$ | 49 |
| Eucalyptus leaves | Essential oils | $CO_2$ | 1 |
| Brocolli | Pesticides | $CO_2$ | 90 |
| Oilseed | Triglycerides | $CO_2$/derivatisation | 85 |
| Brewing hops | Essential oils | $CO_2$ | 60 |
| Lemon peel | Essential oils | $CO_2$ | 3 |
| Spices | Essential oils | $CO_2$ | 62, 80, 92 |
| Swiss cheese | Free fatty acids | $CO_2$ | 101 |
| Wine | Wine aroma components | $CO_2$ | 102 |
| Food packaging | Aroma compounds | $CO_2$ | 100 |

head of the GC column, the extract was refocused prior to GC analysis which enabled good chromatographic resolution of the methyl esters.

A novel application for SFE–GC is the analysis of aroma compounds sorbed by food packing material [101]. It has been well established that food components can be sorbed by polymeric packaging material, which can lead to loss of aroma intensity and damage to the barrier properties of the package. On-column SFE–GC/FID has been used to analyse apple aromas sorbed onto a range of polyethylene, polypropylene and poly(ethylene terephthalate) polymers. The aromas from the polymers are quantitatively extracted with SF-$CO_2$ (80 bar, 40°C, for 15 min) and efficiently retained on a 60 m × 0.25 mm i.d. Supelcowax 10 capillary column at −50°C. The recoveries and the chromatography are comparable to a conventional liquid methylene chloride extraction and GC analysis, but the SFE–GC analysis can be accomplished in 40 min and requires a much smaller amount of sample (1%–2%) than does the liquid extraction method. A further on-line experiment involved the successful determination of the amount of limonene sorbed into poly(ethylene terephthalate) bottles which had orange-flavoured soft drinks stored in them for several weeks [101].

Finally, the analysis of wine aroma was achieved using SFE–GC coupled via a PTV [102]. Red or white wine (1 ml) is placed on filter paper and extracted for 25 min with SF-$CO_2$ (370 atm, 40°C) and the extracted components are retained at −5°C on 15 mg of Thermotrap TA (Chrompack) packed into the glass liner of the PTV. GC analysis was performed by thermally desorbing the extracted analytes into a 30 m × 0.25 mm i.d. fused silica capillary column coated with a 0.25 μm layer of polyethylene glycol. The SFE–GC analysis of the wine proved to be more sensitive and reproducible than did the off-line SFE approach, and the on-line technique was able to recover quantitatively the majority of volatile components in the wine with a detection limit of between 0.08 and 0.34 ng.

## 5.10 Conclusions

On-line SFE–GC is an attractive approach to coupling the extraction and analysis procedures for real-world samples and offers the analyst a rapid, sensitive, quantitative and yet selective analytical method. A quantitative SFE–GC analysis (including the loading and assembly of the extraction vessel, performing the extraction and GC analysis) can be routinely undertaken in less than an hour. Comparable recoveries obtained by conventional solvent extraction usually require several hours.

The optimisation of the SFE–GC method can initially appear to be a complicated procedure and is inherently more involved than off-line SFE. However, the utilisation of the GC column for analyte collection and focusing has sufficient advantages over off-line techniques to warrant the extra time and effort in assembling the on-line system. By optimising the most important experimental parameters – namely, the SFE conditions, the cryogenic trapping temperature and the column stationary phase thickness – quantitative SFE–GC results can quickly be achieved. On-column SFE–GC is recommended when maximum sensitivity is required, whereas split SFE–GC is more suitable for larger sample sizes and high analyte concentrations. External analyte traps are advantageous when the sample requires the use of modified supercritical fluids.

SFE–GC unquestionably offers an excellent alternative to conventional sample preparation and has the potential to become a routine problem-solving tool for the analytical chemist. It is envisaged that the technique will continue to be developed and become a more reliable, simple and automated analytical procedure which could realistically replace purge and trap analysis of volatiles and liquid solvent extraction of semivolatiles.

## References


1. Levy, J.M., Cavalier, R.A., Bosch, T.N. *et al.* (1989) *J. Chromatogr. Sci.*, **27**, 341.
2. Hawthorne, S.B. and Miller, D.J. (1987) *J. Chromatogr.*, **403**, 63.
3. Hawthorne, S.B., Miller, D.J. and Krieger, M.S. (1989) *J. Chromatogr. Sci.*, **27**, 347.
4. Ramalho, S., Hankemeier, T., de Jong, M. *et al.* (1995) *J. Microcol. Sep.*, **7**, 383.
5. Staniewski, J. and Rijks, J.A. (1993) *J. High Resolut. Chromatogr.*, **16**, 182.
6. Rinkema, F.D., Louter, A.J.H. and Brinkman, U.A.T. (1994) *J. Chromatogr.*, **678**, 289.
7. Grob, K. and Müller, E. (1987) *J. Chromatogr.*, **404**, 297.
8. Parr, J.L., Walters, G. and Hoffman, M. (1991) *Hydrocarbon Contaminated Soils and Groundwater*, Enseco Inc., Arvada, CO, Ch. 8.
9. Tekmar (1982) *Technical bulletin*, Tek/Dat B021062, Tekmar, Cincinnati, OH, USA.
10. Hawthorne, S.B. and Miller, D.J. (1986) *J. Chromatogr. Sci.*, **24**, 258.
11. Hawthorne, S.B. and Miller, D.J. (1987) *Anal. Chem.*, **59**, 1705.
12. Wright, B.W., Wright, C.W. and Fruchter, J.S. (1989) *Energy Fuels*, **3**, 474.
13. Wheeler, J.R. and McNally, M.E. (1989) *J. Chromatogr. Sci.*, **27**, 534.
14. Campbell, R.M., Meunier, D.M. and Cortes, H.J. (1989) *J. Microcol. Sep.*, **1**, 302.
15. Lopez-Avila, V., Benicto, J., Dodhiwala, N.S. and Young, R. (1992) *J. Chromatogr. Sci.*, **30**, 335.

16. Eckert-Tilotta, S.E., Hawthorne, S.B. and Miller, D.J. (1993) *Fuel*, **72**, 1015.
17. Field, J.A., Miller, D.J., Field, T.M. *et al.* (1992) *Anal. Chem.*, **64**, 3161.
18. Lin, Y., Brauer, R.D., Laintz, K.E. and Wai, C.M. (1993) *Anal. Chem.*, **65**, 2549.
19. Wang, J. and Marshall, W.D. (1994) *Anal. Chem.*, **66**, 1658.
20. Hirata, Y. and Oikamoto, Y. (1989) *J. Microcol. Sep.*, **1**, 46.
21. Sugiyama, K. and Saito, M. (1988) *J. Chromatogr.*, **442**, 121.
22. Smith, R.M. and Burford, M.D. (1994) *J. High Resolut. Chromatogr.*, **32**, 265.
23. King, J.W. (1989) *J. Chromatogr. Sci.*, **27**, 355.
24. Berg, B.E., Hansen, E.M., Gjorven, S. and Greibrokk, T. (1993) *J. High Resolut. Chromatogr.*, **16**, 358.
25. Schneiderman, M.A., Sharma, A.K. and Locke, D.C. (1988) *J. Chromatogr. Sci.*, **26**, 458.
26. British Patent (1975) Specification 1 388 581.
27. Sugiyama, K., Saito, M., Hondo, T. and Senda, M. (1985) *J. Chromatogr.*, **332**, 107.
28. Levy, J.M. and Rosselli, A.C. (1989) *Chromatographia*, **28**, 613.
29. Nielen, M.W.F., Stäb, J.A., Lingeman, H. and Brinkman, U.A.T. (1991) *Chromatographia*, **32**, 543.
30. Jinno, K., Hoshino, T., Hondo, T. *et al.* (1986) *Anal. Chem.*, **58**, 2696.
31. Smith, R.D., Udseth, H.R. and Wright, B.W. (1986) *J. Chromatogr. Sci.*, **24**, 238.
32. French, S.B. and Novotny, M. (1986) *Anal. Chem.*, **58**, 164.
33. Jinno, K. (1992) Chapter 9, in *Journal of Chromatography Library Series, Volume 53. Hyphenated Techniques in Supercritical Fluid Chromatography and Extraction* (ed. K. Jinno), Elsevier, New York.
34. Baner, L., Bøcherl, T., Ewender, J. and Franz, R. (1992) *J. Supercritical Fluids*, **5**, 213.
35. Ashraf-Khorassani, M., Boyer, D.S. and Levy, J.M. (1991) *J. Chromatogr. Sci.*, **29**, 517.
36. Xie, Q.L., Markides, K.E. and Lee, M.L. (1989) *J. Chromatogr. Sci.*, **27**, 365.
37. Vannoort, R.W., Chervat, J.-P., Lingeman, H. *et al.* (1990) *J. Chromatogr.*, **505**, 45.
38. Liu, M.H., Kapila, S., Nam, K.H. and Elseewi, A.A. (1993) *J. Chromatogr.*, **639**, 151.
39. Davies, I.L., Raynor, M.W., Kithinji, J.P. *et al.* (1988) *Anal. Chem.*, **60**, 683A.
40. Stahl, E., Schilz, W., Schutz, E. and Willing, E. (1978) *Angew. Chem. Int. Ed. Engl.*, **17**, 731.
41. Stalling, D.L., Saim, S., Kuo, K.C. and Stunkel, J.J. (1992) *J. Chromatogr. Sci.*, **30**, 486.
42. Hawthorne, S.B. (1990) *Anal. Chem.*, **60**, 633A.
43. Anderson, M.R., King, J.W. and Hawthorne, S.B. (1990) Chapter 5, in *Analytical Supercritical Fluid Chromatography and Extraction* (eds M.L. Lee and K.E. Markides).
44. Dean, J.R. and Kane, M. (1993) Chapter 3, in *Applications of Supercritical Fluids in Industrial Analysis* (ed. J.R. Dean), CRC Press, Boca Raton, FL.
45. Riekkola, M.-L., Manninen, P. and Hartonene, K. (1992) Chapter 14, in *Journal of Chromatography Library Series, Volume 53. Hyphenated Techniques in Supercritical Fluid Chromatography and Extraction* (ed. K. Jinno), Elsevier, New York.
46. Burford, M.D., Hawthorne, S.B. and Miller, D.J. (1993) *Anal. Chem.*, **65**, 1497.
47. Janda, V., Bartle, K.D. and Clifford, A.A. (1993) *J. Chromatogr.*, **642**, 283.
48. Nielen, M.W.F., Sanderson, J.T., Frei, R.W. and Brinkman, U.A.T. (1989) *J. Chromatogr.*, **474**, 388.
49. Hartonen, K., Jussila, M., Manninen, P. and Riekkola, M.-L. (1992) *J. Microl. Sep.*, **4**, 3.
50. Raymer, J.H. and Velez, G.R. (1991) *J. Chromatogr. Sci.*, **29**, 467.
51. Pankow, J.F. (1988) *Anal. Chem.*, **60**, 950.
52. Liang, X. and Hewitt, C.N. (1992) *J. Chromatogr.*, **627**, 219.
53. Cooper, S.D. and Pellizzari, E.D. (1990) *J. Chromatogr.*, **498**, 41.
54. Burford, M.D., Hawthorne, S.B. and Miller, D.J. (1994) *J. Chromatogr.*, **685**, 79.
55. Lohleit, M. and Bächmann, K. (1990) *J. Chromatogr.*, **505**, 227.
56. Wright, B.W., Frye, S.R., McMinn, D.G. and Smith, R.D. (1987) *Anal. Chem.*, **59**, 640.
57. Onuska, F.I. and Terry, K.A. (1989) *J. High Resolut. Chromatogr.*, **12**, 527.
58. Levy, J. and Rosselli, A.C. (1989) *Chromatographia*, **28**, 613.
59. Leibman, S.A., Levy, E.J., Lurcott, S. *et al.* (1989) *J. Chromatogr. Sci.*, **27**, 118.
60. Hawthorne, S.B., Miller, D.J. and Langenfeld, J.J. (1990) *J. Chromatogr. Sci.*, **28**, 2.
61. Houben, R.J., Janssen, H.-G.M., Leclercq, P.A. *et al.* (1990) *J. High Resolut. Chromatogr.*, **13**, 669.
62. Hawthorne, S.B., Krieger, M.S. and Miller, D.J. (1988) *Anal. Chem.*, **60**, 472.
63. Lou, X., Janssen, H.-G. and Cramers, C.A. (1993) *J. High Resolut. Chromatogr.*, **16**, 425.

64. Pyle, S.M. and Setty, M.M. (1991) *Talanta*, **38**, 1125.
65. Johansen, H.R., Becher, G. and Greibrokk, T. (1992) *Fresenius J. Anal. Chem.*, **344**, 486.
66. Hawthorne, S.B., Miller, D.J., Burford, M.D. *et al.* (1993) *J. Chromatogr.*, **642**, 301.
67. Langenfeld, J.J., Burford, M.D., Hawthorne, S.B. and Miller, D.J. (1992) *J. Chromatogr.*, **594**, 297.
68. Lohleit, M., Hillmann, R. and Bachmann, K. (1991) *Z. Anal. Chem.*, **339**, 470.
69. Levy, J.M. and Guzowski, J.P. (1988) *Fresenius Z. Anal. Chem.*, **330**, 207.
70. Burford, M.D., Hawthorne, S.B. and Miller, D.J. (1994) *J. Chromatogr.*, **685**, 95.
71. Wallace, J.C., Krieger, M.S. and Hites, R.A. (1992) *Anal. Chem.*, **64**, 2655.
72. Burford, M.D., Hawthorne, S.B. and Miller, D.J. (1993) *J. Chromatogr.*, **657**, 413.
73. Hagman, A. and Jacobsson, S. (1988) *J. Chromatogr.*, **448**, 117.
74. Graydon, J.W. and Grob, K. (1983) *J. Chromatogr.*, **254**, 265.
75. Hopper, M.L. and King, J.W. (1991) *J. Assoc. Off. Anal. Chem.*, **74**, 661.
76. Porter, N.L., Rynaski, A.F., Campbell, E.R. *et al.* (1992) *J. Chromatogr. Sci.*, **30**, 367.
77. Burford, M.D., Hawthorne, S.B. and Miller, D.J. (1992) *J. Chromatogr.*, **609**, 321.
78. Smith, R.D., Fulton, J.L., Petersen, R.C. *et al.* (1986) *Anal. Chem.*, **58**, 2057.
79. Bartle, K.D., Clifford, A.A., Jafar, S.A. *et al.* (1991) *J. Phys. Chem. Ref. Data.*, **20**, 713.
80. Hawthorne, S.B., Miller, D.J. and Krieger, M.S. (1989) *J. High Resolut. Chromatogr.*, **12**, 714.
81. Levy, J.M. and Ashraf-Khorassani, M. (1992) Chapter 11, in *Journal of Chromatography Library Series, Volume 53. Hyphenated Techniques in Supercritical Fluid Chromatography and Extraction* (ed. K. Jinno), Elsevier, New York.
82. Levy, J.M., Storozynsky, E. and Ashraf-Khorassani, M. (1991) *Recent Advances in Supercritical Fluid Technology* (ed. M.E. McNally), ACS Symposium Series, American Chemical Society, Washington, DC.
83. Levy, J.M., Dolata, L., Ravey, R.M. *et al.* (1993) *J. High Resolut. Chromatogr.*, **16**, 368.
84. Xianwen, L., Janssen, H.-G. and Cramers, C.A. (1996) *J. Chromatogr.*, **750**, 215.
85. King, J.W., France, J.E. and Snyder, J.M. (1992) *Fresenius J. Anal. Chem.*, **344**, 474.
86. Hillmann, R. and Bachmann, K. (1994) *J. High Resolut. Chromatogr.*, **17**, 350.
87. Hawthorne, S.B. and Miller, D.J. (1994) *Anal. Chem.*, **66**, 4005.
88. Langenfeld, J.J., Hawthorne, S.B., Miller, D.J. and Pawliszyn, J. (1995) *Anal. Chem.*, **67**, 727.
89. Yang, Y., Gharaibeh, A., Hawthorne, S.B. and Miller, D.J. (1995) *Anal. Chem.*, **67**, 641.
90. Levy, J.M., Rosselli, A.C., Storozynsky, E. *et al.* (1992) *LC–GC*, **10**, 386.
91. Pawlyszin, J.B. and Alexandrou, N. (1989) *Water Poll. Res. J. Can.*, **24**, 207.
92. Johansen, H.R., Thorstensen, C., Greibrokk, T. and Becher, G. (1993) *J. High Resolut. Chromatogr.*, **16**, 148.
93. Slack, G.C., Mcnair, H.M., Hawthorne, S.B. and Miller, D.J. (1993) *J. High Resolut. Chromatogr.*, **16**, 473.
94. Francis, E.S., Wu, M., Farnsworth, P.B. and Lee, M.L. (1995) *J. Microcol. Sep.*, **7**, 23.
95. Wu, M., Liu, Z., Farnsworth, P.B. and Lee, M.L. (1993) *Anal. Chem.*, **65**, 2185.
96. Liu, Z., Farnsworth, P.B. and Lee, M.L. (1992) *J. Microcol. Sep.*, **4**, 199.
97. Bartle, K.D., Clifford, A.A., Hawthorne, S.B. *et al.* (1990) *J. Supercrit. Fluids*, **3**, 143.
98. Levy, J.M., Storozynsky, E. and Ravey, R.M. (1991) *J. High Resolut. Chromatogr.*, **14**, 661.
99. Huston, C.K. and Ji, H. (1991) *J. Agric. Food Chem.*, **39**, 1229.
100. Tuomala, T. and Kallio, H. (1996) *Z. Lebensm. Unters Forsch*, **203**, 236.
101. Nielsen, T.J. and Jägerstad, M. (1995) *Spec. Publ. R. Soc. Chem.*, **162**, 59.
102. Blanch, G.P., Reglero, G. and Herraiz, M. (1995) *J. Agric. Food Chem.*, **43**, 1251.

# 6 Coupled supercritical fluid extraction–capillary supercritical fluid chromatography

H.J. VANDENBURG, K.D. BARTLE, N.J. COTTON AND M.W. RAYNOR

## 6.1 Introduction

Supercritical fluids have several advantages over liquids when used as extraction solvents. These are fully detailed in Chapter 1 and include:

- improved extraction efficiency in many cases with respect to Soxhlet extraction, resulting in reduced extraction times:
  - supercritical fluids have more rapid mass-transfer properties than do liquids and therefore penetrate and extract components in sample matrices more quickly and efficiently; in contrast to Soxhlet extraction, which requires several hours or days to perform, supercritical fluid extraction (SFE) can yield quantitative extraction in as little as 10 min [1];
  - on some compound groups SFE is reported to give better recoveries than can liquid–liquid extraction [2];
- the possibility of adjusting the solvent strength of the fluid: as the solvating power of a supercritical fluid is directly related to its density (see Figure 4.1) it can be tuned to the analysis as required; for example, the sample can be extracted exhaustively at constant high density to provide quantitative analysis, or at various pressures or with different mobile phases to fractionate the sample;
- the extraction of heat-sensitive or reactive compounds at low temperatures;
- the use of extraction solvents, such as $CO_2$, which are non-toxic and inexpensive;
- the rapid removal of the supercritical fluid from the extracted analytes on decompression.

Many of the advantages discussed above are also applicable to supercritical fluid chromatography (SFC). Separations are performed in solution at low temperatures and therefore SFC is a particularly good technique for analysing thermally labile, reactive or involatile materials. Solutes have higher diffusion coefficients in supercritical fluids than they do in liquids which allows the use of more efficient open tubular capillary SFC (cSFC) columns.

Further, supercritical fluids such as $CO_2$ are also compatible with flame-based detectors and with Fourier transform infra-red spectroscopy (FTIR) and mass spectrometry (MS), allowing the trace analysis of compounds without a chromophore.

After extraction the sample needs to be transferred to a chromatographic analytical column. This is commonly done off-line by collecting the extract in solvent and injecting an aliquot into the cSFC column. The intrinsic problems with off-line collection are that sample loss and contamination is possible, the process is difficult to automate and the sample must be diluted with solvent to allow transfer, resulting in dilute solutions and hence higher detection limits. cSFC columns are small and easily overloaded, particularly with injection solvent. For example, a 1 μl injection occupies 0.5 m of a 50 μm inner diameter (i.d.) column. Larger injections can easily cause band broadening and peak splitting. The limitation of injection size further increases the detection limit. A logical method of solving the intrinsic problems of off-line collection and cSFC is to link them on-line. This was first reported in 1985 by Sugiyama *et al.* [3]. Tables 6.1 and 6.2 list the SFE–cSFC studies and applications that have been reported in the literature since then. These studies, in keeping with the title of this chapter, are restricted to those that report on-line extraction and cSFC analysis.

## 6.2 Samples for which supercritical fluid extraction–capillary supercritical fluid chromatography is applicable

Table 6.1 shows a comparison of the application areas of SFE–cSFC and SFE–GC (gas chromatography). If the analytes are thermally stable and volatile, GC is the best separation technique to use. Many flavour and fragrance compounds in complex food samples should therefore be analysed by SFE–GC. The same is true of polychlorinated biphenyls (PCBs), pesticides and polycyclic aromatic hydrocarbons (PAHs) in environmental samples.

When the sample contains thermally labile or reactive compounds, SFE–SFC is recommended. The procedure is excellent for thermally unstable polymer additives in commercial plastics or for fatty acids and triglycerides in food, etc., which cannot be analysed by GC very easily without derivatisation. Natural products, such as those containing terpene compounds, or hops, which contain highly reactive bitter compounds such as humulone and lupulone, must also be analysed by SFC or high-performance liquid chromatography (HPLC) as rearrangement can easily occur at elevated temperatures. Speciation studies on organotins, an important environmental pollutant, are difficult using GC or HPLC as derivatisations are required to increase volatility or provide a chromophore [4, 5]. Other application areas specific to cSFC include the analysis of explosives and certain steroids,

**Table 6.1** Comparison of reported application areas of on-line supercritical fluid extraction–capillary supercritical fluid chromatography (SFE–cSFC) and supercritical fluid extraction–gas chromatography (SFE–GC)

| Sample matrix | Analysis | SFE–cSFC | SFE–GC | Reference |
|---|---|---|---|---|
| Industrial samples, smokeless powders | Explosives, propellants | Yes | No | 35 |
| Commercial plastics | Polymer additives, oligomers | Yes | No | 23, 25, 34, 40, 43, 47, 48, 49, 50 |
| Paper | Impregnated substances | Yes | No | 51 |
| Biological samples: | | | | |
| Animal and plant tissue | Steroids, vitamins, drug residues | Yes | Yes | 26, 45, 52, 53, 54 |
| Water, urine | Phenols, phosphonates | Yes | No | 10, 11 |
| Animal faeces | Fatty acids, triglycerides | Yes | No | 55 |
| Foods and natural products: | | | | |
| Butter, cheese, oils | Fatty acids, triglycerides | Yes | No | 56 |
| Coffee | Flavour compounds, caffeine | Yes | No | 3, 31, 56, 57 |
| Spices | Flavour and fragrance compounds | Yes | Yes | 57, 58, 59 |
| Fruit peels, leaves, wood | Terpenes, aldehydes, alcohols | Yes | Yes | 58, 59, 60 |
| Wheat | Fatty acids | Yes | No | 18 |
| Eggs | Cholesterol | Yes | No | 17 |
| Environmental samples: | | | | |
| Coal, coal tar, treated wood | PAHs, sesquiterpenes | Yes | Yes | 57, 61 |
| Soil and sedimentary samples | PCBs, PAHs, pesticides | Yes | Yes | 22, 38, 60, 62 |
| Air and exhaust particulates | PAHs, hydrocarbons, nicotine | No | Yes | 29, 57, 58, 63 |
| Sorbents: | | | | |
| Tenax-GC, charcoal | PCBs, PAHs, hydrocarbons | Yes | Yes | 64, 65, 66 |
| Silica, alumina, LC material | PAHs, hydrocarbons | Yes | Yes | 61, 63, 67, 68 |
| | Prostaglandins | Yes | No | 12 |
| Polyurethane foam | Air, water pollutants, PAHs, phenols, hydrocarbons | No | Yes | 58, 69 |
| Deactivated glass beads | Fatty acids, fatty acid methyl esters, trilaurin | Yes | No | 13 |

Note: LC = liquid chromatography; PAH = polycyclic aromatic hydrocarbon; PCB = polychlorinated biphenyl.

**Table 6.2** Selected applications of supercritical fluid extraction–capillary supercritical fluid chromatography

| Sample and extracts | SFE conditions | Trapping or sampling conditions | cSFC conditions | Detector | Quantitative | Ref. |
|---|---|---|---|---|---|---|
| Fatty acids (standards and from wheat flour) in new microextraction cell | 100–500 atm, 40°C–80°C | Cryo-cooled trap | 100°C, 140–170 atm (112 min). 20 m × 100 μm i.d. SB-cyano-50 (0.25 μm) (2 in series) | FID | Yes | 18 |
| Total extractables and antioxidants from polymers (Biopol/PP) | 20 MPa, 40°C (Biopol) 35 MPa, 40°C (PP) | Aliquot sampling, SFE output measured using FID, peak cut to SFC to calibrate SFE 'chromatogram' | 140°C, 20 MPa (Biopol) 15 MPa (10 min) to 35 MPa (5 min) at 0.3 MPa/min (PP). 10 m × 0.102 mm i.d. 0.5 μm DB 5 | FID | Yes | 23 |
| Fats enzymatically converted to esters in SFE cell. Formic-acid-modified $CO_2$ as extraction solvent and mobile phase | 50°C, 150–300 bar | 1.7 m × 100 μm capillary, 0.1 μm BP-1 held at 25°C | 110°C, 100–400 bar. 10 m × 50 μm i.d. SB-biphenyl-30 (0.25 μm) | FID | No | 16 |
| Edible fat and cholesterol in egg yolk, enzymatic esterification of fats in extraction cell | 50°C, 300–400 bar | 1.8 m × 100 μm fused-silica capillary. Solvent venting by helium purge | 55°C–100°C at 49°C/min, 100–230 bar at 4 bar/min. 10 m × 50 μm i.d. SB-cyanopropyl-50 0.25 μm | FID | Yes | 17 |
| Polymer additives identified by new SFC–MS interface | 15–25 MPa, 60°C | Fisons equipment, no details | 140°C, 10.6 MPa (10 min)–15 MPa (5 min) in 3.5 min, then to 35 MPa at 0.5 MPa/min. 20 m × 100 μm i.d. 0.5 μm DB-5 | FID/MS | No | 49 |
| Hydrocarbons, FAMES, alcohols, pesticides, spiked onto silica and soil. Selective extraction of hydrocarbon, alcohols and FAMES using different pressure or modifier conditions | 60°C–105°C, 200–300 atm, various modifiers | Collection system: 3 m fused-silica capillary. 100 μm (0.5 μm SB-octyl) and 1 m × 0.53 mm capillary (5 μm polydimethyl siloxane). Solvent removed after trapping with $CO_2$ Refocusing system: 1 m × 0.15 mm capillary. 1.2 μm CP-Sil 5CB | 90°C, 10 m × 100 μm i.d. fused-silica capillary HP-5 (0.34 μm) | FID | Yes | 38 |
| Polymer additives and test mix | 150 bar, 45°C, 15 min | Cryogenic fused-silica capillary plus retention gap – had no effect on chromatography | 80–350 bar (20 bar/min), 100°C. 10 m × 50 μm i.d. SB-phenyl-30 | FID | No | 25 |
| Vitamins analysed using micropacked columns | 40°C, 281 atm | Deposited at head of analytical column, cooled by expanding $CO_2$ | 100°C, 90 atm (4 atm/min), 20 m × 0.5 cm i.d. SE-54 and Carbowax 20 | FID | Yes | 33 |

**Table 6.2** (*cont.*)

| Sample and extracts | SFE conditions | Trapping or sampling conditions | cSFC conditions | Detector | Quantitative | Ref. |
|---|---|---|---|---|---|---|
| 11 Prostaglandins from aqueous solutions (SPA–SFE–cSFC). Compares with direct injection SFE–cSFC | Sample loaded onto ODS or XAD-2, SFE 35°C–50°C, 0.8 g/ml $CO_2$ or direct injection (1 μl) | Deactivated cryo-cooled fused-silica capillary (183 μm i.d.) | 100°C, density programmed. 3 m × 50 μm i.d. Oligoethylene-oxide-substituted polysiloxane stationary phase | FID | Yes | 12 |
| Assess various solid phase adsorbents, SPA–SFE–cSFC (triglycerides and FAMES) | 400 atm, 80°C, 10 min | Deactivated or uncoated fused-silica capillary cryo-cooled with $CO_2$ | 100–400 atm, 5 m × 50 μm i.d. SB-Octyl-50 (0.25 μm) | FID | No | 14 |
| Fatty acids and sterols from hamster faeces | 60°C, 400 atm | Cryogenically cooled trap | 140°C–240°C at 5°C/min, 75–350 atm at 5 atm/min. 10 m × 50 μm i.d. SB-methyl-100 (0.25 μm) | FID | No | 55 |
| Odour compounds in *Dalea spinosa*, single seeds analysed | 45°C, 100–400 atm | Cryofocusing trap ($CO_2$ cooled) | Several on different columns, 10 m × 0.1 mm i.d. SB-Phenyl-5, 50 and SB-octyl-50, 0.5 μm | FID | No | 42 |
| Fatty acids and FAMES adsorbed on glass beads | 80°C, 75–400 atm (10 min) | Cryogenic trap, glass-lined or deactivated fused-silica capillary | 100–140 atm at 10 atm/min, 80°C. 5 m × 50 μm i.d. SB-octyl-50, 25 μm | FID | Yes | 13 |
| Differentiate between native and added compounds from paper, e.g. drugs, using FTIR | Details elsewhere | | 20–10 m × 100 μm i.d. DB-5 (4 μm) | FTIR | No | 51 |
| Polycyclic aromatic compounds (PAC) and other aromatic compounds extracted from vacuum-still bottoms | Pentane 210°C, 20–200 atm; static SFE | Aliquot sampling | Pentane, 210°C, 30–45 atm at 0.2 atm/min. 20 m × 50 μm i.d. SB-phenyl-50 | Fluorescence | No | 21 |
| Extractable compounds in cheese, butter, coffee, tobacco and camomile | $CO_2$, 40°C, 100 bar, static SFE | Aliquot sampling | $CO_2$ isobaric. 130 bar, temperature programmed to 220°C. 14 m × 250 μm i.d. Supelcowax 10 | FID | No | 56 |
| PAC extracted from a coal tar pitch | $CO_2$, 35°C–50°C, 70–200 atm, 30 min dynamic SFE | Trapped in rotor groove of cooled injection valve (200 nl) | $CO_2$, 110°C, 0.25–0.74 g/ml at 0.006 g/ml per min, 10 m × 50 μm i.d. SB-biphenyl-30 | FID | No | 30 |

**Table 6.2** (*cont.*)

| Sample and extracts | SFE conditions | Trapping or sampling conditions | cSFC conditions | Detector | Quantitative | Ref. |
|---|---|---|---|---|---|---|
| Caffeine from coffee, oligomeric compounds from polyethylene, hydrocarbons from spiked soil | $CO_2$, 25°C–60°C, 50–250 bar; dynamic SFE | Extracted compounds deposited in 0.1 μl rotor groove of injection valve | $CO_2$, 150°C, 107–423 bar at 7 bar/min, 10 m × 50 μm i.d. biphenyl–methyl–silicone phase | FID | No | 47 |
| Caffeine from decaffeinated coffee sample | $CO_2$, 50°C, 14 MPa, 10 min dynamic extraction | Aliquot sampling | $CO_2$, 80°C, 0.25–0.66 g/ml, 10 m × 50 μm i.d. DB-5 (0.2 μm film) | FID | No | 26 |
| Ouabain in biological assay | $CO_2$, 80°C, 400 atm, 30 min, dynamic extraction | Cryogenic trapping in 150 μm capillary; fraction collection | $CO_2$, 80°C, 200–400 atm, 3 m × 50 μm i.d. SB-methyl | FID | No | 45 |
| Terpenes, limonene carvone, etc. from grapefruit oil | $CO_2$, 70°C. Dynamic extraction at 0.356 g/ml, 12 min | −10°C to −65°C cryogenic trapping in deactivated capillary | $CO_2$, 150°C , 0.1–0.61 g/ml per min, 10 m × 100 μm i.d. SB-phenyl-5 | FID | No | 60 |
| Pesticides (diuron, linuron) in soil and wheat grain | $CO_2$ and $CO_2$ modified with MeOH and EtOH in cell at 120°C, 0.3–0.7 g/ml density, static and dynamic SFE | Off-line collection vessel or 200 nl aliquot injected into cSFC | 120°C, 0.2–0.7 g/ml, 10 m × 100 μm i.d. SB-Cyanopropyl-25 or 100 μm × 1 mm i.d. $C_{18}$ packed column | Liquid scintillation counting and FID | Yes | 22 |
| $C_{12}$–$C_{20}$ hydrocarbons and polymer additives from polypropylene | $CO_2$, 90°C, 300 atm, 30 min, dynamic extraction | −40°C cryogenic trap with: (1) deactivated fused silica; (2) OV-1 coated fused silica; (3) 250 μm fused silica packed with ODS | $CO_2$, 90°C, 80–400 atm at 5 atm/min, 10 m × 100 μm i.d. octyl (0.5 μm) | FID | No | 34 |
| Cuticular hydrocarbons and waxes from fruit beetle. Triglycerides and fatty acids from cottonseed kernel | $CO_2$, 45°C, 200 atm, 1 min dynamic extraction | Cryofocusing using expanding $CO_2$ | $CO_2$, 100°C, 50–290 atm, 10 m × 50 μm i.d. SB, methyl or SB-biphenyl-30 | FID | No | 41 |
| Additives in polyethylene | $CO_2$, 50°C, 100 atm at 150 atm/min, 5 min dynamic extraction | −30°C cryofocusing onto head of column | $CO_2$, 140°C, 100–200 atm at 3 atm/min then to 400 atm; 10 m × 50 μm i.d. SB-biphenyl-30 | FID, FTIR | No | 47 |
| Essential oils, inerol geranyl acetate from dragonhead herb | $CO_2$, 60°C–100°C, 140 atm and 80°C, 0.2–0.65 g/ml; static extraction | Not reported | $CO_2$, 180°C, 0.2–0.67 g/ml at 0.014 g/ml per min; 10 m × 50 μm i.d. SB-octyl-50 | FID | No | 70 |

**Table 6.2** (*cont.*)

| Sample and extracts | SFE conditions | Trapping or sampling conditions | cSFC conditions | Detector | Quantitative | Ref. |
|---|---|---|---|---|---|---|
| Waxes, fatty acids, amides and other coating materials from polyester sewing thread | $CO_2$, 70°C, 35 MPa, 7 min dynamic extraction | Cryogenic trap cooled with expanding $CO_2$ | $CO_2$, 130°C, 11–41 MPa at 0.6 MPa/min, 10 m × 50 μm i.d. SB-biphenyl-30 | FID | No | 71 |
| Polymer additives in polypropylene pellets, nylon pellets and poly(ether–ether–ketone) granules | $CO_2$, 70°C, 50–400 atm dynamic extraction from 2–10 min | Cryogenic trap cooled with expanding $CO_2$ | $CO_2$, 120°C, 100–400 atm at 5 atm/min, 10 m × 50 μm i.d. SB-methyl | FID | Yes | 43 |
| Pesticide traces in rendered fats and soya oil | $CO_2$, 100°C, 0.2–0.6 g/ml over 15 min; dynamic extraction | SF solid phase extraction cryofocusing after SFC using 35 mm × 2 mm i.d. ODS packed column | $CO_2$, 100°C, 0.2–0.6 g/ml, 20 m × 50 μm i.d. SB-phenyl-50 | FID | Yes | 46 |
| Explosives and propellants in smokeless powders and soil (spiked samples) | $CO_2$, 75°C, 350 atm, 150 ml/min flow of $CO_2$; dynamic extraction for 5–60 min | 20 mm × 2 mm i.d. LC pre-column trap packed with amino silica | $CO_2$, 75°C, 60–350 atm at 8 atm/min. 10 m × 50 μm i.d. SB-biphenyl-30 | FID, UV, ECD | No | 35 |
| Oligomers in poly(ethylene terephthalate) films | $CO_2$, 70°C, 400 atm; 30 min dynamic extraction | Cryogenic trapping | $CO_2$, 130°C, 100–400 atm at 10 atm/min, 10 m × 50 μm i.d. SB-biphenyl-30 | FID | Yes | 40 |

Note: cSFC = capillary supercritical fluid chromatography; ECD = electron capture detection; FAMES = fatty acid methyl esters; FID = flame ionisation detection; FTIR = Fourier transform infra-red spectrometry; i.d. = inner diameter; LC = liquid chromatography; MS = mass spectrometry; PAC = polycyclic aromatic compound; PP = polypropylene; SF = supercritical fluid; SFE = supercritical fluid extraction; SPA = solid phase absorption; UV = ultraviolet.

vitamins and other drug residues in biological samples. SFE–cSFC finds important applications in environmental science. The analysis of pollutants in matrices such as soil and sediments, and extraction of sorbents on which pollutants from air or water have been selectively adsorbed, have been reported for this technique.

## 6.3 Influence of the sample matrix

The factors affecting the extractability of analytes during SFE have been fully discussed in Chapters 1 and 3, including the important influence of sample matrix. Particle size and shape, adsorption effects and the presence of water can all have a vital bearing. Although a reduction in particle size leads to more rapid extraction, as predicted by the hot-ball model [6], care must be taken to avoid large pressure drops across the extraction vessel as blockages arising from compaction of the material being extracted may occur.

Matrix–solute interactions can severely reduce extraction efficiency. Clifford *et al.* [7] devised a model for SFE based on three steps:

1. reversible release of analytes from matrix adsorption sites;
2. transfer out of the matrix;
3. removal by fluid.

According to this model, poor desorption from the matrix, slow transport out of the particles and poor solubility may all be to blame for slow extractions or poor recoveries. The rate of readsorption of the analyte onto matrix active sites can reduce the extraction rate almost to zero. Thus modifiers may increase extraction rates by capping active sites in the matrix. Recoveries from adsorbent trapping materials are also highly variable. Wright *et al.* [8] compared the use of XAD-2 resin, polyurethane foam (PUF) and activated carbon as adsorbents for PAHs; high recoveries were obtained for low molecular weight compounds with $CO_2$ from XAD-2 and PUF, but these were progressively reduced with increasing molecular weight. Tenax-GC was found to be a suitable material for the trapping and subsequent recovery by SFE of volatile compounds. Recoveries from carbon were negligible unless the $CO_2$ was modified. The choice of suitable adsorbent materials represents a compromise between the breakthrough volume and the conditions required for subsequent extraction [9].

The presence of water in the matrix presents considerable problems during SFE–cSFC. The direct SFE of aqueous media has been described with coupling to packed column SFC [10, 11]. Thiebaut *et al.* [10] used a phase separator to separate water from a supercritical fluid extract that contained dissolved phenol analytes (further details of this study can be found in section 7.4.2). The smaller capacity of capillary columns has made direct SFE–cSFC on aqueous samples difficult if not impossible. However, the analysis of samples in aqueous and hydrophilic matrices has been developed using solid phase adsorption (SPA) of the samples. The analytes of interest are adsorbed onto a matrix which is dried using a gas purge. The analytes are then flushed off the adsorbent and introduced into the analytical column using SFE. Examples of this technique include analysis of prostaglandins [12], fatty acids and fatty acid methyl esters (FAMES) [13, 14].

The presence in the matrix of interfering compounds that are also soluble in the supercritical fluid may also pose problems. However, by careful control of extraction pressure so as to limit co-extraction it is possible to analyse for low levels of compounds of interest. King [9] has made use of the known solubilities of triglycerides to extract pesticides at the 100 ppm level from lard, and Nam *et al.* [15] demonstrated the extraction of parts-per-billion-level pesticides from a fish–lipid matrix. An interesting approach to the removal of interferences was demonstrated by Berg *et al.* [16, 17]. In the analysis of cholesterol in eggs [17] co-extracted lipids usually interfere with the cholesterol peak in the SFC chromatogram. However, immobilised enzymes in the extraction vessel can esterify the lipids [16, 17] during the SFE step. The cholesterol peak is then eluted free of interfering compounds.

## 6.4 Instrumentation

The instrumentation required for on-line SFE–cSFC can range from relatively simple systems to complicated arrangements involving ovens, switching valves and multiple pumps. However, in most systems there is an extraction vessel which is temperature controlled or held in a thermostated oven, a switching valve, a trap which is usually cryogenically cooled, a chromatographic oven housing the capillary column, and a detector. The fluid may be delivered to both the extraction vessel and capillary column using a single syringe pump. In some cases, when different conditions are required for extraction and chromatography simultaneously, a dual pumping system is required.

## 6.5 Extraction vessels

When considering extraction vessels for SFE–cSFC it is important that they are easy to open and close as well as being leak tight and properly rated for high pressures. The type of vessel used will depend on the sample matrix. Often, HPLC guard cartridges will work well for solid matrices as the end caps are easily screwed on or off by hand. Frits (1–2 μm) are incorporated into the endcaps to ensure that the sample matrix remains within the vessel. For small samples, microextraction vessels can be made from readily

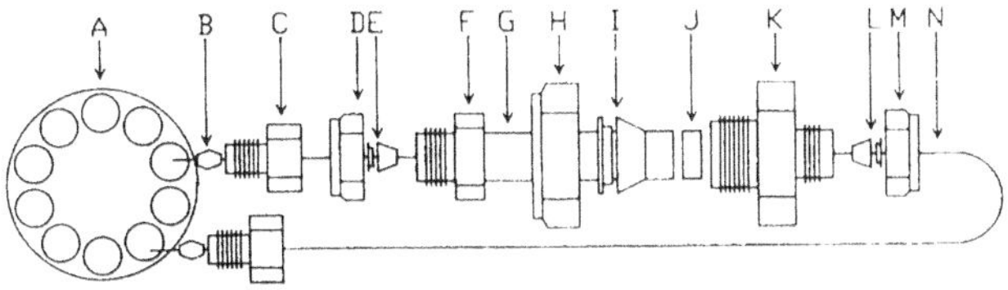


**Figure 6.1** The components of a typical small-scale supercritical fluid extraction (SFE) vessel. A = high-pressure ten-port valve; B = 0.159 cm (1/16″) poly(ether-ether-ketone) (PEEK) double-sided ferrules; C = 10/32 PEEK nut; D–G = tubular stainless-steel vessel components; H–M = stainless-steel reducing union; N = stainless-steel tubing. $CO_2$ flow is from left to right. Reproduced with permission from ref. 18.

available components [18]. The components of a small extraction vessel typically used in analytical SFE [8, 19] are shown in Figure 6.1. Vessels often contain poly(ether-ether-ketone) (PEEK) ferrules, but there are reports of contamination originating from PEEK during microanalysis [16, 20]. Therefore, for trace analysis work, the use of stainless-steel fittings is recommended. Vessels should be designed so that good contact of the supercritical fluid with the matrix occurs. To this end it can be advantageous to mix the sample with inert materials such as glass wool or pelletised Celite®.

Satisfactory recovery of solutes is dependent on analyte solubility and the volume of fluid with which the matrix comes into contact. Obviously, the greater the number of vessel volumes of fluid passing through the vessel per unit of time the better the extraction efficiency. Therefore the aim is to keep the volume of the extraction vessel as small as possible by increasing the sample volume. Extraction can then be carried out as quickly and efficiently as possible without affecting the analyte trapping procedure. With a knowledge of the vessel geometry and the linear velocity ($u$) through the vessel, the total number of vessel volumes per unit time that pass through the extraction vessel can be calculated from

$$u = \frac{F_a \delta_c}{A_c \delta_a} \tag{6.1}$$

where $F_a$ is the gaseous flow rate under ambient conditions, $\delta_c$ is the fluid density within the vessel, $\delta_a$ is the fluid density under ambient conditions and $A_c$ is the cross-sectional area of the extraction vessel.

## 6.6 Supercritical fluid extraction–capillary supercritical fluid chromatography interface

The analytes extracted during the SFE step can be introduced onto the analytical column in two main ways. The SFE extract can be passed through a sample loop and an aliquot directed to the cSFC column, or the analytes can be trapped after SFE and introduced onto the column quantitatively.

### 6.6.1 *Aliquot sampling*

For on-line SFE–cSFC under static conditions, the sample to be extracted is held in a vessel under pressure to equilibrate for a certain period of time. An aliquot of the extract is then introduced into the capillary SFC column. A closed-loop or an open-loop system may be used. Closed-loop static SFE–cSFC involves the sample being sealed in an extraction vessel for a period of static extraction. The extraction vessel is connected to a sample loop into which the analytes diffuse before an aliquot is injected into the cSFC column and the analysis started [21]. The major advantage of this procedure is that small aliquots of the extract can be taken for consecutive analyses with virtually no difference in the extraction profile. However, a major disadvantage is that the solute-containing extraction fluid has to reach equilibrium and diffuse out of the vessel and into the injection valve before sampling is made. Wheeler and McNally [22] have investigated static SFE–cSFC in detail and found that it can take many hours before complete equilibrium is attained. Thus, some sort of recirculating system must be used to reach equilibrium in a shorter time. Unfortunately, recirculating pumps are not suitable for this purpose as they can easily become contaminated. A solution is to incorporate a low-flow restrictor connected to a switching valve in line after the injector, as shown in Figure 6.2. Extraction fluid can then be passed through the loop when required [22]. This is known as open-loop SFE, and with this configuration one also has the opportunity of passing the sample through an ultraviolet/visible (UV/VIS) or flame ionisation detector (FID). At periodic intervals aliquots of the extract can be injected into the cSFC column for analysis.

Usually, aliquot sampling is not suitable for quantitative analysis. However, a method of quantifying open-loop sampling was developed by Baner *et al.* [23]. The SFE extract was passed through a sample loop to an FID and the SFE extraction profile was monitored [Figure 6.3(a)]. The area under the extraction curve gives an indication of the total extractables, but the quantitation of individual species is not possible because their contribution to the total area is not known. In order to determine this, at the peak of the extraction curve an aliquot was directed to the cSFC column, where the quantity of the target analyte could be accurately determined [Figure 6.3(b)]. The directing of this aliquot away from the extraction FID left a negative peak in the extraction profile, and the quantity of the analyte in this negative peak was accurately known from the cSFC data. Therefore the quantity of analyte in the entire extraction profile could be determined by using the negative peak area as a 'calibration' of peak area with quantity of analyte.

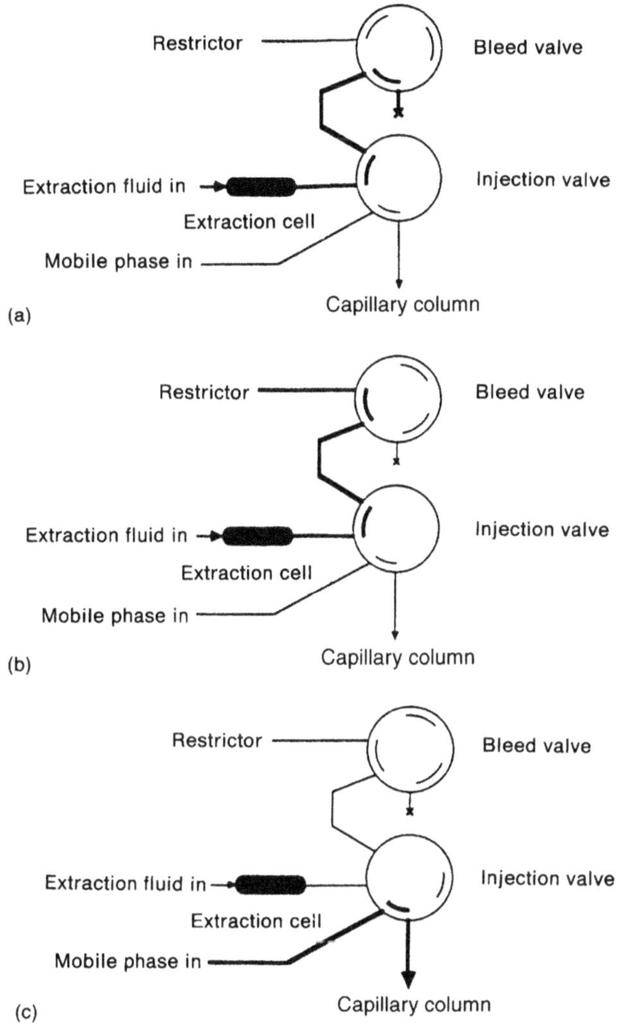


**Figure 6.2** Configurations for static supercritical fluid extraction–capillary supercritical fluid chromatography (SFE–cSFC) with aliquot sampling: (a) static extraction; (b) filling sample loop (open loop or dynamic extraction mode); (c) aliquot sampling SFC analysis. Reproduced with permission from ref. 22.

SFE–cSFC with aliquot sampling is a good technique for basic qualitative investigation and for measuring fundamental parameters such as partition coefficients of solutes in supercritical fluids. However, it is limited in that it is not usually suited to quantitative or trace analysis where analytes in the whole extract must be accumulated prior to chromatographic analysis.

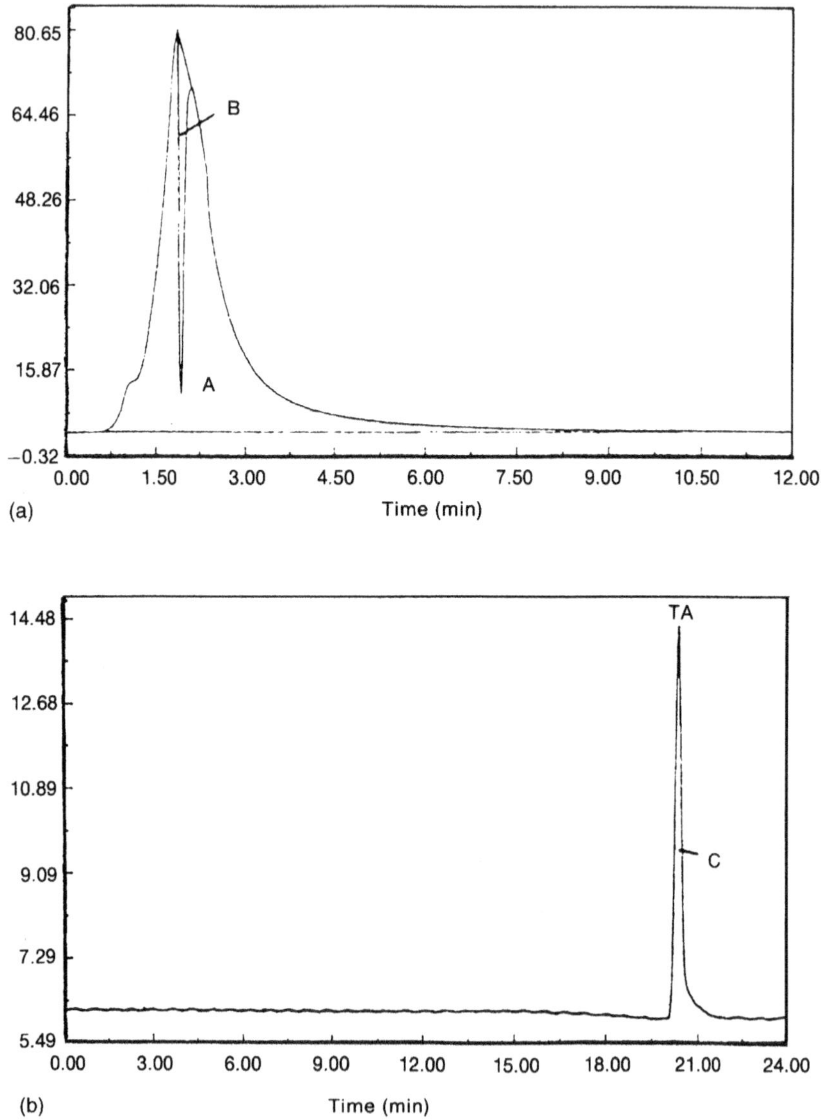


**Figure 6.3** Quantitation of triacetin (TA) from Biopol by open-loop supercritical fluid extraction–capillary supercritical fluid chromatography (SFE–cSFC). (a) Peak A = flame ionisation detector trace of total extractables from the SFE vessel. Negative peak B = aliquot sample diverted to a cSFC column. (b) Peak C = SFC peak attributable only to triacetin. Reproduced with permission from ref. 23.

### 6.6.2 *Trapping of analytes*

In contrast to static extraction with aliquot sampling, dynamic SFE–cSFC operates principally by continuously exposing the analytes to a fresh stream of supercritical fluid. Extracted components are accumulated from this stream in a trap of some kind [22]. Only after extraction is complete are the trapped analytes transferred to the cSFC column for analysis. The major advantages of dynamic SFE–cSFC are that it is much more rapid than static SFE–cSFC and that trace analysis can be performed. Figure 6.4 shows a simple system for use with pure $CO_2$. A high-pressure syringe pump supplies the extraction vessel with fluid. The outlet of the vessel is connected to a capillary flow restrictor which is connected to an accumulating trapping system.

During extraction the depressurised gas from the restrictor passes through the trap and is then vented to the atmosphere. The extracted analytes are concentrated within the trap. After extraction is completed the valve switches the $CO_2$ onto the cSFC column. This raises the pressure within the trap and the $CO_2$ becomes a supercritical fluid, capable of dissolving the trapped analytes and carrying them to the column. If uncoated fused-silica tubing is used to connect the trap to the analytical column (the retention gap) the analytes will in theory be unretained during the transfer. Once they reach the stationary phase film of the cSFC column they become concentrated as a narrow band as a result of differences in analyte migration rates in the two sections of the tubing. If such phase ratio focusing occurs successfully then good chromatographic efficiency is observed during the separation

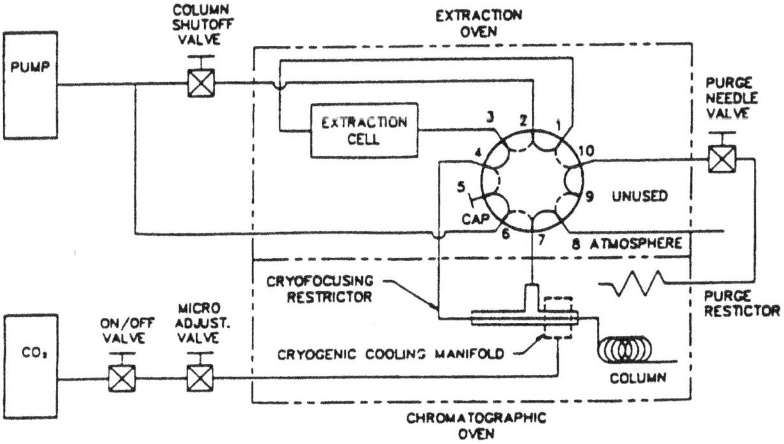


**Figure 6.4** A schematic of the Lee Scientific 602 D supercritical fluid extraction–capillary supercritical fluid chromatography (SFE–cSFC) instrument. Flow is shown in the extract position. 1–10 = ports. Reproduced with permission from Lee Scientific Corporation (now Dionex Corp., Salt Lake City, UT, USA).

[24]. This effect can be achieved with the mobile phase at low density in order to reduce the solubility of compounds in the fluid. Engelhardt *et al.* [25] studied the effect of different retention gaps on the SFC efficiency of a series of standards. They found that i.d. and length of the deactivated fused-silica retention gap had no effect on resolution or peak area, indicating that the solutes were transferred to the column at a density too low to initiate chromatography. Hence the analytes were focused at the head of the column. The chromatography can then be initiated using pressure programming.

Other more complicated systems have been reported using on–off and multiport switching valves to allow continuous extraction or to permit the extraction vessel to be vented during simultaneous chromatographic analysis [26, 27]. The 'plumbing' of such a system can be constructed to any specific requirement. Two different examples are shown in Figures 6.4 and 6.5. When incorporating switching valves into an SFE–cSFC system, attention should be paid to the pressure and temperature ratings of the valves as leaks may develop if they are continuously used at high pressures (more than 400 atm) and temperatures above 80°C. Further, the whole system (the pump, extraction vessel, connecting tubing and all fittings) should be able to withstand high pressures and should be rated at least 1.5 times the maximum intended working pressure [28]. Attention should also be paid to the nature of the supercritical fluid used, particularly with respect to explosion hazards and the corrosion of equipment.

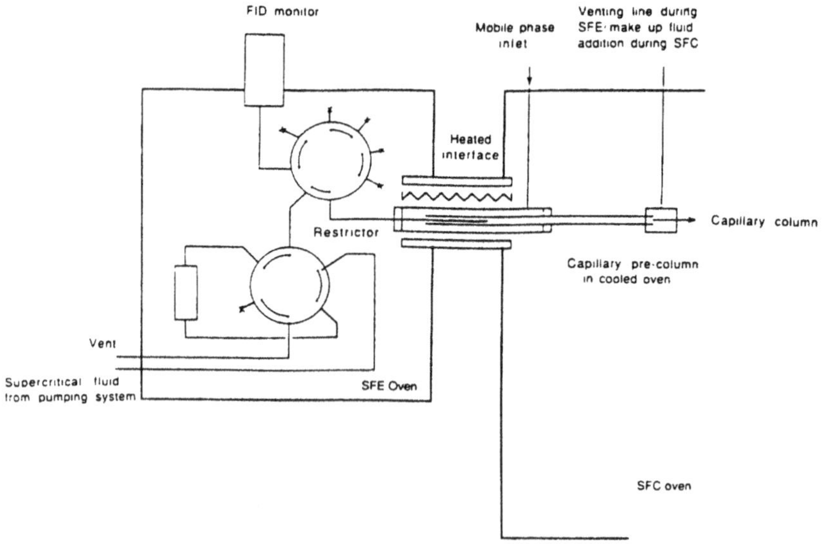


**Figure 6.5** On-line supercritical fluid extraction–capillary supercritical fluid chromatography (SFE–cSFC) system incorporating heated interface housing the SFE restrictor and cooled capillary pre-column for extract trapping. FID = flame ionisation detector.

Since analytical SFE is most often performed with fluids that decompress to gases at ambient conditions (such as carbon dioxide; 1 ml/min liquid $CO_2$ produces a gas flow of approximately 500 ml/min), the success of trapping depends on the efficient recovery of the analytes from the expanded gas. Hawthorne and Miller [29] have shown that expanded flow rates (in most cases expanded $CO_2$) greater than 300 ml/min reduce the effectiveness of the trap so that volatile solutes may be lost through the vent, or band spreading may occur; both effects result in poor chromatographic performance. Thus, for quantitative results, recovery of extracted components should be performed at lower flow rates. The problem of loss of volatile analytes is often not severe in SFE–cSFC, as these are likely to be analysed by SFE–GC. Therefore SFE–cSFC traps generally need to be more concerned with trapping less volatile materials.

## 6.7 Trapping procedures

There are several methods of trapping extracted components from dynamic SFE in preparation for on-line capillary SFC analysis. The first reports of Raynor *et al.* [30] and Anton *et al.* [31] utilised the microinjection valve of the cSFC itself. Extracted components were deposited inside the internal volume rotor or sample loop during the dynamic extraction and the valve was then switched to transfer them into the cSFC column for separation. More recent studies have focused on the use of uncoated or coated fused-silica transfer lines as retention gaps or retaining pre-columns, and on sorbent traps packed with HPLC materials.

### *6.7.1 Trapping on uncoated fused-silica retention gaps*

Figure 6.6 shows a typical arrangement for a cryogenically cooled fused-silica trap. This arrangement is an improvement over earlier designs in that the expanded mobile phase from the extraction vessel is released from an outlet different from that of the incoming $CO_2$ for cSFC. This minimises contamination of the system from previous analyses. The extracted analytes are in contact only with deactivated fused silica after leaving the extraction vessel, which reduces loss of polar analytes by adsorption on metal surfaces.

As was indicated in the previous section, the flow rate of the expanding extraction fluid and the temperature at which analytes are trapped markedly affect the recoveries obtained when uncoated fused-silica tubing is used for accumulating extracted components. In many systems, linear extraction restrictors are used, since they provide the correct flow rate range for on-line coupling with cSFC. They also tend not to plug as quickly as other restrictors when used for SFE. The length and internal diameter of the

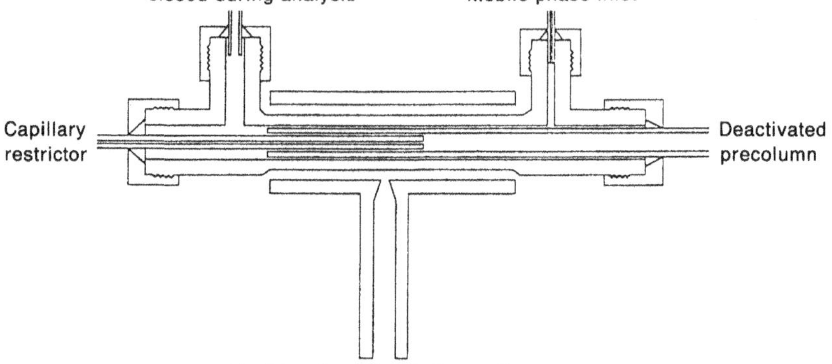


**Figure 6.6** Schematic diagram of cryogenically cooled trap. Reproduced with permission from ref. 12.

capillary restrictor tubing and the pressure at which the extraction is performed should therefore be considered to obtain suitable flow rates.

Restrictors with internal diameters greater than 30 μm result in higher extraction efficiencies, but lower recoveries and significant band broadening of more volatile components. However, restrictors with internal diameters less than 15 μm do not allow sufficient flow for efficient extractions over a short period of time but yield good chromatographic peak shapes. As a rough guide, the gaseous flow rates from 15 cm lengths of 15, 20, 25 and 30 μm i.d. restrictors at a pump pressure of 300 atm are, approximately, 80, 150, 240 and 300 ml/min, respectively. A good compromise therefore is to use a restrictor with a flow rate of 100–200 ml/min. Lengths of capillary tubing of 20 μm or 25 μm i.d. are suitable for most needs.

The trapping efficiency and subsequent chromatographic peak shape are also strongly dependent on the temperature at which analytes are accumulated. This is clearly shown by Figure 6.7 which reports the recoveries of extracted *n*-alkanes as a function of carbon number at different cryotrapping temperatures [32]. Too high a temperature yielded good recoveries and peak shapes for less volatile materials but gave increasingly broad peaks and lower recoveries for more volatile analytes (earlier eluting components). The trap should therefore be cooled in the region of −40°C to −60°C. One problem with cooling the trap, if the restrictor is contained within the cooled zone (Figure 6.4), is that too low a cryofocusing temperature may result in restrictor plugging, or in components such as water freezing in the restrictor. This reduces the rate of extraction and makes it difficult to reproduce analyses. It may therefore be necessary to keep the restrictor hot and deposit the analytes in the transfer line held in a cryogenically cooled oven as shown in Figure 6.5.

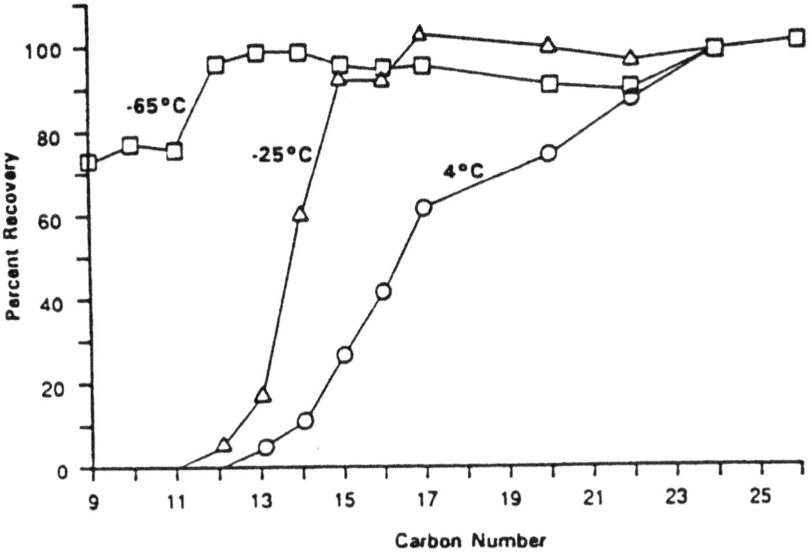


**Figure 6.7** Recoveries of extracted *n*-alkanes as a function of carbon number at different cryotrapping temperatures. □ = − 65°C; △ = − 25°C; ○ = 4°C. Reproduced with permission from ref. 32.

The use of micropacked columns has also been reported [33]. In this case the restrictor can be vented onto the head of the analytical column. The cooling of the expanding gas cools the column and the analytes are deposited on the packing at the start of the column.

### *6.7.2 Trapping on coated fused-silica retaining pre-columns*

An alternative to the cryotrapping method is the use of a coated fused-silica retaining pre-column for concentrating extracted solutes. Although trapping techniques with such columns are continuously developing, Daimon and Hirata [34] have reported results comparing the recoveries of $C_{12}$–$C_{20}$ *n*-alkanes at room temperature using an uncoated capillary, a 3 m × 100 μm i.d. capillary with 0.1 μm OV-1 film and a 3 m × 100 μm capillary with a 0.25 μm OV-1 film. Figure 6.8 summarises the results obtained. Clearly, there is band broadening and solute loss up to $C_{18}$ with an uncoated capillary. However, if a retaining pre-column is used these compounds can be trapped effectively with maintenance of chromatographic peak shape even at room temperature.

### *6.7.3 Trapping on sorbent traps*

Sorbents may also be used as an effective method of trapping after SFE [34]. This entails the use of short traps (usually 2 cm in length) packed with

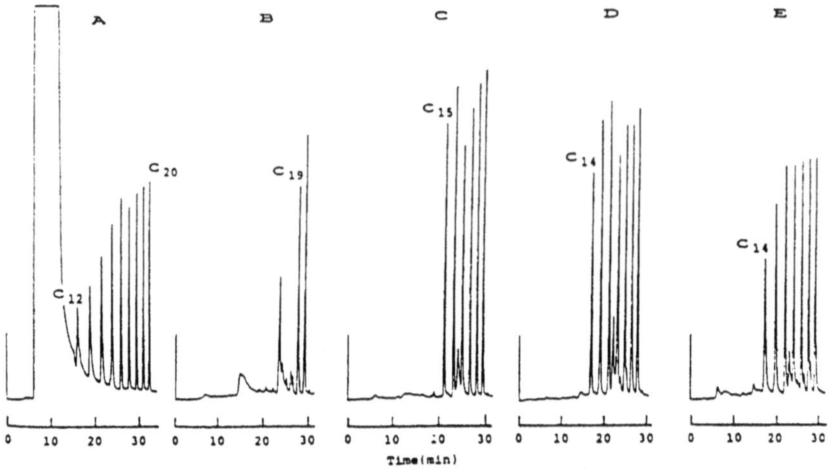


**Figure 6.8** Comparison of sample recoveries using various traps in a supercritical fluid extraction–capillary supercritical fluid chromatography (SFE–cSFC) system. A = obtained by direct injection; B = use of uncoated capillary trap; C = use of capillary trap with a 0.1 μm film; D = use of capillary trap with 0.25 μm film; E = use of uncoated capillary trap plus packed column. Sample: *n*-alkanes ($C_{12}$–$C_{20}$), 0.5 μl of 100 ppm solution, extracted and trapped at room temperature. Column: 10 m × 0.1 mm i.d. octyl, 0.5 μm. Column temperature: 90°C. Column pressure: 80 atm for A and 100 atm for B–E, for 10 min, then programmed at a rate of 5 atm/min. Detection: flame ionisation. Reproduced with permission from ref. 34.

organic sorbents such as Tenax-GC, Carbotrap or with HPLC packing materials. It is important when using such a system that breakthrough of the analytes from the sorbent does not occur and also that the desorption behaviour is suitable for on-line chromatographic analysis. Desorption is performed by increasing the trap temperature or by using the supercritical fluid to desorb the sample. Chromatograph E in Figure 6.8 shows the recovery of $C_{12}$–$C_{20}$ *n*-alkanes from a 2 cm × 0.25 mm trap packed with ODS-300-5 HPLC material in on-line SFE–cSFC [34].

Munder *et al.* [35] have reported the analysis of explosives by SFE–cSFC using a 20 mm × 2 mm i.d. trap, packed with aminopropyl bonded silica material. Explosive residues extracted from soils were accumulated in the sorbent trap and later desorbed by backflushing the components with $CO_2$ onto the analytical column. It was found that the desorption process for each specific component depended on the solvating power of the $CO_2$ and that not all compounds were desorbed under all conditions. Hence, the desorption process is density and temperature dependent. In this regard, supercritical nitrous oxide has been found to be more effective than supercritical carbon dioxide in removing solutes from adsorbents. In some cases this may be useful, since selective desorption could be carried out for sample clean-up or fractionation procedures. This has already been demonstrated off-line by Campbell and Lee [36] who were able to fractionate complex coal-derived mixtures using silica adsorbents prior to GC analysis. Schaeffer *et al.* [37] have also successfully isolated an alkaloid on a cation exchange resin from a supercritical $CO_2$/ethanol fluid.

## 6.8 Use of modifiers and solvent venting

Although $CO_2$ is a versatile extraction solvent, sometimes modifiers are needed to solvate particular analytes or overcome analyte–matrix interactions. This presents a problem in SFE–cSFC with cryogenically cooled traps as the modifier will be trapped and flood the column when the flow is switched to the analytical column. Daimon and Hirata [38] describe a trapping technique on coated capillaries which allows the use of modifiers. The analytes are trapped on two coated capillaries in series, a 3 m × 0.1 mm i.d. capillary coated with a 0.5 μm film of SB-octyl (Dionex) followed by a 1 m length of 0.53 mm i.d. capillary coated with a 5 μm film of polydimethyl siloxane (Quadrex) (Figure 6.9). After extraction the traps were

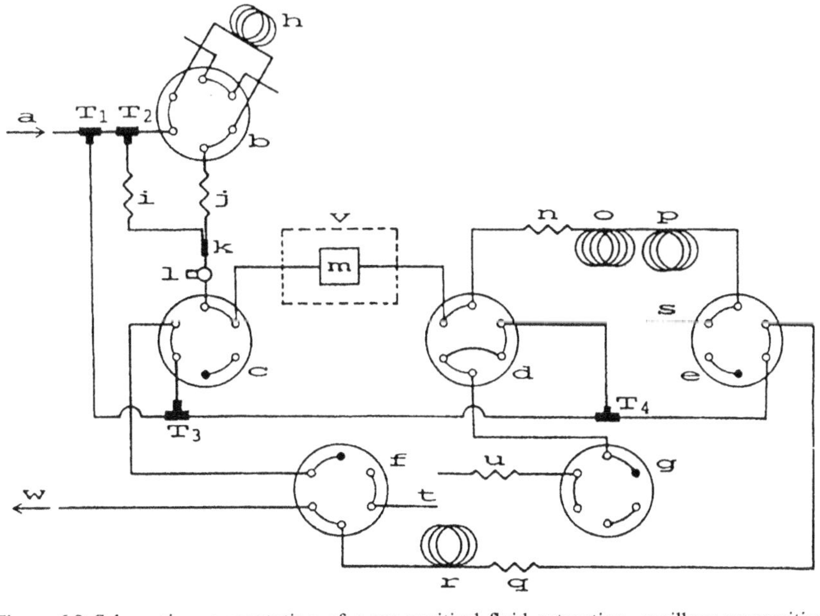


**Figure 6.9** Schematic representation of a supercritical fluid extraction–capillary supercritical fluid chromatography (SFE–cSFC) system for use with modified $CO_2$: a = from pump; b–g = six-way valves; h = modifier loop; i = carbon dioxide line; j = modifier line; k = mixing chamber; l = injector; m = extraction vessel; n, q, u = restrictors; o, p = collecting trap; r = refocusing trap; s, t = venting lines; v = oven; w = separation column; $T_1$–$T_4$ = tee joints. Reproduced with permission from ref. 38.

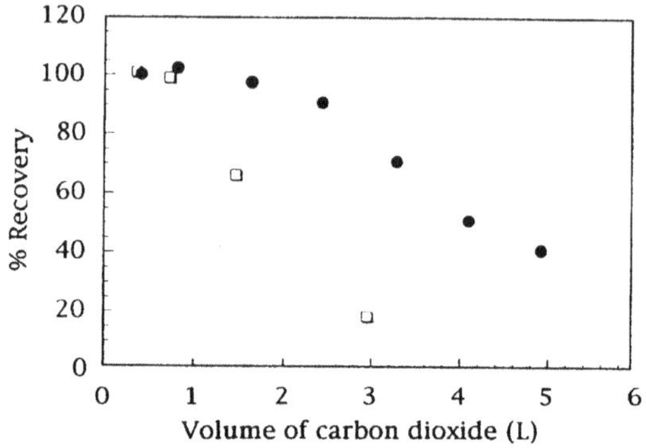


**Figure 6.10** Effect of the extraction volume of methanol-modified supercritical carbon dioxide on the recovery of $n$-$C_{18}$. Collecting trap temperature: 25°C. ● = 5% m/m; □ = 10% m/m. Reproduced with permission from ref. 38.

purged with $CO_2$, typically at 80 ml/min for 3 min to evaporate the excess modifier. The wide bore tube was used to reduce back pressure in the trap, which otherwise caused some of the methanol to pass through as a liquid which flushed the analytes from the trap. However, with a wide bore trap alone, some of the analytes passed through in the expanding gas as an aerosol, therefore the narrow bore trap was used as well. With the two traps the dead volume was too large for effective focusing, so a third, refocusing, trap was installed, consisting of a 30 cm × 20 μm i.d. fused capillary restrictor and a 1 m length of fused-silica capillary coated with a 1.2 μm film of CP-Sil 5CB (Chrompack). After the solvent venting the analytes were transferred to the refocusing trap using $CO_2$ at 250 atm and a 40 ml/min gas flow for 20 min. Finally, the flow was switched to the cSFC column and a pressure programme started. For the modifiers methanol, ethanol and toluene the recovery of a series of $n$-alkanes dropped to zero when the concentration of modifier exceeded the concentrations which saturated $CO_2$ gas at 1 atm and 25°C (14%, 10.7% and 8%, respectively). This was because some of the solvent condensed as a liquid and flushed the analytes from the trap. These concentrations are therefore the upper limits of modifier concentration which can be used with this system. At lower concentrations, $C_{20}$ was quantitatively trapped. Using methanol as modifier, hydrocarbons lower than $C_{20}$ were partially lost from the trap, and the rate of loss was faster with higher levels of methanol (Figure 6.10).

Apart from the use of modifiers other situations occur when large amounts of solvent are trapped along with the analyte. Berg *et al.* [16, 17] used immobilised enzyme to esterify fats in the SFE cell. The fat was introduced to the cell which contained the enzyme and excess alcohol. The alcohol remaining after the esterification reaction was carried to the deactivated fused-silica trap and had to be removed before the SFC analysis. This was achieved with a helium purge at 15 ml/min for 15 min.

## 6.9 Supercritical fluid extraction as a sample introduction technique

As stated previously, one of the problems of cSFC is sample introduction without flooding the column with solvent. SFE can be used as a solventless sample introduction technique to avoid this problem. Cortes [39] showed that 10–100 μl samples could be concentrated and transferred to 50 μm i.d. columns while maintaining high efficiency. The sample was injected into a 15 m × 150 μm i.d. pre-column and the solvent removed by gas purging. The pre-column had a restrictor fitted at the outlet, and thus the whole pre-column could be pressurised, and acted as an SFE vessel. The analytes were dissolved in supercritical $CO_2$ and transferred to the analytical column inlet via the restrictor and redeposited by decompression. The analytes were then redissolved by pressure programming and transferred to the analytical column. This system could be used as an HPLC–SFC interface.

The use of solid sorbents has proved very useful in sample introduction to cSFC. The dissolved analytes are injected onto a sorbent, the solvent can then be removed by evaporation and the analytes transferred to the analytical column using SFE–cSFC. The whole process is called SPA–SFE–cSFC. Koski *et al.* [12] used this technique to analyse 11 prostaglandins in aqueous media. Prostaglandins are difficult to analyse by HPLC as they have no chromophore and require extensive derivatisations to be analysed by GC. Therefore SFC is an ideal technique, but as the analytes are often present in aqueous biological samples the sample introduction is problematical. The authors injected 100 μl of sample onto ODS or XAD-2 adsorbent and evaporated the solvent with SFC grade $CO_2$. The analytes were transferred to the analytical column for on-line SFE–cSFC analysis. Therefore 100 μl samples could be analysed in one injection, compared with 1 μl using a sample loop.

Ullsten *et al.* [13] developed a fully automated SPA–SFE–cSFC system for trace analysis of compounds in polar media. Samples (10 μl) of fatty acids, FAMES, cholesterol and trilaurin in hexane or ethanol were injected with an autosampler and transferred to the adsorbent (50 μm diameter silanised glass beads) in an SFE vessel using a small volume of solvent. The solvent was drained from the extraction vessel which was dried with nitrogen and carbon dioxide. Supercritical $CO_2$ was then used to transfer the analytes to the analytical column. The adsorbent was cleaned with supercritical $CO_2$ during the analytical run. The adsorbent was reusable up to 60 times. Petersson *et al.* [14] extended this work to include other adsorbents. They found that DVB-EVB resin and 5 μm silica particles with 10 nm pores provided the best combination for easy cleaning, high recovery and absence of memory effects.

## 6.10 Optimisation of conditions for supercritical fluid extraction–capillary supercritical fluid chromatography

A number of parameters must be optimised for successful analysis by coupled SFE–cSFC. Principal among these are the conditions for quantitative extraction. This should begin with a determination of the supercritical fluid extractability of the analyte(s) from non-sorptive matrices (filter paper, etc.) to assess the appropriate solvent, density and temperature conditions. Trial runs on spiked samples then allow investigations of matrix–solute interactions; if necessary these may be overcome by a period of static extraction. The kinetics of extraction must then be determined in order to define the required extraction time. A full discussion of these considerations is presented in Chapters 1 and 3.

Factors affecting the efficiency of intermediate trapping must then be addressed. The nature of the analyte is crucial, while the possible presence of co-extracted, interfering compounds demands either selectivity during extraction or the trapping on an adsorbent from which selective desorption into the cSFC column is possible. The sample size must be carefully chosen so that the capacity of the cSFC column is not exceeded, and the extracting supercritical solvent must be of sufficient purity to avoid introduction of extraneous material into the column [40]. Finally, the conditions for efficient cSFC analysis must be optimised, preferably off-line. Correct choice of column, temperature and pressure/density programme are vital. Compromises may be inevitable if the extracted analytes have a range of polarities.

## 6.11 Selected applications of supercritical fluid extraction–capillary supercritical fluid chromatography

Table 6.2 lists selected applications of SFE–cSFC. Many of these are real problems in the analysis of fuels, foods and food contaminants and in polymer chemistry.

The simplified procedure of trapping the $CO_2$ SFE extract in the rotor groove of the cooled injection valve was applied by Raynor *et al.* [30] to 1 mg samples of coal pitch. Switching the contents of the loop by means of the $CO_2$ mobile phase then permitted the analysis by cSFC; fractions containing individual hydrocarbon peaks were collected on a potassium bromide disc for subsequent identification by FTIR microspectroscopy, a procedure that allowed isomer identification not possible by cSFC–MS. King [41] applied a

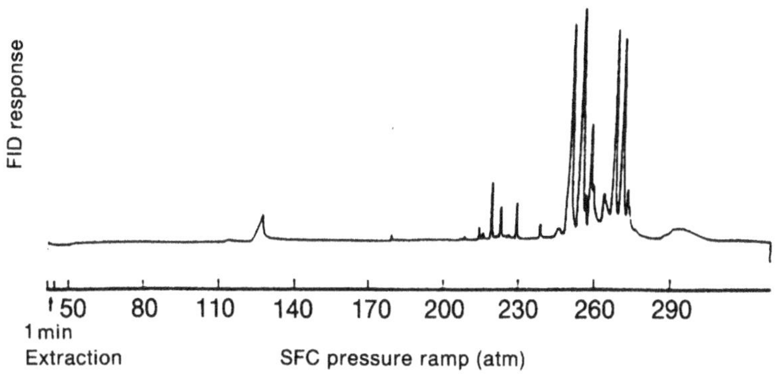


**Figure 6.11** Supercritical fluid extraction–capillary supercritical fluid chromatography (SFE–cSFC) separation of the on-line $CO_2$ extract from a single dried fruit beetle (live specimen). Extraction pressure = 200 atm; temperature = 45°C; time = 1 min, extract cryo-focused. FID = flame ionisation detector. Reproduced with permission from ref. 41.

commercial SFE–cSFC system to a variety of food-related analyses. Thus, the triglycerides and fatty acids of single seeds were characterised to give information required during genetic engineering studies. The cSFC chromatogram after SFE of less than 1 mg of a cottonseed kernel at 200°C and 45°C showed the $C_{50}$–$C_{54}$ triglycerides of cottonseed oil which were injected after cryofocusing in the 'tee' cooled by external $CO_2$. Figure 6.11 is the cSFC chromatogram of the on-line SFE extract obtained by the same procedure from a single dried-fruit beetle (*Carcophilus hemipterus*) showing the high molecular weight wax components. Single seeds of *Dalea spinosa* were analysed by Taylor *et al.* [42] in order to determine in which parts of the plant important odour compounds were most concentrated.

A similar technique was employed by Cotton *et al.* [43] to the quantitative extraction and separation of additives and oligomers from polypropylene and a number of other polymers. Quantitative extraction of additives from polypropylene was investigated at five different extraction pressures, at a constant flow rate and temperature. Below 50 atm, extraction was negligible; between 50 atm and 200 atm, Tinuvin 326 and 770 were extracted, along with small quantities of oligomers, although not in the ratio anticipated from their relative concentrations in the material. Higher pressures led to the extraction of all the additives present (Figure 6.12). The integrated peak areas corresponded well with the actual concentrations.

Raynor and Bartle [44] used a similar SFE–cSFC method of establishing the presence of stabilisers in automotive clear coatings. A thermosetting acrylic resin which had been cross-linked with an isocyanate curing agent to form a hard brittle film was chosen for the study. A sample (5 mg) was extracted with $CO_2$ at 75°C and 400 atm for 10 min. The extract was deposited in a cryogenically cooled pre-column before analysis by cSFC.

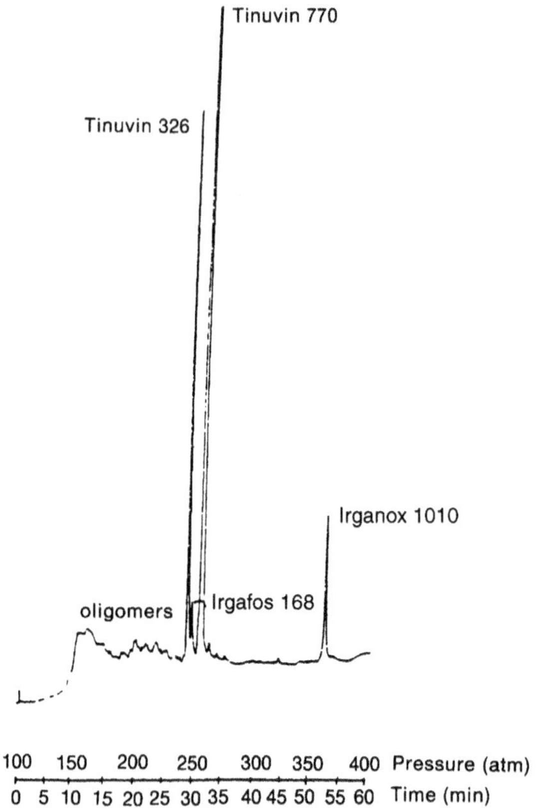


**Figure 6.12** On-line supercritical fluid extraction–capillary supercritical fluid chromatography (SFE–cSFC) of polypropylene pellets. Conditions: extraction with supercritical $CO_2$ at 70°C, 300 atm for 4 min. Chromatography with $CO_2$ at 100°C. Pressure programmed from 100 atm to 400 atm (at 5 atm/min). Column: 10 m × 50 μm inner diameter SB-biphenyl; 6 mg introduced into the extraction vessel.

The presence of Tinuvin 326 and Tinuvin 770 was confirmed using retention time data from standards. Quantitative extraction from the cross-linked sample was more difficult than from polypropylene, presumably because of the cross-links preventing swelling.

The oligomer tris-(ethylene terephthalate), a cyclic trimer of poly(ethylene terephthalate) (PET) monomer, was identified [40] in the supercritical fluid extract of a commercially available PET film. The coefficient of variation for extraction and chromatography of the extracted cyclic trimer was determined for two extraction times to be approximately 8%, as shown in Table 6.3.

Baner *et al.* [23] used an open-loop with aliquot sampling technique to quantify the plasticiser, triacetin, in Biopol (a biodegradable polymer) and antioxidants Irganox 1010 and Irganox 1076 in polypropylene.

**Table 6.3** Coefficient of variation for the analysis of the cyclic trimer of poly(ethylene terephthalate) (PET) by supercritical fluid extraction–capillary supercritical fluid chromatography

| Duration of extraction (min) | Number of replicates | Mass of PET extracted (mg) | Coefficient of variation (%RSD)[a] |
|---|---|---|---|
| 10 | 5 | $15.1 \pm 0.02$ | 8.2 |
| 25 | 18 | $15.2 \pm 0.02$ | 7.8 |

[a] Percentage relative standard deviation.

Extraction rates from polymers increase strongly at the glass transition temperature ($T_g$). However, higher temperatures in combination with $CO_2$ have been reported to cause crystallisation in the polymer, which locks the target molecules inside the matrix. Melting is undesirable too as this reduces the surface area of the sample. Hence the optimum SFE temperature for polymers should be just below the melting point and crystallisation point but above the glass transition point.

Xie *et al.* [45] employed a cryogenic trap in which the supercritical fluid extract was decompressed through a small internal diameter linear restrictor and deposited in a deactivated capillary solute concentrator. The efficient operation of the system was demonstrated by the extraction (at 80°C and 400 atm), trapping (at −40°C to −50°C) and cSFC analysis of the digitalis cardioglycoside ouabain, 3-[(6-deoxy-$\alpha$-*l*-mannopyranosyl)oxyl]-1,5,11a,14,19-pentahydroxycard-20(22)-enolide (Figure 6.13). High recoveries of ouabain were deduced from biological assays of collected fractions.

Firearm propellants were characterised through their extractable organic constituents by SFE–cSFC by Munder *et al.* [35]. Detection limits of 100 pg were achieved by coupling three different detectors in line (UV/VIS, FID and electron capture detection) and injecting the whole extract rather than a portion. Precipitation pre-concentration without cryotrapping was performed by expanding the fluid containing the analytes into a short sorbent packed column from which they were eluted by supercritical $CO_2$. Figure 6.14 illustrates the SFE traces which allowed the propellants present in trace gunshot residues remaining on expended cartridge cases to be identified.

Murugaverl and Voorhees [46] analysed trace amounts of pesticides in fats and oils by modifying the SFE–cSFC procedure to permit extraction and clean-up in one step. Extraction and clean-up were monitored by a photodiode array detector. The clean-up stage was achieved by selectively retaining extracted lipids on a packed column while depositing analytes of interest in a cryogenic trap. Three pesticides were found in the cSFC chromatogram of the first fraction obtained by extraction of parts-per-million level spiked soybean oil; the lipids are found in the second and subsequent fractions.

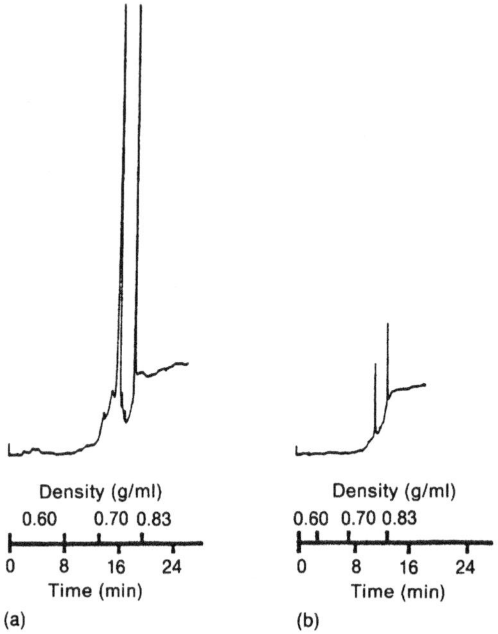


**Figure 6.13** Supercritical fluid extraction–capillary supercritical fluid chromatography (SFE–cSFC) chromatograms of oubain and $n$-$C_{38}$: (a) after 30 min extraction; (b) after a second 30 min extraction. Reproduced with permission from ref. 35.

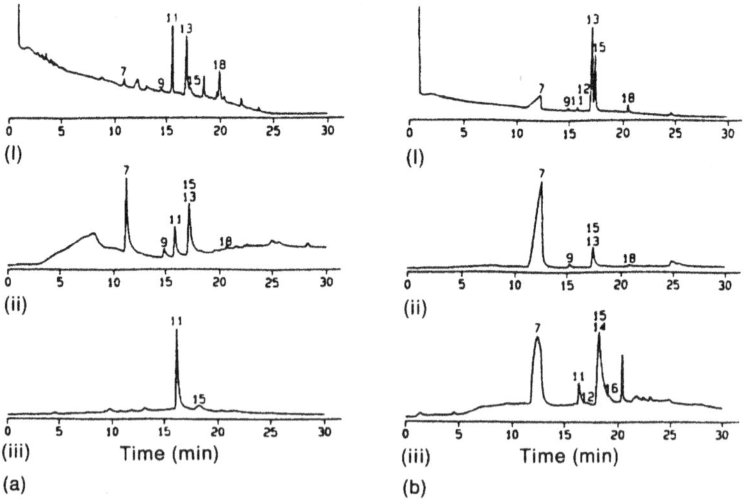


**Figure 6.14** Characterisation of gunshot residue. Direct $CO_2$ extraction of the residue left in the cartridge case. Residue from (a) 12 grains of 680 BR; (b) 7 grains of 630 P. (i) Absorption spectrum at 240 nm; (ii) electron capture detection response; (iii) flame ionisation detection response. Peaks are not identified in the current context. Reproduced with permission from ref. 35.

An alternative method of cleaning up samples containing fats was demonstrated by Berg *et al.* [17]. In the analysis of cholesterol in eggs the presence of lipids co-eluting with the analyte makes quantitation difficult. However, the fats can be esterified in the extraction cell by the use of immobilised enzymes and a small amount of alcohol. The resulting esters have much shorter retention times than the underivatised lipids, resulting in the cholesterol peak being free of interferences (a more detailed account of this study can be found in section 4.6).

Daimon and Hirata [38] used different pressures and modifiers selectively to extract paraffins, FAMES and alcohols from a silica matrix. Extraction with pure $CO_2$ at 200 atm extracted only paraffins; increasing the pressure to 300 atm extracted the FAMES; addition of 10% methanol extracted the alcohols (Figure 6.15). Use of different modifiers also selectively extracted different pesticides from soils. At 60°C and 300 atm and 5% modifier only isodrin and mirex were extracted with toluene, hexane and dichloromethane as modifier. Methanol-modified and ethanol-modified $CO_2$ also extracted malathion and linuron at 5% addition, and methanol at 10% also extracted bensulphide. Therefore, by altering modifier identity and concentration as well as temperature and pressure very selective extractions are possible.

## 6.12 Conclusions

On-line SFE–cSFC has shown itself over the past few years to be an invaluable and powerful coupled analytical procedure combining a sample preparation stage with a high-resolution chromatographic technique. Commercial instrumentation has appeared on the market and is now in routine use in a number of individual analytical laboratories.

The success of on-line SFE–cSFC can be attributed to three main factors: its sensitivity, reproducibility and speed of analysis. As is usual in commercial instruments, the whole of the extract is transferred to the analytical column giving very low detectable quantities. Sensitivity of this order is difficult and in many cases is impossible to obtain using off-line two-stage techniques. Compared with the classical solid–liquid extraction methods, SFE has been widely found to be more reproducible for a number of reasons, one being the ease with which the experimental conditions can be duplicated. This is also true for this on-line procedure.

Not all samples are suitable for SFE–cSFC. For example, if an analyte can be analysed by GC then SFE–GC would be the preferred technique. Moreover, samples with large amounts of extractable compounds or water present are suitable, at the moment, only for off-line analysis; however, as we have shown, there are a whole variety of analytical problems where SFE–cSFC can provide the answer.

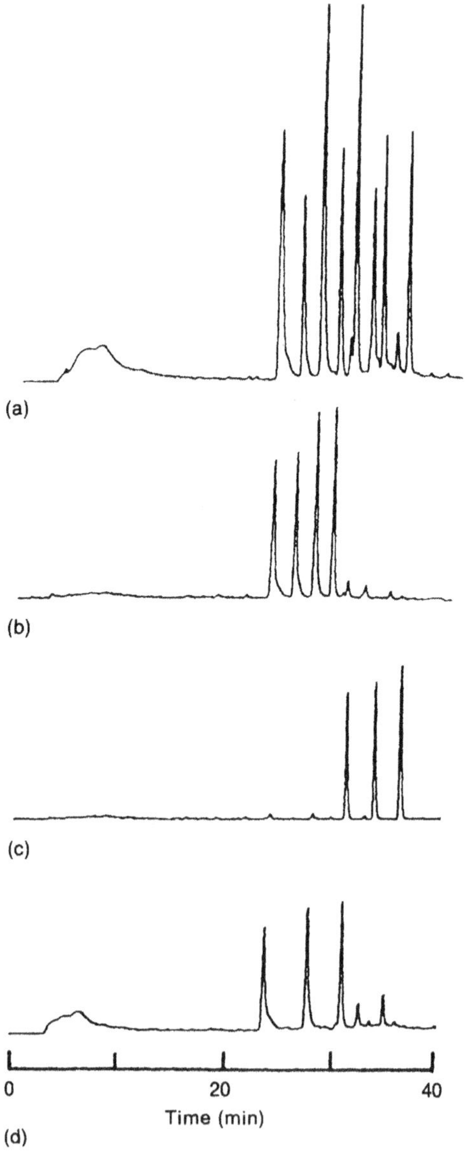


**Figure 6.15** Chromatograms obtained for the selective extraction of *n*-alkanes, fatty acid methyl esters (FAMES) and alcohols from a silica matrix: (a) 300 atm, $CO_2$ modified with 10% methanol; (b) first extraction, 200 atm, unmodified $CO_2$ (*n*-alkanes extracted); (c) second extraction, 300 atm, unmodified $CO_2$ (FAMES extracted); (d) third extraction, 300 atm, $CO_2$ modified with 10% methanol (alcohols extracted). Extraction temperature = 60°C; collecting temperature = 25°C. Reproduced with permission from ref. 38.

## References


1. Saito, M., Hondo, T. and Senda, M. (1990) *Multidimensional Chromatography – Techniques and Applications* (ed. H.J. Cortes), Chromatographic Science Series Vol. 50, Marcel Dekker, New York, pp. 331–57.
2. Lee, M.L. and Markides, K.E. (eds) (1990) *Analytical Supercritical Fluid Chromatography and Extraction*, Chromatography Conferences Inc., Provo, UT, pp. 344–62.
3. Sugiyama, K., Saito, M., Hondo, T. and Senda, M. (1985) *J. Chromatogr.*, **332**, 107.
4. Bayona, J.M. and Cai, Y. (1994) *TRAC*, **13**, 327–32.
5. Oudsema, J.W. and Poole, C.F. (1993) *J. High Res. Chromatogr.*, **16**, 198.
6. Bartle, K.D., Clifford, A.A., Hawthorne, S.B. *et al.* (1990) *J. Supercrit. Fluids*, **3**, 143.
7. Clifford, A.A., Burford, M.D., Hawthorne, S.B. *et al.* (1995) *J. Chem. Soc. Faraday Trans.*, **91**, 1333.
8. Wright, B.W., Wright, C.W., Gale, R.W. and Smith, R.D. (1987) *Anal. Chem.*, **59**, 38.
9. King, J.W. (1989) *J. Chromatogr. Sci.*, **27**, 355.
10. Thiebaut, D., Chervet, I.P., Vannoort, R.W. *et al.* (1989) *J. Chromatogr.*, **477**, 151.
11. Hedrick, J. and Taylor, L.T. (1989) *Anal. Chem.*, **61**, 1986.
12. Koski, I.J., Jansson, B.A., Markides, K.E. and Lee, M.L. (1991) *J. Pharmaceut. Biomed. Anal.*, **9**, 281.
13. Ullsten, U. and Markides, K.E. (1994) *J. Microcol. Sep.*, **6**, 385.
14. Petersson, U. and Markides, K.E. (1996) *J. Chromatogr. A*, **734**, 311.
15. Nam, K.S., Kapila, S., Pieczonka, G. *et al.* (1988) *Proc. Int. Symp. on Supercritical Fluids*, p. 743.
16. Berg, B.E., Hansen, E.M., Gjorven, S. and Greibokk, T. (1993) *J. High Res. Chromatogr.*, **16**, 358.
17. Berg, B.E., Lund, H.S. and Greibokk, T. (1997) *Chromatographia*, **44**, 399.
18. Artz, W.E. and Sauer, R.M. (1992) *JAOCS*, **69**, 309.
19. Lee, M.L. and Markides, K.E. (eds) (1990) *Analytical Supercritical Fluid Chromatography and Extraction*, Chromatography Conferences Inc., Provo, UT, p. 349.
20. Wallace, J.C., Krieger, M.S. and Hites, R.A. (1992) *Anal. Chem.*, **64**, 2655.
21. Jackson, W.P., Markides, K.E. and Lee, M.L. (1986) *J. High Resolut. Chromatogr.*, **9**, 213.
22. Wheeler, J.R. and McNally, M.E. (1989) *J. Chromatogr. Sci.*, **27**, 534.
23. Baner, L., Bucherl, T., Ewender, J. and Franz, R. (1992) *J. Supercrit. Fluids*, **5**, 213.
24. Schomburg, G. and Roeder, W. (1989) *J. High Resolut. Chromatogr.*, **12**, 218.
25. Engelhardt, H., Zapp, J. and Kolla, P. (1991) *Chromatographia*, **32**, 527.
26. Mapelli, G., Pigozzo, F., Raynor, M.W. and Tristianu, S. (1991) *13th Int. Symp. on Capillary Chromatography* (ed. P. Sandra), Riva del Garda, Italy, p. 1383.
27. Lee Scientific (1988) *Series 600-D Supercritical Fluid/Gas Chromatograph User's Manual*, Dionex Corp., Salt Lake City Technical Center, 1515 West 2200 South, Salt Lake City, UT 84119, USA.
28. Saito, M., Hondo, T. and Yamauchi, Y. (1988) *Supercritical Fluid Chromatography* (ed. R.M. Smith), Royal Society of Chemistry, Cambridge, pp. 203–30.
29. Hawthorne, S B. and Miller, D.J. (1987) *J. Chromatogr.*, **403**, 63.
30. Raynor, M.W., Davis, I.L., Bartle, K.D. *et al.* (1988) *J. High Resolut. Chromatogr.*, **11**, 766.
31. Anton, K., Menes, R. and Widmer, H.M. (1988) *Chromatographia*, **26**, 221.
32. Anderson, M.R., Swanson, J.T., Porter, N.L. and Richter, B.E. (1989) *J. Chromatogr. Sci.*, **27**, 371.
33. Ibanez, E., Herraiz, M. and Reglero, G. (1995) *J. High Resol. Chromatogr.*, **18**, 507.
34. Daimon, H. and Hirata, Y. (1991) *Chromatographia*, **32**, 549.
35. Munder, A., Christensen, R.G. and Wise, S.A. (1991) *J. Microcolumn Sep.*, **3**, 127.
36. Cambell, R.M. and Lee, M.L. (1986) *Anal. Chem.*, **58**, 2247.
37. Schaeffer, S.T., Zalkov, L.H. and Teja, A.S. (1989) *Ind. Eng. Chem. Res.*, **28**, 1017.
38. Daimon, H. and Hirata, Y. (1994) *J. High Resol. Chromatogr.*, **17**, 809.
39. Cortes, H.J., Campbell, R.M., Himes, R.P. and Pfeiffer, C.D. (1992) *J. Microcol. Sep.*, **4**, 239.
40. Bartle, K.D., Boddington, T., Clifford, A.A. *et al.* (1991) *Anal. Chem.*, **63**, 2371.
41. King, J.W. (1990) *J. Chromatogr. Sci.*, **28**, 9.

42. Taylor, S.L., King, J.W. and Snyder, J.M. (1994) *J. Microcol. Sep.*, **6**, 467.
43. Cotton, N.J., Bartle, K.D., Clifford, A.A. *et al.* (1991) *J. High Resol. Chromatogr.*, **14**, 164.
44. Raynor, M.W. and Bartle, K.D. (1992) *J. Supercrit. Fluids*, **6**, 39.
45. Xie, Q.L., Markides, K.E. and Lee, M.L. (1989) *J. Chromatogr. Sci.*, **27**, 365.
46. Murugaverl, B. and Voorhees, K.J. (1991) *J. Microcol. Sep.*, **3**, 11.
47. Wiebolt, R.C., Kemptfert, K.D. and Dalrymple, D.C. (1990) *Appl. Spectrosc.*, **44**, 1028.
48. Hirata, Y., Nakata, F. and Horihata, M. (1988) *J. High Resolut. Chromatogr.*, **11**, 81.
49. Bucherl, T., Eschler, M., Gruner, A. and Wolff, E. (1994) *J. High Resolut. Chromatogr.*, **17**, 765.
50. Jordan, S.L., Taylor, L.T., Seemuth, P.D. and Miller, R.J. (1997) *Textile Chemist and Colorist*, **29**, 25.
51. Yang, J., Hasenoehrl, E.J. and Griffiths, P.R. (1997) *Vibrational Spectroscopy*, **14**, 1.
52. Ramsey, E.D., Perkins, J.R., Games, D.E. and Startin, J.R. (1989) *J. Chromatogr.*, **464**, 353.
53. Saito, M., Yamauchi, Y., Inomata, K. and Kottkamp, W. (1989) *J. Chromatogr. Sci.*, **27**, 79.
54. Englehardt, H. and Gross, A. (1988) *J. High Resolut. Chromatogr.*, **11**, 38.
55. Pinkstone, J.D., Delaney, T.E., Bowling, D.J. and Chester, T.L. (1991) *J. High Resolut. Chromatogr.*, **6**, 401.
56. Gmuer, W., Bosset, J.O. and Plattner, E. (1987) *J. Chromatogr.*, **388**, 335.
57. Skelton, R.J., Johnson, C.C. and Taylor, L.T. (1986) *Chromatographia*, **21**, 3.
58. Hawthorne, S.B., Miller, D.J. and Krieger, M.S. (1989) *J. Chromatogr. Sci.*, **27**, 7.
59. Hawthorne, S.B., Krieger, M.S. and Miller, D.J. (1988) *Anal. Chem.*, **60**, 472.
60. Anderson, M.R., Swanson, J.T., Porter, N.L. and Richter, B.E. (1989) *J. Chromatogr. Sci.*, **27**, 371.
61. Jahn, K.R. and Wenclawiak, B. (1988) *Chromatographia*, **26**, 345.
62. Onuska, F.I. and Terry, K.A. (1989) *J. High Resol. Chromatogr.*, **12**, 527.
63. Levy, J.M., Cavalier, R.A., Bosch, T.N. *et al.* (1989) *J. Chromatogr. Sci.*, **27**, 341.
64. Hawthorne, S.B. and Miller, D.J. (1986) *J. Chromatogr. Sci.*, **24**, 258.
65. Nielen, M.W.E., Sanderson, J.T., Frei, R.W. and Brinkman, U.A.T. (1989) *J. Chromatogr.*, **474**, 388.
66. Liebman, S.A., Levy, E.J., Lurcott, S. *et al.* (1989) *J. Chromatogr. Sci.*, **27**, 118.
67. Levy, J.M. and Guzowski, J.P. (1989) *Fresenius Z. Anal. Chem.*, **330**, 207.
68. Levy, J.M., Guzowski, J.R. and Huhak, W.E. (1987) *J. High Resolut. Chromatogr.*, **10**, 7.
69. Hawthorne, S.B., Krieger, M.S. and Miller, D.J. (1989) *Anal. Chem.*, **61**, 736.
70. Manninen, R., Holm, Y., Hittlinen, R. and Riekkola, M.L. (1990) *11th Int. Symp. on Capillary Chromatography* (eds P. Sandra and G. Redant), Monterey, CA, p. 676.
71. Hofler, E. and Alt, G. (1990) *11th Int. Symp. on Capillary Chromatography* (eds P. Sandra and G. Redant), Monterey, CA, p. 682.

# 7 Supercritical fluid extraction coupled to packed column supercritical fluid chromatography

I.G.M. ANDERSON

## 7.1 Introduction

Although the solvating ability of supercritical fluids was first recognised by Hannay and Horgarth working with supercritical ethanol in 1897 [1], supercritical fluids found little application as extraction media until the early 1960s when Zosel [2] working at the Max Planck Institute for Coal Research introduced supercritical fluid extraction (SFE) as an industrial process, although Zhuze and co-workers had described an earlier process for the extraction of lanolin from wool grease in 1958 [3].

Supercritial fluids, such as supercritical carbon dioxide, were soon found to be highly efficient extraction media, chiefly on account of their high solvating power, their low viscosities (intermediate between a gas and a liquid) and their low surface tensions which enable their penetration deep into the extraction matrix.

SFE used in isolation is generally not selective enough to isolate specific solutes from the extraction matrix without further clean-up or resolution from co-extracted species prior to qualitative and quantitative analysis. Consequently, for analytical applications, SFE is generally used in conjunction with chromatographic techniques to improve the overall selectivity of the process in isolating specific solutes.

The first combined use of SFE with chromatographic techniques was in 1976 when Stahl and Schiltz described the use of a micro supercritical fluid extractor with thin-layer chromatography (TLC) [4]. Since then SFE has found ever increasing application with chromatographic techniques such as gas chromatography (GC) [5, 6], high-performance liquid chromatography (HPLC) [7] and supercritical fluid chromatography (SFC).

SFE combined with chromatography can be either 'off-line' or 'on-line'. In the off-line process supercritical fluid extraction takes place as a separate and isolated process to the chromatography: extracted solutes are trapped or collected, often in a suitable solvent for later injection into GC, HPLC or SFC instrumentation. In the on-line process SFE and chromatography are coupled to form an integrated process: following SFE the extracted species are passed directly to the chromatograph, usually via a trap or sample loop and a valve-switching device.

Chromatography, where a supercritical fluid is used as the mobile phase, was introduced by Klesper and co-workers in 1962 [8]. Advances in instrument technology, sophistication and computation over the next 20 years led to the development of increased on-line coupling of chromatographic (and other analytical) techniques. In the mid-1980s Sugiyama *et al.* [9–11] and, separately, Skelton *et al.* [12] introduced SFE which was directly coupled to SFC, combining the advantages of both techniques.

Supercritical fluid chromatography can be performed on either packed or capillary columns. Although Sugiyama *et al.* [10] used a packed HPLC column in their initial work on directly coupled SFE–SFC to extract and resolve caffeine from coffee beans, capillary column SFC was more widely used until the late 1980s when there was a renewed interest in packed column SFC. This interest has been generally maintained to the present day, and more recently because of the increased interest in packed column SFC for the chiral separation of biologically active and pharmaceutical compounds [13]. This increased interest in packed columns may be fuelled even further by developments and advancements in on-line $^1H$ nuclear magnetic resonance (NMR) in particular, where the use of supercritical carbon dioxide eliminates the problems related to high background signals arising from the use of protonated solvents and minimises the requirement for deuterated solvents whilst providing detailed structural information [14].

However, where analytical SFE has been coupled with other techniques there has been a trend away from packed column SFE–SFC over the past few years towards coupling with capillary gas chromatography and open tubular and packed column capillary SFC, probably, in part, because of the routine use of capillary GC in most laboratories and in particular the increased use and availability of reliable bench top gas chromatography–mass spectrometry (GC–MS) instrumentation.

This chapter reviews the advantages and disadvantages of packed columns and their packing materials when compared with capillary columns, particularly for SFE coupled to packed column SFC. SFE coupled to packed column SFC is discussed with introductory sections on supercritical fluid mobile phase, SFE and SFC. Details of instrumental aspects of SFE coupled with packed column SFC are reviewed, highlighting some relevant applications to illustrate their use.

Instrumental aspects discussed are: back pressure regulators, extraction vessels, on-line solute trapping, collection and concentration (focusing) devices, on-line solute introduction, columns, detectors and fraction collection. Specific details of commercially available pumps for supercritical fluid delivery and of commercial detectors used for SFC will not be given here since detailed information is readily available from the instrument manufacturers.

Following the review of instrumental components, applications of SFE coupled with packed column SFC are detailed and discussed followed by some thoughts on the future of the coupled technique.

## 7.2 Supercritical fluid chromatography: packed versus capillary columns

In general, columns used for SFC may be broadly classified as capillary, packed capillary or packed. The common physical differences between these three types of column, in terms of column diameter and length, are summarised in Figure 7.1 and Table 7.1.

Additionally, since 1984, several authors have evaluated the use of micro-packed capillary columns [15–20]. More recently Ibáñez and co-workers [21–23] have advocated the use of micropacked columns having large particle diameters of up to 100–120 μm, with shorter column lengths (20 cm) and broader column diameters (0.5 mm) than used for conventional packed capillary columns, with the objective of minimising pressure drop through the length of the column. For ease of discussion, in this review chapter, packed and micropacked columns are included with packed capillary and microbore columns under the general term 'packed columns'.

In their initial application of SFE directly coupled to SFC, Sugiyama *et al.* [10] used packed column SFC. However, the majority of workers in the field

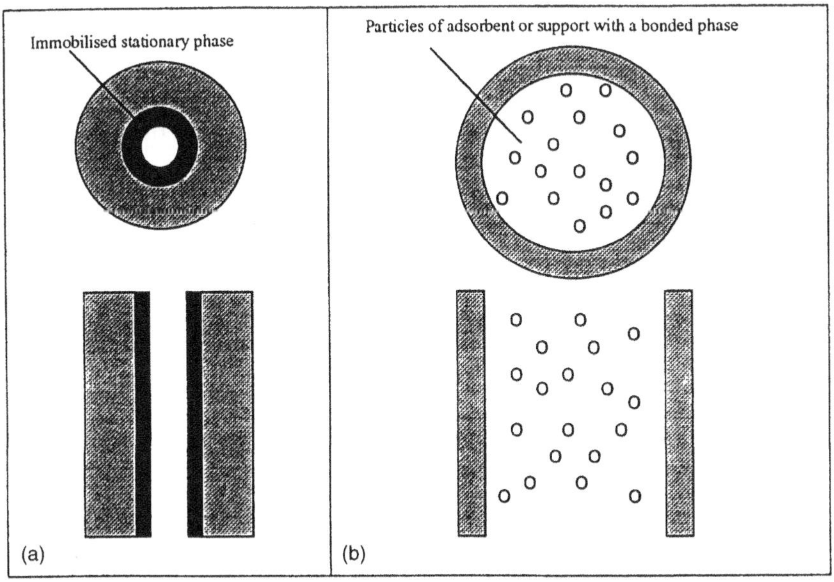


**Figure 7.1** Types of column used for supercritical fluid chromatography: (a) capillary columns and (b) packed capillary columns and packed columns. For dimensions, see Table 7.1.

**Table 7.1** Typical dimensions of column used in supercritical fluid chromatography. Columns are illustrated in Figure 7.1

| Nomenclature | Capillary | Packed capillary | Packed |
|---|---|---|---|
| Internal diameter | 25–100 μm | 0.25–1 mm | 2–4.6 mm |
| Length | 3–20 mm | 10–100 cm | 10–25 cm |
| Film thickness | 0.05–1 μm | | |
| Particle size | | 3–10 μm | 3–10 μm |
| Stationary phase | Methylsilicon | Silica | Silica |
| | Phenyl | $C_{18}$–$C_8$ | $C_{18}$–$C_8$ |
| | Cyanopropyl | Diol | Diol |
| | Biphenyl | Amino | Amino |
| | Poly(ethylene glycol) | Cyano | Cyano |
| | Chiral phases | Chiral phases | Chiral phases |
| | Liquid phases | | |

used capillary columns until the late 1980s, partially for two historical reasons. First, SFC was developed mainly from high-pressure GC technology [8] and many of the early workers had originated from a gas chromatographic background and, second, open tubular capillary columns had been introduced by Novotny and co-workers [24] in 1981, coincidental with Hewlett-Packard's launch of commercial SFC instrumentation at that year's Pittsburg Conference.

Apart from these historical reasons, there were two major practical factors which influenced the early preference for the use of capillary over packed columns. These were the problem of pressure drop over the length of the column, emphasised by Novotny *et al.* [24], and the inertness of surface coatings used for capillary columns when compared with packed column SFC, where the majority of packing materials are based on bonded silica with high surface activity because of residual silanol groups. Both these problems will be discussed.

Differences, advantages and disadvantages of packed and capillary columns and between packed and capillary SFC have been reviewed by a number of authors. For example, Wheeler and McNally [25] compared packed and capillary SFC with HPLC using representative herbicides and pesticides to evaluate the use of the supercritical fluid techniques as an alternative to HPLC. Schoenmakers [26] used a mathematical approach to compare packed columns with open tubular capillary columns for SFC in terms of column efficiencies, column pressure drop, influence of film thickness, speed of analysis and the maximum number of theoretical plates available for each of the column types. Further papers where packed and capillary columns have been compared include those of Petersen [27], who dealt with packed columns in SFC, and Schoenmakers *et al.* [28] in which a number of stationary phases were evaluated and compared for packed column SFC, principally on the basis of packing material inertness. Griebrokk *et al.* [29] and Taylor and Chang [30] discussed the development and uses of packed columns in SFC up to 1990. More recently Ibáñez and co-workers [21] compared the performance of micropacked, open capillary and packed columns in SFC. These papers identify the major advantages and disadvantages of packed and capillary columns as outlined in the following sections.

### 7.2.1 *Efficiency*

Approximately identical numbers of theoretical plates are obtained per unit length of packed and capillary columns when the particle size of the packed column is equivalent to the internal diameter of the capillary column [31]. Excluding micropacked capillary columns such as those described by Ibáñez *et al.* [21] the average internal diameter of capillary columns used for SFC is 50 μm and, as particle sizes for packed columns are in the range 3–10 μm, so the number of theoretical plates per unit length of packed column is higher than that for a capillary column. However, set against this is the large difference in permeability between packed and capillary columns, since permeability is obviously greater for an open tubular column than for a packed column.

Permeability influences the pressure drop along the length of the column, and as the pressure drop increases the supercritical fluid mobile phase density gradient increases. Density is directly linked to the capacity factor ($k'$) and therefore a capacity factor gradient exists having a low value of ($k'$) at the column inlet and a higher value at the column outlet. Under certain circumstances this negative capacity factor gradient can lead to the practical result of a decrease in the chromatographic resolution with column length. For a given pressure drop over the column, capillary column SFC will allow 100 to 500 times more theoretical plates than will packed column SFC with 5 μm or 10 μm diameter particles [26].

For difficult separations, it is generally better to use the resolving power of capillary SFC. However, for most practical purposes short packed and micropacked SFC columns will be able to provide enough theoretical plates for most separations when other factors, such as advantages of packed column selectivity, are taken into account.

### 7.2.2 *Selectivity*

The phase ratio, $V_s/V_m$ (where $V_s$ is the volume of the stationary phase in the column and $V_m$ is the volume of the mobile phase in the column) is higher for packed than it is for capillary columns, which means that interactions between the stationary phase and the sample solutes are of more significance for packed than they are for capillary columns. This provides an advantage for packed column SFC over capillary columns in terms of their selectivity towards specific solutes, since advantage can be taken of the wide range of packing material selectivities commercially available for conventional HPLC and SFC columns compared with a narrower range of surface coatings available for capillary columns.

Additionally, chiral stationary phases, such as cyclodextrins and bonded chiral diamines, are more readily available for packed columns where they may be used to separate optically active compounds. These chiral phases have in part been responsible for the increased interest in packed column SFC in recent years, with numerous reports published and presented at meetings of superior separations, easier and faster method development and adoption by industry [13]. Much of this success is a result of the higher ordering of the packing materials under SFC conditions in comparison with the same materials used under gas chromatographic conditions [13].

A problem with packed columns, especially those based on bonded silica, is that of the surface activity of the packing materials, principally arising from residual silanol groups as a consequence of incomplete or patchy coverage of the porous silica particles with the bonded phase material. With polar compounds this may cause asymmetric peak shape and sometimes severe tailing as a result of strong hydrogen bonding and ionic and dipole interactions with the stationary phase, particularly with acidic and basic compounds, which can result in their partial retention on the chromatographic column through irreversible absorption.

This surface activity, particularly for early packing materials for packed columns, was the second and perhaps most serious reason why early SFC workers preferred capillary columns, since the internal surface area of capillary columns is covered with sufficient film thickness of stationary phase surface coating to cover the silanol groups present, rendering the columns inert. For this reason, capillary column SFC is often employed for the analysis of polar solutes, whereas packed column SFC is more commonly used for non-polar solutes such as polynuclear aromatic hydrocarbons. Elution of polar materials from the commonly used non-polar $C_{18}$ bonded phase silicas is difficult with supercritical carbon dioxide because of its lack of polarity and the interaction of polar solutes with the surface activity of the silica-based materials.

Over the past decade there has been much effort to reduce the surface activity of silica-based packing materials. This, combined with the introduction of newer packing materials, the introduction of smaller diameter and more uniform particles, together with the ability to pack narrower fused-silica columns, helped to initiate increased interest in packed column SFC in the mid-to-late 1980s [29].

Partial reduction in surface activity has been achieved by 'end-capping' bonded silicas and by other surface treatments aimed at shielding residual silanols. Hydrosilyloxane polymers have been used to shield residual silanol groups by their chemical bonding to porous silica particles using a coating technique similar to that developed for capillary GC [32]. In this technique a less active surface coating is produced by using a cross-linking agent to combine the polysiloxane chains which mask the silanol groups.

Several non-silica-based packing materials have been used which overcome the silica-based packed column limitation of surface activity. Better results were obtained for the separation of iso-octylphenoxyl polyethoxy ethanol oligomers from Triton X-100 surfactant by SFC on a reversed phase PRP polymer column than on a $C_{18}$ bonded silica column [33], although there may be some problems with using polymer resin columns under the high pressures required for SFC [30].

Schoenmakers *et al.* [28] looked at a number of different stationary phases for packed column SFC in terms of their surface activities. These included modified and unmodified porous graphitic carbon and modified and unmodified poly(styrene–divinylbenzene) copolymers as well as chemically modified silica and polysiloxane-coated silica. The compatibility of the phases with supercritical carbon dioxide mobile phase, and a number of test solutes of different polarities, was established from qualitative examination of their eluted peak shapes. Polysiloxane-coated silica packing materials performed better than $C_{18}$ bonded phase silica for certain solutes but some peak shapes and efficiencies were still poor. Of the packing materials tested under a uniform set of conditions the unmodified carbon column proved to be most stable. By experimentally coating poly(ethylene glycol) onto porous graphitic carbon a 'carbonwax' column was prepared which provided the best, non-active, packing material for the SFC of polar solutes.

Supercritical carbon dioxide is the most commonly used mobile phase for SFC on account of the ease with which supercriticality may be accomplished ($T_c = 31°C$, $P_c = 73$ bar) its availability in a pure form, its non-toxic nature, its cheapness and its compatibility with a number of chromatographic detectors. However, it provides an essentially non-polar mobile phase which has poor eluting characteristics for polar materials on most commonly used non-polar stationary phases used for packed column SFC.

Although recent developments, such as polyethylenimine-coated silica particles [34] and the commercial development of fluorinated silica packing materials [35], have resulted in increasingly inert surface activities for SFC packing materials, there is still some way to go in terms of reducing surface silanol activity for most silica-based material and its interaction with polar solutes. Because of the latter, polar materials have tended to chromatograph poorly and have relatively long retention times with packed columns and, unless a more polar supercritical fluid mobile phase is used, some other means must be found to deactivate the packing material. Of the supercritical fluids available the polar supercritical ammonia may be a possible choice as an alternative to the non-polar carbon dioxide. However, it would be obnoxious to use, and a better way of improving the polarity of the mobile phase is to add a small amount of a polar organic solvent, commonly methanol, to the supercritical carbon dioxide as a modifier. The addition of polar modifier increases the mobile phase polarity and also deactivates surface silanol groups, resulting in better peak shapes and reduced retention time for polar solutes. The addition of a small amount of modifier also increases the critical temperature and hence the density of the mixture. This increases the solvation power of the mobile phase, further reducing the retention of polar solutes.

Columns for capillary SFC are more inert than are packed columns, but less selectivity towards particular solutes is available because of a smaller range of commercial stationary phase materials. The inertness of capillary columns provides better chromatography for polar compounds without the need for polar modifiers, although these can be added to the supercritical fluid at source. However, GC-type detectors used with SFC capillary columns are not compatible with mixed carbon dioxide–modifier phases and this tends to restrict the use of polar modifiers to packed columns where, for example, ultraviolet (UV) detectors compatible with both supercritical carbon dioxide and most polar modifiers are mainly used.

### *7.2.3 Sample capacity*

Because of their higher phase ratio, packed columns also have a greater sample capacity than do capillary columns. This means that greater amounts of material can be added directly to the column without the need of sample splitting, commonly required for capillary SFC, which simplifies sample introduction by injection or sample valve. Higher sample load onto a packed column allows smaller concentrations of materials to be detected and as a consequence detection limits are often better from packed than from capillary columns, particularly where small microbore (1–2 mm inner diameter columns) are used. However, sensitivity is also dependent on detector characteristics.

The ability to load significantly more material onto a packed column is an important consideration and a major advantage of packed column SFC directly linked with SFE, since the process of solute transfer from SFE is easier without sample splitting and ideal when an initially large sample is required for SFC followed by component collection or fractionation.

The large sample capacity of packed columns is also an important consideration if transfer to semi-preparative or preparative SFC is being considered.

### *7.2.4 Detectors*

The low mass flow from capillary columns means they are amenable for use with the gas detectors used for GC, such as flame ionisation, flame photometric, electron capture and the nitrogen phosphorus specific detectors. Capillary columns are also directly compatible with mass spectrometry

(MS) and Fourier transform infra-red (FTIR) detectors and may be adapted for use with most common analytical detection systems.

With greater supercritical mobile phase flow, packed columns are more directly compatible with liquid detectors traditionally used for HPLC, in particular fixed or variable wavelength UV detectors, and multiwavelength or photo-diode array detectors are commonly used with SFC. Packed capillary columns have the major advantage of being compatible with gas and liquid detectors used for SFC and are useful in cases where SFE coupled with SFC requires the introduction of relatively large samples to the SFC column followed by FTIR, MS or NMR analysis.

In a review, Saito and Yamauchi [36] determined the frequency of application for detectors used for packed and capillary column SFC, reported in publications between 1988 and the first half of 1992, which clearly demonstrated that UV detection was the most commonly used detection system for packed column SFC, whereas the GC flame ionisation detection (FID) was most commonly used for capillary SFC. At that time, the frequency of detectors used for packed column and capillary SFC were as shown in Table 7.2, taken from Saito and Yamauchi's review.

More recently, an increasing number of detection systems have been applied to both capillary and, in particular, packed column SFC; among those which have recently been applied to packed column SFC are chemiluminescence [37], fluorescence [38], electrochemical [39], radiochemical [40], evaporative light scattering detection [41–43], $^{1}$H NMR [14], plasma emission spectrometry with MS and inductively coupled plasma MS detection systems [44], the latter being used for the ultratrace determination of organometallics. The use of these detectors, in the literature, is briefly described in a review by Chester *et al.* [13]. Overall, the applicability of such a large range of detectors is seen as one of the main attractions of SFC and of packed column SFC in particular.

**Table 7.2** Frequency of detector use for packed column and capillary column supercritical fluid chromatography, appearing in publications 1988–mid 1992. Source: ref. 36

| Detector | Packed column | Capillary column |
|---|---|---|
| Absorption | 90 | 14 |
| Flame ionisation | 23 | 100 |
| Mass spectrometer | 15 | 30 |
| Fourier transform infra-red spectrometer | 4 | 12 |
| Evaporative light scattering | 5 | – |
| Atomic emission spectrometer | 7 | 8 |
| Nuclear magnetic resonance spectrometer | 1 | – |
| Electron capture | 1 | 2 |
| Photo ionisation | 1 | – |
| Thermionic | 1 | 8 |
| Chemiluminescent | 1 | 4 |
| Ion mobility | – | 3 |
| Flame photometric | – | 3 |

### 7.2.5 *Analysis times*

Schoenmakers [26] developed simple mathematical expressions to compare the speed of analysis in an open capillary with a packed SFC column and concluded that in terms of speed of analysis capillary SFC columns could not compete with packed column SFC, unless the column diameters were greatly reduced. Hence, for speed of analysis, packed SFC columns are better, and because of this, are often more suited to method development [28].

### 7.2.6 *Restrictors*

For capillary column SFC simple capillary restrictors are generally used to control supercritical fluid pressure. These are prepared from short lengths of narrow-bore fused-silica tubing. Pressure regulation is achieved through a combination of flow resistance brought about by the length and the diameter of the capillary tube. The longer and narrower the tube the greater the pressure maintained. Several types of capillary restrictor have been used having linear, crimped, tapered, converging or diverging, integral or fritted ends (section 2.7.1). The main advantage of these restrictors is cheapness; however, restrictor-to-restrictor back pressure reproducibility is difficult and the restrictors are prone to blockage, often because of condensation of moisture from the sample, which becomes frozen at the tip of the restrictor on evaporation of the carbon dioxide eluent. More recently, heated capillary restrictors, which reduce some of these problems, have become available from commercial manufacturers, such as Isco-Suprex [45].

More robust back pressure regulators, which are often heated, do not clog and are able to maintain a more reproducible pressure in the SFC system, are generally used for packed column SFC systems. For this reason, packed column SFC can provide more stable and reproducible results when coupled to SFE than can capillary column SFC.

### 7.2.7 *Temperature*

As capillary SFC columns were generally developed from GC stationary phases they can be used at higher temperatures than can traditional packed columns, which were originally developed for use at the ambient and low temperatures used for HPLC. An increase in temperature decreases the fluid density and the solvating ability of the mobile phases, which in turn will increase the retention time for most compounds. However, it is generally the case that an increase in temperature in capillary SFC decreases retention times because of the increased volatility of volatile solutes. Therefore retention time behaviour in capillary SFC can be complex, with some compounds increasing and some decreasing in retention with increasing temperature.

At increased temperatures with silica packing materials there may be some problems with the use of supercritical carbon dioxide, which because of its high solvating power dissolves the silica backbone and/or bonded phases. The effects of high temperatures on conventional HPLC bonded phase materials and polymer-coated bonded phase packings especially developed for SFC have been studied by Taylor and Chang [30] who compared a number of commercially available conventional $C_{18}$, aminopropyl and silica packings using retention times and peak shapes from a number of test solutes. Possible loss of packing materials during heat treatments was examined by thermal gravimetric analysis (TGA) linked to mass spectrometry (TGA–MS). After heat treatment, retention time was reduced and peak shape improved for $C_{18}$ material. TGA–MS revealed that only water had been lost from the column at temperatures up to 150°C. However, aminopropyl columns treated in the same way gave increased retention times and poorer chromatographic results. TGA–MS revealed loss of water and some aminopropyl stationary phase, which illustrates that the loss of stationary phase from conventional HPLC packing materials is dependent on the exact nature of the bonded phase and, of course, the conditions used.

Taylor and Chang [30] also compared a number of polymer-coated hydroxysiloxane packing materials, including polymer-coated cyanopropyl and phenyl, specially developed for SFC with conventional bonded phase packing materials, and concluded that the polymer-coated materials had less surface activity and were more thermally stable than were conventional phases.

In summary, the cross-linked stationary phases used for capillary SFC columns are generally stable at elevated temperatures under SFC conditions. Although conventional HPLC packing materials employed for packed column SFC are generally unstable at elevated temperatures, the newer polymer-coated packing materials exhibit much greater thermal stability, and consequently both capillary and packed column SFC can make use of elevated temperatures to reduce retention times of solutes, provided that the stability of the packing material is checked before use.

Apart from the major differences and similarities between packed and capillary SFC columns discussed above, Wheeler and McNally [25] found packed column SFC easier to use, having no need for injection splitters, and that the back pressure regulators used with packed columns were easier to use than were the capillary restrictors used with capillary columns. Furthermore, better injection-to-injection reproducibility with less sample carry-over was found for packed column SFC.

## 7.3 Supercritical fluid extraction coupled to packed column supercritical fluid chromatography

The major differences between packed and capillary columns for SFC have been reviewed, and, where applicable, comments have been made on the benefits of packed column for SFE–SFC. We will now deal with equipment and techniques used for the direct 'on-line' coupling of SFE with packed column SFC.

### 7.3.1 *Supercritical fluid mobile phase*

The most commonly used supercritical fluid for SFE, SFC and for SFE coupled with SFC is carbon dioxide on account of its ease of supercriticality ($T_c = 31°C$, $P_c = 73$ atm), its availability at high purity, its inertness and its lack of toxicity and flammability. At typical working pressures (80–600 atm) supercritical fluid carbon dioxide is an excellent solvent for non-polar solutes such as alkanes and terpenes and is a reasonably good solvent for moderately polar compounds such as polynuclear aromatic hydrocarbons (PAHs), polychlorinated biphenyls (PCBs), aldehydes, esters, alcohols, organochlorine pesticides (OCPs) and fats but is less useful for polar compounds [46]. Polar compounds, however, may be successfully extracted by the addition of a small amount of polar modifier to the sample or supercritical carbon dioxide mobile phase or by *in situ* derivatisation [47] of targeted solutes to reduce their polarity. As a rule of thumb, solutes that can be determined by conventional gas chromatographic techniques can be quantitatively extracted using supercritical carbon dioxide [48].

A number of other substances can be used as supercritical fluids, such as the more polar nitrous oxide, ammonia, methanol, water, ethane and ethylene. However, these other compounds are more difficult to use in terms of handling and tend to be used only for more specialist applications, where supercritical carbon dioxide cannot be used alone or with an appropriate modifier.

### 7.3.2 *Supercritical fluid extraction*

In its simplest form a supercritical extractor consists of a high-pressure pump, similar to or adapted from pumps used for HPLC, an extraction vessel, a back pressure regulator and a separation vessel or trap where the extracted solutes of interest are collected. There are three basic types of SFE system characterised by the way in which the solutes are separated from the supercritical fluid extraction media used [49, 50]. In the first type, solutes are separated from the extraction media based on pressure reduction which causes a solubility decrease. In the second type, a temperature change is used to bring about a decrease in solubility from the extraction media. In the third type the solutes are adsorbed onto an appropriate adsorbate. Often a combination of the first and second is used, where after extraction the supercritical fluid is simply evaporated to leave the solutes of interest.

Supercritical fluid extraction can be either 'static' or 'dynamic'. In static SFE the sample and supercritical fluid are introduced into the extraction vessel and the fluid is left, under continued supercritical conditions, at a specific pressure and temperature while extraction takes place for a period of time, after which the fluid is removed from the vessel and the solutes of interest collected. If extraction is not complete in the first pass subsequent cycles of extraction with fresh supercritical fluid may be used. In dynamic SFE the sample in the extraction vessel is continually flushed or swept with supercritical fluid so that extraction is a continuous process during the extraction period.

In 'off-line' SFE, the solutes of interest are recovered either by evaporation of the carbon dioxide (possibly with cryogenic trapping) or by bubbling the depressurised carbon dioxide containing the solutes through a suitable trapping solvent. Alternatively, the solutes of interest may be trapped onto an adsorbent medium such as silica, solid phase extraction materials such as $C_{18}$ silica, Tenax or activated charcoal. Further details of supercritical extraction and of collection and trapping methods used by workers in the field will be discussed later.

### 7.3.3 *Supercritical fluid chromatography*

A packed column SFC system is very similar to an HPLC system except that there is a back pressure regulator, which pressurises the mobile phase above its supercritical pressure, and a column heater, which is employed to keep the mobile phase above its critical temperature. Generally, the back pressure regulator maintains the system pressure through the detector for packed column SFC, unlike most cases for capillary SFC where use is made of a flame ionisation detector which does not operate under the high pressures used for SFC. Figure 7.2 shows a typical schematic for a packed column SFC system.

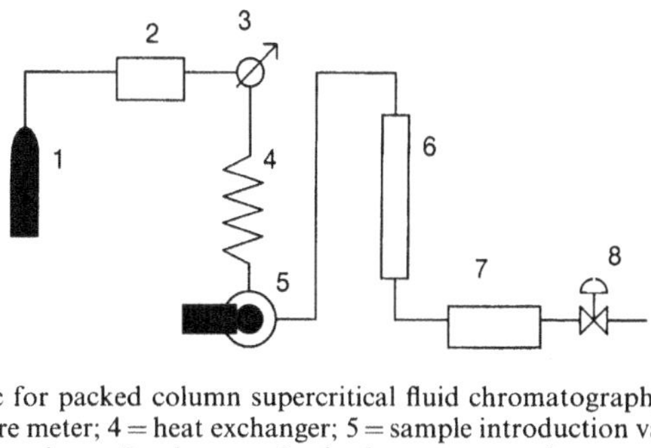


**Figure 7.2** Schematic for packed column supercritical fluid chromatography. 1 = gas cylinder; 2 = pump; 3 = pressure meter; 4 = heat exchanger; 5 = sample introduction valve; 6 = separation column, 7 = detector, 8 = back pressure regulator.

### *7.3.4 Supercritical fluid extraction coupled to packed column supercritical fluid chromatography*

On-line SFE–SFC can provide an extremely powerful analytical tool. The efficient, fast and selective extraction capabilities of supercritical fluids allows the quantitative extraction and direct transfer of the selected solutes of interest to be accomplished to the column, often without the need for further sample treatment or clean-up.

Extraction selectivity is usually achieved by adjusting the pressure of the supercritical fluid at constant temperature or, less often, by changing the temperature of the supercritical fluid at constant pressure.

The extraction selectivity of the supercritical fluid used may also be changed by the addition of small amounts of modifier, typically water or methanol, directly to the sample matrix as an easier alternative to adding the modifier at a constant concentration to the supercritical fluid itself. For supercritical carbon dioxide the addition of small amounts of polar modifier to the sample matrix will increase the selectivity towards polar solutes. Selectivity towards the target solutes may therefore be enhanced by the use of both a specific pressure/temperature regime and the addition of polar modifier to the sample matrix or mobile phase.

A further approach, which has been applied to SFE–GC, capillary SFC (cSFC) and SFE-cSFC, but has undoubted future application to packed column SFE, is to perform *in situ* modification, or derivatisation, of targeted solute molecules to reduce their polarity and increase their extraction efficiency in supercritical carbon dioxide [47, 51]. Berg *et al.* [52] used *in situ* reaction and simultaneous SFE with the immobilised enzyme lipase to synthesise methyl and butyl esters from a sample containing fatty acids, from triglycerides, followed by direct transfer to capillary SFC. Further details of this study can be found in section 4.6.

Major advantages of SFE directly coupled to packed column SFC are:

- its suitability for trace analysis – quantitative transfer of solutes between extraction and chromatography with little or no loss of (particularly) volatile solutes compared with off-line SFE with SFC or other instrumental techniques;
- ease of selective sample preparation which is easy to automate with the minimum of operator involvement;
- less contamination potential for trace level solutes as a result of the direct coupling of extractor and chromatography;
- the ability rapidly to extract and directly analyse unstable and oxidation-sensitive solutes such as metal carbonyl compounds [53], which decompose on contact with most solvents, for example tocopherols from wheatgerm samples [54] and polyunsaturated fatty acids [55] sensitive to oxidation. For these compounds, off-line SFE followed by SFC would be difficult, if not impossible.

The direct coupling of SFE to SFC is nearly always achieved through the on-line linkage of an SFE vessel through a valve switching system, although other means for the direct introduction of SFE extracts such as the thermal modulation interface, developed by Mitra and Wilson [56], have been used for capillary SFC. The thermal modulator, which is interfaced between the SFE vessel and supercritical fluid chromatograph, is made of a short length of fused-silica capillary tubing containing stationary phase on the inside and painted on the outside with an electrically conducting paint. As the SFE medium flows through the modulator the solutes of interest are trapped and concentrated in the stationary phase. The solutes are then rapidly released, like an injection, to the SFC column when the modulator is rapidly heated by passage of electrical current through the paint.

Commercial SFE systems which include autosamplers, high-pressure pumps, extraction cells, back pressure regulators and valves for off-line solute collection and on-line solute transfer, together with the necessary heating units to maintain the supercritical fluids at temperatures above critical, are available through a number of commercial manufacturers. Additionally, many SFE–SFC systems have been constructed by workers using component parts from commercial systems, together with commercially available sample switching valves and extraction vessels. The extraction vessels may be adapted from HPLC columns or specially prepared and developed for SFE for a particular sample matrix.

A major advantage of using packed column SFC in conjunction with SFE is the sample loading capacity of the columns. This in turn enables enough material to be chromatographed to make fraction collection of the components of interest a viable option. Fraction collection is often added to, or an option for, SFE directly coupled to packed column SFC.

Typical configurations for SFE coupled to packed column SFC are shown in Figure 7.3(a) and 7.3(b). These were chosen to illustrate typical configurations for an SFE–SFC system in terms of the supercritical flow path.

The figures show the position of post-SFE trapping (or pre-concentration as shown in the figures) and post-SFC fractionation. Figure 7.3(a) illustrates the mobile phase flow path during the extraction and pre-concentration stage, whereas Figure 7.3(b) shows the flow path during SFC (and fractionation). In the figures given, UV light is used for solute detection. As discussed previously, UV detection is commonly used for packed column SFC; however, IR and FTIR are also frequently used, and GC detectors such as the FID can be used when an eluent splitter is introduced post-column. In Figure 7.3 the dashed lines indicate the three different sections of the SFE–SFC system: fluid delivery (carbon dioxide and modifier cylinders plus fluid delivery pumps); extraction, including pre-concentration and chromatography; and fractionation.

During the extraction process liquefied carbon dioxide, indicated by the thick flow line in Figure 7.3(a), is pumped via the high-pressure pump

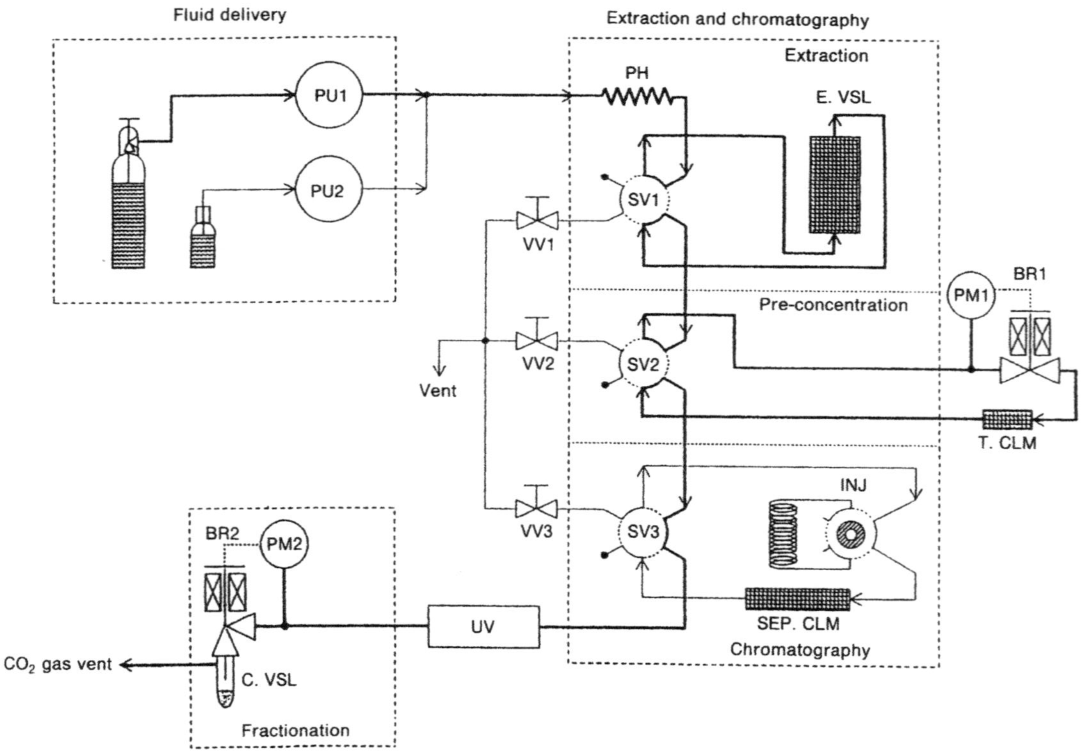

Fluid delivery
PU1
PU2
Extraction and chromatography
Extraction
PH
E. VSL
SV1
VV1
Pre-concentration
PM1
BR1
SV2
VV2
Vent
T. CLM
INJ
SV3
VV3
SEP. CLM
Chromatography
UV
BR2
PM2
$CO_2$ gas vent
C. VSL
Fractionation


(a)

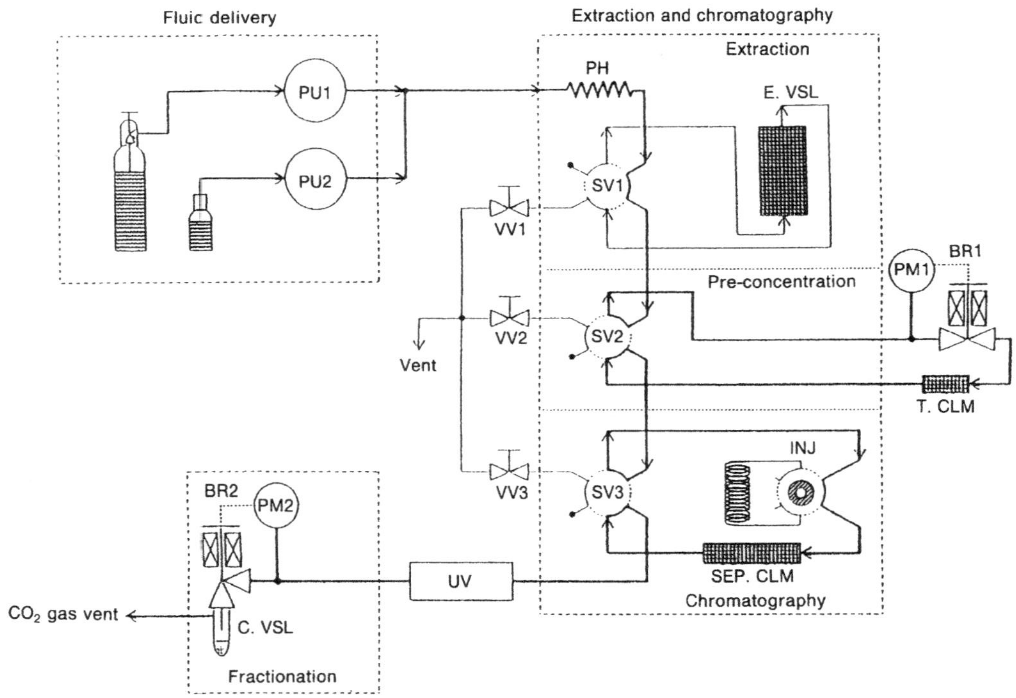


(b)

**Figure 7.3.** Typical instrumental set-up for supercritical fluid extraction–supercritical fluid chromatography (SFE–SFC) with fractionation: (a) flow line for extraction and preconcentration; (b) flow line for SFC and fractionation. Dashed lines delimit the three sections of the apparatus (fluid delivery; extraction and chromatography; fractionation); solid line = $CO_2$ flow; PU1 = high-pressure pump for $CO_2$; PU2 = pump for modifier; PH = preheat coil; E.VSL = extraction vessel; BR1, BR2 = back pressure regulators; T.CLM = trap column; SV1–SV3 = switching valves; SEP.CLM = separator column; PM1, PM2 = pressure transducers; VV1–VV3 = vent valves; C.VSL = collection vessel; INJ = injector. Reproduced with permission from ref. 36.

(PU1) through the preheat coil (PH) where the carbon dioxide is heated to a temperature above critical. The pressure of the pumped carbon dioxide then builds up to the pressure to be applied to the extraction vessel (E.VSL), above the critical pressure where it is controlled by the back pressure regulator (BR1), whilst the pressure of the trap column (T.CLM), placed downstream of the regulator, is kept at a lower pressure by the second back pressure regulator (BR2). On passage from the extraction vessel to the trap column the density of the now supercritical fluid carbon dioxide is decreased rapidly, which results in a decrease of its solvating power, with consequent deposition of solutes onto the packed bed of the trap column.

On completion of extraction and trapping of the solutes, the extraction vessel is bypassed by means of the switching valve (SV1), and supercritical fluid, which can also contain a modifier added via pump (PU2), is passed through the trap column to flush the solutes in a concentrated band to the separator column (SEP.CLM) via the first back pressure valve which is now completely open. The flow path for SFC is indicated by the thick flow line in Figure 7.3(b). During chromatography, the pressure of the carbon dioxide in the entire flow line from the pumps through to the detector is controlled by back pressure regulator (BR2).

After SFC the pressure of the column eluent is reduced to atmospheric on passing through the regulator (BR2) which drastically reduces the solubility of the solutes in the carbon dioxide, enabling them to be collected by an appropriate collection device, such as the one shown.

In SFE–SFC the separation and trapping are housed within an oven so that temperatures above the critical temperature may be maintained. The SFE vessel may be housed in the same oven used for the separator column but for versatility the extraction vessel is usually kept in a separate oven with independent temperature control. Alternatively, the extraction vessel may be thermostated in some other way to maintain a constant temperature during extraction.

Selectivity of a supercritical fluid towards target solutes may be at least partially achieved through variation of its solvating power through the control of pressure (density) in the system. The higher the density of supercritical fluid carbon dioxide the greater its solvating power. Change in solvating power may also be achieved by variation of extraction temperature, though this is less commonly used, by the addition of polar modifiers or by *in situ* derivatisation of targeted solutes to reduce polarity.

Coupled SFE–SFC is often configured with an analyte concentration device prior to valve switching onto the SFC. The objective of this device is to concentrate the analytes of interest into a small band for 'injection' onto the supercritical fluid chromatograph. There are two main approaches to this concentration stage.

The first approach is to use an 'extractor' column, that is generally packed with adsorbent material such as silica, on which the solutes are trapped. Saito and Yamauchi [54] used this type of column to collect extracted tocopherols from wheatgerm, whereas Yarita *et al.* [57] trapped extracted thiolcarbamate herbicides from soil matrices on $C_{18}$ silica. Alternatively, Jordan *et al.* [58] used a short length of stainless steel column held, without adsorbent, at 20°C to trap and concentrate caprolactam from nylon copolymers before SFC.

The second approach is the use of cryofocusing. This technique has been applied by a number of groups, in particular Ashraf-Khorassani *et al.* [59–61], who used the technique to trap polymer additives prior to valve switching onto SFC, Andersen *et al.* [62], who used it to extract cold-pressed grapefruit oil flavorants, and Xie *et al.* [63], who used SFE with cryogenic trapping for the extraction and collection of a standard hydrocarbon (octatriacontane) and the biologically active compound oubain in conjunction with packed column SFC and a fraction collector. Cryofocusing of SFE extracted solutes may also be achieved by the direct on-line depressurisation of supercritical carbon dioxide onto the SFC column head through the use of a capillary restrictor, an approach recently employed by Ibáñez *et al.* [23] for the SFE–micropacked column SFC of a test vitamin mix.

## 7.4 Instrumental aspects

### *7.4.1 Back pressure regulators*

In SFE and SFC back pressure regulators are required to maintain the pressure of the mobile phase above its critical point. Two main types of back pressure regulators are used, the simple fixed capillary tube restrictor and the mechanical regulator which may be pneumatically or electrically controlled.

The simple fixed restrictor, discussed more fully in section 7.2.6, achieves pressure regulation through a combination of flow resistance brought about by the length and the diameter of the capillary tube. The longer and narrower the tube the greater the pressure maintained. Capillary restrictors are useful for the direct on-line coupling of SFE to capillary, packed capillary or micropacked SFC or gas chromatography, in which the restrictor can be used as an on-column injection device through the injection port and directly into the separator column [64]. In addition capillary restrictors may be used in conjunction with small SFE vessels [23].

Mechanical restrictors are better suited to the higher mass flow rate of packed column systems and therefore to the demands of SFE coupled with packed column SFC, and for use with UV detectors. Whereas capillary restrictors for capillary SFC are situated before GC type detectors, mechanical restrictors are usually situated behind the UV detection cell to maintain pressure throughout the whole SFC (or SFE–SFC) system. Mechanical restrictors consist of a pressure sensing device, together with a needle valve, which is usually heated to prevent blockage from ice or extracted material build up resulting from cooling from adiabatic extraction of the carbon dioxide. These regulators, which may be controlled manually or electronically, are able to control back pressure irrespective of mass flow rate of the supercritical fluid. Consequently, they are more reliable than are simple fixed capillary restrictors, for both SFE and SFC, where extraction or chromatographic reproducibility is required as a function of pressure.

Mechanical restrictor valves control the flow resistance of the fluid by changing the gap between the needle valve and seat. Saito *et al.* [50] developed a regulator specifically for the exacting needs of SFE, having low dead volume and high precision. This uses a needle valve attached to a solenoid (see Figure 4.7) which operates at high speed, switching the fluid flow by opening and closing the flow path and preventing blockages from precipitated solutes and dry ice which are continually trapped and forced to pass through the valve.

More recently, instrument manufacturers such as Isco-Suprex [45] have developed new heated capillary and packed column restrictors for SFE and SFC to overcome the reproducibility and blockage problems encountered with simple capillary restrictors and to improve the performance of the older, sometimes unheated, mechanical restrictors used for packed column SFE–SFC. For example, the Isco-Suprex coaxially heated adjustable needle valve (illustrated in Figure 2.13) consists of a restrictor which has its flow-regulating nozzle at the tip of a slender probe so that it may be positioned at the bottom of a solvent-filled vial for solute collection. With this restrictor, flow rates may be manually regulated to between 0.5 and 10 ml/min. Blockage problems are eliminated since the restrictor is pressurised and heated to the tip of the nozzle.

### 7.4.2 *Extraction vessels*

There are a number of commercial analytical supercritical extraction systems marketed by Jasco, Isco-Suprex, Dionex, Hewlett-Packard and Carlo Erba which may be purchased as an integrated instrument with SFC or bought separately and linked by the user to SFC.

Commercial SFE instrumentation employs custom-made vessels, with volumes ranging from a few microlitres to tens of millilitres, made of stainless steel or poly(ether-ether-ketone) (PEEK) or other polymeric materials. However, many workers perform supercritical fluid extractions in pressure-rated HPLC columns, with perhaps the most popular size being the standard 250 mm × 4.6 mm inner diameter (i.d.) standard analytical HPLC column.

The advantage of using HPLC columns as extraction vessels for SFE are that they are readily obtainable and can be packed and unpacked so that they can be reused. McNally and Wheeler [65] have used a variety of HPLC column sizes as extraction vessels for coupled SFE–SFC: commonly 250 mm × 4.6 mm i.d. and 100 mm × 4.6 mm i.d. columns. They also used 250 mm × 1.27 mm i.d hollow stainless steel tubes and a stainless steel bolted closure pressure vessel to extract soil, plant materials and cell culture medium. The size of the vessels used depends on the quantity of material to be extracted, the detection limits required and the back pressures obtained during extraction. Vessels having volumes over 3 ml were pressure rated to withstand pressures up to 4200 psi at 100°C. Standard HPLC columns have been used by many other workers, including Xie *et al.* [63] and Ramsey *et al.* [66]. In their paper on supercritical fluid extraction as a means of sample introduction for chromatography, Andersen *et al.* [62] discuss volume considerations for SFE extraction vessels, particularly in terms of requirements for environmental analysis. In their laboratory they used extraction vessels with volumes ranging from 0.3 to 5 ml, but they felt there was a requirement to build extraction vessels capable of extracting 50 g of material specified in most current regulatory methods for environmental analysis. Vessels having an extraction capability of 50 g of leaf sample have been built and are referred to in the paper.

During extraction, the sample, usually ground for solid material, is packed into the extraction vessel and retained by frits. Where liquids or oils are to be extracted they may be deposited on a packing material such as silica, quartz wool, cellulose, starch or filter paper. Filter paper was used by Nishikawa [67] to examine extraction parameters for diniconazole pesticide standards, using filter paper spiked with the pesticide prior to SFE–SFC. Fibreglass paper was used by Weixi *et al.* [68] as a support for standard PAHs in methylene chloride prior to coupled SFE–SFC.

Microextraction and miniextraction vessels having void volumes of 3–4 µl and 85 µl, respectively, were built by Jahn and Wenclawiak [53] for direct on-line coupling of SFE with packed column SFC for use under subcritical and supercritical conditions. The advantages of using such small extraction volumes were seen as rapid equilibrium between the analyte samples and the supercritical extraction fluid, and minimal disturbance of the column packing material on direct transfer to the analytical column. The microextractors and miniextractors used are illustrated in Figures 7.4(a) and 7.4(b), respectively.

The microextractor [Figure 7.4 (a)] was constructed from a Rheodyne in-line solvent filter. Short connections were made from 0.1 mm i.d. tubing and $\frac{1}{16}''$ (1.6 mm) fittings. The 5 µm frit used in the microextractor was prepared by compression with a conical tool to create a small depression for the sample material. A filter was placed into the frit holder to prevent access of sample material to the column during SFE. Polytetrafluoroethene (PTFE) and cellulose acetate (pore sizes 0.2 µm and 0.45 µm) were found suitable as filter materials since they could withstand supercritical carbon dioxide at temperatures up to 80°C. The microextractor was resistively heated via the capillary extractors (0.4 V, 180 W, adjustable) and used to

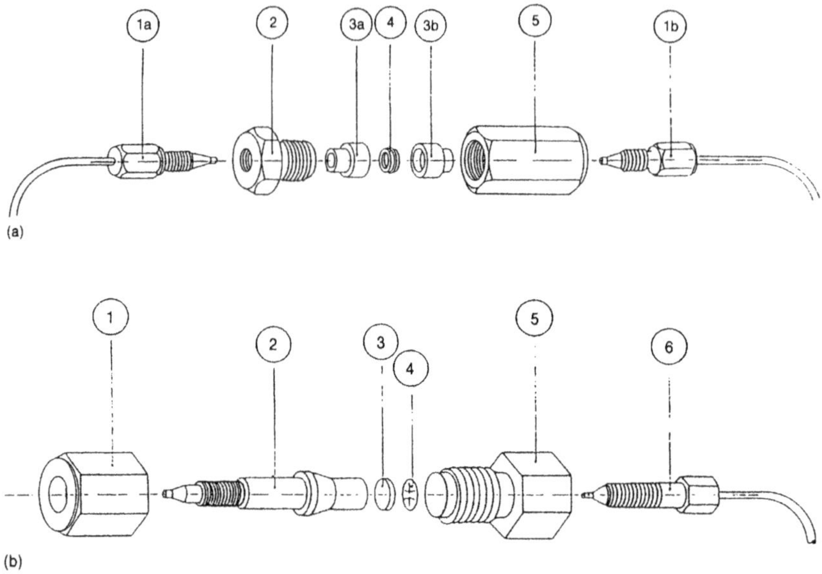


**Figure 7.4** (a) Microextractor: 1a, 1b = male nuts with ferrules; 2, 5 = casing; 3a, 3b = female connection and frit holder; 4 = frit. (b) Miniextractor: 1 = nut; 2 = extraction cell; 3 = frit; 4 = stream distributor; 5 = bolt; 6 = extra long (1/16″) bushing. Reproduced with permission from ref. 53.

extract highly viscous and solid samples, such as used engine oil, benzene, toluene and naphthalene, and solid chromium, molybdenum and tungsten mesitylene carbonyl compounds which react with normal solvents and are also thermally unstable.

The miniextraction vessel [Figure 7.4(b)] was constructed from a guard column having an internal diameter of 2.1 mm, a length of 25 mm and an internal volume of approximately 85 μl. The miniextraction vessel was used mainly for SFE of solid materials such as soil, soot or liquids deposited onto adsorbent materials, such as $C_{18}$ silica. The miniextractor was fitted with a filter plate on the inner side of the frit. After filling the extraction vessel with the material to be extracted the vessel was screwed directly to the sampling valve via capillary tubing. The miniextractor was heated by the attachment of an aluminium clamp containing a 70 W GC detector heater.

As can be seen from the above, there is little difficulty in extracting solid samples by SFE. Viscous samples, such as used engine oil, may also be introduced directly for SFE–SFC provided the sample is very small.

Liquid samples present more of a problem. The usual solution is to place the liquid in an extraction vessel by injection onto solid adsorbent material, either in small, microlitre, volumes or larger volumes if a volatile organic solvent is used, and allowed to evaporate before extraction.

Aqueous samples (and samples containing a high proportion of water) present special problems because of the difficulties of water freezing in the SFE–SFC flow path creating potential blockages and because of the effect of water, acting as a polar modifier, on analyte retention time in SFC using chemically bonded silicas. Thus for SFE and SFE–SFC it is important that the majority of water is removed.

This may be achieved before SFE by evaporation, by use of solid adsorbents such as anhydrous magnesium or sodium sulphate, molecular sieves or Hydromatrix or by freeze drying, which is commonly used for biological tissues [66].

Recently, Zegers *et al.* [20] and Sandra [69, 70] have elegantly overcome water and SFE–SFC incompatibility difficulties by using pre-columns or cartridges containing solid phase extraction (SPE) material such as $C_{18}$ or poly(styrene divinylbenzene) for the indirect supercritical extraction of aqueous analytes, such as pesticides in water. Using this approach, analytes are concentrated onto an SPE pre-column or cartridge which is then dried by passage of dry nitrogen through the SPE material before SFE extraction.

In recent years SFE of aqueous samples has assumed an increased importance for aqueous environmental samples, for the analysis of pesticides and herbicides from field drainage and municipal wastewater and for the pharmaceutical industry for the analysis of water-soluble drug formulations and metabolites. Indirect (SPE) extraction, illustrated by Zegers *et al.* and Sandra *et al.*, is one approach. However, for the direct introduction of aqueous samples for SFE or SFE–packed column SFC, two initial approaches were either to use an on-line aqueous extraction vessel, as reported by Taylor and Hedrick [71], or to use a phase separator to remove water post-SFE, pre-SFC, as described by Thiebault *et al.* [72].

Hedrick and Taylor extracted a phosphonate [diisopropyl methylphosphonate (DIMP)] from aqueous media using an extraction vessel obtained from the Suprex Corporation (Pittsburgh, PA, USA) as previously shown in Figure 4.3.

A total of 3 ml of aqueous sample is introduced into the 100 mm × 10 mm i.d. extraction vessel, which has an internal volume of 8 ml. With the extraction vessel located in an oven, supercritical carbon dioxide is passed into the bottom of the vessel where it bubbles through the aqueous medium to extract the analytes of interest. As a result of the lower density of supercritical carbon dioxide compared with water, the supercritical fluid rises to the top of the extraction vessel where it is recirculated through the aqueous media for the determined extraction time.

A 20 μl extraction aliquot is then transferred to the SFC column via a sample loop for quantitation. With this method of on-line extraction for aqueous samples Taylor and Hedrick were able to obtain a four-point linear calibration graph over the concentration range of 834 ppb to 834 ppm for DIMP, using FTIR detection. The rate of extraction of

DIMP from water was determined using SFC/FID response: an increase in amount of material extracted with time was found up to equilibrium at 1.5 h. However, addition of 0.1 mg of sodium chloride to the aqueous extraction medium reduced equilibrium time to less than 5 min.

Use of the aqueous extraction vessel was also extended to the qualitative study of various hydrochloride salts of drug compounds, including triprolidine and pseudoephedrine hydrochloride by SFE–SFC [73]. Further 'off-line' attempts at aqueous extraction for organic bases were reported by the same authors [74] but this was not successful since during direct extraction with supercritical carbon dioxide the pH of the water decreased to 3.5, causing protonation of the bases with consequential increased water solubility.

In the second approach to direct on-line extraction of analytes from aqueous media Thiebaut *et al.* [72] used a phase separator to demonstrate the potential of segmented flow systems for on-line liquid SFE and SFE–SFC, as shown in Figure 7.5. This figure illustrates the location of the separator in the flow path for coupled SFE–SFC. Interior details of the phase separator are shown at the bottom of the figure. The aqueous extract together with supercritical carbon dioxide pass along the groove in the separator, which has an upper hydrophobic and a lower hydrophilic (usually stainless steel) surface where separation of water and supercritical carbon dioxide occurs based on the respective 'wettability' of the hydrophilic stainless steel part and hydrophobic part of the phase separator's construction. For SFE, the disc material functioning as the hydrophobic surface in the separation process had to be carefully selected to withstand the pressure conditions of SFE. PTFE was found unsuitable, but good stability was achieved with Delrin® and PVDF®, which did not show any signs of deterioration after several months. Initially the system was used to extract phenol and 4-chlorophenol from water, and phenol from urine. Extraction efficiency of the test compounds was over 85%, with a repeatability of about 8% relative standard deviation (RSD) for eight samples for the total SFE–phase separation–SFC system.

In 1996, Janda *et al.* [75] reviewed the direct supercritical extraction of organic compounds from aqueous samples; however, apart from the examples cited above, all involved the simpler off-line approach – trapping extracted solutes in solvent or on solid phase adsorbents.

Against this trend, Ramsey *et al.* [76] have recently reported a new approach for the direct extraction of organic contaminants in water, demonstrating the feasibility of their technique by the extraction and detection of three phenols at low parts-per-billion level, using SFE–SFC–MS with full-scan negative-ion atmospheric chemical ionisation (APCI). To achieve this a high-pressure multivalve switching arrangement incorporating two amino columns was built, the first column to act as an analyte trap and the second as a separation column for SFC.

Direct aqueous extraction of phenols was performed in a patented, pressure-tested stainless steel extraction vessel which was housed within a

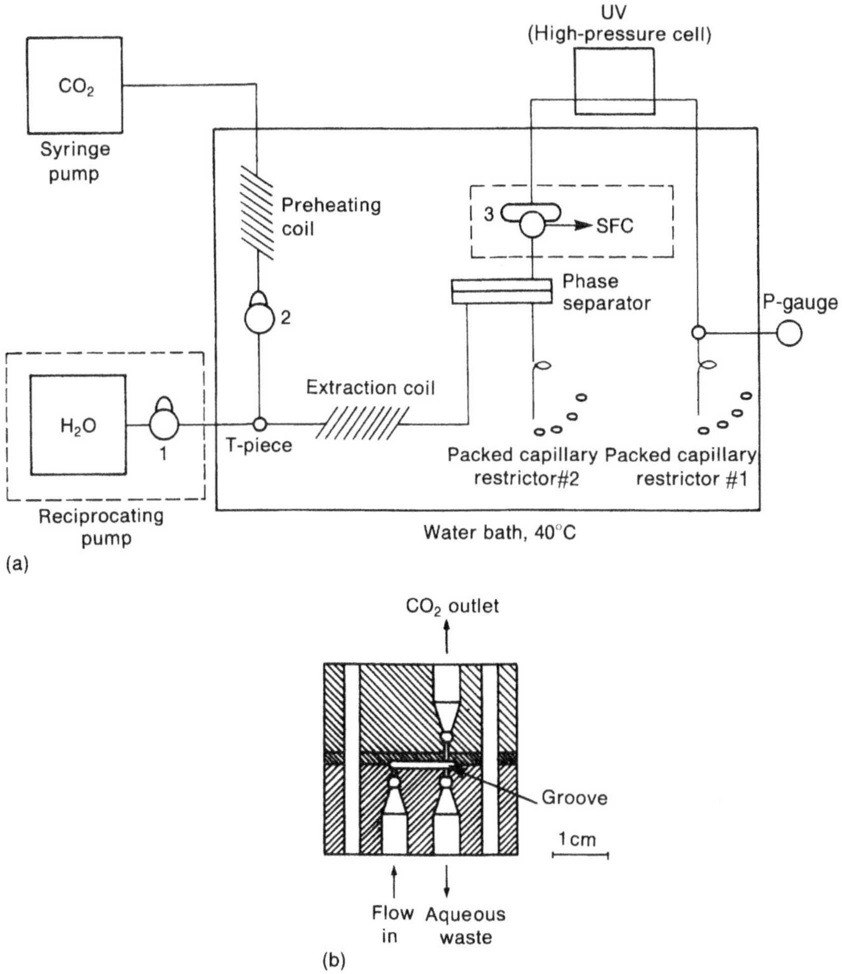


**Figure 7.5** (a) Schematic of supercritical fluid extraction (SFE)–phase separation–supercritical fluid chromatography (SFC) to show the position of the phase separator; (b) expanded interior view of the phase separator. Redrawn with permission from ref. 72.

GC oven for temperature control. A spiked acidic aqueous solution of three phenols, 2,4-dibromophenol, 2,4,5-trichlorophenol and 4-nitrophenol, was prepared, with a total loading of 10 μg of each phenol in a volume of 250 ml, and placed in the extraction vessel. Using the multivalve system shown in Figure 7.6(a), direct aqueous SFE was performed at 65°C over 45 min, at a pressure of 32.33 MPa and a flow of 4 ml/min. After extraction the multivalve system was switched, as shown in Figure 7.6(b), to enable SFC–MS.

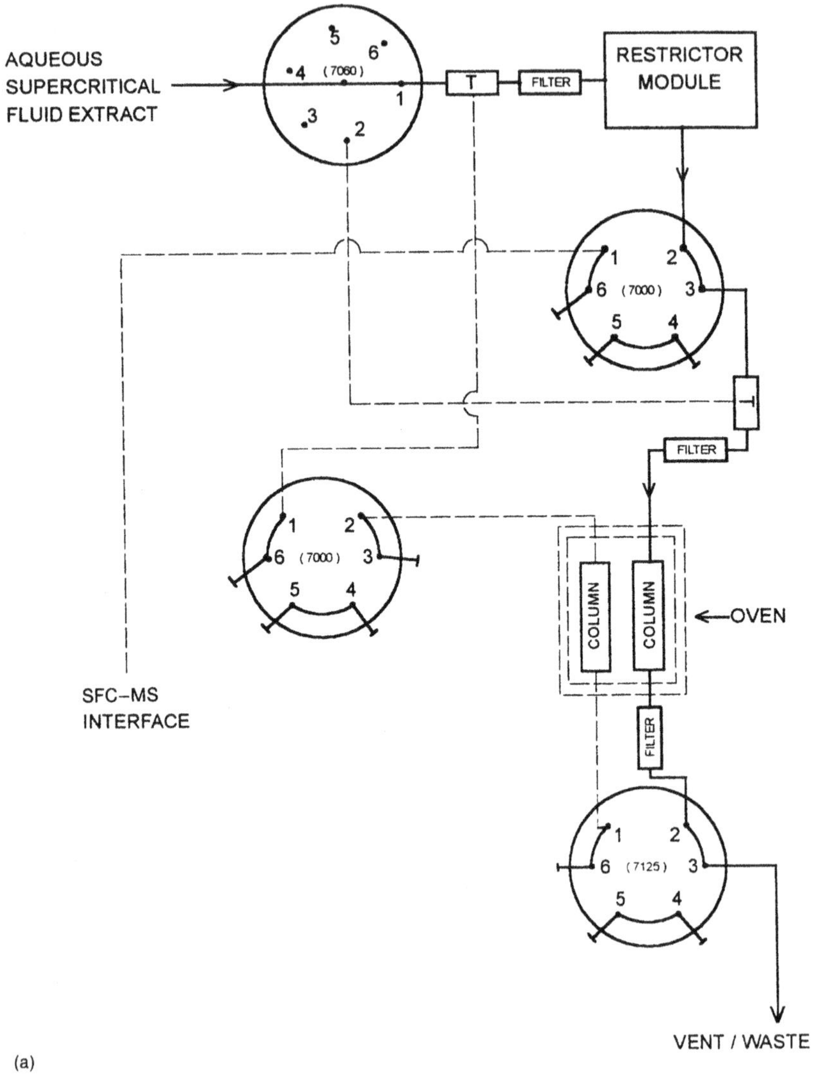


**Figure 7.6** Configuration of high-pressure multivalve system used for (a) direct aqueous supercritical fluid extraction; (b) supercritical fluid chromatography–mass spectrometry (SFC–MS). Reproduced with permission from ref. 76.

During dynamic aqueous SFE the restrictor module was situated pre-collection-column, by valve selection, and the phenols were collected and trapped on the first of the two amino columns, as shown in the flow diagram, Figure 7.6(a). After extraction, with the extraction vessel isolated, the valve configuration was switched, as shown in Figure 7.6(b), so that the

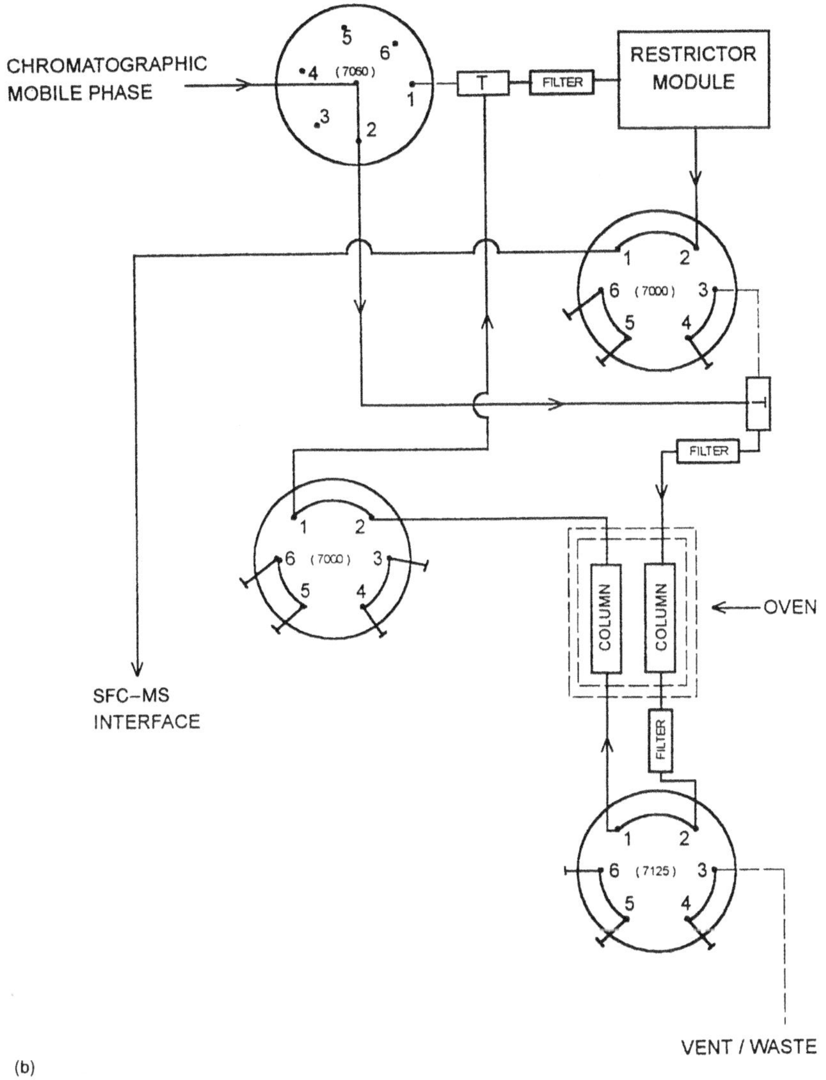


(b)

restrictor module became situated post-separation column, in the flow path, to enable SFC–MS of the extracted phenols.

Initially, studies were performed to determine whether a single amino column would be sufficient for both the trapping and the chromatographic separation of phenols; however, the studies revealed degradation of the chromatographic performance in comparison with results determined off-line for SFC, and an additional column was introduced to restore chromatographic integrity. Further details of this study are presented in section 9.6.

### *7.4.3 On-line analyte trapping and concentration*

After SFE there is often a need to further concentrate the analytes of interest before sample introduction to SFC. This may be achieved by repeated evaporation of supercritical carbon dioxide in the sample loop prior to injection so a greater concentration of material builds up, or by trapping the solutes onto a suitable adsorbent material packed into a sample loop or small column or accumulator column in the flow path for SFE–SFC prior to SFC as used by several workers, including Saito and Yamauchi [54], who used silica as a trapping material for trapping tocopherols, and Yarita *et al.* [57], to trap thiolcarbamate herbicides from soil matrices on $C_{18}$ silica, and by Jordan *et al.* [58] who used a similar approach to trap monomers and polymer additives between SFE and SFC.

A third and popular approach of increasing sample concentration is to use on-line cryogenic trapping. The main advantage of using cryogenic trapping for coupled SFE–SFC is the ability to concentrate the analytes of interest into a discrete narrow band, before injection via a sampling valve, to give quantitative and reproducible results with better detection limits. Cryofocusing also reduces possible interaction between solutes and adsorbent when the latter concentration system is used.

Cryogenic trapping, or focusing, has been used by a number of workers, including Ashraf-Khorassani and Levy [59], who demonstrated the potential of SFE–SFC with cryotrapping for the quantitative analysis of polymer additives in low-density polyethylene using a commercial Suprex system consisting of an extraction vessel, two electronically actuated valves, a zero-volume 'T' piece, a cryogenic trap and a packed SFC column. The stainless steel cryogenic trap measured 30 mm × 1.0 mm i.d. and had the potential for rapid cooling and ballistic heating.

On extraction the SFE extract is passed through a 50 μm i.d. fused restrictor and into the cryogenic trap, which can be cooled to as low as −50°C using industrial carbon dioxide, to collect the solutes. During extraction depressurised carbon dioxide mobile phase is vented through the ten-port valve in the chromatography/extraction oven.

Following extraction the pressure automatically changes from that set for extraction to that required for commencement of SFC. During this period the extracted solutes remain in the cryogenic collection trap until pressure equilibrium for chromatography is reached, when the ten-port and four-port selector valves automatically switch to the injection mode. The temperature of the cryogenic collection trap is then rapidly heated to the injection temperature at a rate of 250°C/min. The supercritical carbon dioxide from the pump then passes through the cryogenic trap when the extracted and concentrated components are transferred to the SFC column for quantitation. Using the system described, Ashraf-Khorassini and Levy were able to obtain linear calibration curves for four different polymer additive standards added to polymer pellets – BHT, Isonox 129, Irganox 1076 and Irganox 1010 – at the concentrations given in Figure 7.7.

In a second paper Ashraf-Khorassani *et al.* [60] evaluated SFE–SFC cryogenic trapping with two different hydrocarbon standard mixtures, and in a third [61] the effects of temperature, matrix configuration and extraction time were studied.

Andersen *et al.* [62] discussed aspects of the use of cryogenic traps, particularly with respect to trapping temperature for volatile solutes and some of the problems involved. It was generally observed that trapping becomes more difficult with increasing solute volatility, showing that the efficiency at which the solute is trapped is a function of temperature. Cryofocusing of extracts containing significant water content require special precautions to prevent ice crystals accumulating in the trap. Aspects of cryofocusing were demonstrated with essential oils.

Xie *et al.* [63] constructed an SFE capillary column SFC system with cryogenic trapping (and fraction collection) to evaluate its use with a biologically active compound and a standard hydrocarbon, normal octatriacontane. These workers also found that, in general, the lower the cryofocusing temperature the better the result. By altering trapping temperature selective trapping of solutes was possible.

### *7.4.4 On-line sample introduction*

For SFE directly coupled to packed column SFC, introduction of SFE extracted solutes to the packed SFC column after they have been trapped in a sampling loop, concentrated onto an adsorbent (accumulator) column or by cryogenic trapping is invariably achieved by the use of a high-pressure rotary sample introduction valve, having either manual or, far more commonly, electronic or pneumatic actuation. In the case of cryogenic trapping the trap may also be heated rapidly to ensure the rapid transfer of a narrow band of analytes to the head of the SFC column through the open sampling valve.

An increasingly popular technique for the introduction of supercritical fluid extracted analytes on-column, following cryogenic or adsorbent trapping, is to make use of a post-trap and post-valve capillary restrictor, which is inserted directly 'on-column'. Following extraction and trapping of analytes they are desorbed and passed through an open valve via a capillary restrictor onto the head of the SFC column where, initially, depressurisation and simultaneous solute cryofocusing, due to adiabatic cooling, takes place. Once all the extracted solutes have been focused on the column head the sample valve is switched and the SFC column repressurised with supercritical fluid mobile phase, followed by rapid heating to release the solutes and to initiate chromatography.

Direct transfer of solutes from SFE to SFC using supercritical carbon dioxide is a major attraction of using SFE directly coupled to SFC. Off-line

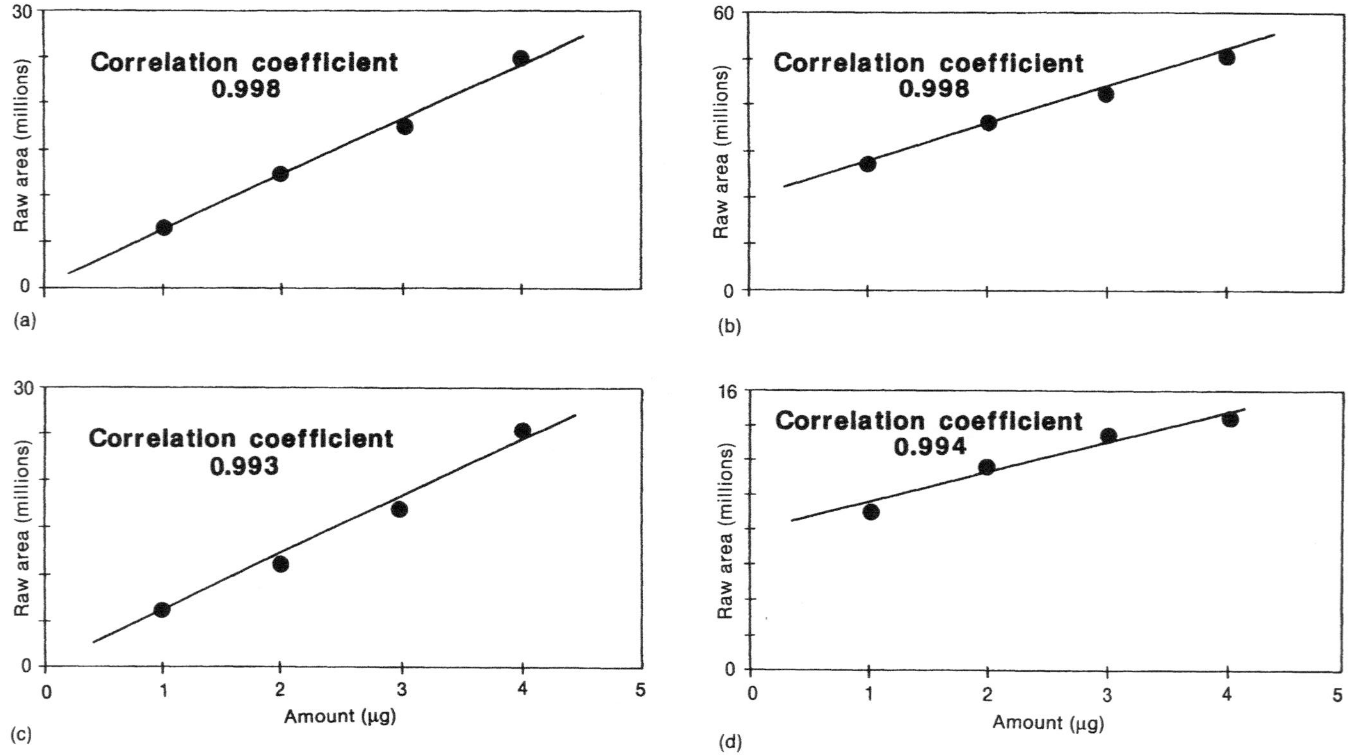


**Figure 7.7** Supercritical fluid extraction–supercritical fluid chromatography calibration curves for (a) BHT; (b) Isonox 129; (c) Irganox 1076; (d) Irganox 1010, at spike levels from 1000 to 4000 ppm. Correlation coefficients: (a) 0.998; (b) 0.998; (c) 0.993; (d) 0.994. Redrawn with permission from ref. 59.

SFE usually involves collection of the solutes in an organic solvent, which needs to be reintroduced into the supercritical fluid for SFC. This often results in the incomplete mixing of solvent and supercritical fluid which leads to band broadening and loss of chromatographic efficiency. In addition, traces of the organic solvent containing the solutes may be left in the SFC sampling valve, which can result in memory effects. Direct coupling of SFE with SFC avoids these incompatibility problems and results in greater chromatographic efficiency, especially for trace analysis.

As many commercial sampling valves are available which may be used under the pressure and temperature conditions for SFE–SFC analysis, they will not be discussed further here; however, it is important in SFE–SFC, as with all other forms of chromatography, to ensure that dead volumes are minimised, or as far as possible eliminated, if good reliable chromatographic results are to be obtained.

### *7.4.5 Columns*

Materials for packed columns used for SFC were discussed in section 7.2.2. For SFC most workers still use conventional HPLC packing materials based on silica which, as previously discussed, may give some problems with polar solutes because of the surface activity of these materials. To overcome this problem, increasingly inert column packing materials for SFC are being developed and used [34, 35]. The increasing availability of a wide range of chiral selective packing materials [13], together with new packing materials being developed for SFC, provides an advantage of packed over capillary SFC where a smaller range of surface coating materials are available.

Virtually all standard analytical HPLC column sizes and types can be used for SFC applications, provided that the columns have been sufficiently pressure rated to withstand the intended SFC operating pressures. Where resolution is required, longer, narrower or packed capillary (or microbore) columns may be used. When sample loading is an important consideration larger or semipreparative columns may be used, particularly when fraction collection of eluted solutes is required.

### *7.4.6 Detectors*

Detectors used for packed column SFC were discussed in section 7.2.4. As with packed column SFC, SFE coupled with packed column SFC lends itself to detection systems capable of accepting high mass flow – the liquid-type (HPLC) detectors. Ultraviolet (UV), diode-array (DAD) and multiwavelength UV detectors are used by many workers, as demonstrated by Saito and Yamauchi in their review of SFC detector usage between 1988 and mid-1992 [36]. UV detectors have the advantage of being compatible with the use of polar modifiers, such as methanol, which causes a high background signal with the gas chromatographic FID when added to the supercritical carbon dioxide mobile phase. Infra-red or FTIR detectors are also fairly commonly used for packed column SFC. GC detectors may also be used, after a column effluent splitter to reduce the flow, particularly FID which is also fairly commonly used for packed column SFC, although other GC detectors may be used. Nam and King [77] used electron capture (ECD) and nitrogen phosphorus specific (NPD) GC detectors for the determination of organochlorine and organophosphorus pesticide residues, respectively, following on-line SFE–SFC.

Mass spectrometry has been used for the on-line detection of eluted solutes from packed column SFC by Ramsey *et al.* [66] where MS–MS was used to detect veterinary drug residues extracted by SFE from freeze-dried pig's kidney, and in a more recent application [76] for the direct on-line determination of phenols from aqueous samples, using full-scan negative-ion atmospheric pressure chemical ionisation.

In recent years, much effort has been expended into interfacing most types of modern analytical detectors to SFC and in particular to packed column SFC, as discussed in section 7.2.4 and more fully in the 1996 review paper by Chester *et al.* [13].

### *7.4.7 Fraction collection*

A major consideration in using SFE combined with packed column SFC is that of the greater sample capacity (discussed in section 7.2.3) of packed compared with capillary columns, particularly when fraction collection of eluted analytes is required for further investigation. Saito and Yamauchi have described and illustrated a number of fraction collection devices for SFC, which may be operated at atmospheric pressure or under pressure, in their review of SFC instrumentation [36]. A system for fraction collection at atmospheric pressure is described in their paper on enrichment of tocopherols in wheatgerm [54]. The thick line in Figure 7.3(b) taken from this paper, shows the flow path through an SFE–SFC system for fraction collection. The extraction vessel is bypassed by means of the switching valve (SV1), and supercritical carbon dioxide, in this case containing modifier, flows through the trap and separation column (SEP.CLM) and through the UV detector via the first back pressure regulator (BR1), which is now open. The second back pressure regulator (BR2), which is used to maintain the pressure throughout the flow path, including the trap and separation columns and UV detector, is used to release the pressure of the column effluent to atmospheric pressure, which causes a sharp decrease in solute solubility, enabling the collection of individual solutes into glass collection vessels, which in this case were changed manually.

Collection of fractions under pressure, generally restricted to industrial applications, requires a greater degree of instrumental sophistication [36].

Jentoft and Gouw [78] demonstrated a fraction collector for use under pressure, which is suited for small-scale SFE and preparative SFC systems. The fraction collector consisted of a turntable holding six glass collection tubes enclosed within a high-pressure vessel, pressurised by nitrogen, to allow fractionation collection in a similar way to preparative HPLC [36].

McNally and Wheeler [65] investigated a number of collection devices for SFE coupled with packed column SFC in their determination and fractionation of sulphonyl urea herbicides and their metabolites. During initial SFE–SFC experiments, radiolabelled metabolites were extracted from various matrices, collected and then detected in a liquid scintillation counter. As radiolabelled compounds were used it was necessary to achieve quantitative recovery of eluted species, and a number of collection devices were investigated, including two HPLC guard columns in series containing $C_{18}$ packing material, a hollow stainless steel U-tube placed in ice or a dry-ice acetone bath and a U-tube containing copper shot or stainless steel balls. The $C_{18}$ guard columns proved ineffective, and less than 25% radioactive recovery was achieved from the U-tube devices. McNally and Wheeler's solution was to use a simple glass collection vessel with which collection efficiencies of up to 100% were obtained. Supercritical effluent from the separator column was passed through a final back pressure regulator, where the pressure was reduced to decrease the solubility of the analytes, and through a (solvent reservoir) frit into a glass reservoir where the radioactive materials were collected. The reservoir was fitted with a side port to ensure pressure release during operation, which was plugged with glass wool as a precaution against radioactivity loss.

## 7.5 Selected applications

With the continued growth of SFE and SFC, SFE coupled with packed column SFC has found many applications in recent years and has found particular application in the determination of food compounds, drug residues, environmental contaminants, herbicide and pesticide residues, PAHs and polymer additives. Some of these have been briefly discussed in previous sections of this chapter to illustrate specific aspects or uses of the technique. These, plus additional applications, will now be discussed further.

Sugiyama *et al.* [10] initially demonstrated the technique of SFE coupled to packed column SFC by extracting and determining caffeine from green coffee beans with supercritical carbon dioxide. After extraction of the powdered beans (with 20% added water) at 200 bar and 40°C for 15 min, extracts were chromatographed on a $C_{18}$ HPLC column. Caffeine was detected using a multiwavelength UV detector.

Tocopherols in wheatgerm were investigated by Saito and Yamauchi [54] using the SFE–SFC system illustrated in Figure 7.3, in which tocopherols were extracted from 3 g of wheatgerm containing approximately 0.03% tocopherols and 10% oil by weight. Extraction was performed at 250 bar and 40°C. Separation of tocopherols was achieved using a 200 mm × 20 mm i.d. column packed with silica gel with supercritical carbon dioxide plus ethanol modifier as mobile phase. The eluent was monitored with a multiwavelength UV detector and the tocopherol content in each fraction collected was determined.

A test mixture of pure vitamins (A, A-acetate, D2, D3, E, E-acetate, K1 and K3) was used by Ibáñez *et al.* [23] to demonstrate the performance of an on-line SFE–SFC system, using two coupled micropacked columns containing SE-54 and Carbowax 20M phases coated on unsilanised silica. A total of 0.1 μl of test vitamins in dichloromethane were added to a packed bed of silanised glass wool in a 50 μl stainless steel extraction vessel. Extraction was performed at 40°C with supercritical carbon dioxide at a density of 0.9 g/ml for 30 s. After extraction, the extracted vitamins were flushed through a valve system and a fused-silica integral restrictor to the head of the micropacked column, where depressurisation resulted in cryofocusing before SFC. Chromatographic results from direct injection of the vitamin test mix onto SFC and from SFE–micropacked column SFC gave comparative results in terms of resolution and peak distortion. A table of qualitative results and data describing the performance of the system showed that the relative standard deviations of the normalised areas were acceptable and that retention times were very reproducible. With recoveries of *c.* 60% further optimisation was planned.

Engelhardt and Gross [79] used SFE–SFC to extract and determine compounds from food products and spices, such as from curry leaves, and carvone and terpene from caraway seeds. A modified FID was used where part of the column effluent fluid is expanded through a small orifice directly into the flame. The short expansion path from column to flame prevented solute condensation and consequent detector spiking. Some 1–2 g of finely ground material was dry packed into the extraction vessel and extracted with supercritical carbon dioxide. After extraction the solutes were passed via a sample loop/valve to the packed SFC column which was either filled with silica or $C_{18}$ packing material. Using the system described, completeness of extraction and extraction kinetics of carvone and terpene from caraway seeds were studied.

The applicability of SFE packed column SFC to the determination of lipid peroxide levels in foods and fats was investigated by Sugiyama *et al.* [80]. Good agreement was found between results from SFE–SFC and potentiometry, previously used for hydroperoxide compounds extracted from samples of ground peanuts. SFE–SFC also had the advantage of providing an oxygen-free environment for the analysis of these sensitive, thermally labile compounds.

Fatty acids, fatty acid esters and tocopherols were extracted and analysed by Ikushima *et al.* [81] using SFE–SFC and FTIR detection. After extraction, the solutes were separated on a packed column containing silver nitrate on silica. A heated transfer line was used to transfer the separated compounds to the FTIR detector using Ga–Gd garnet windows specially selected for the work at supercritical carbon dioxide pressures. Separation and efficiency determined for this system were correlated with a parameter derived from the solubility parameter concept.

Higashidate *et al.* [55] used SFE coupled with packed column SFC, containing silica gel coated with silver nitrate, to enrich and fractionate fish oils, rich in polyunsaturated fatty acids, especially eicosapentaenoic acid (EPA) and docosahexanoic acid (DHA). Following fractionation by SFC, the methylated oils were studied off-line by GC. SFE with carbon dioxide enabled extraction and fractionation at moderate temperatures without exposure to oxygen, preventing thermal degradation or oxidation. A sample of 0.5 g of fish oil was placed in an extraction vessel connected to the separation column by a switching valve. After extraction at 8 MPa and 40°C for 20 min, the extract was directly introduced into the SFC column, where separation was performed using programmed elution during which the pressure was increased stepwise to 8 MPa, 12 MPa and 20 MPa, and with, finally, ethanol added as a modifier. Using this programme, five fractions were obtained for further study by using off-line GC/FID. During SFC the column effluent was monitored using diode-array UV detection. Results showed that the programmed extraction/elution system was capable of providing highly purified EPA and DHA methyl esters.

Veterinary drug residues, including trimethoprim, hexestrol, diethylstilbestrol and dienestrol, were determined by SFE–SFC–MS–MS from pig's kidney by Ramsey *et al.* [66]. As a result of the quantity of endogenous material extracted from powdered freeze-dried kidney, SFE–MS or SFE–MS–MS was considered impractical because of the likelihood of rapid contamination of the mass spectrometer. To avoid this difficulty, an SFE–SFC procedure was developed to provide additional clean-up before MS–MS. SFC was performed on a polar amino bonded column, initially using 100% supercritical carbon dioxide followed by a sharp methanol gradient to 80:20 $CO_2$ : MeOH in 0.5 min. The final composition was then maintained for 10 min. The methanolic (modifier) gradient was developed to exploit the difference in polarity between polar drug residues and the generally non-polar endogenous material co-extracted during SFE to obtain maximum chromatographic separation. Although extraction efficiencies for the investigated drugs were not established, reanalysis of the dried extraction residue showed no further extractable drugs. Further details of this study are presented in section 9.6.

The potential of SFE packed column SFC for the determination of prostaglandins in hydroxypropyl methylcellulose (HPMC) dispersions widely used for controlled-release drug formulations was demonstrated by Roston [82]. Quality-control procedures for formulated drug substances usually involve lengthy clean-up, to remove the drug of interest from the formulation matrix. Using SFE–SFC the only sample preparation necessary was to weigh the material into the extraction vessel. After extraction the extracts were cryofocused and submitted to SFC using a cyano column. Extraction efficiency for 4 min for misoprostol was estimated at 65%. Total analysis time for SFE–SFC was 20 min. With optimisation of experimental parameters it was felt that extraction efficiency could be improved and that SFE–SFC had potential for quality-control procedures.

The determination of phenol and 4-chlorophenol in aqueous samples and phenol in urine was investigated by Thiebaut *et al.* [72] using a segmented flow system and phase separator after SFE and prior to SFE–SFC.

A number of environmental samples have been determined by SFE–SFC. Using a patented extraction vessel and a multivalve switching system in conjunction with two packed amino columns, Ramsey *et al.* [76] successfully extracted and detected three phenols, 2,4-dibromophenol, 2,4,5-trichlorophenol and 4-nitrophenol at the 40 ppb level from a large-volume sample (250 ml) of acidified water using SFE–SFC–MS, demonstrating the applicability of the system to the analysis of organic contaminants in water, as described in section 7.4.2.

Hedrick and Taylor [71] directly extracted and analysed a phosphonate from aqueous media using a specially adapted extraction vessel.

An alternative on-line technique for the extraction and trace enrichment of pesticides from aqueous riverwater samples was used by Zegers *et al.* [20] prior to SFE–SFC. Three organophosphate pesticides (OPPs) of differing polarities were used to evaluate the procedure: diazinon, pyrazophos and azinphos-methyl.

Using a combination of valve and restrictor switching techniques large (chromatographic) aqueous sample volumes (i.e. 75 µl) were pumped into an HPLC capillary pre-column packed with $C_{18}$ silica onto which the pesticides were adsorbed, enabling sample enrichment.

Water was removed from the pre-column by passage of dry nitrogen through the column. Once dry, the pesticides were desorbed with supercritical carbon dioxide at high pressure to aid desorption. The analytes were then passed to the SFC column which was open ended to an NPD. As a consequence, a pressure gradient was developed between the pre-column and analytical column, causing a drop in supercritical fluid density which in turn enhanced the retention of the pesticides on the analytical column where they were focused during the desorption process.

Using this approach, two difficulties associated with large-volume aqueous sample injection for SFC were overcome by Zegers *et al.* First, the drying procedure eliminated possible detector compatibility problems associated with the introduction of solvents or water into the supercritical carbon dioxide mobile phase stream; second, large aqueous sample volumes could be introduced to improve detectability substantially. With the SPE/SFE–SFC/NPD set-up described, OPPs could be determined to below the threshold values of 0.1–1 $\mu g\,l^{-1}$ set up by the European Community for drinking and surface waters.

Sandra *et al.* [69, 70] used a similar approach to evaluate off-line and on-line solid phase extraction (SPE)/SFE for aqueous analyte concentration and extraction, including an automated SPE/SFE–SFC/DAD system, which was used to extract and determine pesticides in water samples, and xanthines in iced tea. In addition to allowing a large aqueous sample size, advantages of using SPE materials in conjunction with SFE were seen as enhanced selectivity, which could be yet further enhanced by variation of extraction pressure and temperature and through the use of modified supercritical carbon dioxide. Efficient drying after SPE, supercritical fluid desorption time and selection of the SPE material were also seen as crucial to SFE–SFC performance.

Two instrumental approaches were used by Sandra *et al.* [69, 70]: a single cartridge system, or, for automation, a Gilson Model 233XL sample preparation unit which was directly coupled to an SFC instrument. The Gilson sample preparation instrument consisted of two high-pressure valves and a dispensing unit, which enabled the sequential automatic conditioning, loading, drying, desorption and reconditioning of solid phase cartridges.

McNally and Wheeler [65] used SFE coupled with packed column SFC to determine sulphonylurea herbicides and metabolites from complex matrices, including soil, plant materials and cell culture media. Near quantitative or quantitative recovery was achieved for radiolabelled metabolites eluted from SFC and collected in a special glass collection vessel. McNally and Wheeler [83] also described the use of SFE packed column SFC for the analysis of the herbicides linuron and diuron in soil, studying the extraction and chromatographic behaviour of these compounds under varying conditions of temperature, pressure and modifier (methanol, ethanol and acetonitrile). In a further paper, Wheeler and McNally [84] extracted linuron and diuron from wheat and soil, using both SFE coupled with a microbore packed column and SFE with capillary SFC.

SFE–SFC has been used for the direct extraction, clean-up and determination of pesticide residues, from agricultural products, by Warner and Yancey [85]. SFE–SFC, using supercritical carbon dioxide, has advantages over conventional methods of pesticide residue analysis, principally because supercritical carbon dioxide is an efficient solvent for the non-polar or moderately polar pesticide residues. In contrast, supercritical carbon dioxide is a poor solvent for the highly polar carbohydrate and proteinaceous materials, which tend to be co-extracted and interfere with conventional GC–MS pesticide residue analysis from non-oily agricultural crops. SFE–SFC was seen as being amenable for automation and that many non-volatile and thermally labile pesticides had the potential to be determined without derivatisation using selective ionisation GC detectors or MS.

Nishikawa [67] applied SFE–SFC to pesticide residue analysis from plant samples and soils. Three representative categories of pesticides were investigated: an organophosphorus pesticide, a synthetic pyrethroid and an azole compound. Extraction conditions, pressure, temperature and time were optimised by spiking standard compounds onto filter paper prior to SFE. Although good reproducibility was obtained for the compounds from filter paper, initial results from soil were variable. Investigation showed that the results were dependent on sample homogeneity and preparation. Extraction of fenitrothion spiked on powdered and from tabulated soil gave relatively low RSD values, but the tablet method improved extraction efficiency as a result of the homogenous fixing of the pesticide. However, extraction efficiencies were lower than those obtained using filter paper. It was observed that the pesticides investigated gave individual extraction efficiencies which were dependent upon their electronic structures. SFE–SFC of pesticide residues from straw, green tea and vegetables was also investigated.

Using SFE coupled to packed column SFC, with unmodified supercritical carbon dioxide, Yarita *et al.* [57] developed a rapid method for the determination of the thiocarbamate herbicides, thiobencarb and molinate. Optimum conditions for extraction pressure, temperature, restrictor flow rates and extraction time were determined, and the influence of moisture and soil types on extraction efficiency were studied. Recovery rates over 90%, using pressures above 15 MPa, were obtained. Extraction temperature, between 40°C and 90°C had little effect on recovery, so most extractions were optimised at 60°C. Soil moisture levels of between 5% and 10%, obtained from air drying the samples in an oven, had little influence on the extraction efficiencies for the herbicides studied. At higher moisture levels the recovery of molinate reduced as a result of its increasing solubility in water when compared with thiobencarb. Thiobencarb and molinate were extracted satisfactorily from all soil types tested, containing varying percentages of sand, silt and clay; however, reproducibility was poor, probably because of the small sample size used. Following extraction and passage through a capillary restrictor, the solutes were trapped on a $C_{18}$ silica gel column before desorption and SFC on an analytical column of the same material. Figure 7.8 shows a chromatogram of thiocarbamate herbicides in a soil matrix, taken from the paper, illustrating excellent peak shape with no chromatographic interferences for the herbicides studied.

Nam *et al.* [86] developed a multiple on-line extraction system, capable of handling large sample sizes, up to 20 g, for the determination of polychlorinated organics, such as organochlorine pesticides (OCPs) and PCBs in complex biological matrices, such as fish tissue, milk and blood. Static extraction was performed for an hour using supercritical carbon dioxide, after which the extract was sampled using a fixed-volume sampling loop and cryogenically trapped, before initial analyte fractionation by SFC using a 150 mm × 1 mm i.d. stainless steel column packed with 5 μm silica

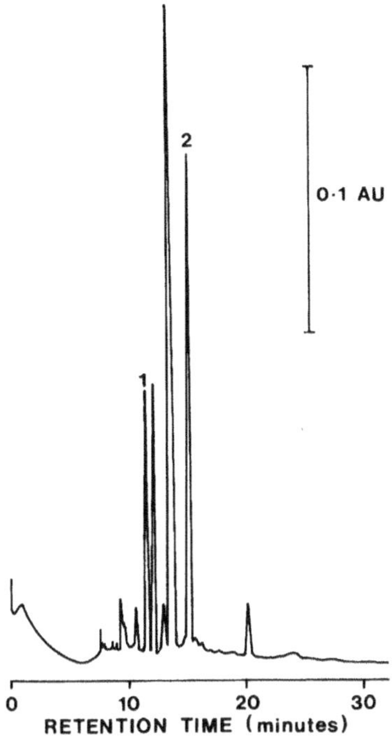


**Figure 7.8** Chromatogram of thiolcarbamate herbicides in a soil matrix obtained by on-line supercritical fluid extraction–supercritical fluid chromatography. Peaks: 1 = molinate; 2 = thiobencarb. Reproduced with permission from ref. 57.

beads chemically bonded with $C_8$ siloxane. Following SFC over 2–15 min at 170 atm and 50°C the analytes were again cryogenically trapped and thermally desorbed before final separation and detection by capillary gas chromatography, with ECD or ion trap detection. On-line SFE–SFC–GC compared very favourably with analytical procedures based on conventional Soxhlet extraction of the polychlorinated organics, especially in terms of the time: 3 h compared with 20 h. Quantitative recoveries were obtained for the OCPs and PCBs studied.

Nam and King [77] later used a similar approach to couple SFE–SFC–GC for the trace analysis of organochlorine and, separately, organophosphorus pesticide residues in fatty food samples: chicken fat, ground beef and lard. In the method, 0.5–1 g of fatty sample, together with an internal standard, was mixed with either Hydromatrix or anhydrous sodium sulphate in the extraction vessel and extracted in the static mode, using a reciprocating pump to circulate supercritical carbon dioxide through the sample matrix during the extraction period. Following extraction, the solutes plus co-extracted material, passed onto the SFC column for initial separation at 200 atm and 50°C. The solutes were then flushed with the mobile phase through a

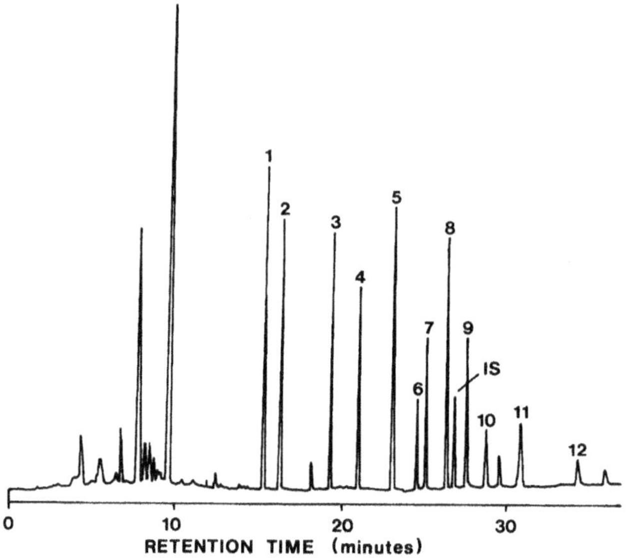


**Figure 7.9** Chromatogram of organochlorine pesticides spiked at 0.5 ppm in chicken fat; analyses on-line by supercritical fluid extraction/supercritical fluid chromatography/gas chromatography–electron capture detector. Peaks: 1 = hexachlorobenzine; 2 = $\gamma$-benzenehexachloride; 3 = heptachlor; 4 = aldrin; 5 = heptachlor epoxide; 6 = $o,p'$-dichlorodiphenyldichloroethylene; 7 = $\alpha$-chlordane; 8 = dieldrin; 9 = eldrin; 10 = $p,p'$-dichlorodiphenyldichloroethane; (11) $p,p'$-dichlorodiphenyltrichloroethane; 12 = methoxychlor. Reproduced with permission from ref. 77.

restrictor and into a fused-silica capillary cold tap. After valve switching, to enable solute transfer to the GC, the trap was rapidly heated to release the solutes into the helium carrier gas and onto the capillary GC column, for final high resolution separation and ECD, or NPD, as illustrated by Figure 7.9, which shows a chromatogram resulting from organochlorine pesticides spiked onto chicken fat after SFE–SFC–GC/ECD, together with a key for the pesticides determined. Using the procedure described, pesticide residues were quantitatively recovered from the fatty matrices studied, having improved limits of detection over the levels used by the Food Safety and Inspection Service of the US Department of Agriculture. As with their earlier paper [86] the authors reported that using the on-line approach of SFE–SFC–GC was considerably faster and less laborious than are conventional analytical procedures for the determination of pesticides in fatty foods.

Weixi *et al.* [68] developed a rapid, simple SFE–SFC method, with FID, for the on-line determination of PAHs, based on the US Environmental Protection Agency list for priority monitoring and applied their system to the determination of PAHs in air particulate samples, taken from near a coke oven and traffic island. Overall, SFE–SFC proved much faster than did the conventional approach for PAH analysis using Soxhlet extraction, with only 0.5 h needed for SFE compared with between 8 h and 24 h for Soxhlet extraction. Recoveries and chromatographic behaviour were determined from PAH standards spiked onto fibreglass paper, as support, before extraction. Average recovery for 16 PAHs studied was nearly 90%, having RSDs of between 1.9% and 6.0%.

A variety of polymer additives have been quantitatively investigated by Ashraf-Khorassani *et al.* [59–61] using the technique of SFE–SFC with intermediate cryogenic trapping. Extraction efficiencies of polymer additives from a number of different matrices were examined by varying the extraction conditions. Polymer additives from low-density polyethylene (LDPE) as well as propellant stabilisers and normal hydrocarbons have been examined by the hyphenated technique. SFC chromatograms of propellant stabilisers from a single base propellant and from standards, reproduced from one of the papers by Ashraf-Khorassani *et al.* [60] are shown in Figure 7.10.

Polyethylene and polypropylene samples from several manufacturers were extracted by SFE coupled to SFC by Ryan *et al.* [87]. Extraction efficiencies of greater than 92% were obtained for ten different additives, including butylhydroxytoluene (BHT) and Irganox 1010. SFE–SFC had the advantage over conventional determinations for polymer analysis of speed and efficiency. Off-line liquid extraction of polymer additives involved many time-consuming steps, including lengthy Soxhlet extraction often followed by concentration, clean-up and reconstitution for instrumental analysis by GC, HPLC or SFC. Coupled SFE–SFC for most polymer additives can generally be accomplished in less than an hour using supercritical carbon dioxide. After extraction, the additives were trapped on an accumulator column containing cyano trapping material. Separations were achieved using a packed $C_8$ column. Both FID and UV detection were used. Linear calibration curves for quantitation were constructed by extracting standards spiked onto quartz wool in the extraction vessel.

On-line SFE–packed capillary SFC systems with FTIR and flame ionisation detection for the identification and quantitation of additives and monomers in nylon and polystyrene have been developed and evaluated by Jordan *et al.* [58] to demonstrate the potential for SFE–SFC within the polymer and textile industries.

Polymer pellets were initially ground and sieved under liquid nitrogen to maximise surface area. Liquid nitrogen was used since heat generated during grinding causes degradation of polymer additives. Ground polymer was mixed in a 1:1 ratio with Celite in the extraction vessel to prevent solid plug formation during extraction, if traces of methanol as modifier were added. For identification and quantitation, of monomer or polymer additives, standards prepared in dichloromethane were spiked onto Celite and the dichloromethane dried prior to extraction. Stainless steel vessels of either 0.16 ml or 0.88 ml were used.

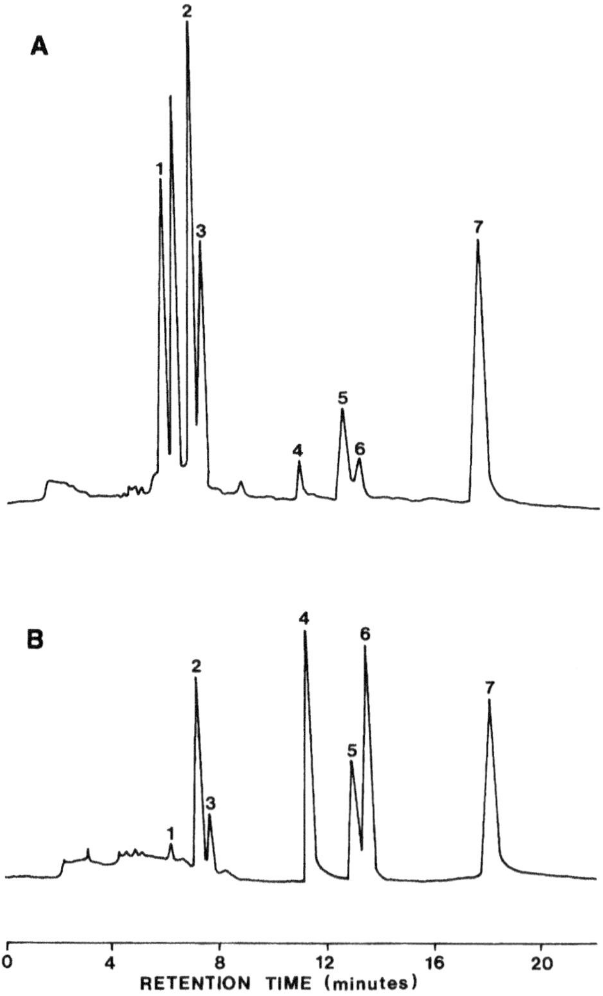


**Figure 7.10** Extraction and separation of different propellant stabilisers: (a) from M-10 single-base propellant; (b) from standards. Peaks: 1 = 2-nitrodiphenylamine; 2 = *n*-nitrosodiphenylamine; 3 = diphenylamine; 4 = 2,4-dinitrodiphenylamine; 5 = 2,4′-dinitrodiphenylamine; 6 = 2,2′-dinitrodiphenylamine; 7 = 4-nitrodiphenylamine. Reproduced with permission from ref. 60.

Systems for determination of monomers and additives from polystyrene and Nylon 6,6/Nylon 6 copolymers were evaluated, taking into account such factors as access of supercritical carbon dioxide to the polymer matrix and analyte solubility in pure or modified supercritical carbon dioxide. To improve sensitivity, analytes were generally trapped and concentrated between SFE and SFC. A variety of 'column' traps were used: stainless steel, deactivated fused silica and portions of methyl and cyano open tubular columns, which were either cooled to $-10°C$ for polystyrene extractions or kept at 20°C for nylon copolymer extractions. Trapped analytes were then heat transfered to a $100\,mm \times 1\,mm$ i.d. cyano packed capillary column. FTIR or flame ionisation detection were used for identification and quantitation. Further details of Jordan *et al.*'s studies are presented in section 9.2.3.

Jahn and Wenclawiak [53] used their microextractors and miniextractors, discussed in section 7.4.2, for the direct SFE–SFC of pyrolysed wood and viscous materials, such as used engine oil, PAHs and oxygen-sensitive and reactive compounds such as metal carbonyls, demonstrating the usefulness of the technique for these categories of materials.

## 7.6 Future prospects

The future for SFE coupled to packed column SFC continues to look bright, although there appears to have been a reduction in the number of papers covering coupled SFE–packed column SFC applications in favour of papers dealing with coupled SFE–capillary column or SFE–GC in the past five years. This trend most probably reflects the ease with which capillary SFE restrictors can be interfaced 'on-column' to capillary SFC or GC and the availability of capillary GC instrumentation as 'the norm' in research and industrial laboratories, together with the continued growth and availability of 'bench-top' GC–MS instrumentation optimised for capillary systems.

However, this current trend towards capillary SFC or GC may be reversed, or at least slowed, by increased interest in chiral separations, particularly in the pharmaceutical sector, for the separation of biologically active compounds, made possible by the increasing availability of new chiral packing materials for packed column SFC and HPLC. Furthermore, advances in the on-line coupling of detection systems for packed column SFC, especially for $^{1}H$ NMR coupled with SFC, will increase the attraction of using the chiral packed columns with information-rich detectors such as NMR, which in turn may be further 'hyphenated' with MS to provide an extremely powerful tool for drug search.

The many and varied applications discussed in this chapter are testimony to the established use of SFE–packed column SFC in the food industry, in agriculture, for environmental analysis of waste and drainage waters and for herbicides, pesticides and PAH analysis in soil and other environmental samples, in the pharmaceutical, petrochemical and polymer industries, and in many other areas where qualitative and quantitative trace analysis is required, particularly from difficult matrices, for quality-control, regulatory and research purposes.

Automation and improved sample handling techniques, especially for the extraction of analytes from aqueous samples, either through direct contact with supercritical carbon dioxide or through indirect extraction using SPE/SFE will ensure increased application for SFE coupled to packed column SFC for the determination of herbicides and pesticides in the environment and foods, and for PAHs and organic contaminants in the environment, driven by increasing legislation and regulation.

SFE–SFC using supercritical carbon dioxide has, because of its inertness, great advantage and continued potential for use with reactive and heat labile compounds, such as metal carbonyls which react with most organic solvents, and for substances liable to oxidation, such as polyunsaturated fish oils, which would be difficult to extract and analyse by other means.

Combined with the increased availability of pre-configured commercial instrumentation and automation, the advantages of efficiency, speed, functional group selectivity and simplicity of SFE, coupled with the ease of solute concentration and/or direct transfer to SFC and GC, will continue to make SFE coupled with packed column SFC very attractive compared with many off-line processes.

## Acknowledgement

The author wishes to express thanks to British American Tobacco, Research & Development for their support and use of facilities.

## References

1. Hannay, J.B and Horgarth, J. (1897) *Proc. Royal. Soc. (London)*, **29**, 324.
2. Zosel, K. (1963) *Australian patent Appl.*, 16 April.
3. Zhuze, T.P., Jushkevic, G.N. and Gekker, J.E. (1958) *Maslo – Zhir. Promst.*, **24**, 34.
4. Stahl, E. and Schiltz, W. (1976) Extraction with supercritical gases in coupling with thin layer chromatography. *Anal Chem.*, **280**, 99.
5. Smith, R.D., Udseth, H.R. and Wright, B.W. (1985) Chapter in *Supercritical Chromatography* (eds J.M.L. Penninger, M. Radosz, M.A. McHugh and V.J. Krukonis), Elsevier, Amsterdam, pp. 191–223.
6. Hawthorne, S.B. and Miller, D.L. (1986) Extraction and recovery of organic pollutants from environmental solids and Tenax-GC using supercritical carbon dioxide. *J. Chromatogr. Sci.*, **24**, 258.
7. Unger, K.K. and Raumeliotis, P. (1983) On-line high pressure extraction–high performance liquid chromatography. *J. Chromatogr.*, **282**, 519.
8. Klesper, E., Corwin, A.H., and Turner, D.A. (1962) High pressure gas chromatography above critical temperature. *J. Org. Chem.*, **27**, 700.
9. Sugiyama, K., Saito, M. and Wada, A. (1984) *Jap. Patent*, 58-117773.
10. Sugiyama, K., Saito, M., Hondo, T. and Seda, M. (1985) New double-stage analysis method. Directly coupled laboratory-scale supercritical fluid extraction–supercritical fluid chromatography, monitored with a multiwavelength ultraviolet detector. *J. Chromatogr.*, **332**, 107.
11. Sugiyama, K., Saito, M. and Wada, A. (1986) *US Patent*, 4597,943.

12. Skelton, R.J. Jr, Johnson, C.C. and Taylor, L.T. (1986) Sampling considerations in supercritical fluid chromatography. *Chromatographia*, **21**, 3.
13. Chester, T.L., Pinkston, J.D. and Raynie, D.E. (1996) Supercritical fluid chromatography and extraction. *Anal. Chem.*, **68**, 487R.
14. Albert, K., Braumann, U., Streck, R. *et al.* (1995) Application of direct on-line coupling of HPLC and SFC with $^1$H NMR spectroscopy for the investigation of monomeric acrylates. *Fresenius J. Anal. Chem.*, **352**, 521.
15. Hirata, Y. and Nakata, F. (1984) Supercritical fluid chromatography with fused-silica packed columns. *J. Chromatogr.*, **295**, 315.
16. Hirata, Y., Nakata, F. and Kawasaki, M. (1986) Supercritical fluid chromatography with a slurry-packed capillary column. *J. High Resolut. Chromatogr./Chromatogr. Commun.*, **9**, 633.
17. Payne, K.M., Tarbet, B.J., Bradshaw, J.S. *et al.* (1990) Simultaneous deactivation and coating of porous silica particles for microcolumn supercritical fluid chromatography. *Anal. Chem.*, **62**, 1379.
18. Steenmackers, D. and Sandra, P. (1991) Experiments with micropacked columns towards unified chromatography, *J. High Resolut. Chromatogr.*, **14**, 842.
19. Payne, K.M., Davies, I.L., Bartle, K.D. *et al.* (1989) Multidimensional packed capillary column supercritical-fluid chromatography using a flow-switching interface. *J. Chromatogr.*, **477**, 161.
20. Zegers, B.N., de Geus, H.-J., Wildenburg, S.H.J. *et al.* (1994) Large-volume injection in packed-capillary supercritical fluid chromatography. *J. Chromatogr. A*, **677**, 141.
21. Ibáñez, E., Alvarez, P.J.M., Reglero, G. and Herraiz, M. (1993) Large particle micropacked columns in supercritical fluid chromatography. *J. Microcol. Sepn.*, **5**, 371.
22. Ibáñez, E., Tabera, J., Herraiz, M. and Reglero, G. (1994) Effect of temperature and density on the performance of micropacked columns and supercritical fluid chromatography. *J. Chromatogr. A.*, **667**, 249.
23. Ibáñez, E., Herraiz, M. and Reglero, G. (1995) On-line SFE–SFC coupling using micropacked columns. *J. High Resol. Chromatogr.*, **18**, 507.
24. Novotny, M., Springston, S.R., Peaden, P.A. *et al.* (1981) Capillary supercritical fluid chromatography. *Anal. Chem.*, **53**, 407A.
25. Wheeler, J.R. and McNally, M.E. (1987) Comparison between packed and capillary supercritical fluid chromatography and high performance liquid chromatography using representative herbicides and pesticides as typical moderate polarity and molecular weight range molecules. *J. Chromatogr.*, **410**, 343.
26. Schoenmakers, P.J. (1988) Supercritical-fluid chromatography: open columns vs packed columns. *J. High Resol. Chromatogr. & Chromatogr. Commun.*, **11**, 278.
27. Petersen, M. (1990) Review: packed columns in supercritical fluid chromatography. *J. Chromatogr.*, **505**, 3.
28. Schoenmakers, P.J., Uunk, L.G.M. and Janssen, H.-G. (1990) Comparison of stationary phases for packed-column supercritical fluid chromatography. *J. Chromatogr.*, **506**, 563.
29. Greibrokk, T., Doehl, J. and Lundones, E. (1989) Current use of packed columns in SFC, in *Progress in HPLC*, Volume 4 (eds Yoshioka *et al.*), pp. 53–71.
30. Taylor, L.T. and Chang, H.-C. (1990) Packed column development in supercritical fluid chromatography. *J. Chromatogr. Sci.*, **28**, 357.
31. Guiochon, G. and Guillemin, C.L. (1988) *Quantitative Gas Chromatography* from Journal of Chromatography Library Volume 42, Elsevier, Amsterdam.
32. Schomburg, G., Deege, A., Kohler, J. and Bien-Vogelsang, V. (1983) Immobilisation of stationary liquids in reversed-phase and normal-phase liquid chromatography. *J. Chromatogr.*, **282**, 27.
33. Gere, D.R. (1983) Oligomer separations by supercritical fluid chromatography. Application Note AN 800-3, Hewlett-Packard, Avondale, PA.
34. Shen, Y. and Lee, M.I. (1996) Performance of polyethylenimine-coated particles in packed capillary column supercritical fluid chromatography. *J. Microcolumn Separations*, **8**, 519.
35. Li, W., Ibáñez, E., Malik, A. and Lee, M.L. (1996) Evaluation of a fluorocarbon bonded silica using packed capillary column supercritical fluid chromatography. *J. Microcolumn Separations*, **8**, 175.

36. Saito, M. and Yamauchi, Y. (1994) Chapter 4, in *Fractionation by Packed Column SFC and SFE: Principles and Applications* (eds M. Saito, Y. Yamauchi and T. Okuyama), VCH Publishers, New York.
37. Shi, H., Taylor, L.T. and Fujinari, E.M. (1997) Chemiluminescence nitrogen detection for packed-column supercritical fluid chromatography with methanol modified carbon dioxide. *J. Chromatogr. A*, **757**, 181.
38. Höner, A., Arnold, M., Hüsers, N. and Kleiböhmer, W. (1995) Monitoring polycyclic aromatic hydrocarbons in waste gases. *J. Chromatogr. A*, **710**, 129.
39. Wallenborg, S.R., Markides, K.E. and Nyholm, L. (1997) Development of an amperometric detector for packed capillary column supercritical fluid chromatography. *Anal Chem.*, **69**, 439.
40. Lembke, P., Engelhardt, H. and Ecker, R. (1994) On-line coupling of packed column supercritical fluid chromatography with multiple wavelength detector and radioactivity flow-through detector. *Chromatographia*, **38**(7/8), 491.
41. Carraud, P., Thiebaut, D., Caude, M. *et al.* (1987) Supercritical fluid chromatography/light-scattering detector: a promising coupling for polar compounds analysis with packed columns. *J. Chromatogr. Sci.*, **25**, 395.
42. Berry, A.J., Ramsey, E.D., Newby, M. and Games, D.E. (1996) Applications of packed-column SFC using light scattering detection. *J. Chromatogr. Sci.*, **34**, 245.
43. Thompson, J., Stode, B. III and Taylor, L.T. (1996) Evaporative light scattering detection for supercritical fluid chromatography. *J. Chromatogr. Sci.*, **34**, 261.
44. Long, G.L., Ducatte, G.R. and Lancaster, E.D. (1994) Helium microwave-induced plasmas for element specific detection in chromatography. *Spectrochimica Acta*, **48**(1), 75.
45. Isco (1994) *Isco Technical Bulletin 77: SFE Restrictors*, Isco Inc. Europe, New Road, Hengoed, Mid Glamorgan GF8 8AU, UK.
46. Poole, C.F. and Poole, S.K. (1991) Supercritical fluid chromatography, in *Chromatography Today*, Elsevier, Amsterdam, pp. 601–48.
47. Hawthorne, S.B., Miller, D.J. and Langenfeld, J.J. (1992) Advances in analytical supercritical fluid extraction (SFE), in *Hyphenated Techniques in Supercritical Chromatography and Extraction* (ed. K. Jinno), Journal of Chromatography Library Series, Volume 53, Elsevier, Amsterdam.
48. Hawthorne, S.B. (1990) Analytical scale supercritical fluid extraction. *Anal. Chem.*, **62**, 633A.
49. Saito, M., Hondo, T. and Yamauchi, Y. (1989) Fractionation by coupled micro-supercritical fluid extraction and supercritical fluid chromatography, in *Supercritical Fluid Chromatography* (ed. R.M. Smith), RSC Chemistry Monographs, Royal Society of Chemistry, London, pp. 203–29.
50. Saito, M., Hondo, T. and Senda, M. (1990) Coupled supercritical fluid extraction–supercritical fluid chromatography. *Chromatogr. Sci.*, **50**, 331.
51. Hawthorne, S.B., Miller, D.J., Nivens, D.E. and White, D.C. (1992) Supercritical fluid extraction of polar analytes using *in situ* chemical derivatisation. *Anal. Chem.*, **64**, 405.
52. Berg, B.E., Henson, E.M., Gjørven, S. and Greibrokk, T. (1993) On-line enzymatic reaction, extraction and chromatography of fatty acids and triglycerides with supercritical carbon dioxide. *J. High Resol. Chromatogr.*, **16**, 358.
53. Jahn, K.R. and Wenclawiak, B. (1988) Direct on-line coupling of small subcritical and supercritical fluid extractors with packed column supercritical fluid chromatography. *Chromatographia*, **26**, 345.
54. Saito, M. and Yamauchi, Y. (1989) Enrichment of tocopherols in wheat germ by directly coupled supercritical fluid extraction with semi-preparative supercritical fluid chromatography. *J. Chromatogr. Sci.*, **27**, 79.
55. Higashidate, S., Yamauchi, Y. and Saito, M. (1994) Fractionation of eicosapentaenoic acid and docosahexanoic acid esters from esterified fish oil by coupled supercritical fluid extraction–chromatography, in *Fractionation by Packed-column SFC and SFE: Principles and Applications* (eds M. Saito, Y. Yamauchi and T. Okuyama), VCH Publishers, New York, ch. 9.
56. Mitra, S. and Wilson, N.K. (1990) Thermal modulation interface between supercritical fluid extraction and supercritical fluid chromatography. *J. Chromatogr. Sci.*, **28**, 182.

57. Yarita, T., Norma, A. Horimoto, Y. and Gonda, S. (1996) On-line coupled supercritical fluid extraction and chromatography for the determination of thiolcarbamate herbicides in soil matrix. *J. Chromatogr. A*, **750**, 175.
58. Jordan, S.L., Taylor, L.T., Seemuth, P.D. and Miller, R.J. (1997) Analysis of additives and monomers in nylon and polystyrene. *Textile Chemist and Colorist*, **29**(2), 25.
59. Ashraf-Khorassani, M. and Levy, J.M. (1990) Quantitative analysis of polymer additives in low density polyethylene using supercritical fluid extraction/supercritical fluid chromatography. *J. High Resol. Chromatogr.*, **13**, 742.
60. Ashraf-Khorassani, M., Kumar, M.L., Koebler, D.J. and Williams, G.P. (1990) Evaluation of coupled supercritical fluid extraction–cryogenic collection–supercritical fluid chromatography (SFE–CC–SFC) for quantitative and qualitative analysis. *J. Chromatogr. Sci.*, **28**, 599.
61. Ashraf-Khorassani, M., Boyer, D.S. and Levy, J.M. (1991) Optimisation of experimental parameters for the determination of polymer additives using on-line SFE–SFC. *J. Chromatogr. Sci.*, **29**, 517.
62. Andersen, M.R., Swanson, J.T., Porter, N.L. and Richter, B.E. (1989) Supercritical fluid extraction as a sample introduction method for chromatography. *J. Chromatogr. Sci.*, **27**, 371.
63. Xie, Q.L., Markides, K.E. and Lee, M.L. (1989) Supercritical fluid extraction–supercritical fluid chromatography with fraction collection for sensitive analysis. *J. Chromatogr. Sci.*, **27**, 365.
64. Davis, I.L., Raynor, M.W., Kithinji, J.P. *et al.* (1988) LC, SFE, GC, SFC interfacing. *Anal. Chem.*, **60**, 683A.
65. McNally, M.E. and Wheeler, J.A. (1988) Supercritical fluid extraction coupled with supercritical fluid chromatography for the separation of sulfonylurea herbicides and their metabolites from complex matrices. *J. Chromatogr.*, **435**, 63.
66. Ramsey, E.D., Perkins, J.R., Games, D.E. and Startin, J.R. (1989) Analysis of drug residues in tissue by combined supercritical fluid extraction–supercritical fluid chromatography–mass spectrometry–mass spectrometry. *J. Chromatogr.*, **464**, 353.
67. Nishikawa, Y. (1991) Application of supercritical fluid extraction/supercritical fluid chromatography to pesticide residue analysis. *Anal. Sciences*, **7**, 567.
68. Weixi, Y., Spitzauer, P. and Kettrup, A. (1995) Determination of polycyclic hydrocarbons with supercritical fluid extraction and chromatography (SFE/SFC). *J. Environ. Sci.*, **7**(2), 224.
69. Sandra, P., Kot, A., Medvedovici, A. and David, F. (1995) Selected applications of the use of supercritical fluids in coupled systems. *J. Chromatogr. A*, **703**, 467.
70. Sandra, P., Medvedovici, A., Kot, A. *et al.* (1996) SPE–SFC–DAD: a new hyphenated system for monitoring organic micropollutants in aqueous samples. *LC–GC International* (September), 504.
71. Hedrick, J. and Taylor, L.T. (1989) Quantitative supercritical fluid extraction/supercritical fluid chromatography of a phosphonate from aqueous media., *Anal Chem.*, **61**, 1986.
72. Thiebaut, D., Chervet, J.-P., Vannoort, R.W. *et al.* (1989) Supercritical-fluid extraction of aqueous samples and on-line coupling to supercritical-fluid chromatography. *J. Chromatogr.*, **477**, 151.
73. Hedrick, J. and Taylor, L.T. (1990) Supercritical fluid extraction strategies of aqueous based matrices. *J. High Resol. Chromatogr.*, **13**, 312.
74. Hedrick, J. and Taylor, L.T. (1992) Direct supercritical fluid extraction of nitrogenous bases from aqueous solutions. *J. High Resol. Chromatogr.*, **15**, 151.
75. Janda, V., Mikešová, M. and Vejrosta, J. (1996) Direct supercritical fluid extraction of water based matrices. *J. Chromatogr. A*, **733**, 35.
76. Ramsey, E.D., Minty, B., McCullogh, M.A. *et al.* (1997) Analysis of phenols in water at ppb level using direct supercritical fluid extraction of aqueous samples combined on-line with supercritical fluid chromatography–mass spectrometry. *Anal. Communications*, **34**, 3.
77. Nam, K.-S. and King, J.W. (1994) Coupled SFE/SFC/GC for the trace analysis of pesticide residues in fatty food samples. *J. High Resol. Chromatogr.*, **17**, 577.
78. Jentoft, R.E. and Gouw, T.H. (1972) Apparatus for supercritical fluid chromatography with carbon dioxide as the mobile phase. *Anal. Chem.*, **44**, 336.

79. Engelhardt, H. and Gross, A. (1988) On-line extraction and separation by supercritical chromatography with packed columns. *J. High Resolution Chromatogr. & Chromatogr. Commun.*, **11**, 38.
80. Sugiyama, K., Shiokawa, T. and Moriya, T. (1990) Application of supercritical fluid chromatography and supercritical fluid extraction to the measurement of hydroperoxides in foods. *J. Chromatogr.*, **515**, 555.
81. Ikushima, Y., Saito, N., Hatakeda, K. *et al.* (1989) Development of extraction and separation monitoring *in-situ* with on-line SFE–SFC/FTIR system. *Chemistry Letters*, **10**, 1707.
82. Roston, D.A. (1992) Supercritical fluid extraction–supercritical fluid chromatography for analysis of a prostaglandin: HPMC dispersion. *Drug Development and Industrial Pharmacy*, **18**(2), 245.
83. McNally, M.E. and Wheeler, J.R. (1988) Increasing extraction efficiency in supercritical fluid extraction from complex matrices. Predicting extraction efficiency of diuron and linuron in supercritical fluid extraction using supercritical fluid chromatographic retention. *J. Chromatogr.*, **447**, 53.
84. Wheeler, J.R. and McNally, M.E. (1989) Supercritical fluid extraction and chromatography of representative agricultural products with capillary and microbore columns. *J. Chromatogr. Sci.*, **27**, 534.
85. Warner, J.S. and Yancey, M.F. (1989) Application of supercritical fluid extraction and supercritical fluid chromatography to residue analysis. *Brighton Crop Protection Conference – Weeds. Part 1*, 267.
86. Nam, K.S., Kapila, S., Yanders, A.F. and Puri, R.K. (1991) A multiple sample extraction and on-line system for the analysis of chlorinated compounds. *Chemosphere*, **23**(8–10), 1109.
87. Ryan, T.W., Yocklovich, S.G., Watkins, J.C. and Levy, J.E. (1990) Quantitative analysis of additives in polymers using coupled supercritical fluid extraction–supercritical fluid chromatography. *J. Chromatogr.*, **505**, 273.

# 8 Supercritical fluid extraction for off-line and on-line high-performance liquid chromatographic analysis

A.T. REES

## 8.1 Introduction

The on-line coupling of supercritical fluid extraction (SFE) with gas chromatography (GC) and supercritical fluid chromatography (SFC) has been described in previous chapters. In the case of GC and capillary SFC (cSFC) a commonly employed interfacing procedure involves decompressing the SFE extract on to the top of the capillary column. This results in the analyte(s) becoming band focused prior to the onset of the chromatographic stage. In packed column SFC the interfacing procedures are again relatively straightforward, generally involving either trapping the extracted analyte(s) on to the top of the analytical column (whose retentive capacity has been carefully selected) prior to supercritical decompression or decompressing the SFE extract into a trapping column which is subsequently linked to an analytical column. The on-line coupling of SFE with high-performance liquid chromatography (SFE–HPLC) is technically more challenging. This situation arises from the fundamental incompatibility of a sample preparation procedure that ultimately produces a large volume of gas with a chromatographic technique which uses a liquid mobile phase. Maintenance of a gas-free HPLC mobile phase is generally considered mandatory in order to obtain optimum pump and detector performance. Admission of gas into an HPLC system can lead to erratic pump performance and complete loss of detector stability. Despite these problems the development of SFE–HPLC has potential to enhance the analyst's capabilities in situations where, for example, the analyte(s) of interest:

- are involatile and/or thermally labile;
- are light-sensitive and/or air-sensitive;
- are present at ultratrace levels, requiring minimum sample handling;
- are to be analysed using a combined liquid chromatography–mass spectrometry (LC–MS) procedure.

Howard and Taylor [1] have published a dedicated review concerning on-line and off-line applications of SFE with HPLC up to 1993. Extensive reviews which are either dedicated or deal with the on-line combination of SFE with chromatographic procedures, including HPLC, have also been published, amongst which are those by Vannoort *et al.* [2], Jinno and Saito [3], Chester *et al.* [4] and Stuart *et al.* [5]. Off-line techniques have been extended to the use of micellar electrokinetic chromatography (MEKC) [6] for specific target analytes. Since 1990 there have been only about 12 papers published concerning the use of SFE–HPLC. However, studies performed by Ashraf-Khorassani *et al.* [7] indicate that with careful instrument set-up the previously described problems can be overcome. The results of their investigation demonstrate that the degree of detector interference is dependent on the solubility of carbon dioxide, the percentage of water and the flow rate of the mobile phase.

The objective of this chapter is to focus on selected work performed since the beginning of the 1990s to illustrate:

- that off-line SFE can be used to extract compounds whose polarity and/or high relative molecular mass makes HPLC rather than GC or SFC the method of choice for analysis;
- whether off-line SFE can provide clean extracts for HPLC analysis;
- the application of on-line SFE–HPLC either as a sample preparation technique or as a direct analytical procedure.

## 8.2 Supercritical fluid extraction–high-performance liquid chromatography

Since the work carried out by Unger and Roumeliotis [8] little has changed in the general set-up of an SFE–HPLC system, comprising a series of switching valves, an extraction vessel and a trap for the analytes. It is how the analytes are trapped and subsequently analysed that defines whether the process is on-line or off-line. Li *et al.* [9] have described a systematic approach for the optimisation of the SFE process where they use off-line HPLC analysis to determine the extraction efficiency of vitamin E from spiked corn oil. A review of on-line and off-line strategies covering method development, chromatographic methodology and recoveries are described herein. The purpose of this approach is to give the reader a better understanding of the practical issues surrounding the technique and provide an insight into the diversity of analytes that have been successfully determined in recent years. Armed with this understanding, and that from other chapters in this volume, it is envisaged that those new to the technique will be able to set up a working system with a reasonable degree of confidence.

An article by Ashraf-Khorassani *et al.* [10] describes the effects of pressure, temperature, modifier, modifier concentration and sample matrix on the extraction process which will be of use to the reader. In addition to this Thompson *et al.* [11] describe the trapping efficiencies of various collection solvents. The results of this study indicated that mixed collection solvents provided better analyte recovery and that no correlation between recovery and the physical constants of the solvent existed (boiling point, density, viscosity, surface tension, etc.). Hyphenated SFE techniques have not escaped the attention of automation. Gere and co-workers [12] describe an overview of techniques including SFE. They illustrate the use of SFE as a means of sample preparation for a series of polycyclic aromatic hydrocarbons (PAHs) for subsequent analysis by an off-line HPLC method.

## 8.3 Supercritical fluid extraction followed by off-line high-performance liquid chromatography analysis

Owing to the relative simplicity of off-line SFE analyte collection, the majority of reports published since 1990 dealing with SFE sample preparation for HPLC analysis have used this approach. The two main methods for off-line analyte collection are decompression into an appropriate solvent (or solvents) and decompression onto a solid phase sorbent trap. In addition to the extraction-trapping process some reported methods require additional sample cleanup by solid phase extraction (SPE) prior to HPLC analysis. The primary drivers for all of the reported work is either regulatory (outlaw of solvents, etc.) or with a view to reducing method times. At present no applications have described an analyte–matrix combination where combined SFE–HPLC has been the natural choice for the analysis.

The vast majority of investigations fall into defined categories: pharmaceutical, environmental, food products, fermentation or flora and regulatory. The pharmaceutical industry is subject to a wide range of stringent regulations. Anything that does or is likely to affect the physiology of the user is subject to these regulations. As a consequence, assay of the components in a preparation and the accurate quantification and identification of impurities and metabolites is of the utmost importance. Scalia and co-workers [13–16] have performed extensive investigations on various matrices for a variety of target analytes, namely, parabens in cosmetic products [13], triclosan in deodorant sticks [14], the assay of vitamin A palmitate and vitamin E acetate in cosmetic creams and lotions [15] and vitamin A, vitamin E and their esters in tablet preparations [16]. These investigations were carried out between 1992 and 1995. The initial investigation with Games [13] centred around an SFE method which used supercritical carbon dioxide at an extraction density of 0.95 g/ml and a flow rate of 2 ml/min, trapping onto stainless-steel balls (0.36–0.43 mm in diameter). The method development process involved spiking methyl, ethyl, propyl and butyl parabens at the 0.04% m/m level into a hand cream product and subsequently adsorbing the mixture onto filter paper. Initial analyte recoveries were low, determined at approximately 20% by off-line HPLC. Additional experiments indicated that temperature was the most significant SFE parameter. Appropriate modification of the SFE parameters resulted in quantitative recovery of the parabens. Recovery

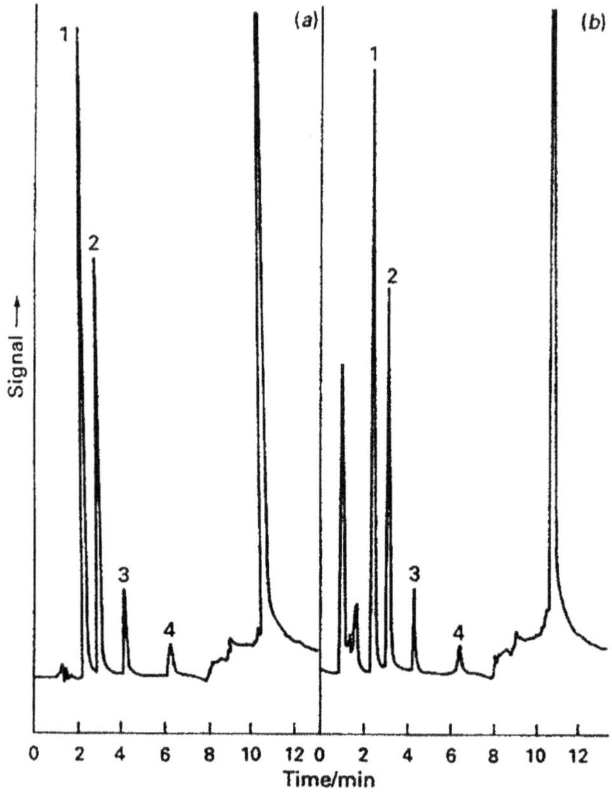


**Figure 8.1** High-performance liquid chromatography (HPLC) chromatograms obtained for a suncream product extracted by: (a) supercritical fluid extraction; (b) liquid extraction. HPLC conditions: Spherisorb® ODS column using a water:acetonitrile gradient. Peaks: 1 = methyl paraben; 2 = ethyl paraben; 3 = propyl paraben; 4 = butyl paraben. Reproduced with permission from ref. 13.

of the parabens from the trap was achieved with methanol, and off-line reverse phase HPLC was used for quantification. A representative chromatogram comparing a standard solvent extraction method with the SFE process is shown in Figure 8.1. The authors also found that because of the selectivity of the SFE process column longevity was significantly improved, the cosmetic preparation SFE fraction being sufficiently clean to be analysed by HPLC without further cleanup. As shown in Figure 8.2, the SFE procedure produced a day cream extract with significantly lower levels of early eluting peaks relative to that obtained using conventional liquid extraction.

The result of this comprehensive study indicated that the parabens limit of quantification was a factor of at least 10 below the levels normally used in the cosmetic formulations. The study was extended to the analysis of three different cosmetic preparations (containing no detectable parabens), spiked

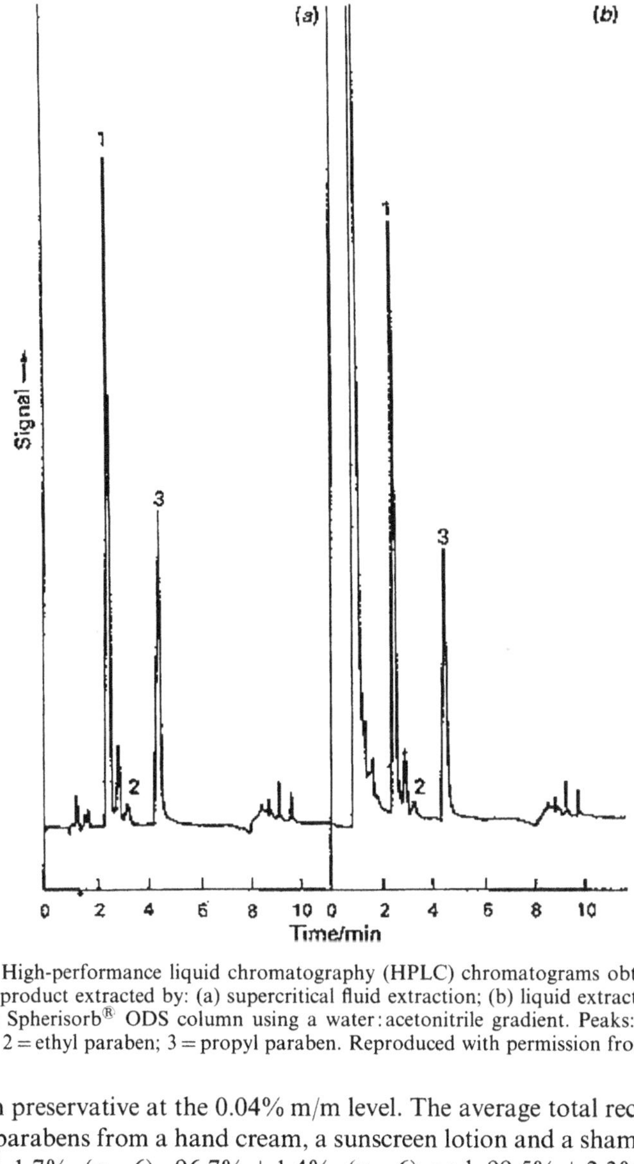


**Figure 8.2** High-performance liquid chromatography (HPLC) chromatograms obtained for a day cream product extracted by: (a) supercritical fluid extraction; (b) liquid extraction. HPLC conditions: Spherisorb® ODS column using a water:acetonitrile gradient. Peaks: 1 = methyl paraben; 2 = ethyl paraben; 3 = propyl paraben. Reproduced with permission from ref. 13.

with each preservative at the 0.04% m/m level. The average total recovery for the four parabens from a hand cream, a sunscreen lotion and a shampoo were 100.3% ± 1.7% ($n = 6$), 96.7% ± 1.4% ($n = 6$) and 99.5% ± 2.3% ($n = 6$), respectively. Table 8.1 demonstrates the efficacy of SFE compared with classical solvent extraction for five cosmetic products. These results indicate that SFE is a superior sample preparation method than is solvent extraction, offering higher recoveries of these compounds in cleaner fractions.

In a series of subsequent investigations, Scalia *et al.* [14–16] used SFE and off-line HLPC for the analysis of various cosmetic products. The

**Table 8.1** Comparison of total concentrations of parabens in cosmetic products recovered by supercritical fluid extraction (SFE) or liquid extraction. Source: ref. 13

| Sample | Concentration (% m/m) | |
|---|---|---|
| | SFE | Liquid extraction |
| Suncream | 0.164 | 0.150 |
| Moisturising lotion | 0.179 | 0.169 |
| Cleansing milk | 0.207 | 0.197 |
| Skin cream | 0.237 | 0.219 |
| Day cream | 0.058 | 0.052 |

deodorant and soap product matrices [14] produced results that were entirely consistent with the initial work [13] in terms of column longevity and the superior recoveries of triclosan over liquid extraction. Temperature was again found to be an important factor. In order to obtain the same SFE recoveries for triclosan, initially studied by spiking onto filter paper, it was necessary to increase the extraction temperature. The SFE procedure produced a much cleaner sample for HPLC analysis than did an established liquid extraction method since a co-extracted compound giving rise to interference with the latter approach was found to be insoluble in supercritical carbon dioxide. Detection limits were determined at the 0.01% m/m level with recoveries detailed in Table 8.2.

In a related study Scalia *et al.* [15] used SFE as a means of sample preparation in the analysis of cosmetic creams and lotions for vitamin A palmitate and vitamin E acetate. The optimised SFE conditions were found to be a pressure of 3625 psi at 40°C for 30 min. The extracted analytes were trapped in a glass vial containing tetrahydrofuran–methanol (4:1 vol/vol) cooled to 0°C. Reversed phase HPLC [μBondapak® 10 μm ODS 300 mm × 3.9 mm

**Table 8.2** Comparison of triclosan levels in deodorant products recovered by supercritical fluid extraction (SFE) or liquid extraction. Source: ref. 14

| Sample | Concentration (% m/m)[a] | |
|---|---|---|
| | SFE | Liquid extraction |
| Deodorant stick | | |
| A | 0.154 (1.2) | 0.149 (1.3) |
| B | 0.025 (4.0) | 0.023 (7.4) |
| C | 0.199 (2.1) | 0.190 (6.3) |
| D | 0.160 (3.7) | 0.157 (4.3) |
| Deodorant soap | | |
| A | 0.288 (1.5) | 0.266 (4.9) |
| B | 0.234 (0.6) | 0.208 (1.2) |

[a] Percentages are given in parentheses.

inner diameter (i.d.) column eluted with methanol–acetonitrile (75:25 vol/vol) at 1.5 ml/min] was used to assay the SFE extract. Recoveries were found to be low when filter paper was used as the matrix support, yielding 16.4% for the vitamin E acetate and 19.3% for the vitamin A palmitate. However, finely dividing the cream matrix with Celite® increased recoveries for both analytes to over 93%. Depending on the product under investigation relative standard deviations (RSDs) ranging from 4.0% to 6.1% were achieved.

Scalia *et al.* [16] continued the work on SFE as means of sample preparation for HPLC assay in the analysis of tablet preparations for vitamin A, vitamin E and their esters. The analytes were determined by off-line non-aqueous reverse phase HPLC using a Lichrospher® CH-8 5 μm, 250 mm × 4.0 mm i.d. column coupled to a 4 mm × 4 mm i.d. guard column and eluted with methanol–acetonitrile (75:25 vol/vol). The optimised SFE conditions were found to be a pressure of 3625 psi at 40°C for 15 min. The extracted analytes were trapped in a glass vial containing 6 ml of tetrahydrofuran at 0°C. Under these conditions recoveries were found to be in excess of 95.6%, with RSDs in the range 1.5 to 3.9, depending upon the product under investigation.

D'Odorico and co-workers [17] have described the use of off-line SFE for the HPLC determination of putative carcinogens and anticarcinogens in faecal samples. The authors of this application found that the main problem in developing an SFE method was with the ability of the carbon dioxide to penetrate the faecal mass. The nub of the problem was the compaction of the small sample sizes of lyophilised material which was overcome by dispersal in anhydrous sodium sulphate. As one might expect the more polar carotenoids were poorly extracted with neat carbon dioxide, resulting in the development of a static SFE procedure using benzene or toluene as modifiers. Highly polar bile acids could be extracted only by using methanol-modified carbon dioxide. In this application, pressure was found to be the most important extraction parameter, with 7500 psi needed to extract carotenes and carotene-epoxides. In this preliminary study no extraction efficiencies were reported and method precision and detection limits were not determined. Off-line HPLC analysis was performed using a gradient reverse phase system of 0%–100% ethyl acetate in acetonitrile/water (9:1 vol/vol) at 1 ml/min with a Spherisorb® octadecyl silane (ODS)2 250 mm × 4.6 mm i.d. column. Unfortunately, a representative chromatogram was not presented.

One of the most topical areas in analytical chemistry at present is that of pollution and the environment. The whole range of analytical techniques in the analyst's armoury is being used and SFE is no exception. Wenclawiak *et al.* [18] and Meyer *et al.* [19] have described the use of off-line HPLC for the analysis of PAHs in soils following sample preparation using SFE. The use of SFE, including HPLC, for the analysis of compounds of agricultural significance has been reviewed by Stuart *et al.* [5]. Gere *et al.* [12] have used SFE in conjunction with GC, GC–MS and HPLC in their work on environmental samples.

In their studies, Wenclawiak *et al.* [18] performed SFE using carbon dioxide (5800 psi, 100°C, extraction time 60 min). The soil extracts required an additional two-stage column chromatography cleanup prior to HPLC analysis because of non-selective extraction of the PAHs, as shown in Figure 8.3. The authors report detection at the parts-per-billion level using fluorescence detection. Off-line HPLC was performed using a gradient solvent system of acetonitrile/water at 1 ml/min using an ODS Knauer Eurospher 80-5 , 250 mm × 4 mm i.d. column at 30°C. Recovery of the internal standard, 2,2-binaphthyl, was found to be 97%, but the two-ring and three-ring PAHs – naphthalene, acenaphthylene, fluorene and phenanthrene – gave recoveries of only 30%. The three-to-six-ring PAHs have apparent recoveries of between 89% and 115%. The authors account for these figures as sample loss rather than incomplete extraction, owing to the high volatility of two-to-three-ring PAHs and the need for additional sample cleanup prior to quantification.

The primary objective of the work by Meyer *et al.* [19] was to compare SFE with traditional Soxhlet extraction for the recovery of PAHs from soil. The analytes were trapped on a liquid/solid trap of petroleum ether/silica following SFE and were recovered by washing with petroleum ether/toluene. Diethyl phthalate was then added and the whole extract was evaporated under a stream of nitrogen. The purpose of the diethyl phthalate, being less volatile than petroleum ether/toluene, was to retain the volatile PAHs prior to dilution with acetonitrile. Analysis was performed using reverse phase HPLC with a programmable ultraviolet/visible (UV/VIS) detector. Figure 8.4 clearly demonstrates the precise nature of the extraction/trapping process by the quality of the chromatography with co-extracted compounds being well resolved. Table 8.3 provides a summary of the extraction efficiencies achieved in 75 min compared with 24 h for the Soxhlet determinations. For some PAHs, Meyer suggests that poor SFE recoveries were attributable to inefficient extraction and/or incomplete trapping.

Sewage treatment and the screening of environmentally hazardous compounds in sewage sludge is an important issue in today's society. Fernandez *et al.* [20, 21] and Breen *et al.* [22] have conducted studies involving SFE for the off-line HPLC determination of surfactants in sewage sludge and sediments. In their initial study Fernandez *et al.* [20] developed an SFE procedure for ditallowdimethylammonium chloride (DTDMAC) in sewage sludge and compared this method with a conventional liquid extraction technique. The SFE method provided equivalent extraction efficiencies but required an additional SPE cleanup step using an anion-exchange column prior to HPLC analysis. Analyses were performed using normal phase HPLC with post-column ion-pair extraction using aqueous

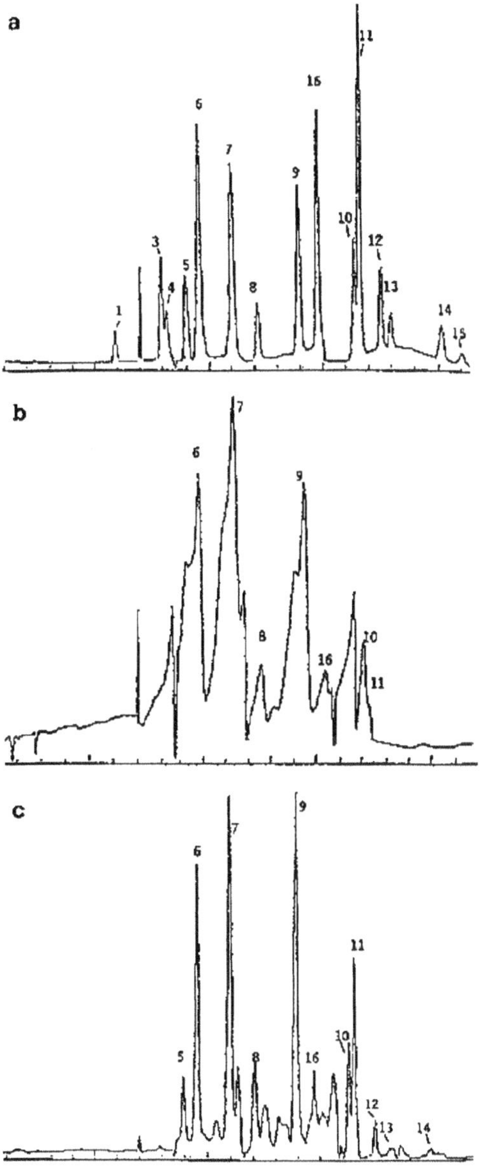


**Figure 8.3** High-performance liquid chromatography (HPLC) chromatograms obtained for polycyclic aromatic hydrocarbon (PAH) standards recovered from soil: (a) PAH standards; (b) supercritical fluid extraction (SFE) extract without cleanup; (c) SFE extract following cleanup. HPLC conditions: Knauer Eurospher 80-5 ODS column using a water : acetonitrile gradient. Peaks: 1 = naphthalene; 2 = acenaphthylene; 3 = fluorene; 4 = acetnaphthene; 5 = phenanthrene; 6 = anthracene; 7 = fluoranthene; 8 = pyrene; 9 = chrysene; 10 = benzo[*b*]fluoranthene; 11 = benzo[*k*]fluoranthene; 12 = benzo[*a*]pyrene; 13 = dibenz[*a*,*h*]anthracene; 14 = indeno[1,2,3-*cd*]-pyrene; 15 = benzo[*ghi*]perylene; 16 = 2,2′-binaphthyl. Reproduced with permission from ref. 18.

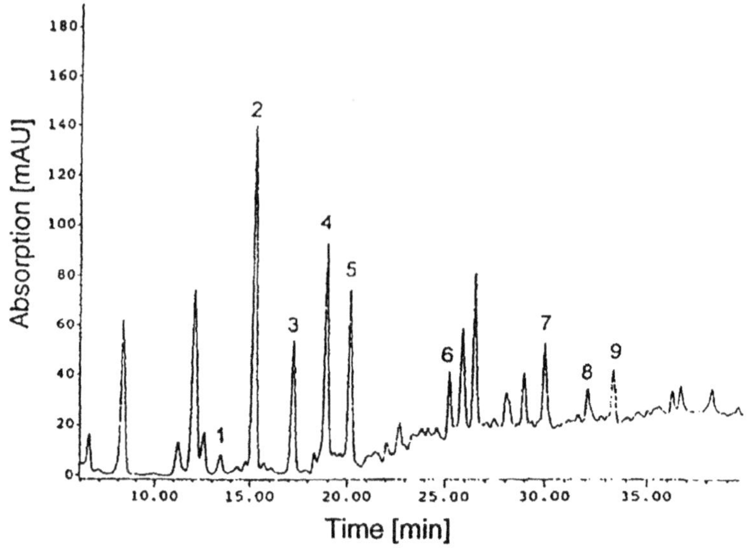


**Figure 8.4** High-performance liquid chromatography (HPLC) analysis of supercritical fluid extraction (SFE) soil extract. HPLC conditions: Bakerbond® Widepore 250 mm × 4.6 mm i.d. 5 µm ODS column using a water:acetonitrile gradient at 1 ml/min. Peaks: 1 = fluorene; 2 = phenanthrene; 3 = anthracene; 4 = fluoranthene; 5 = pyrene; 6 = benz[*a*]anthracene; 7 = benzo[*b*]fluoranthene; 8 = benzo[*k*]fluoranthene; 9 = benzo[*a*]pyrene. Reproduced with permission from ref. 19.

methyl orange. This complex was detected using UV/VIS at 420 nm following phase separation. Figure 8.5 shows good separation of the DTDMAC and an internal standard (didecyldimethylammonium bromide); the former is only partially resolved from a co-extracted species. The authors report detection limits of 70–100 ng for DTDMAC using the HPLC method.

In 1996 Fernandez *et al.* [21] reported the results of further SFE studies of DTDMAC using the same HPLC assay procedure described above. In addition to dried sewage samples, SFE was also applied to marine sediments. These investigations were more comprehensive than those reported in 1993 [20], with the authors determining limits of quantitation and instrument detection limits for both sample matrices. Electrospray ionisation–mass spectrometry (ESI–MS) was also used to confirm the identity of the analytes. The SFE recoveries of DTDMAC were compared with those obtained by liquid extraction for both sewage and marine sediment samples and are summarised in Table 8.4.

This research makes no comment on the quality of the chromatographic determination and unfortunately no representative chromatogram was reported. The most attractive feature of this work is the time saving afforded by SFE without any loss in analyte recovery compared with solvent extraction. The solvent extraction method was reported to take about one week for four samples, compared with 30–40 min for SFE sample preparation.

**Table 8.3** Comparison between supercritical fluid extraction (SFE) and Soxhlet extraction results (in μg/g). Source: ref. 19

| Polycyclic aromatic hydrocarbon | SFE (60°C)[a] | SFE (140°C)[a] | SFE (toluene, 140°C)[b] | SFE (5% toluene, 140°C)[c] | Soxhlet extraction[d] |
|---|---|---|---|---|---|
| Fluorene | 4.6 ± 0.3 | 11.3 ± 0.6 | 10.1 ± 0.8 | 12.0 ± 1.3 | 8.4 ± 1.8 |
| Phenanthrene | 27.0 ± 1.3 | 55.7 ± 5.6 | 79.1 ± 2.0 | 92.9 ± 1.4 | 52.3 ± 2.9 |
| Anthracene | 4.6 ± 0.3 | 11.8 ± 1.3 | 15.8 ± 0.8 | 20.1 ± 0.1 | 10.6 ± 0.7 |
| Fluoranthene | 33.1 ± 1.0 | 43.9 ± 1.4 | 53.4 ± 1.9 | 60.8 ± 1.8 | 47.1 ± 2.5 |
| Pyrene | 26.0 ± 0.7 | 36.2 ± 0.6 | 48.4 ± 2.3 | 46.3 ± 0.5 | 41.4 ± 2.2 |
| Benz[*a*]anthracene | 9.8 ± 0.6 | 9.7 ± 1.5 | 8.3 ± 2.3 | 9.6 ± 1.2 | 11.3 ± 2.7 |
| Benzo[*b*]fluoranthane | 9.4 ± 1.5 | 5.4 ± 2.0 | 9.0 ± 2.7 | 16.3 ± 1.9 | 16.6 ± 2.2 |
| Benzo[*k*]fluoranthene | 4.4 ± 0.6 | 3.7 ± 0.7 | 3.8 ± 0.4 | 4.8 ± 0.7 | 5.5 ± 2.1 |
| Benzo[*a*]pyrene | 9.0 ± 1.1 | 7.7 ± 1.4 | 7.6 ± 2.7 | 7.5 ± 1.2 | 12.9 ± 0.3 |

[a] Extraction parameters: $p = 5800$ psi, $t = 75$ min, $n = 3$.
[b] Equilibration: $V_{toluene} = 1$ ml; $T_0 = 120$°C; $t_0 = 15$ min. SFE: $p = 5800$ psi, $t = 75$ min, $n = 5$.
[c] Equilibration and SFE conditions as in footnote b; carbon dioxide was continuously modified with 5% toluene, $n = 3$.
[d] Soxhlet extraction with *n*-hexane/acetone (1 : 1 vol/vol), $t = 24$ h, $n = 5$.

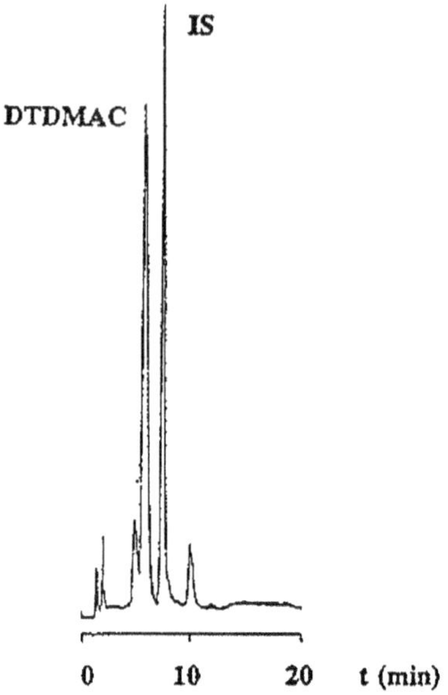


**Figure 8.5** Normal phase high-performance liquid chromatography (HPLC) analysis of ditallowdimethylammonium chloride (DTDMAC) extracted from digested sewage sludge by supercritical fluid extraction (SFE) with didecyldimethylammonium bromide acting as internal standard (IS). HPLC conditions: aminopropyl 100 mm × 4.6 mm i.d. 3 μm column using a chloroform:methanol:acetic acid gradient. Post-column ion-pair extraction with methyl orange. $t$ = time. Reproduced with permission from ref. 20.

Breen *et al.* [22] have also evaluated off-line SFE for the analysis of cationic surfactants–dihardenetallowdimethylammonium chloride (DHTDMAC) in dried sewage sludge as an alternative to the time-consuming methanolic hydrochloric acid extraction procedure. They demonstrated that methanol-modified carbon dioxide afforded the best recoveries for this analyte. Conditions for their work were an extraction density of 0.8068 g/ml at a pressure of 5877 psi and temperature of 85°C. With these SFE conditions additional sample cleanup was needed prior to HPLC assay using a combination of liquid/liquid extraction and anion-exchange chromatography. The reference method involved extraction with methanolic hydrochloric acid of the dried

**Table 8.4** Comparison of ditallowdimethylammonium chloride (DTDMAC) recoveries by liquid extraction with those by supercritical fluid extraction (SFE) from dried sewage sludge and marine sediment (in g/kg). Source: ref. 21

| Extraction technique | Sample source | | |
|---|---|---|---|
| | Zurich-Glatt | Bulach | Marine sediment |
| SFE | 3.29 ± 0.06 | 3.17 ± 0.06 | 42.3 ± 0.06 |
| Solvent extraction | 3.30 ± 0.06 | 3.08 ± 0.06 | 30.3 ± 0.06 |

sludge followed by solvent extraction and anion-exchange chromatography prior to assay. Both extraction methods were compared using HPLC (Figure 8.6) and a colorimetric method using disulphine blue. The colorimetric method was used to estimate the DHTDMAC since the method is non-specific, as any long-chain amine and compounds that can be protonated in acid conditions or that contain a quaternary nitrogen group may form interfering extractable complexes. Table 8.5 demonstrates the efficacy of the method leading to the conclusion that SFE is at least capable of qualitative extraction of DHTDMAC from sewage sludge.

The modern consumer demands more and more from our food industry. Production techniques have evolved in a variety of ways in response to the demand for higher yields, better quality and reliability of supply. The use of pesticides is now fundamental to the success of most suppliers. Unfortunately, although we need the benefits of pesticides they are detrimental to the well being of the consumer [23]. A considerable amount of effort has been invested in the development of analytical procedures for the determination of pesticide residues, and SFE shows considerable potential. Howard *et al.* [24] have used SFE in conjunction with a tandem trapping system consisting of a solid phase trap and a methanol solvent trap in a feasibility study for the analysis of methomyl, methiocarb and eptam in apples. HPLC with UV detection was found to be the assay method of choice when compared with sulphur chemiluminescence and GC with flame ionisation detection (FID). Two types of solid phase trap were evaluated in this research: ODS and stainless-steel beads. The authors found that neither of the traps by themselves were effective at trapping the analytes. Following their initial studies, the authors used stainless-steel beads in serial combination with a methanol trap, and a summary of the results is shown in Table 8.6. This method development involved the use of a spiked cotton swab using HPLC/UV analysis to assay the analytes.

On one occasion whilst recovering the analytes with methanol from the solid phase trap (an apple sample having been extracted) a small amount of solution dropped onto the bench. After some of the solvent had evaporated the remaining 'solution' was sticky to the touch. The authors suggest that this material consisted of sugars co-extracted with the analytes. At first glance this is an unusual conclusion since sugars are highly polar and insoluble in non-polar supercritical carbon dioxide, possessing only marginal solubility in methanol. The most plausible explanation for this observation is that water or apple juice containing dissolved sugars flashed over from the extraction vessel. In their initial studies involving direct liquid SFE, Ramsey and co-workers (unpublished development work) incorporated an empty semi-preparative HPLC column acting as a 'demister' into their SFE system in order to trap out aqueous flashover prior to analyte collection.

The reverse phase HPLC method used for the analysis of the three pesticides indicated, apart from the additional peak at $t_0$, that SFE yields a

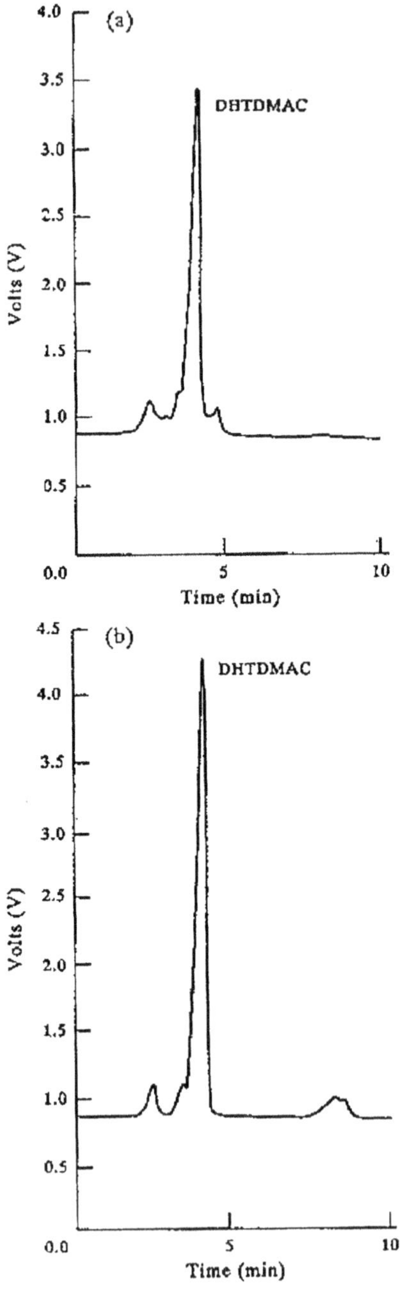


**Figure 8.6** High-performance liquid chromatography (HPLC) analyses of: (a) dihardenetallow-dimethylammonium chloride (DHTDMAC) extracted from sewage sludge by supercritical fluid extraction (SFE); (b) 2 μg DHTDMAC standard. Reproduced with permission from ref. 22.

**Table 8.5** Supercritical fluid extraction (SFE) of disulphine blue active substances and dihardenetallowdimethylammonium chloride (DHTDMAC) from sewage sludge. Ref. = reference method. Source: ref. 22

| | SFE conditions | | Disulphine blue active substances | | | DHTDMAC | | |
|---|---|---|---|---|---|---|---|---|
| Sludge sample | Methanol (%) | Time (min) | Extracted by SFE (mg/g) | Extracted by Ref. (mg/g) | SFE/Ref. (%) | Extracted by SFE | Extracted by Ref. | SFE/Ref. (%) |
| 1 | 0 | 60 | 0.2 | 5.0 | 4 | – | 2.5 | – |
| 1 | 10 | 45 | 4.3 | 5.0 | 86 | 3.0 | 2.5 | 120 |
| 2 | 10 | 30 | 2.6 | 4.6 | 57 | 2.0 | 2.3 | 87 |
| 2 | 10 | 60 | 2.4 | 4.6 | 52 | 1.6 | 2.3 | 70 |
| 2 | 50 | 45 | 3.7 | 4.6 | 80 | 2.3 | 2.3 | 100 |
| 3 | 50 | 30 | 2.8 | 5.8 | 48 | 2.1 | 2.9 | 72 |
| 4 | 25 | 30 | 3.4 | 5.6 | 61 | 2.8 | 2.8 | 100 |
| 5 | 25 | 40 | 3.8 | 5.8 | 66 | 2.3 | 2.9 | 79 |

**Table 8.6** Tandem solid-phase–liquid thiocarbamate supercritical fluid extraction (SFE) trapping recoveries. Source: ref. 24

| Trap material and temperature | Analyte recovery | | |
|---|---|---|---|
| | Solid phase trap[a] | Liquid trap[b] | Total |
| Stainless-steel beads, 15°C | | | |
| Methomyl | 35.8 | 46.8 | 82.6 |
| Methiocarb | 31.4 | 56.8 | 88.3 |
| Eptam | 0 | 77.1 | 77.1 |
| Stainless-steel beads, −30°C | | | |
| Methomyl | 5.3 | 92.1 | 97.5 |
| Methiocarb | 4.9 | 104.8 | 109.8 |
| Eptam | 0 | 93.5 | 93.5 |
| ODS, −30°C | | | |
| Methomyl | 8.8 | 77.0 | 85.8 |
| Methiocarb | 13.3 | 90.5 | 103.7 |
| Eptam | 12.9 | 84.2 | 97.2 |

[a] Solid phase trap rinsed with 4 ml of methanol.
[b] 7 ml methanol.

relatively clean extract resulting in the retention of chromatographic performance. Comparison of the chromatograms for the SFE extracts obtained from spiked cotton swab and apple clearly demonstrate this [Figures 8.7(a) and 8.7(b)]. The peaks 1–3 in the chromatogram are, respectively, methomyl, methiocarb and eptam, with thianaphthene acting as internal standard. Using the SFE procedure recoveries of the pesticides at the 2 ppm level were demonstrated. The authors concluded that sub-ppm detection limits should be possible for the pesticides, provided an additional concentration step is used following SFE.

Atienza *et al.* [25] used SFE with HPLC/UV for the determintion of fluvalinate residues in honey and compared this technique with a standard solvent extraction–thin-layer liquid chromatography (TLC)–gas chromatography method. Fluvalinate or chlordimeform are used for the chemical treatment of varroatosis, an external form of parasitosis caused by the mite *Varroa jacobsoni*. Prior to SFE the honey was mixed with water and cellulose powder prior to lypophilising. A mixture of benzene:isopropyl alcohol (IPA) was added to the loaded extraction vessel to act as modifier. The SFE conditions were as follows: 0.45 g/ml supercritical carbon dioxide fluid density, 2000 psi pressure at 70°C, a flow rate of 0.8 ml/min, with an extraction time of 20 min. The extracted analytes were collected on a trap packed with stainless-steel balls at a nozzle temperature of 75°C, subsequently recovered with 1 ml of methanol. Analysis of the extract involved an off-line reverse phase HPLC procedure using a 150 mm × 4.6 mm i.d. NovaPak® ODS column with acetonitrile:water:acetic acid (80:20:14) as mobile phase at 1.5 ml/min; UV detection was at 254 nm. The results using SFE for a range of fortified honey samples, summarised

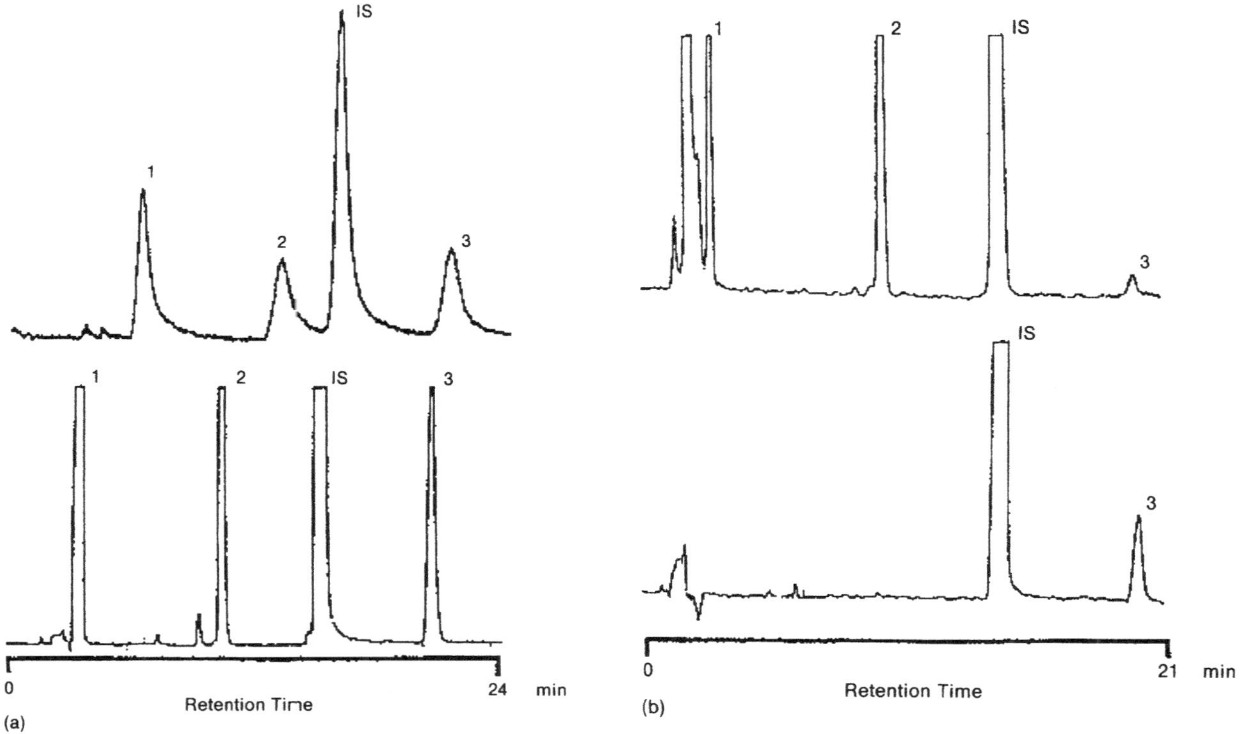


**Figure 8.7** Analysis of a supercritical fluid extraction (SFE) extract obtained from: (a) a cotton swab; (b) an apple, spiked with methomyl (peak 1), methiocarb (peak 2), eptam (peak 3) and internal standard (IS), thianaphthene. High-performance liquid chromatography (HPLC) conditions: Hypersil® 250 mm × 4.6 mm inner diameter ODS column using water : acetonitrile (50 : 50) at 1 ml/min. The upper chromatograms were obtained with sulphur chemiluminescence detection and the lower chromatograms with ultraviolet detection. Reproduced with permission from ref. 24.

**Table 8.7** Comparison of fluvalinate recoveries by supercritical fluid extraction (SFE) with high-performance liquid chromatography (HPLC) and the standard solvent extraction technique, with thin-layer chromatography (TLC) and gas chromatography (GC). Source: ref. 25

| Fortified amount (mg/kg) | SFE | | Standard solvent extraction | |
|---|---|---|---|---|
| | Recovery (%) | $\sigma_{n-1}$ | Recovery (%) | $\sigma_{n-1}$ |
| 10 | 53 | 1.6 | 49 | 3.0 |
| 5 | 80 | 1.6 | 71 | 2.7 |
| 1 | 84 | 1.5 | 77 | 2.5 |
| 0.5 | 94 | 1.3 | 88 | 2.5 |

in Table 8.7, proved to be superior to those obtained using the standard method. It should be noted that the stability of the fluvalinate presents a significant problem to the analyst using either method of sample preparation. After 12 days HPLC analysis of a standard solution in methanol showed the fluvalinate to have degraded to one of its metabolites and to other decomposition products. The GC analysis used to assay the fluvalinate following the solvent extraction–TLC method showed the presence of even more decomposition products.

As a consequence of the fluvalinate instability the authors determined the metabolite concentration and used that to back calculate the original fluvalinate concentration in honey samples. The limit of detection for the HPLC/UV method was found to be 0.02 mg/kg of honey. The authors did not publish any chromatograms, hence it is difficult to comment on the effects of any co-extractives and/or the selectivity of SFE for fluvalinate. Owing to the instability of fluvalinate, the authors did not extend their study for the analysis of real samples of beehive honey.

Continuing the theme of flora-related foodstuffs, Aharonson *et al.* [26] turned their attention to the SFE of residual benzimidazole fungicides from fruit and vegetable matter. Thiabendazole (TBZ), benomyl, carbendazim (MBC) and thiophanate-methyl were chosen for the study as they are the most commonly used benzimidazole fungicides applied in pre-harvest and post-harvest treatment of many different crops. Experiments to determine optimum supercritical carbon dioxide density, extraction temperature and extraction volume of carbon dioxide were performed. The conclusion of these experiments led to the use of reasonably moderate conditions: an extraction pressure of 5800 psi at 60°C, with an ODS trap at 25°C for analyte collection. The addition of 3% methanol modifier improved recoveries only within the standard deviation of the method. Methanol proved to be superior to both acetone and ethyl acetate as it gave fewer matrix interferences in the HPLC/UV chromatograms. The analytes were recovered by washing the trap with methanol. Owing to the relatively low levels of the analytes an additional concentration step was required prior to assay using a reverse phase ion-pair HPLC method. An important observation from this work came from the use of Hydromatrix as a solid support to suppress the effects of sample-based water on the extraction process. It was postulated that water behaves as a modifier for supercritical carbon dioxide, resulting in a more efficient extraction of polar analytes. This is entirely consistent with those findings reached by Howard *et al.* [24] in their work on pesticide determination from apples. The authors found that no additional sample cleanup was necessary following SFE, resulting in little or no serious matrix interference at 285 nm with analyte concentrations above 0.1 μg/ml of MBC and 0.05 μg/ml TBZ (Figure 8.8). The fungicide recoveries using the SFE method compared favourably with those obtained using solvent extraction and in some cases were better, as is evident from Table 8.8. Using UV, the limits of detection were found to be 0.041 μg/g for MBC and 0.024 μg/g for TBZ from the potato matrix.

Shun Huang and co-workers [27] used SFE and off-line HPLC for the quantification of SALTRIM 23CA, a member of a family of triglycerides developed by R.J.R. Nabisco for use in food items. SFE conditions were as follows: 0.8 g/ml supercritical carbon dioxide at a pressure of 5293 psi and a temperature of 80°C, with a fluid flow of 4 ml/min. A period of 2 min static SFE was used followed by 30 min dynamic SFE. The extracted analytes were trapped onto stainless-steel beads and were recovered by washing with hexane. Assay of the resulting solution was achieved using an HPLC method that required two Supelcosil™ ODS 150 mm × 4.6 mm i.d. columns connected in series, eluted with a multistage gradient starting from 100% acetonitrile to acetonitrile/*tert*-butyl methyl ether (75:25 vol/vol) at 1 ml/min. Detection was performed using evaporative light scattering. The quality of the chromatography is excellent, with little or no interference from co-extractives; example chromatograms obtained for the SFE extracts of several food products are shown in Figure 8.9.

Table 8.9 clearly shows that the fat content of several foods is determined by SFE and HPLC with good precision, and the average determined fat content is in excellent agreement with the theoretical value. An important point to note is that the light-scattering detector did not provide a linear response. The authors created a non-linear model demonstrating that an exponential fit based on the expression $y = -0.28 + 0.55x^{1.31}$, where $y$ is the peak area ratio of total SALATRIM 23CA peak areas to the internal standard trinonanoin peak area, and $x$ is the weight ratio of SALATRIM 23CA to trinonanoin, provided the best quantitative analysis.

Animal-based and fish-based products represent other important classes of foodstuffs monitored in the food industry. Matrices ranging from meat, eggs and fish products have all been analysed using SFE and HPLC. Owing to the complexity of these matrices, additional cleanup procedures, both pre-SFE and post-SFE, are generally required to provide extracts sufficiently clean for HPLC analysis. Argauer *et al.* [28] used a two-stage pre-extraction procedure involving extraction of the beef with acetonitrile and liquid/liquid

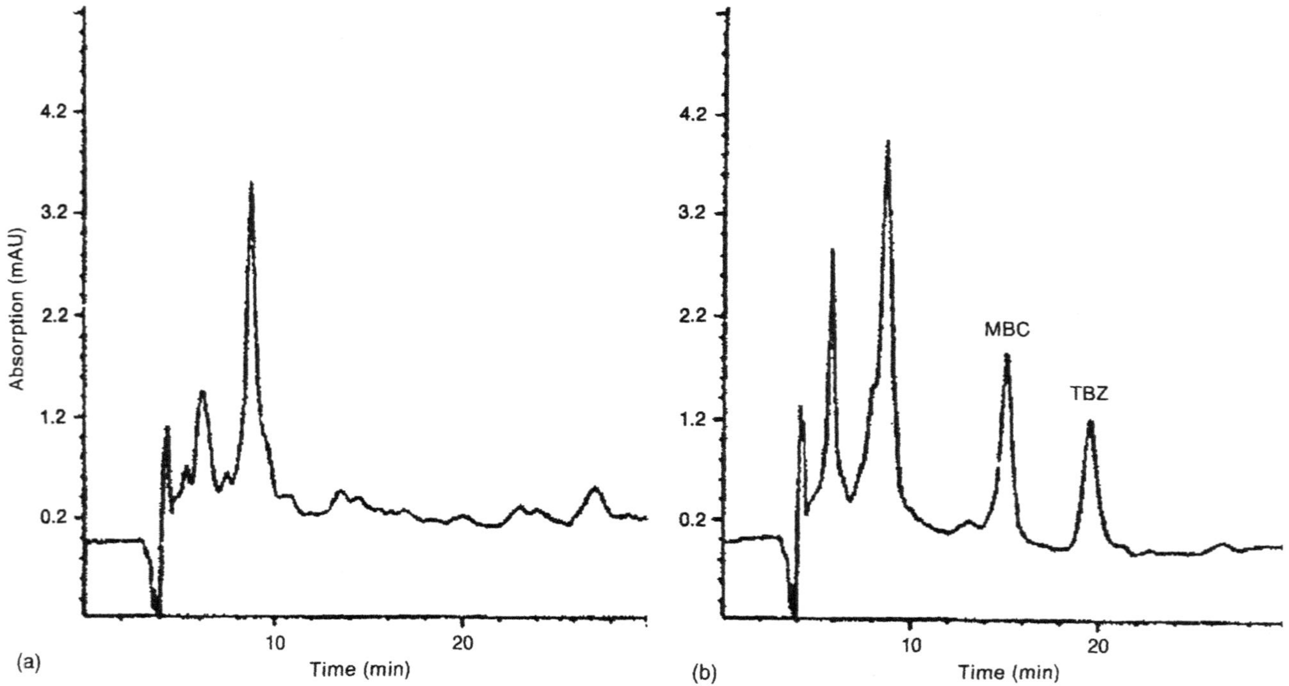


**Figure 8.8** Analysis of supercritical fluid extraction extracts obtained from: (a) blank potato; (b) potato fortified with 0.5 μg/g carbendazim (MBC) and 0.25 μg/g thiabendazole (TBZ). High-performance liquid chromatography (HPLC) conditions: Ultrasphere® 250 mm × 4.6 mm inner diameter ODS column using an isocratic ion-pair mobile phase system consisting of methanol with sodium decanesulphonate, triethylamine and phosphoric acid at 0.8 ml/min. Reproduced with permission from ref. 26.

**Table 8.8** Comparison of the analytical results, with high performance liquid chromatography (HPLC), for thiabendazole (TBZ) in the incurred potato and banana samples using the supercritical fluid extraction (SFE) method and the traditional solvent-based method. Source: ref. 26

| Sample | TBZ concentration, solvent-based extraction (μg/g) | TBZ concentration, SFE (μg/g) | Internal standard recovery, SFE (%) |
|---|---|---|---|
| Banana 1 | 0.33 | $0.29 \pm 0.05$ | $80 \pm 16$ |
| Banana 2 | 0.17 | $0.13 \pm 0.02$ | $73 \pm 29$ |
| Potato 1 | 0.26 | $0.61 \pm 0.11$ | 87.9 |
| Potato 2 | 0.17 | $0.49 \pm 0.04$ | 85.1 |
| Potato 3 | 0.13 | $0.33 \pm 0.05$ | 92.8 |

extraction of the filtered acetonitrile with hexane saturated with acetonitrile. This provided a means of extracting and concentrating the target analytes, propoxur and other carbamate pesticides from a very complex matrix. The hexane extract was mixed with Hydromatrix to adsorb the propoxur and other carbamate pesticides and the solvent evaporated by aqueous aspiration prior to SFE. The target analytes were collected on a Hypersil® ODS trap and recovered with 1.5 ml of acetonitrile. This cleanup process in conjunction with SFE resulted in 100 g of ground beef providing 50 mg of co-extractives plus the analytes. A series of experiments were performed to determine the influence of supercritical fluid carbon dioxide density on SFE efficiency, with the results summarised in Figure 8.10. Poor SFE efficiencies were noted at a fluid density of 0.3 g/ml, the optimum density being 0.85 g/ml. Consistent with other reports dealing with SFE of meat products they found that lipid extraction rapidly increased for fluid densities of 0.4–0.6 g/ml. Reverse phase HPLC/fluorescence along with GC–MS was used to assay the analytes. The HPLC system employed a Hypersil® Green Env C-8 5 μm column eluted with acetonitrile/water (40:60 vol/vol). Fluorescence detection was achieved by post-column addition of a fluorescence reagent consisting of *o*-phthalaldehyde, Thiofluor and *N*,*N*-dimethyl-2-mercapto-propionic acid in 0.05 M sodium hydroxide. The authors report recoveries of 95% from a sample equivalent to 1 g of meat fortified at the 0.1 ppm level with propoxur. The success of this solvent cleanup procedure prior to SFE is demonstrated by the chromatograms shown in Figure 8.11.

Järvenpää *et al.* [29] have also described the need for an additional cleanup stage in their SFE study of PAHs in smoked and broiled fish. The crude PAH extract was isolated using 20 min dynamic SFE with supercritical carbon dioxide at 5075 psi, 70°C and a flow of 1.5–1.7 ml/min. The extract was then subjected to an SPE cleanup procedure to provide a fraction for HPLC/UV analysis. Despite this, the presence of co-extracted compounds still affected the quality of the chromatograms, as shown in Figure 8.12.

Din *et al.* [30] have conducted an SFE study dealing with the analysis of trenbolone, a growth-promoting anabolic steroid, in beef. A major

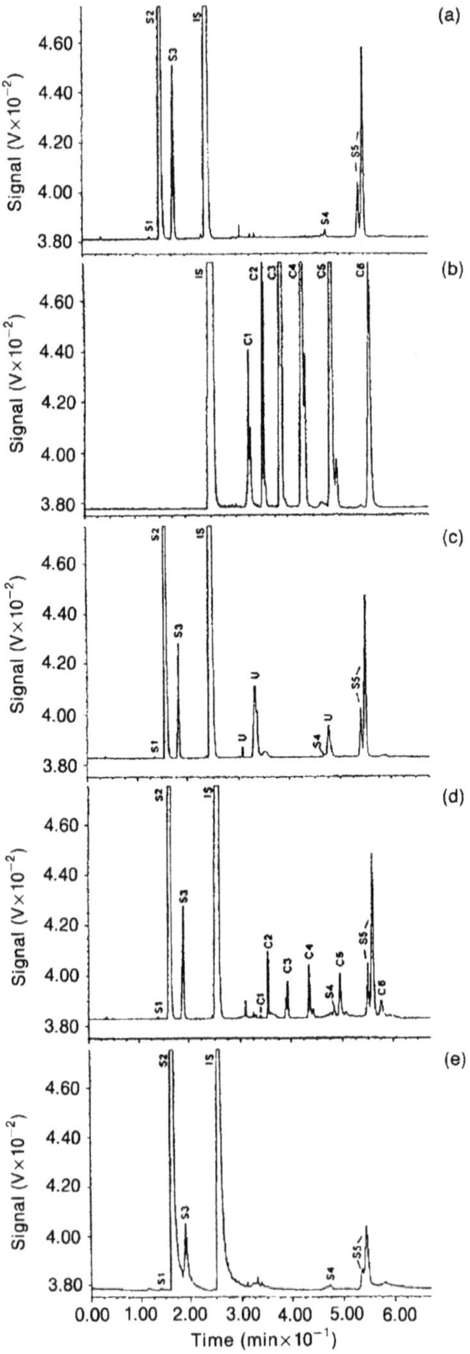

(a)
(b)
(c)
(d)
(e)
S1
S2
S3
S4
S5
IS
C1
C2
C3
C4
C5
C6
U
Signal (V×10⁻²)
4.60
4.40
4.20
4.00
3.80
0.00
1.00
2.00
3.00
4.00
5.00
6.00
Time (min×10⁻¹)

**Table 8.9** Calculated SALATRIM 23CA contents and original added values in foods (percentage by weight). Source: ref. 27

| Sample source | Run 1 | Run 2 | Run 3 | Average | Theoretical values from the original formulation |
|---|---|---|---|---|---|
| Experimental sandwich cookies | 21.7 | 23.2 | 22.8 | 22.6 ± 0.7 | 21.2–23.1 |
| Experimental compound coating candy | 33.3 | 34.6 | 30.5 | 32.8 ± 2.1 | 30.1–35.0 |
| Experimental ice cream | 10.1 | 10.8 | 10.7 | 10.5 ± 0.4 | 10.0 |

feature of this investigation was the experimental demonstration that careful selection of the extraction conditions led to the maximum recovery of trenbolone whilst minimising the effects of co-extractives on the HPLC assay procedure [using a Spherisorb® S5 ODS1, 100 mm × 4.6 mm i.d. column eluted with a methanol/water (54:46 vol/vol) mobile phase]. Figure 8.13(a) illustrates the importance of temperature on extraction efficiency, and Figure 8.13(b) shows the increase in the levels of co-extractives that accompanied the recovery of the analyte. The chromatograms in Figure 8.14 demonstrate the influence that co-extracted fats have on the assay of trenbolone. A further point of interest from this study was that residual water reduced trenbolone extraction efficiency whereas other reports have described the addition of a small percentage of water as being beneficial to promote the SFE recoveries of a range of analytes from various matrices. It was concluded that the SFE procedure developed was capable of providing 98% m/m recovery of trenbolone with minimum co-extraction of fats using 60 min dynamic SFE with a carbon dioxide pressure of 5800 psi at 75°C.

Huopalahti and Henion [31] have evaluated the feasibility of SFE for the isolation of a range of seven growth-promoting veterinary drugs from spiked bovine tissue samples. The estrogenic and anabolic agents studied were: dexamethasone (DEX), diethylstibestrol (DES), medroxyprogesterone (MDP), melengestrol acetate (MGA), trenbolone (TBOH), triamcinolone acetonide (TACA) and zeranol (ZER). The first stage of the study involved the optimisation of four SFE experimental parameters (temperature, pressure, extraction time and restrictor type) with respect to recovery and repeatability.

**Figure 8.9** High-performance liquid chromatography (HPLC) profiles of: (a) SALATRIM 23CA; (b) Hydrol 92 (a refined hydrogenated coconut oil); (c) supercritical fluid extraction (SFE) extract of a sandwich cookie sample; (d) SFE extract of a compound coating a bonbon sample; (e) SFE extract of an ice cream sample. HPLC conditions: two Supelcosil™ LC-18 150 mm × 4.6 mm inner diameter columns connected in series using an acetonitrile: *tert*-butyl methyl ether gradient with light-scattering detection. Peaks: S1 = acetylstearoylglycerol plus diacetylpalmitoylglycerol plus acetylpropionylpalmitoylglycerol (not separated by HPLC); S2 = diacetylstearoylglycerol; S3 = acetylpropionylstearoylglycerol; S4 = acetylpalmitoylstearoylglycerol; S5 = distearoylacetylglycerol; C1 = dilauroylcapryloglycerol; C2 = dilauroylcaproylglycerol; C3 = trilaurin; C4 = dilauroylmyristoylglycerol; C5 = dimyristoyllauroylglycerol; C6 = trimyristin; internal standard (IS) = trinonanoin; U = unknown compounds. Reproduced with permission from ref. 27.

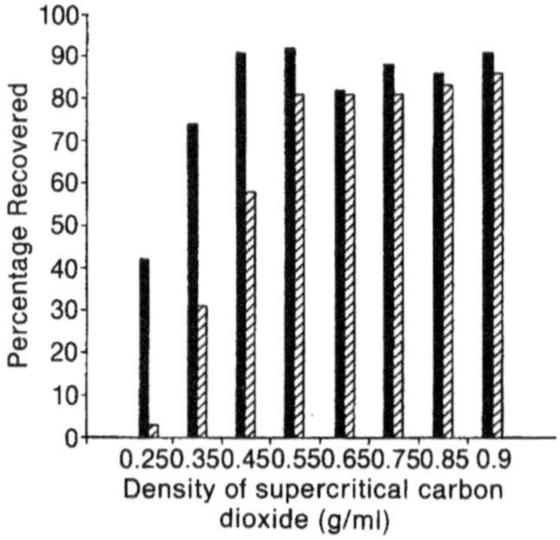


**Figure 8.10** Supercritical fluid extraction (SFE) recoveries of propoxur (dark shading) and carbaryl (hatched shading) standards at various supercritical fluid carbon dioxide densities. An ODS trap and acetonitrile rinse was used for analyte collection. Reproduced with permission from ref. 28.

These investigations involved the use of either a standard solution of the drugs loaded onto filter paper supports or spiked freeze-dried bovine tissue samples as matrices. Following SFE, off-line HPLC/UV and LC–MS analyses were performed. The LC analyses involved the use of a Supelcosil® 50 mm × 4.6 mm i.d. column with a linear gradient programme initially of methanol/acetonitrile:ammonium formate (20 mM) (5:95 vol/vol) then to methanol/acetonitrile:ammonium formate (20 mM) (95:5 vol/vol). Also on-line SFE–cSFC was used to help determine the optimum extraction time. It was demonstrated that in order to obtain optimum SFE recoveries for all seven drugs a compromise was necessary (carbon dioxide at 5802 psi, 60°C, with an extraction time of 60 min, using a linear restrictor). With these SFE conditions, the mean recovery of the seven drugs was 69% with an RSD of 19.7%. The more polar drugs (DEX, TACA and ZER) had SFE recoveries ranging from 44% to 58% whereas less polar drugs (MGA, MDP and DES) had SFE recoveries ranging from 83% to 91%. The SFE recovery of TBOH could not be determined in these experiments because of interference from co-eluting compounds. The use of an SFE *in situ* cleanup agent, such as alumina, to retain co-extracted lipids from tissue samples was not investigated in these studies. The SFE procedure in conjunction with off-line LC–MS resulted in detection limits (using selected ion monitoring atmospheric pressure chemical ionisation) for each of the drugs in spiked tissue samples in the region of 100 ppb.

To continue the theme of SFE for the HPLC analysis of veterinary drug residues in meat products, Cross *et al.* [32] have described a procedure for the determination of a range of sulphonamides in chicken liver and swine muscle tissue. This group of synthetic compounds has been widely used in

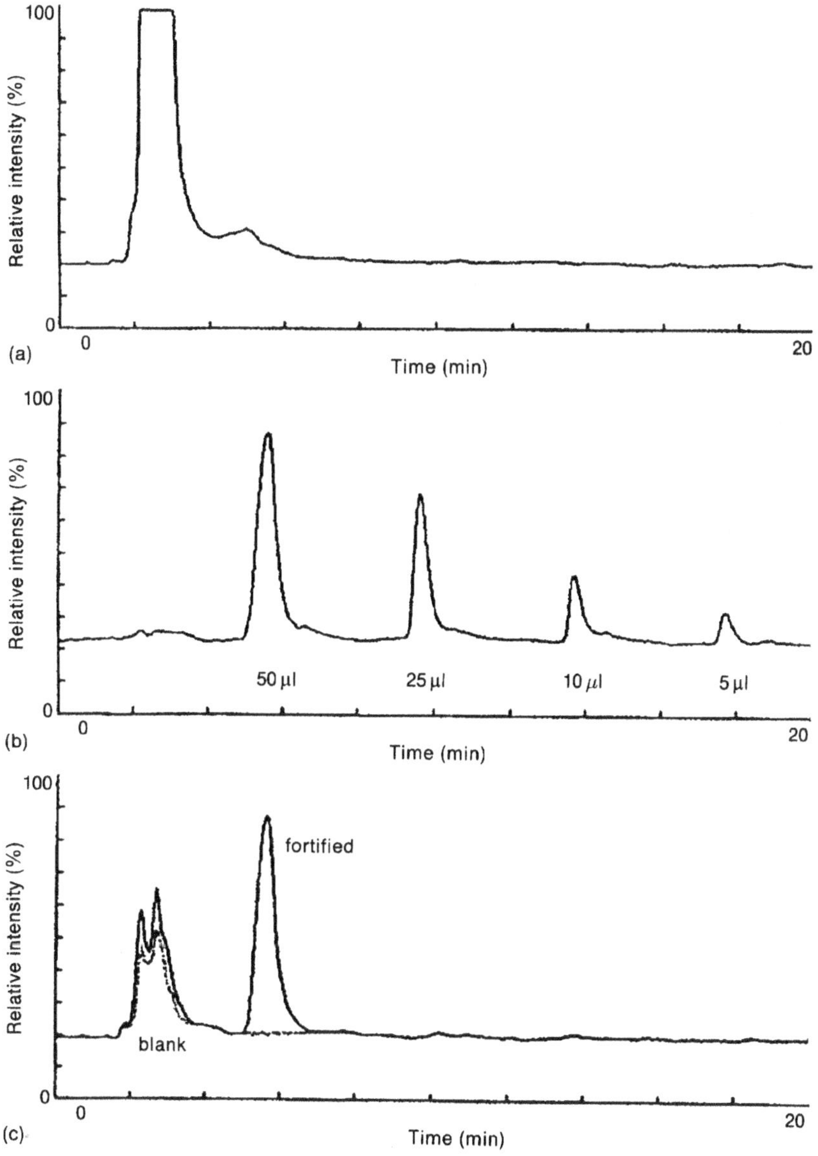


**Figure 8.11** High-performance liquid chromatography (HPLC) analyses of propoxur: (a) hexane-washed acetonitrile extract of spiked ground beef; (b) propoxur standards (2 µg/ml injections at $t = 0$, 4, 8 and 12 min); and (c) supercritical fluid extraction (SFE) extract of hexane fraction (from the liquid extraction procedure for spiked ground beef) loaded onto Hydromatrix. Reproduced with permission from ref. 28.

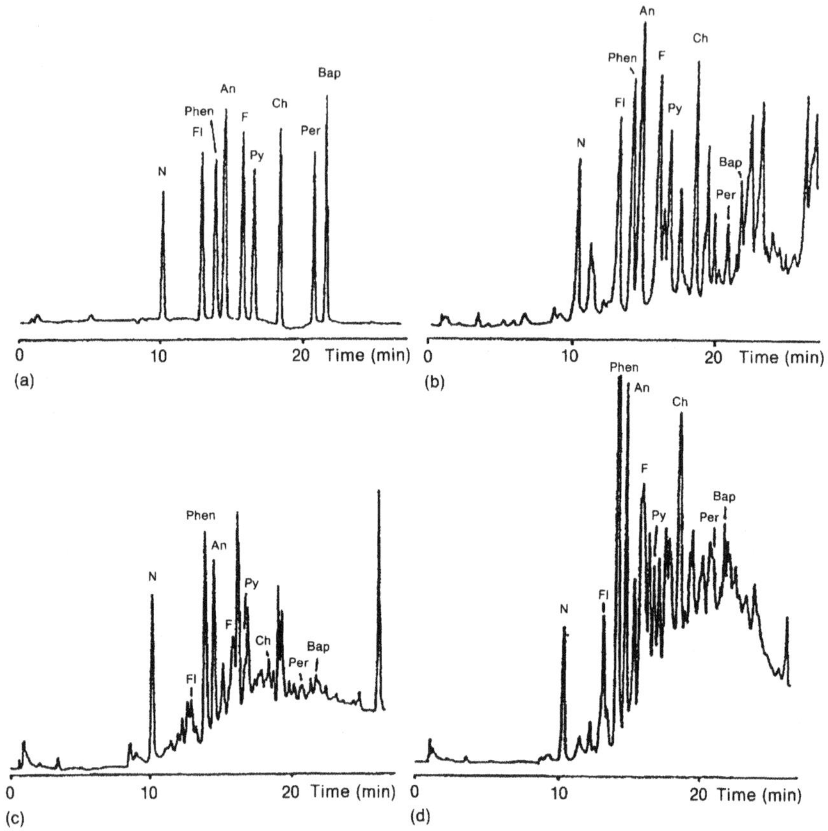


**Figure 8.12** Chromatograms obtained for the analysis of polycyclic aromatic hydrocarbons (PAHs) in fish samples. High-performance liquid chromatography (HPLC) conditions: Lichrosphere 125 mm × 4 mm inner diameter 5 μm ODS column using a water : acetonitrile gradient at 0.8 ml/min: (a) PAH standards: fluoranthene (Fl), phenanthrene (Phen), anthracene (An), fluorene (F), pyrene (Py), chrysene (Ch), perylene (Per), benzo[*a*]pyrene (BeP); (b) supercritical fluid extraction (SFE) extract of spiked baltic herring; (c) SFE extract of spiked smoked baltic herring; (d) SFE extract of spiked broiled herring. Reproduced with permission from ref. 29.

veterinary medicine for the treatment of disease caused by bacterial infection. Owing to their toxicity and carcinogenic properties, the sulphonamides have now been largely replaced by other antibiotics. The following sulphonamides were studied: sulphamerazine (SMR), sulphamethizole (SMI), sulphamethazine (SMZ), sulphamethoxypyridazine (SMOP), sulphamethoxazole (SMOZ) and the major metabolite $N^4$-acetylsulphamethoxazole (N4). This comprehensive study investigated the role of several extraction parameters, including the concentration of methanol acting as modifier, temperature and pressure on the sulphonamide SFE efficiencies. It was demonstrated that the nature of sample presentation to the supercritical fluid was also

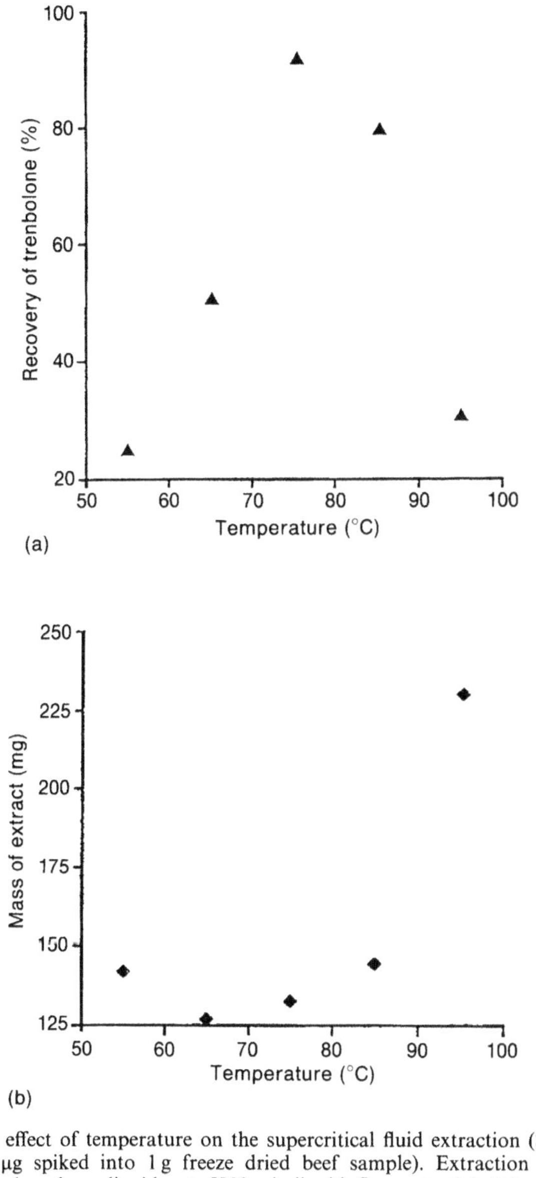


**Figure 8.13** The effect of temperature on the supercritical fluid extraction (SFE) recovery of trenbolone (800 μg spiked into 1 g freeze dried beef sample). Extraction conditions: 10% methanol-modified carbon dioxide at 5800 psi, liquid flow rate 0.3–0.8 ml/min, extraction time 2 h. (b) The effect of temperature on the variation of the endogenous material extracted from freeze dried beef. SFE conditions: 10% methanol-modified carbon dioxide at 5800 psi, liquid flow rate 0.7–1.0 ml/min, extraction time 1 h. Reproduced with permission from ref. 30.

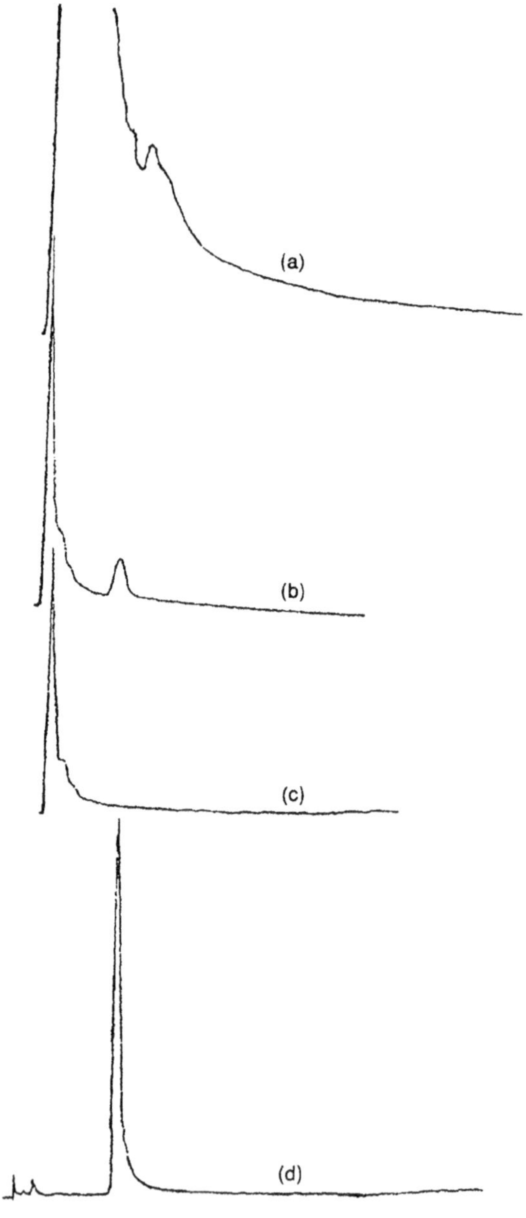


**Figure 8.14** High-performance liquid chromatography (HPLC) chromatograms obtained for supercritical fluid extraction (SFE) extracts of trenbolone (a) at 95°C; (b) at 75°C; (c) blank extract at 75°C; (d) trenbolone standard. Reproduced with permission from ref. 30.

important. This resulted in the conclusion that finally divided material mixed with an inert support, to maximise surface area and hence exposure to the supercritical fluid, led to improved SFE recoveries. This conclusion has been arrived at by many workers dealing with matrices that are particularly prone to compaction under high pressure. Sulphonamide recoveries were determined using reverse phase HPLC after the crude SFE extract had been subjected to SPE cleanup using deactivated alumina. Despite this cleanup stage the presence of interfering co-extractives is still evident in the HPLC analysis obtained for spiked liver extracts (Figure 8.15). Depending on the analyte, recoveries vary from as low as 27% in spiked swine muscle to as high as 97% in spiked liver samples.

In a related study, Parks *et al.* [33], investigating sulphamethazine (SMZ), sulphadimethoxine (SDM) and sulphaquinoxaline (SQX) from chicken liver, breast tissue and thigh muscle, reported similar extraction efficiencies to those previously obtained for the liver samples and were able to detect the analytes at the 100 ppb level. In order to achieve this they used an SFE vessel with an in-line SPE trap (neutral alumina) under the following SFE conditions: pressure, 10 000 psi; temperature, 40°C; density, 1.042 g/ml; flow, 3 l/min expanded gas; extraction time, 40 min. A consequence of using this configuration was that the extract needed no further cleanup prior to off-line LC analysis. The in-line SPE trapping procedure was compared with an off-line SPE cleanup; the analyte recoveries using both techniques are summarised in Table 8.10.

The off-line HPLC employed a reverse phase ion-pairing system using a Supelcosil™ LC–ODS column (5 μm film) with a mobile phase consisting of 0.05 M phosphate buffer : methanol (65 : 35% vol/vol) with 0.1% tetrabutyl ammonium hydroxide at pH 7.2. The quality of the chromatography obtained using the in-line SPE cleanup procedure was good, with little interference from co-extractives, as shown in Figure 8.16.

In an extension to the above study Pensabene *et al.* [34] investigated the merits of SFE using on-line and off-line trapping for the analysis of sulphonamides (SMZ, SDM, SQX) in whole egg samples. The method involved collecting the analytes on alumina, recovering them by solvent extraction, with reverse phase HPLC being used for quantitation. The extraction conditions established in the work by Parks *et al.* [33] on chicken tissues were used to optimise the volume of supercritical fluid carbon dioxide needed for the extraction. Interestingly, the SFE efficiency, equivalent to the use of 80 l gaseous carbon dioxide, was greater than for 85 l although in both cases high levels of co-extractives were also obtained. However, increasing the in-line extraction volume, equivalent to the use of 120 l gaseous carbon dioxide, the chromatograms were much cleaner. This observation supported the theory that the co-extractives were effectively eluted from the alumina by supercritical fluid chromatography. The recoveries at the 0.1 ppm level were: sulphamethazine, 99.5% ± 2.2%; sulphadimethoxine, 87.8% ± 6.0%;

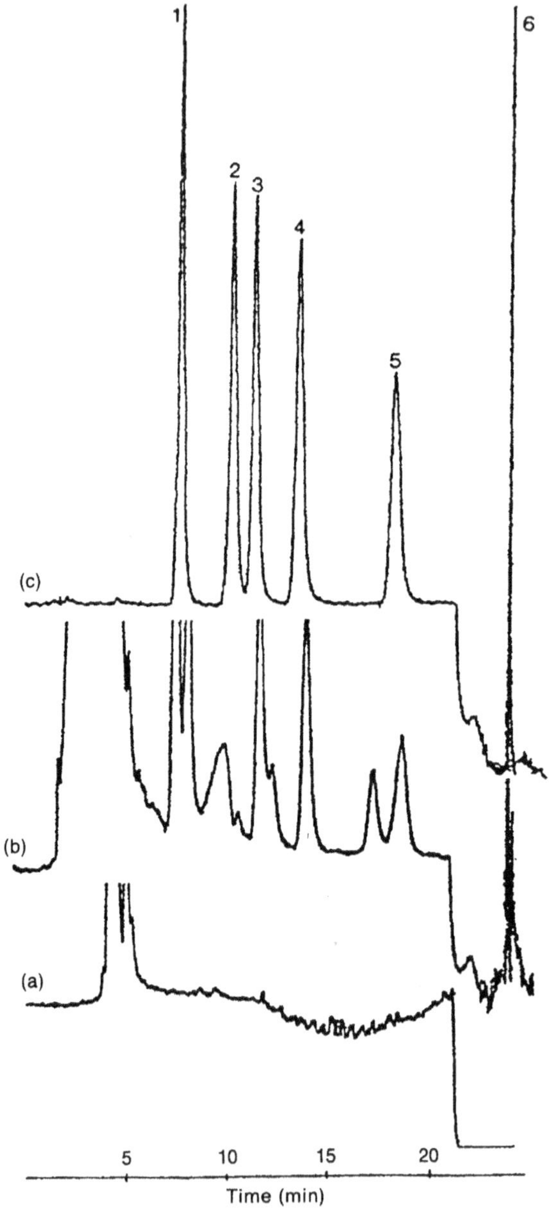


**Figure 8.15** High-performance liquid chromatography (HPLC) chromatograms obtained for the supercritical fluid extraction (SFE) extracts of liver samples after an alumina cleanup stage: (a) blank liver; (b) liver spiked with sulphonamide standards; (c) sulphonamide standards. Peaks: 1 = sulphamerazine (SMR); 2 = sulphamethizole (SMI); 3 = sulphametazine (SMZ); 4 = sulphamethoxypyridazine (SMOP); 5 = sulphamethoxazole (SMOZ); 6 = $N^4$-acetylsulphamethoxazole (N4). Reproduced with permission from ref. 32.

**Table 8.10** Supercritical fluid extraction (SFE) recoveries of sulphonamides from fortified chicken tissues ($n = 5$). Source: ref. 33

| Tissue | SPE | Mean ± standard deviation (%) | | |
|---|---|---|---|---|
| | | Sulphamethazine | Sulphadimethoxine | Sulphaquinoxaline |
| Liver | OL | 60.4 ± 3.3 | 69.3 ± 7.0 | 54.1 ± 2.5 |
| | IL | 89.9 ± 2.3 | 96.9 ± 1.2 | 76.4 ± 3.0 |
| Breast | OL | 66.7 ± 4.1 | 79.3 ± 5.6 | 64.4 ± 5.8 |
| | IL | 86.4 ± 2.5 | 92.0 ± 2.3 | 75.2 ± 4.1 |
| Thigh | OL | 74.9 ± 5.1 | 83.9 ± 9.5 | 72.5 ± 7.7 |
| | IL | 90.8 ± 3.8 | 94.8 ± 2.6 | 79.9 ± 3.2 |

Note: OL = 2 g off-line alumina trap; IL = 2 g in-line alumina trap.

sulphaquinoxaline, 97.6% ± 2.5%. Owing to the nature of the cleanup, the in-line method provided a much cleaner chromatogram with less interference from co-extractives, as shown in Figure 8.17.

Parks *et al.* [35] have also studied the effect of matrix dehydration using SFE for the recovery of drug residues in chicken liver. The drug studied was 3,5-dinitro-*O*-toluamide and its metabolites, extracted with water-modified supercritical carbon dioxide at 14 993 psi at 60°C, consuming 31 of expanded gas. In contrast to Din *et al.*'s study [30], it was demonstrated that the addition of water to a mixture of tissue/$Na_2SO_4$ improved SFE recoveries by 15%–20%. Recoveries between 66% and 92% were achieved for the drug and its three metabolites. Analysis was performed using reverse phase chromatography (Supelcosil™ LC-18, 5 μm, 250 mm × 4.6 mm i.d. column eluted isocratically with 0.05 M phosphate buffer (pH 7.0): methanol (70:30 vol/vol)); unfortunately, no example chromatograms were provided.

In a further study involving the HPLC analysis of veterinary drugs, Din and co-workers [36] investigated the use of SFE for the analysis of sulphamethazine (SMZ) and five of its metabolites – $N^4$-acetylsulphamethazine (AC), desaminosulphamethazine (DES), $N^4$-glucuronylsulphamethazine (GR) and $N^4$-sulphatosulphamethazine ($SO_4$) – from swine liver and kidney. The highlight of this work was the use of *in situ* ion pairing with tetramethylammonium (TMA) hydrochloride in conjunction with methanol as modifier to improve extraction efficiency. The SFE procedure involved the use of an initial static extraction for 30 min at 5800 psi followed by a dynamic extraction, again at 5800 psi (a series of experiments were carried varying temperature in the range 40°C–140°C and with methanol concentrations in the range 0–0.5 mol%). This approach increased recoveries of the ionic metabolites by up to 70%. An additional liquid/liquid extraction cleanup step was found to be necessary prior to the assay procedure. The result of this cleanup meant that the remaining co-extractives were well resolved from the target analytes. As shown in Figures 8.18(a) and 8.18(b), the chromatograms obtained for the tissue extracts compared favourably with that obtained for a standard solution of drugs.

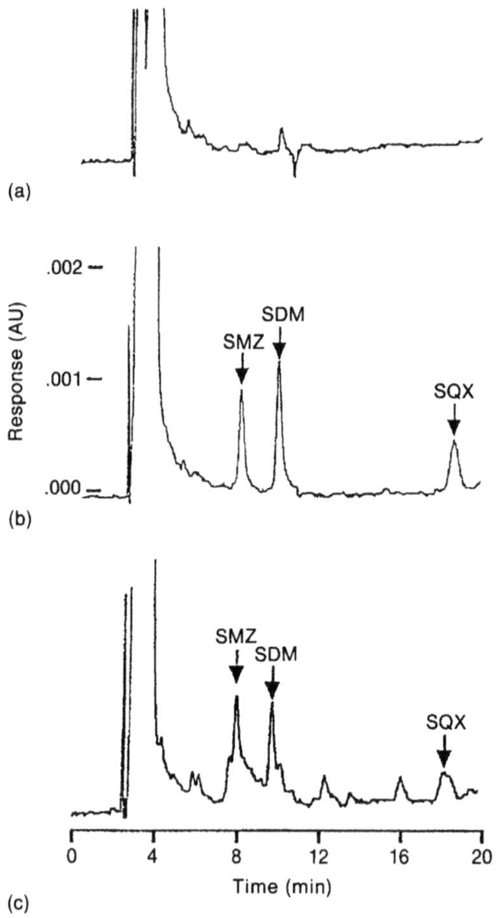


**Figure 8.16** High-performance liquid chromatography (HPLC) chromatograms obtained for the supercritical fluid extraction (SFE) extracts of liver samples with additional solid phase extraction (SPE) cleanup: (a) blank liver; (b) in-line SPE cleanup of liver sample spiked with 100 ppb sulphonamides; (c) off-line SPE cleanup of liver sample spiked with 100 ppb sulphonamides. SMZ = sulphamethazine; SDM = sulphadimethoxine; SQX = sulphaquinoxaline. Reproduced with permission from ref. 33.

Hedrick and Taylor [37] have used SFE for the direct extraction of a range of nitrogenous bases, including sulphamethazine from a range of aqueous samples. This study provides an interesting contrast between relatively simple aqueous-based samples and much more complex matrices such as animal tissues. The reverse phase HPLC analysis of sulphamethazine extracted from whole milk at the 60 ppm level provided good resolution from a low quantity of co-extractives. The sulphamethazine was collected on a diol silica trap at −20°C and recovered using 3 ml of methanol: water (1:10) prior to final HPLC analysis. No salt or base was added to the milk

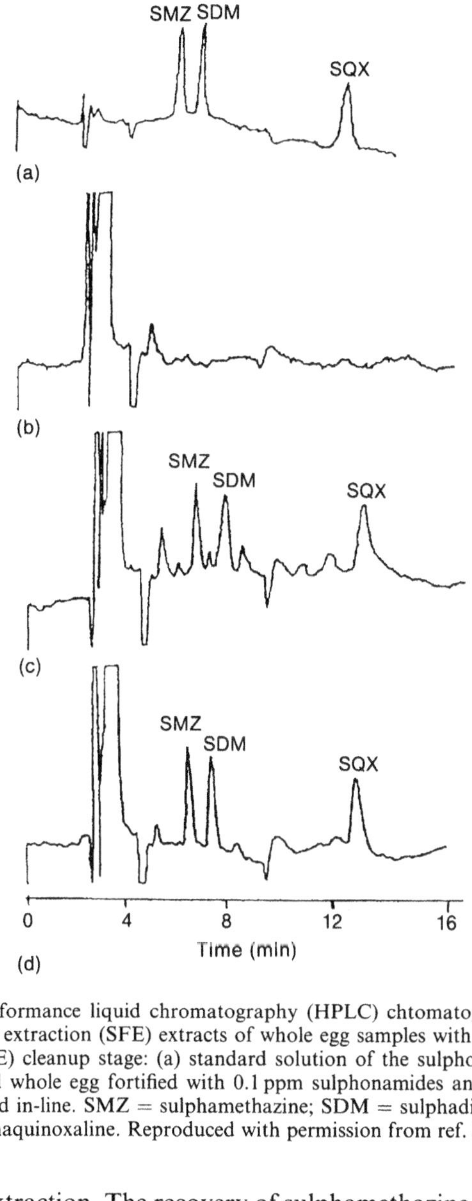


**Figure 8.17** High-performance liquid chromatography (HPLC) chtomatograms obtained for the supercritical fluid extraction (SFE) extracts of whole egg samples with an additional solid phase extraction (SPE) cleanup stage: (a) standard solution of the sulphonamides; (b) blank egg extract; (c) liquid whole egg fortified with 0.1 ppm sulphonamides and collected off-line; (d) as (c) but collected in-line. SMZ = sulphamethazine; SDM = sulphadimethoxine; SQX = sulphaquinoxaline. Reproduced with permission from ref. 34.

sample prior to extraction. The recovery of sulphamethazine was reported to be greater than 95%.

A wide variety of foodstuffs other than meat products have also been analysed using SFE for sample preparation. Verschuere *et al.* [6] evaluated supercritical fluids for the enrichment and fractionation of the essential oil and bitter flavouring components of hops. For sample analysis the authors

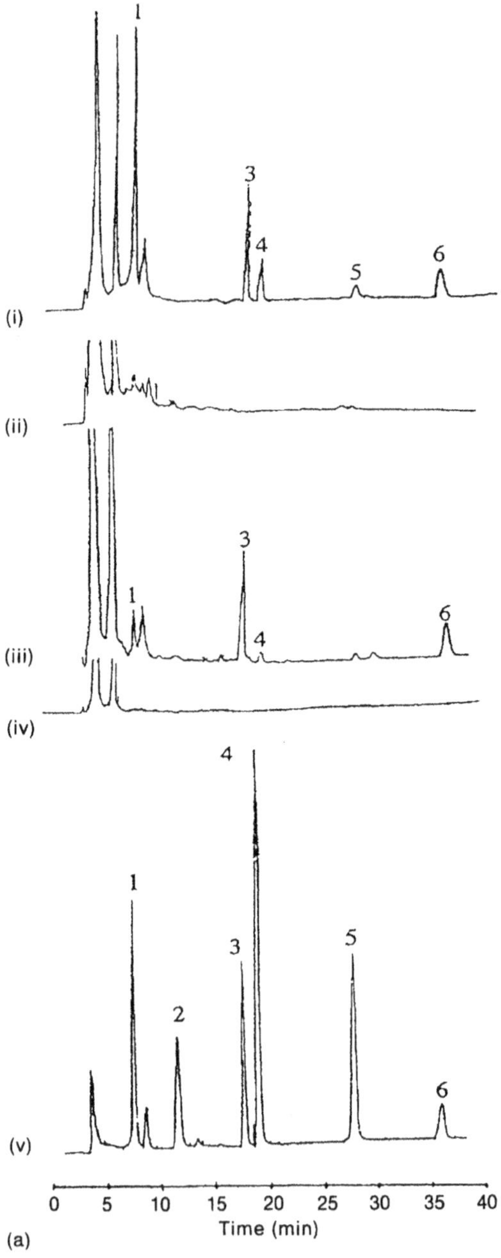


**Figure 8.18** High-performance liquid chromatography (HPLC) chromatograms obtained for (a) swine liver supercritical fluid extraction (SFE) extracts: (i) spiked wet liver extract (after cleanup); (ii) wet liver, blank extract; (iii) spiked freeze-dried liver extract; (iv) freeze-dried liver, blank extract; (v) standard; (b) swine kidney SFE extracts: (i) spiked wet kidney extract;

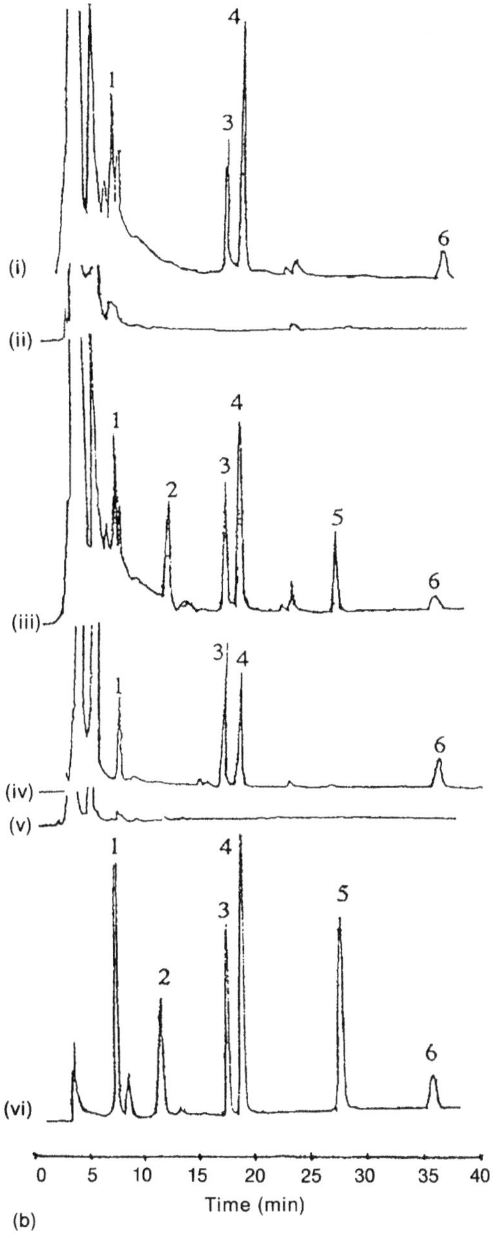


(ii) wet kidney, blank extract; (iii) spiked freeze-dried kidney extract (tetramethylammonium hydrochloride); (iv) spiked freeze-dried kidney extract; (v) freeze-dried kidney, blank extract; (vi) standard. Peaks: 1 = $N^4$-glucosylsulphamethazine; 2 = $N^4$-glucuronylsulphamethazine; 3 = $N^4$-acetylsulphamethazine; 4 = sulphamethazine; 5 = $N^4$-sulphatosulphamethazine; 6 = desaminosulphamethazine. Reproduced with permission from reference 36.

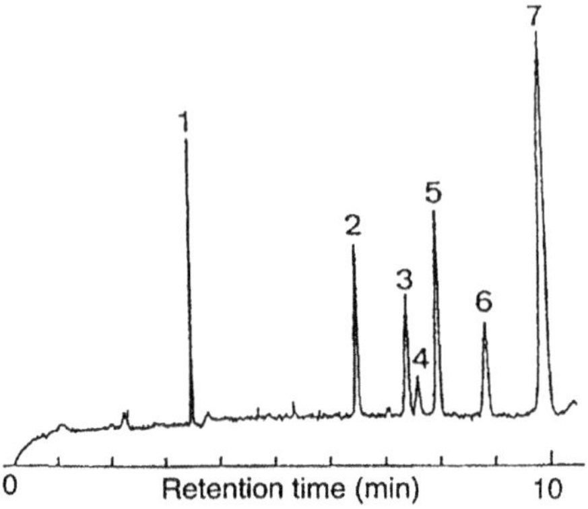


**Figure 8.19** Micellar electrokinetic chromatography analysis of $\alpha$-acids and $\beta$-acids extracted from hops using supercritical fluid extraction. Peaks: 1 = formamide; 2 = co-lupulone; 3 = lupulone; 4 = ad-lupulone; 5 = co-humulone; 6 = ad-humulone; 7 = humulone. Reproduced with permission from ref. 6.

used sequential extractions of 15 min and 30 min, increasing the supercritical carbon dioxide density from 0.05 to 0.1 g/ml at a temperature of 50°C. The essence of this work was the use of MEKC, microcolumn HPLC and GC to differentiate and characterise the analytes. Figure 8.19 is an example of a chromatogram obtained for an SFE extract analysed using MEKC. The chromatogram shows good separation of $\alpha$-acids and $\beta$-acids, with the presence of few co-extractives. In this work they exploit the advantages of off-line analysis, which allows the analyst to choose the most appropriate assay technique for the analyte of interest. Consistent with the findings of other workers, it was reported that the use of modified supercritical fluid carbon dioxide provided less selectivity, resulting in the hop SFE extracts becoming more coloured as a result of increased levels of co-extracted chlorophylls. Metal ions which tend to form complexes with iso-$\alpha$-acids are a problem with HPLC methods for hop analysis. MEKC eliminates any system metal ion problems experienced by traditional HPLC whilst maintaining resolution with a sixfold reduction in analysis time.

Engelhardt and Haas in their SFE study concerning the analysis of aflatoxin B1 [38] opted to use pressure as a means of selectively extracting the target analyte from the lipid fractions found in peanuts. Although methanol-modified supercritical fluid carbon dioxide made the aflatoxin more amenable to SFE, reduction in selectivity resulted in higher levels of interfering co-extractives for HPLC analysis. [The extracts were treated with trifluoroacetic acid (1:1 vol/vol) to convert aflatoxin B1 to aflatoxin B2a, as aflatoxin B2a is more sensitive to fluorescence detection than is aflatoxin B1.] These findings are illustrated by the chromatograms shown in Figure 8.20. The HPLC assay utilised a Kromasil® RP18 5 μm, 250 mm × 4.6 mm i.d. column eluted with methanol/water/acetonitrile

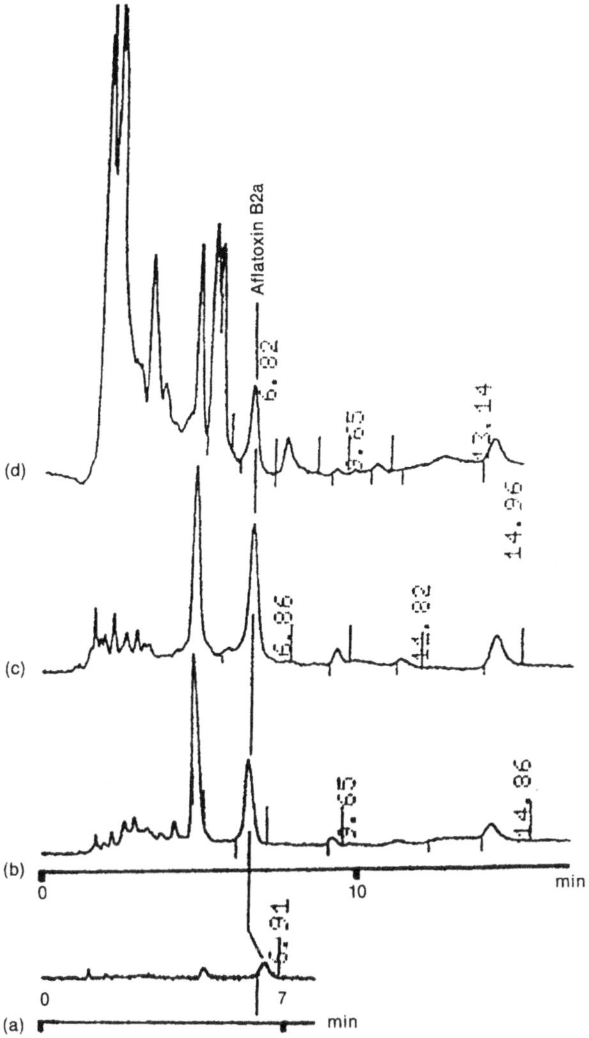


**Figure 8.20** High-performance liquid chromatography chromatograms showing the selectivity of supercritical fluid extraction for the recovery of aflatoxin B1 from peanuts, with increasing modifier concentration using: (a) neat carbon dioxide; (b) 2.9% methanol-modified carbon dioxide; (c) 5.6% methanol-modified carbon dioxide; (d) 30% methanol-modified carbon dioxide. Reproduced with permission from ref. 38.

(16:14:70 wt/wt/wt). These SFE studies were performed at 40°C using a pressure programme (2900 psi for 25 min, linear to 6525 psi in 2.5 min then held at 6525 psi for 60 min) to extract a 1 g sample of peanut meal (43.2 μg/kg aflatoxin B1 content determined by the classical extraction procedure). Aflatoxin B1 recoveries for four different reference samples were in the range of 5.3%–14.6%, with RSDs of 33.1%–79.2% for six determinations. Although solvent extraction data were not presented the authors conclude that SFE is not the ideal method for this particular application.

In contrast to the previous two studies, Yao *et al.* [39] were able successfully to exploit the use of 10% methanol-modified supercritical fluid carbon dioxide for the determination of total flavanoid content in ginko biloba leaves. Recoveries were reported at approximately 87%, with an RSD of 5.4% and detection limits of 28 ppb. No interference due to co-extractives was reported as being present when the SFE extracts were analysed using reverse phase HPLC (no chromatograms were published). The general conclusion that arises from this comparison is that it is difficult to define a prescriptive set of rules for SFE method development. Depending upon specific analyte/matrix combinations it may be difficult to isolate the analytes in high yield in clean SFE extracts, especially when modifiers are used.

Fermentation is a means of producing a wide range of products, from radiolabelled amino acids and carbohydrates to drugs and foodstuffs. Cocks and co-workers [40] investigated the merits of SFE and liquid extraction in combination with HPLC to fractionate and purify microbial fermentation products. Their procedure employed a dual SFE method followed by a dual solvent extraction procedure on the remaining biomass. Each microbial fermentation biomass sample (isolated by centrifugation and ground prior to SFE) was extracted for 30 min using 20% vol/vol methanol-modified supercritical fluid carbon dioxide. The biomass samples were then sequentially extracted with dichloromethane and then methanol. A reverse phase HPLC system utilising a Nova-Pak® ODS radial compression cartridge (100 mm × 8 mm i.d.) was used to assay the analytes. All of the extracts were analysed using a linear water–acetonitrile gradient. The results obtained for the extraction of *Chaetoglobosin A* ($C_{23}H_{26}N_2O_5$, $M_r = 528$) from a biomass of *Penicillium expansum* are shown in Figure 8.21. The majority of *Chaetoglobosin* was present in the modified SFE extract [Figure 8.21(b)] with small quantities in the unmodified SFE extract [Figure 8.21(a)]. Trace quantities remained after SFE being observed in the dichloromethane extract [Figure 8.21(c)]. The final methanol extract [Figure 8.21(d)] contained very few metabolites. Similar results were obtained for the following biologically active compounds extracted from other fermentation biomasses: *mycolutein* ($C_{22}H_{23}NO_6$, $M_r = 397$), *luteoreticulin* ($C_{19}H_{19}NO_5$, $M_r = 341$), 7,8-dihydro-7,8-epoxy-1-hydroxy-3-hydroxymethylxanthone-8-carboxylic acid methyl ester ($C_{16}H_{12}O_7$, $M_r = 316$), *sydowinin B* ($C_{16}H_{12}O_7$, $M_r = 316$) and *elaiophylin* ($C_{54}H_{88}O_{18}$, $M_r = 1025$).

Gawdzik *et al.* [41] have compared SFE and Soxhlet extraction for the isolation of furanocoumarins from the fruits of *Archangelica off.* Hoffm. These compounds are commonly found in the higher plants from the Umbelliferae and Rutaceae families and are used in clinical treatment of skin lesions. The HPLC assay was carried out using an isocratic reverse phase system using a Bakerbond® 5 μm ODS 250 mm × 1 mm column with

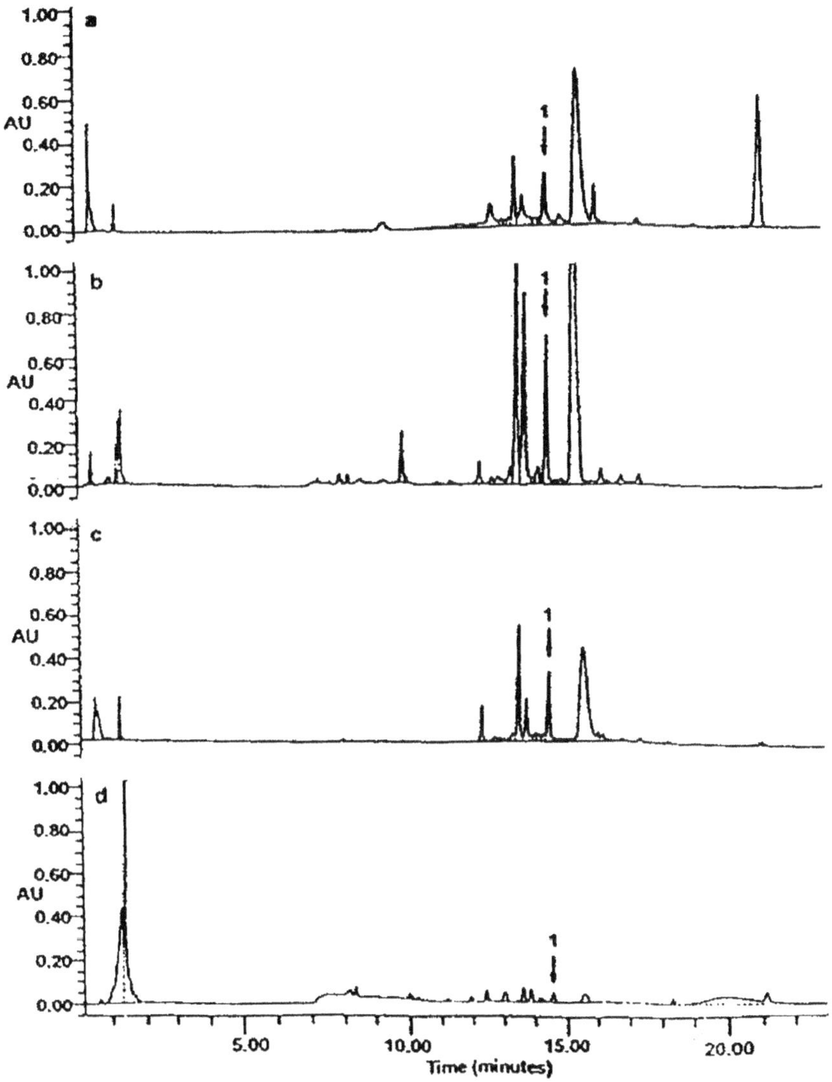


**Figure 8.21** High-performance liquid chromatography chromatograms obtained for the analysis of *Chaetoglobosin A* (peak 1) isolated from *Penicillium expansum* biomass by supercritical fluid extraction (SFE) using (a) neat carbon dioxide; (b) methanol-modified carbon dioxide; (c) dichloromethane extraction; (d) methanol extraction (solvent extractions sequentially performed on the biomass after SFE). Reproduced with kind permission from reference 40.

methanol:water (70:30) at 50 µl/min. The results are shown in Figure 8.22. SFE proved to be more selective than did Soxhlet extraction since analysis of its extract [Figure 8.22(b)] indicated the presence of fewer co-extractives. Relative to Soxhlet extraction (24 h with chloroform) fractionating SFE at

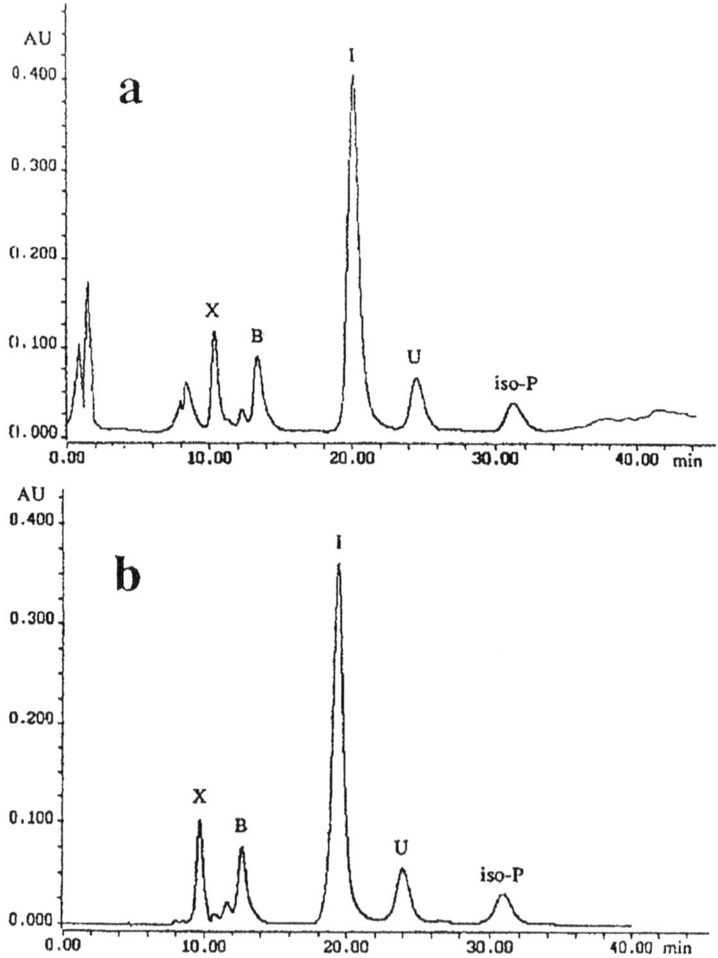


**Figure 8.22** High-performance liquid chromatography chromatograms obtained for furanocoumarins isolated from fruits of *Archangelica off.* Hoffm. using (a) Soxhlet extraction; (b) supercritical fluid extraction. Peaks: X = xanthotoxin; B = bergapten; I = imperatorin; P = pimpinellin; iso-P = isopimpinellin; U = unknown. Reproduced with permission from ref. 41.

constant density and temperatures of 40°C, 60°C, 80°C and 100°C yielded cumulative recoveries relative to those obtained by Soxhlet extraction of: xanthotoxin, 120%; bergapten, 78%; imperatorin, 95%. Recoveries for pimpinellin and isopimpinellin were not reported.

Working with similar types of matrices Marsili and Callahan [42] compared a conventional solvent extraction technique with SFE for the determination of $\alpha$-carotene and $\beta$-carotene in a variety of different vegetables. Extractions involved the use of ethanol-modified supercritical fluid

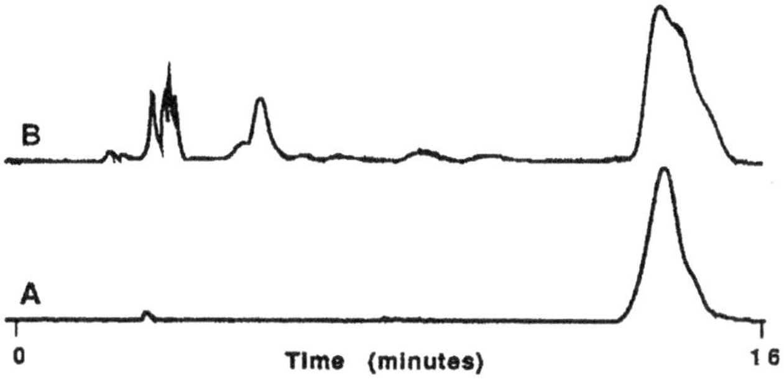


**Figure 8.23** High-performance liquid chromatography chromatograms showing the degradation and isomerisation of β-carotene in hexane used to rinse the ODS sorbent trap following supercritical fluid extraction: (A) elution with fresh hexane; (B) eluted with hexane after exposure to air and light for 24 h. Reproduced with permission from ref. 42.

carbon dioxide and pentane for solvent extraction. Analyses were performed using a non-aqueous HPLC assay. The SFE process took only one third of the time of the solvent extraction, with superior extraction efficiencies. Both extraction techniques provided clean fractions, although problems arose with the chromatography owing to the ability of the pump to portion the eluent and the instability of the carotene standards. Recovering the analytes from the ODS trap following SFE presented problems in that the solvent of choice for carotenes, dichloromethane, also induced analyte decomposition. As a consequence hexane was evaluated as an alternative trap rinse solvent and was found to extract the carotenes quantitatively without the problems associated with dichloromethane. The HPLC assay was carried out using a Vydac® 201TP 5 µm, 250 mm × 4.6 mm i.d. column eluted with a multisolvent mobile phase of methanol:acetonitrile: dichloromethane:hexane (65:27:4:4 vol/vol/vol/vol). However, caution must be exercised since hexane appears to be photosensitive and one of its degradation products may have caused the carotene decomposition shown in Figure 8.23.

Bicking [43] used off-line HPLC to determine the recovery of 3-chloro-*p*-toluidine hydrochloride (CPTH), an ionic compound which is highly soluble in water and polar organic solvents. This compound is an effective agent in the control of starlings (*Sturnus vulgaris*). The US Environmental Protection Agency requires the determination of avian dietary $LD_{50}$ values in quail and ducklings. This study involved the use of a factorial design procedure for optimisation of the SFE conditions used for the recovery of CPTH from spiked bird seed. The optimised SFE method utilised supercritical carbon dioxide–methanol (90:10) at a pressure of 4640 psi at 45°C for 60 min and was effective for the analysis of spiked samples at the 5 µg/g level. Above this level, as shown in Table 8.11, the SFE recoveries of CPTH fell, reflecting the limited solubility of this polar analyte in the extraction fluid. Although

**Table 8.11** Supercritical fluid extraction recoveries of 3-chloro-*p*-toluidine hydrochloride from spiked feed samples. Extraction conditions: 60 min at 45°C at 4640 psi with carbon dioxide:methanol (90:10). Source: ref. 43

| Spike level | Recovery (%) |
|---|---|
| 5 μg/g | |
| replicate 1 | 90.7 |
| replicate 2 | 102 |
| average | 96 |
| standard deviation | 8 |
| 10 μg/g | |
| replicate 1 | 84.3 |
| replicate 2 | 95.5 |
| average | 90 |
| standard deviation | 8 |
| 100 μg/g | |
| replicate 1 | 64.6 |
| replicate 2 | 75.0 |
| replicate 3 | 83.4 |
| average | 74 |
| standard deviation | 9 |

the example chromatogram (Figure 8.24) contains a considerable number of unidentified components, the analyte is well resolved and free from interfering co-extractives.

Messer and Taylor [44] have also developed an SFE procedure involving the use of HPLC for the analysis of a polar drug applied to rat feed. The optimised SFE method used supercritical carbon dioxide at a density of 0.9 g/ml and a temperature of 50°C for 30 min dynamic extraction. Blank samples were initially extracted to ensure that co-extractives would

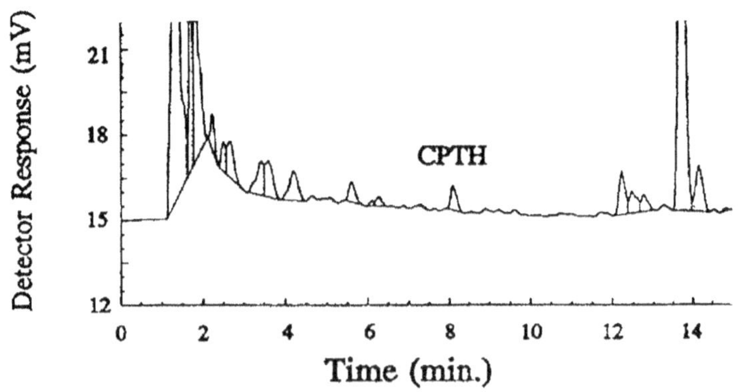


**Figure 8.24** High-performance liquid chromatography chromatogram obtained for the supercritical fluid extraction extract of bird seed spiked with 3-chloro-*p*-toluidine hydrochloride (CPTH) at the 10 μg/g CPTH level. Reproduced with permission from ref. 43.

not complicate the final analysis. Extracted particulate matter was removed by filtration. The authors report good chromatography, although example chromatograms were not published. Experiments were performed to determine the best trap and elution solvent. The SFE efficiency of 4-trifluoromethyl-2-biphenyl carboxylic acid analyte, using a stainless-steel bead trap and acetonitrile as rinse solvent, was 95.3% with an RSD of 4.6% ($n = 3$).

Sauvage *et al.* [45] have reported the use of nitrous oxide for the SFE of pharmaceutical compounds from animal feed. This study involved the determination of a halogenated aromatic derivative of urea (HAU) and a halogenated aromatic phenoxy derivative of an aliphatic alkane (HAPA) in rodent and dog feed. The method development included studying the influence of fluid flow rate on SFE efficiencies. The drugs were individually assayed using different columns and mobile phases. The authors did not report any chromatographic problems, providing the results of analyses obtained for blank SFE extracts and standard solutions of drugs.

An extremely important area, subject to many and varied regulations, is that of pesticides and herbicides. Barnabas and co-workers [46] investigated the use of SFE to determine pesticides and herbicides from aqueous samples. The main feature of this study was the use of SFE selectivity to isolate each class of compounds for different assay procedures. An indirect aqueous SFE procedure, involving the use of a $C_{18}$ SPE disc, was used for sample preparation. The authors used neat supercritical carbon dioxide to extract the organochlorine pesticides (OCPs) and then used modified supercritical carbon dioxide to extract the herbicides. Recoveries in each case were *c.* 90% with *c.* 5% herbicide being co-extracted with the pesticides. Reverse phase HPLC was the method of choice for herbicide determination. All herbicide extracts were collected in the LC eluent since other trapping solvents were found to degrade the chromatographic process. The HPLC analysis obtained for the herbicide extract is shown in Figure 8.25. Figure 8.26 shows the successful analysis of the extracted OCPs using GC–MS.

Howard and Taylor [47] have also described an indirect aqueous SFE procedure for the analysis of two herbicides: sulphometron methyl and chlorsulphon. A reverse phase HPLC procedure was used to assay each herbicide. Full details of this study are presented in section 4.8.

Venema *et al.* [48] have used SFE for the isolation of caprolactam and its oligomers. The extraction was performed using methanol-modified supercritical carbon dioxide, with the analytes being collected on a $C_{18}$ trap. The analytes were eluted from the trap using methanol as the rinse solvent and were then analysed using non-aqueous reverse phase HPLC. As shown in Figure 8.27, SFE enabled the extraction of oligomers up to nine units. It was reported that for this matrix, sample grinding and particle size were important parameters affecting extraction efficiency.

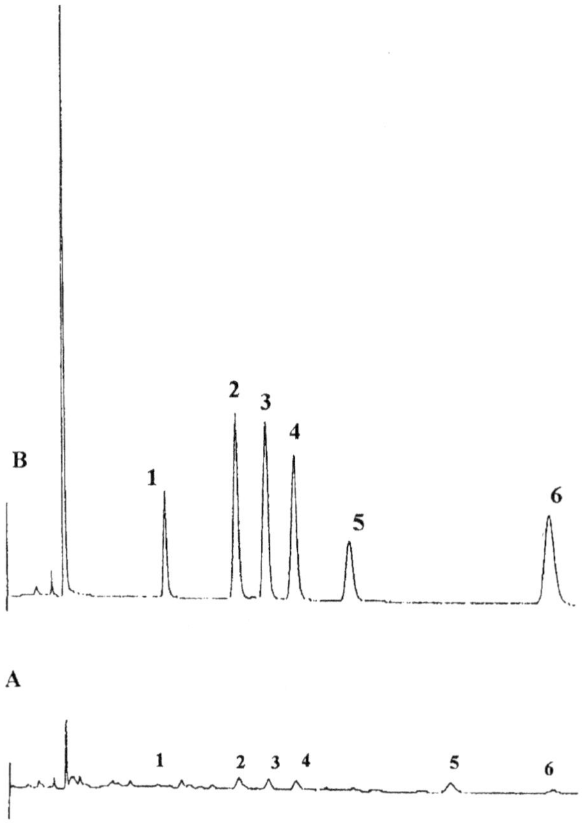


**Figure 8.25** High-performance liquid chromatography chromatograms obtained for herbicides recovered from aqueous solution using an indirect SFE procedure: (A) extract obtained using neat carbon dioxide; (B) extract obtained using methanol-modified carbon dioxide. Peaks: 1 = simazine; 2 = chloroluron; 3 = isoproturon; 4 = diuron; 5 = propazine; 6 = trietazine. Reproduced with permission from ref. 46.

## 8.4 On-line supercritical fluid extraction–high-performance liquid chromatography sample preparation techniques

In general on-line chromatographic analyses have proven more challenging than have those performed off-line, particularly for HPLC because of the reasons cited in the introduction to this chapter. Consequently, it is recommended that prior to any on-line SFE chromatography combination experience should be gained with the analyte(s) of interest using off-line SFE.

Cortes *et al.* [49] have described a method using SFE coupled with multi-dimensional microcolumn HPLC–GC. This study involved the determination

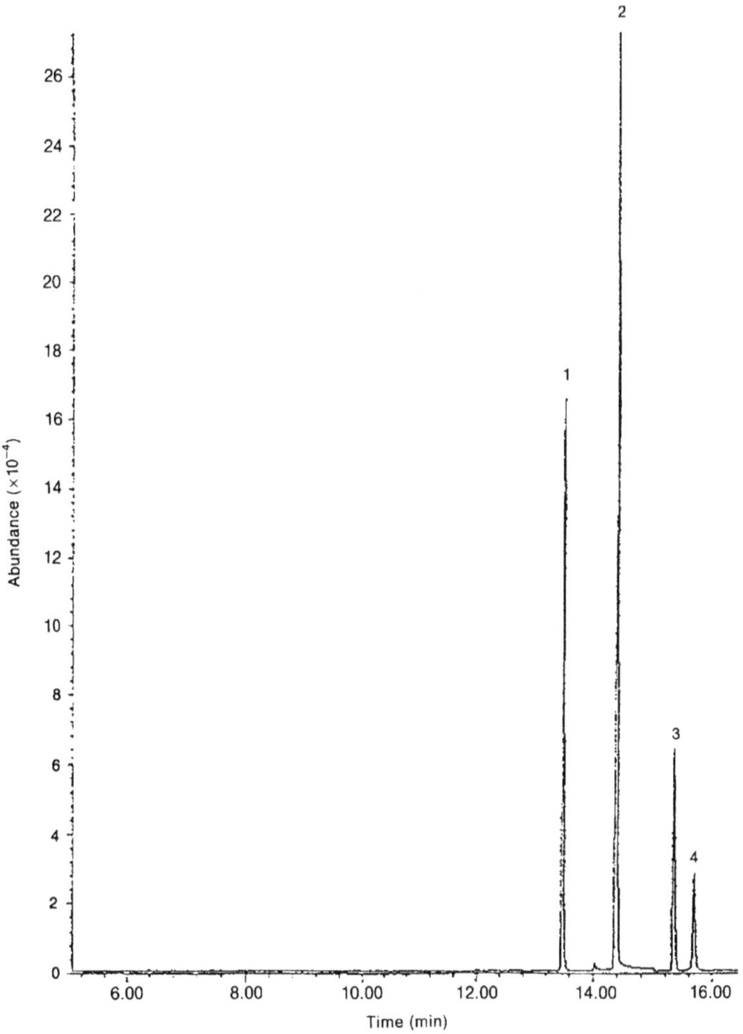


**Figure 8.26** Gas chromatography–mass spectrometry analysis of organochlorine pesticides recovered from aqueous solution using an indirect supercritical fluid extraction procedure (extract obtained using neat carbon dioxide). Peaks: 1 = heptachlor; 2 = isodrin; 3 = dieldrin; 4 = internal standard. Reproduced with permission from ref. 46.

of chlorpyrifos insecticide [*O,O*-diethyl-*O*-(3,5,6-trichloro-2-pyridinyl)phosphothioate] at the parts-per-billion level in grass samples, a challenging proposition in terms of both matrix and interfacing. This on-line system was found to yield a standard deviation of 10.8% on actual samples at a concentration of 160 ng/g. The micro LC was used to provide an on-line extract cleanup stage for final GC analysis. The authors used an impactor interface which acted

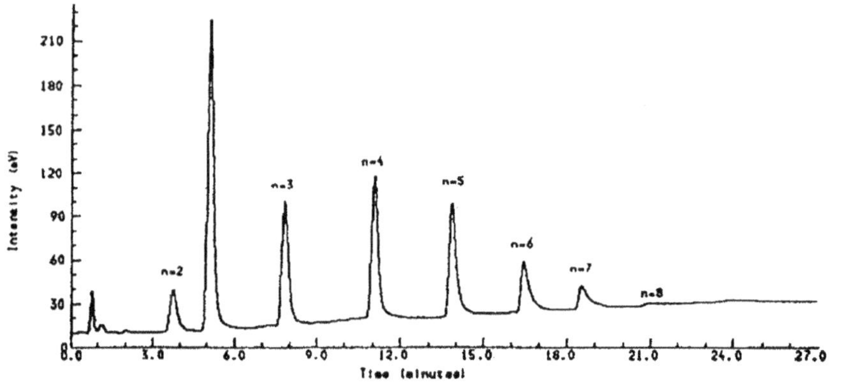


**Figure 8.27** Non-aqueous reverse phase high-performance liquid chromatography analysis obtained for caprolactam oligomers extracted from nylon 6 using supercritical fluid extraction. Caprolactam elutes at *c.* 5.2 min. Reproduced with permission from ref. 48.

as a solid phase trap (Figure 8.28). During SFE appropriate valve settings (not shown) were selected which isolate the impactor 'T' piece from the HPLC system, preventing back diffusion of carbon dioxide to the LC mobile phase.

The design of this interface was chosen since it concentrated the analytes and minimised band broadening when the predictions from the single-dimension compressible flow model are taken into consideration [50]. Sample sizes were orders of magnitude smaller for on-line studies; typically 5 mg compared with 0.5–1.0 g for off-line analyses. The authors compared the use of micro size exclusion chromatography (SEC) and micro reverse phase chromatography. The relative non-specificity of SEC introduced a significant number of impurities into the GC system, as shown in Figures 8.29(a) and 8.30(a). This resulted in poor GC quantitation.

The on-line reverse phase HPLC cleanup stage led to improved specificity, as shown in Figures 8.29(b) and 8.30(b). The result of on-line GC

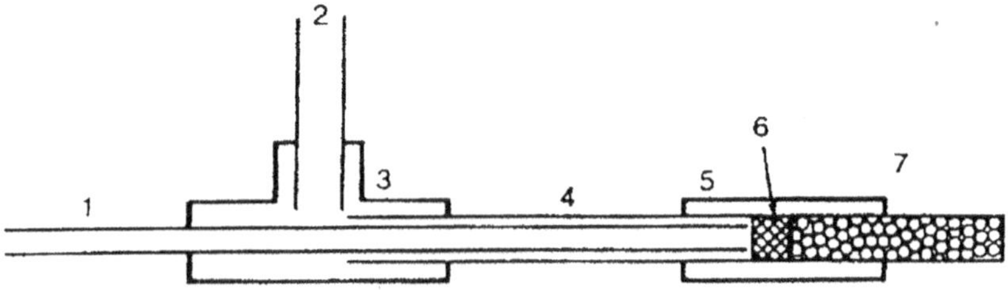


**Figure 8.28** Supercritical fluid extraction–high-performance liquid chromatography impactor interface. Components: 1 = linear restrictor; 2 = liquid chromatography (LC) eluent inlet/carbon dioxide vent; 3 = low dead volume 'T' piece; 4 = impactor tube; 5 = low dead volume union; 6 = impactor; 7 = micro LC column. Reproduced with permission from ref. 49.

quantitation using the reverse phase HPLC cleanup stage are given in Table 8.12.

The aim of the SFE–HPLC study performed by Johansen *et al.* [51] was to develop a simple, rapid, on-line procedure for the quantitative extraction and cleanup of mono-substituted, *ortho*-substituted, non-*ortho*-substituted polychlorinated biphenyls (PCBs) prior to GC/ECD or GC–MS analysis. Method development involved the use of three different biological matrices of different fat content: blood serum, human milk and crab hepatopancreas.

Each matrix was spiked with the following internal standards: PCB-159 for mono-*ortho*-substituted PCBs; $^{13}C$-labelled analogues of PCBs 77, 126 and 169 for non-*ortho*-substituted PCBs. The blood serum and human milk samples were loaded onto $C_{18}$ sorbent for SFE (an indirect liquid SFE procedure, see chapter 4) whereas the crab hepatopancreas was mixed with anhydrous sodium sulphate (1:4 m/m) prior to extraction. In order to prevent plugging of the $C_{18}$ sorbent pores, the milk sample was first treated with acetonitrile (1:1 vol/vol) to precipitate the proteins prior to sample loading. In order to retain extracted lipids that would otherwise interfere with the analyses (reported to reduce PCB normal phase HPLC retention times), activated alumina was loaded into the outlet end of the extraction vessel which was then connected to an alumina column. An impactor interface was used to trap the analytes. When the collector was packed with either silica or Florisil®, HPLC band broadening became significant. The analytes were therefore collected in the absence of a sorbent since Joule–Thomson cooling afforded a means of collecting the analytes as a concentrated band. In order to test the trapping system, six PCB congeners spiked onto glass fibre filter paper were analysed using on-line SFE–HPLC. Normal phase HPLC involved the use of two 150 mm × 4.6 mm i.d. Cosmosil® 5-PYE columns with hexane as mobile phase. Figure 8.31 shows the results of the SFE–HPLC analysis with the congener peaks being labelled. Compared with off-line direct injection, the SFE–HPLC procedure gave 7%–12% lower peak areas, with 12%–36% loss in the number of theoretical plates, depending upon the identity of the congener.

Since HPLC is not sufficiently sensitive for routine PCB studies, analyses are generally performed using either GC/ECD or GC–MS. In order to facilitate such studies the SFE–HPLC eluent was cut into fractions, based upon the retention times of PCB standards. As shown in Table 8.13 the SFE–HPLC procedure provided slightly higher recoveries of PCBs from crab hepatopancreas and human blood serum compared with a conventional solvent extraction procedure.

The average recoveries of the planar non-*ortho*-substituted PCB congeners 77, 126 and 169 from crab hepatopancreas were 71%–101%, with the highest recovery for PCB-169. For human blood serum and milk, recoveries of the three congeners ranged from 35% to 57% (serum) and 76% to 87% (milk), with the lowest recovery being for PCB-169. The RSDs for the

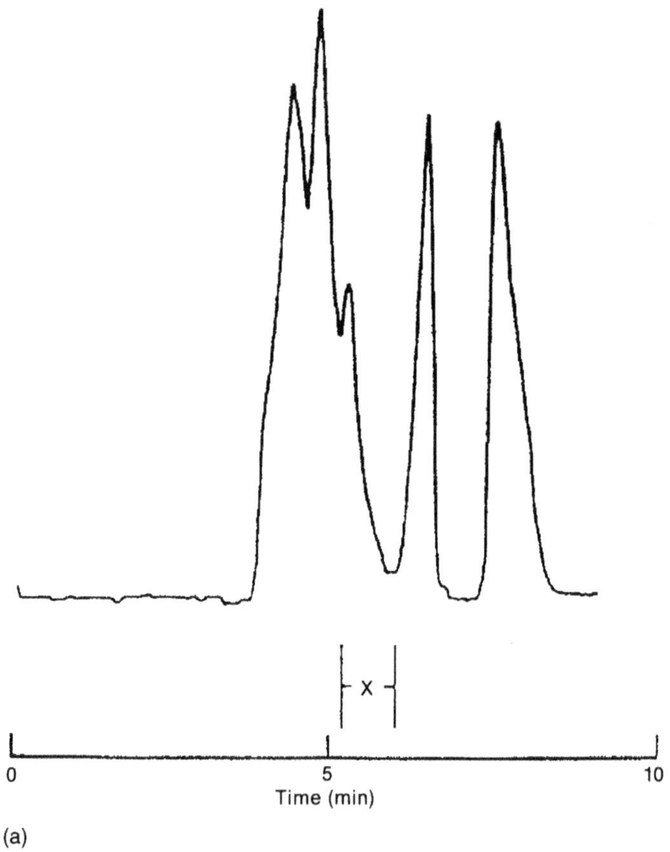


**Figure 8.29** (a) Micro size exclusion chromatography (SEC) chromatogram of grass extract obtained by off-line supercritical fluid extraction. Fraction X was analysed by gas chromatography (GC) for the determination of chlorpyrifos. SEC conditions: PL GEL™ 700 cm × 250 μm inner diameter (i.d.), 500 nm pore size using tetrahydrofuran at 3 μl/min. (b) Micro liquid chromatography chromatogram obtained from a grass sample using on-line supercritical fluid extraction–high-performance liquid chromatography (SFE–HPLC). Fraction X was analysed by GC for the determination of chlorpyrifos. HPLC conditions: Spherisorb® 8 cm × 250 μm i.d.; ODS column using methanol:water (80:20) at 4 μl/min. Reproduced with permission from ref. 49.

complete analyses were determined to be 5%–16%. The authors concluded that on-line SFE–HPLC was much faster than was conventional sample preparation procedures for PCB analysis since several off-line sample manipulations were eliminated.

A further example of an SFE–LC sample preparation procedure has been described by Ahmad [52] for the determination of pesticide residues in vegetable oils using off-line GC–MS. The technique involves the use of an HP 7680A SFE instrument, a diagram of which is shown in Figure 4.8.

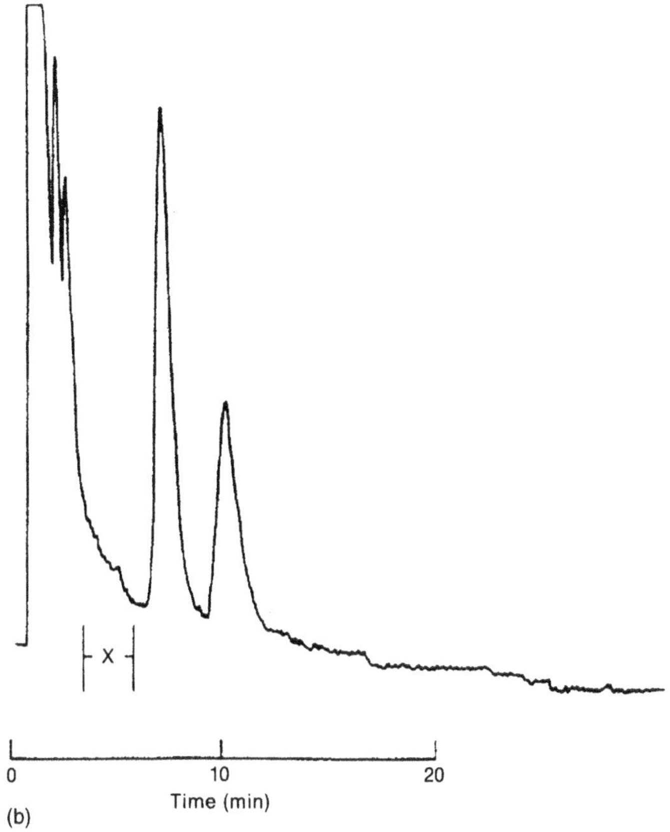


As reported by King [53], SFE does not provide sufficient selectivity for the direct isolation of a clean pesticide extract from vegetable oils. As a consequence, both Nam *et al.* [54] and Murugaverl *et al.* [55] have described the use of SFE–SFC–GC for the determination of pesticides in vegetable oils. Ahmad used an indirect liquid SFE procedure whereby spiked vegetable oil samples were loaded onto SPE cartridges which in turn were placed into the SFE vessel (see Figure 4.4). Using supercritical fluid carbon dioxide (based on a previous study, Ahmed [52] selected a density of 0.75 g/ml at 70°C to obtain a good compromise for pesticide extraction whilst limiting the extraction of vegetable oil), the vegetable oil retention of silica and ODS SPE cartridges were determined using an SFE off-line gravimetric procedure. As shown in Figure 8.32, the silica SPE cartridge provided higher vegetable oil retention than did the ODS cartridge. However, the ODS SPE cartridge was selected since it provided higher pesticide SFE recoveries, and the extracted oil (*c.* 47%) was effectively retained on the ODS trap using a suitable trap rinse cycle. The complete SFE–LC procedure

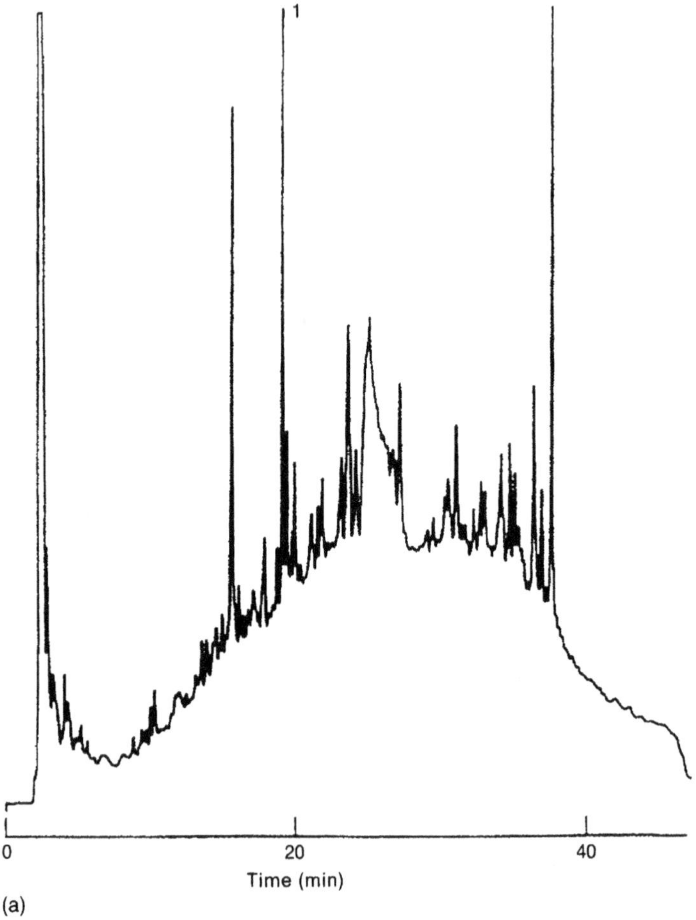


**Figure 8.30** (a) Chromatogram obtained for the gas chromatography (GC) analysis of fraction X [see Figure 8.29(a)]. (b) Chromatogram obtained for the GC analysis of SFE–HPLC fraction X [see Figure 8.29(b)]. Peak 1 = chlorpyrifos. Reproduced with permission from ref. 49.

is summarised in Figure 8.33, the acetonitrile fraction containing the pesticides and the dichloromethane fraction containing the extracted oil.

During the acetonitrile ODS trap rinse the extracted pesticides were collected into vials containing [$^{13}C_6$]-hexachlorocyclohexane which acted as internal standard for quantitative selected ion monitoring (SIM) GC–MS analysis. With the high level of automation the complete SFE–LC sample preparation time was approximately 15 min. An example of a SIM GC–MS analysis obtained for seven organophosphate pesticides (OPPs) recovered from spiked vegetable oil is shown in Figure 8.34.

In general the SFE–LC procedure was reported to provide higher recoveries for OCPs than it does for OPPs. The presence of retentive silanol

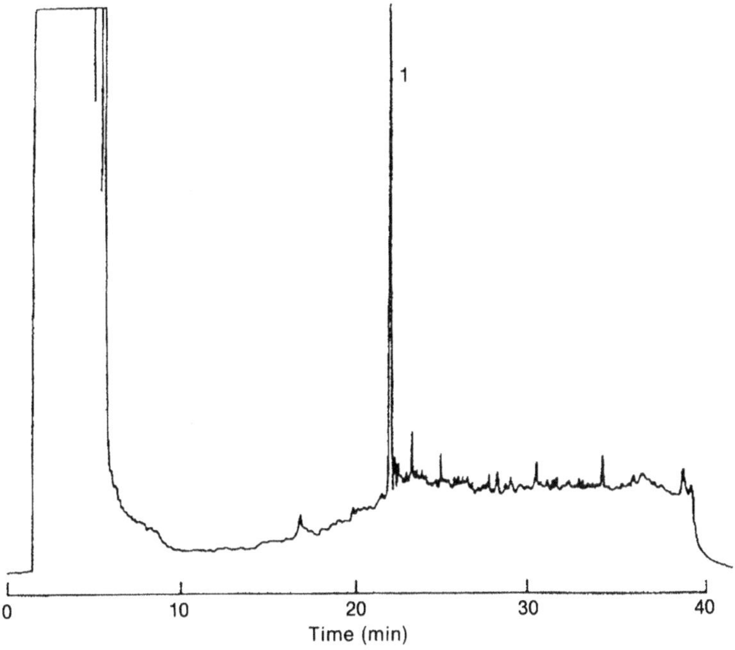


(b)

groups on the SPE ODS sorbent resulted in poor SFE recoveries for the more polar pesticides such as diazinon.

Ahmad *et al.* [56] used the SFE–LC procedure described above for the GC–MS analysis of pesticide residues in authentic vegetable oil and seed samples. An example of these results is shown in Table 8.14.

**Table 8.12** Recovery data for the determination of chlorpyrifos insecticide in grass. Source: ref. 49

| Sample weight (mg) | Chlorpyrifos recovered (ng/g) | Recovery (%)[a] |
|---|---|---|
| 5.08 | 104 | 87.5 |
| 5.09 | 168 | 105 |
| 5.28 | 158 | 98.7 |
| 5.31 | 161 | 101 |
| 4.70 | 188 | 118 |
| 5.01 | 150 | 93.8 |
| 5.71 | 142 | 88.8 |
| 4.64 | 140 | 87.5 |
| mean | | 97.5 |
| standard deviation | | 10.5 |
| relative standard deviation (%) | | 10.8 |

[a] Recoveries were calculated based on 160 ng/g concentrations determined by both off-line supercritical fluid extraction and conventional sample pre-treatment procedures.

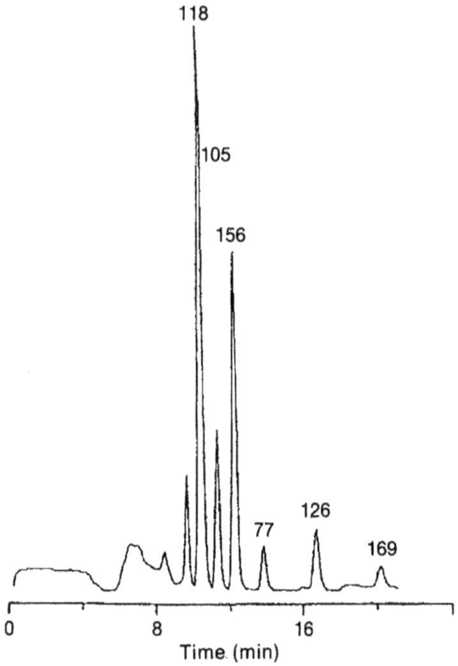


**Figure 8.31** On-line supercritical fluid extraction–high-performance liquid chromatography chromatogram obtained for six polychlorinated biphenyl congeners extracted from filter paper. Identities: 105 = (2,3,3′,4,4′); 118 = (2,3′,4,4′,5); 77 = (3,3′,4,4′); 126 = (3,3′,4,4′,5); 156 = (2,3,3′,4,4′,5); 169 = (3,3′,4,4′,5,5′). Reproduced with permission from reference 51.

**Table 8.13** Quantitation of mono-substituted and non-*ortho*-substituted polychlorinated biphenyls (PCBs) in crab heptopancreas and human blood serum: supercritical fluid extraction (SFE) compared with solvent extraction. Source: ref. 51

| PCB congener | Crab heptopancreas (ng/g) | | Human blood serum (ng/g) | |
|---|---|---|---|---|
| | SFE | Solvent extraction | SFE | Solvent extraction |
| 105[a] | 6.1 | 5.2 | 0.3 | 0.2 |
| 118[a] | 17.2 | 19.0 | 1.1 | 0.8 |
| 156[a] | nd | nd | 0.7 | 0.5 |
| 77[b] | $179.1 \times 10^{-3}$ | $189.8 \times 10^{-3}$ | $3.9 \times 10^{-3}$ | $2.8 \times 10^{-3}$ |
| 126[b] | $50.1 \times 10^{-3}$ | $54.7 \times 10^{-3}$ | $7.4 \times 10^{-3}$ | $5.7 \times 10^{-3}$ |
| 169[b] | $16.3 \times 10^{-3}$ | $14.0 \times 10^{-3}$ | $1.8 \times 10^{-3}$ | $1.2 \times 10^{-3}$ |

nd = not determined.
[a] Determined by gas chromatography–electron capture detection.
[b] Determined by gas chromatography–mass spectrometry.

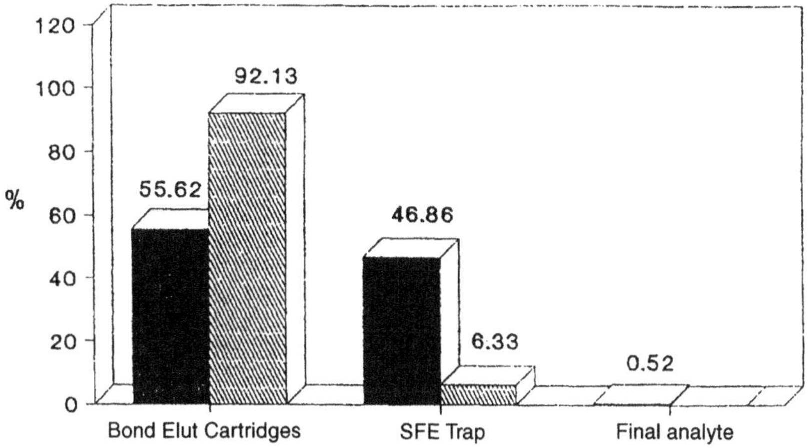


**Figure 8.32** Retention characteristics of vegetable oil retained on solid phase extraction cartridges and extracted oil retained on HP 7680 A supercritical fluid extraction sorbent traps. Solid shading = $C_{18}$ cartridge; hashed shading = silica cartridge. Reproduced from ref. 52.

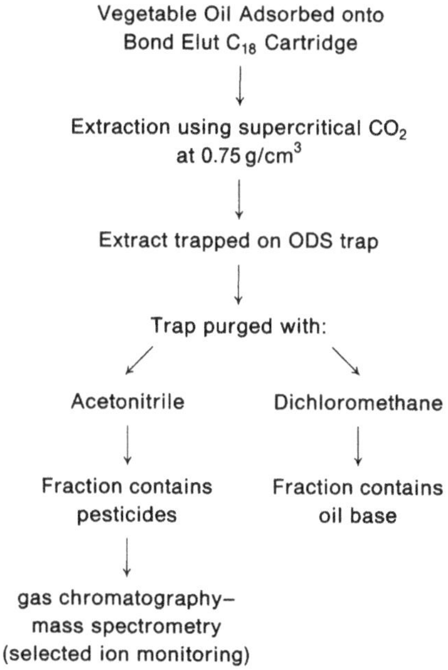


**Figure 8.33** Flow diagram showing the supercritical fluid extraction–liquid chromatography (SFE–LC) procedure for the extraction and cleanup of pesticide residues in vegetable oil using an HP 7680 A SFE instrument. Reproduced from ref. 52.

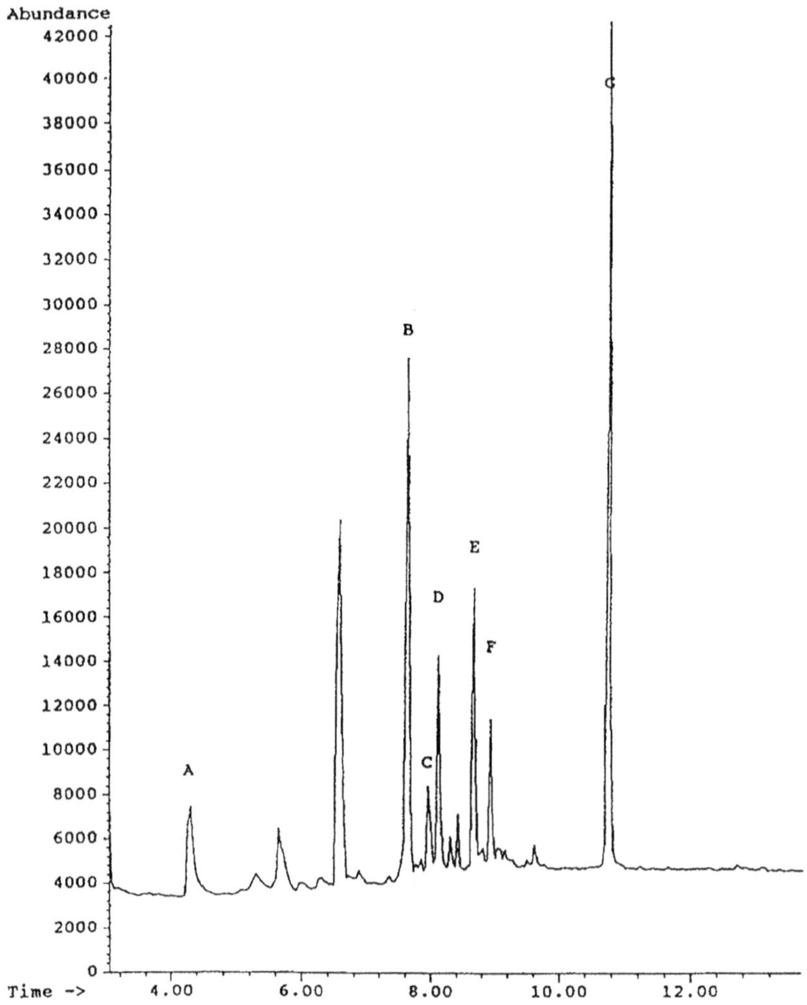


**Figure 8.34** Selected ion monitoring gas chromatography–mass spectrometry chromatogram obtained for organophosphate pesticides extracted by supercritical fluid extraction–liquid chromatography. Spiked vegetable oil was loaded onto a Bond Elut® ODS solid phase extraction cartridge prior to extraction and cleanup. Peaks: A = dichlorvos (DDVP); B = phorate; C = dimethoate; D = diazinon; E = methyl parathion; F = malathion; G = ethion. Reproduced with permission from ref. 52.

## 8.5 Selected analyses performed using on-line supercritical fluid extraction–high-performance liquid chromatography

Ischi and Haerdi [57] and Ashraf-Khorassani *et al.* [7] have described the use of SFE–HPLC for the analysis of spiked matrices serving as model samples.

**Table 8.14** Analysis of pesticide residues in various vegetable oils by supercritical fluid extraction–liquid chromatography and selected ion monitoring gas chromatography–mass spectrometry. Source: ref. 56

| Pesticide type | Crude cotton seed oil (ng/g) | Corn oil (ng/g) | Sunflower oil (ng/g) |
|---|---|---|---|
| *o,p*-DDT | 28.23 | 55.20 | – |
| *o,p*-DDD | 40.00 | 75.82 | – |
| *p,p*-DDD | – | 16.84 | – |
| *o,p*-DDE | – | – | 17.53 |
| α-HCH | – | 114.49 | – |
| β-HCH | – | 201.17 | – |
| Heptachlor | – | 72.16 | 72.79 |
| Methyl parathion | 165.07 | – | – |

Note: DDD = dichlorodiphenyldichloroethane; DDE = dichlorodiphenylethane; DDT = dichlorodiphenyltrichloroethane; HCH = hexachlorohexane.

Ischi and Haerdi [57] selected anthracene, naphthalene and benzoic acid as target analytes which they spiked onto silica and recovered using supercritical fluid carbon dioxide (4350 psi, 50°C, 30 min extraction with a flow rate of 1 ml/min). These investigations were directed towards the study of some key SFE experimental parameters, including restrictor performance and selection of appropriate solid adsorbent for analyte trapping. Two different on-line HPLC procedures were used for determining the extracted analytes. The first involved the use of an anion-exchange column, for analyte pre-concentration prior to reverse phase analysis being performed using a coupled ODS column. The second approach involved the elution from the ODS trap directly onto the ODS analytical column for reverse phase analysis. Both SFE–HPLC procedures provided good quality chromatography with respect to chromatographic peak shape, with no detector instability arising from carbon dioxide.

The main objective of the SFE–HPLC study performed by Ashraf-Khorassani *et al.* [7] was to investigate the effect carbon dioxide dissolved in the LC mobile phase has upon chromatographic performance and detector stability. These studies involved the extraction of two different classes of analytes spiked onto sand: PAHs (naphthalene, anthracene, fluorene and pyrene); five linear alkylbenzene sulphonates ($C_{10}$–$C_{14}$). As shown in Figures 8.35(a) and 8.35(b) good chromatographic performance was obtained for the two different classes of analytes. Extractions were performed using carbon dioxide at 4000 psi, 40°C.

Liu *et al.* [58] have described an SFE–HPLC system that enabled the analysis of chlorinated phenols. An ODS column (5 μm, 250 mm × 4.6 mm i.d.) was used for analyte trapping and HPLC analysis. In common with other workers, problems concerning restrictor plugging and detector stability were addressed. In contrast to other approaches, the restrictor was fitted to the outlet of a high-pressure UV/VIS cell such that carbon dioxide remained dissolved within the HPLC mobile phase. Figure 8.36 shows the

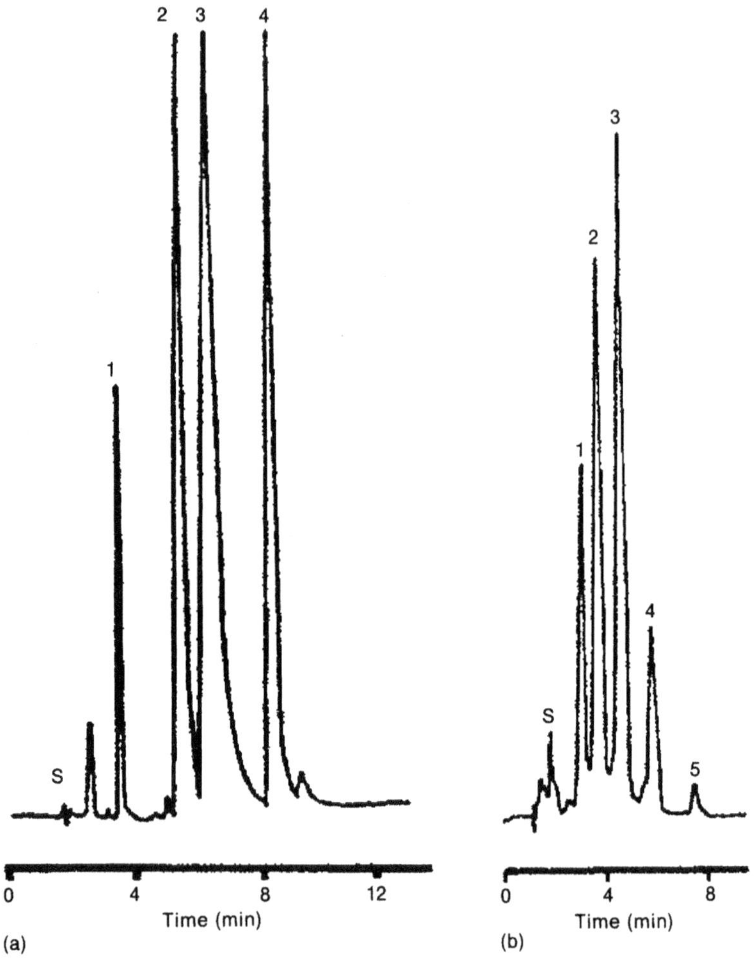


**Figure 8.35** (a) On-line supercritical fluid extraction–high-performance liquid chromatography (SFE–HPLC) analysis of polycyclic aromatic hydrocarbons spiked onto sand. HPLC conditions: ODS column using methanol:water (80:20) at 1 ml/min. Peaks: 1 = naphthalene; 2 = anthracene; 3 = fluorene; 4 = pyrene. S = internal standard. (b) On-line SFE–HPLC analysis of linear alkylbenzene sulphonates. HPLC conditions: ODS column using methanol:aqueous 0.1 M $NaClO_4$ (80:20) at 1 ml/min. Peaks: 1 = $C_{10}$; 2 = $C_{11}$; 3 = $C_{12}$; 4 = $C_{13}$; 5 = $C_{14}$; S = internal standard. Reproduced with permission from ref. 7.

erratic UV/VIS response without the outlet restrictor. The UV/VIS chromatogram obtained for the SFE–HPLC analysis of chlorinated phenols that had been spiked onto glass beads is shown in Figure 8.37 and serves to illustrate the effectiveness of the restrictor arrangement.

For the matrices investigated a linear response was recorded over the concentration range 1–500 ppm with analysis times of 1.5 h compared with

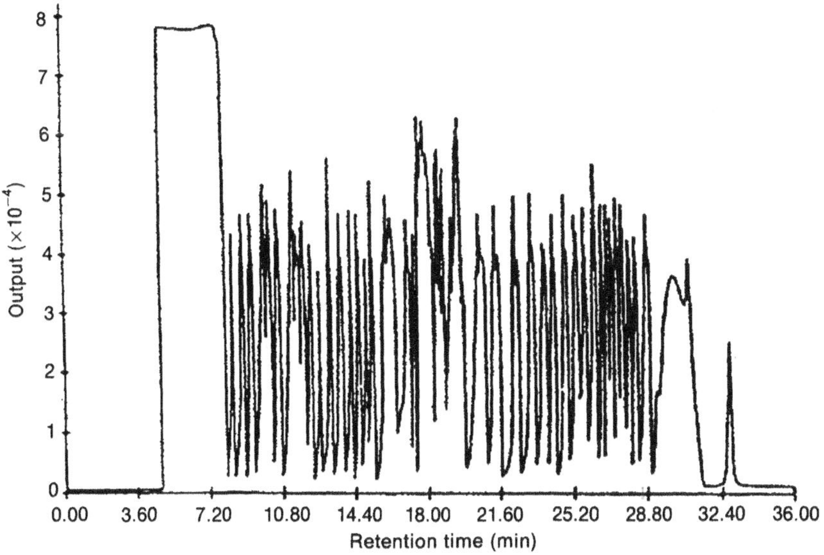


**Figure 8.36** Output of the ultraviolet/visible (UV/VIS) detector without an outlet restrictor. Reproduced with permission from ref. 58.

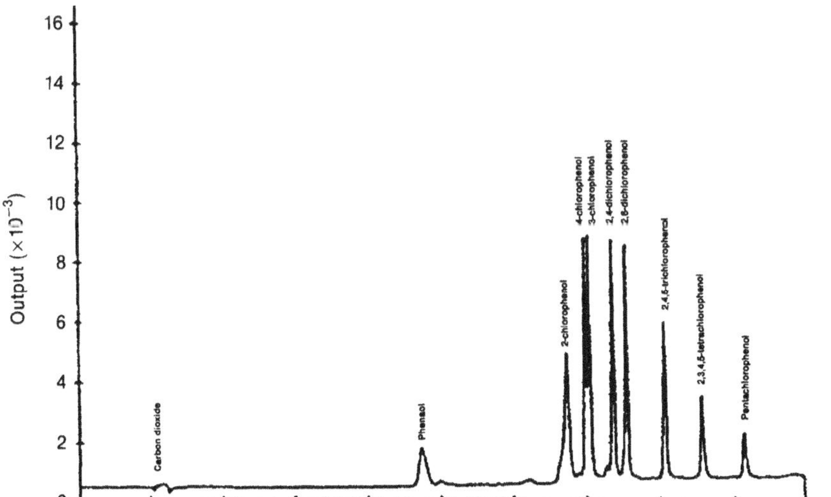


**Figure 8.37** On-line supercritical fluid extraction–high-performance liquid chromatography analysis of chlorinated phenols, extracted from glass beads, with a restrictor connected to the ultraviolet/visible (UV/VIS) detector outlet. Reproduced with permission from ref. 58.

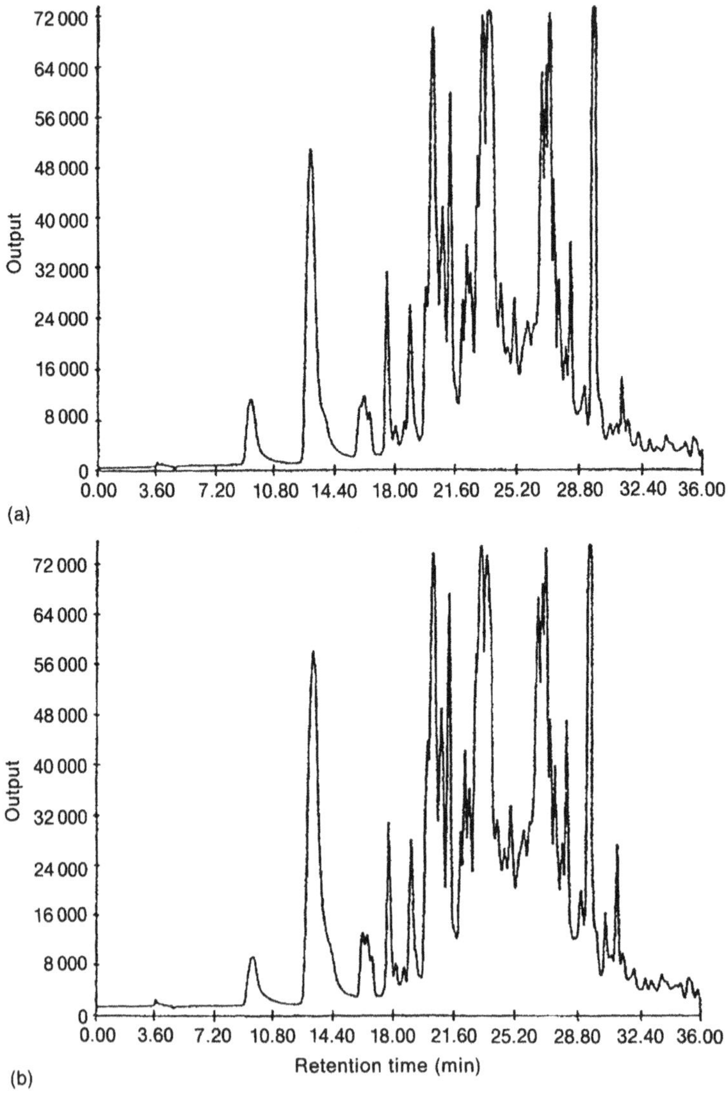


around 15 h for traditional methodologies. In order to evaluate the selectivity of the SFE–HPLC procedure, samples of phenol-spiked wood were analysed on-line and off-line following a traditional solvent extraction procedure. As shown in Figure 8.38(a) and 8.38(b), the off-line technique produced a complex extract such that the presence of phenols could not be readily established. In contrast, as shown in Figure 8.38(c), on-line SFE–HPLC provided a relatively clean extract such that phenols were readily detectable.

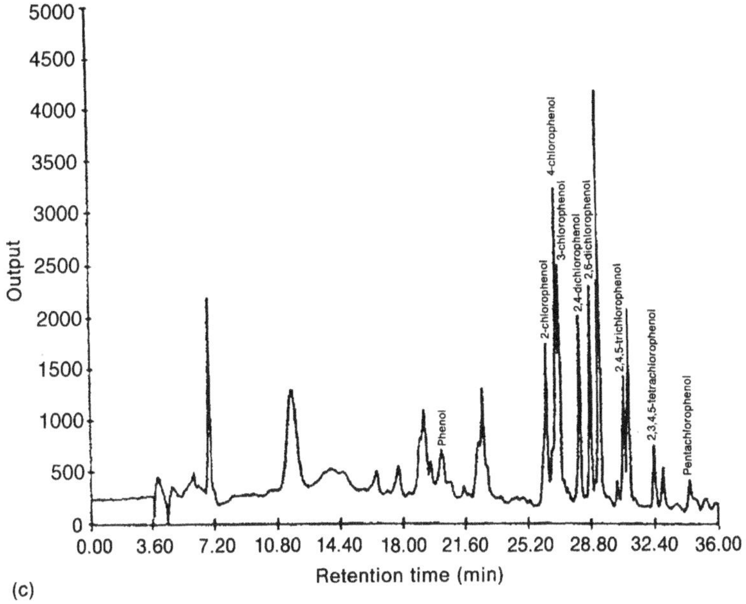


**Figure 8.38** (a) High-performance liquid chromatography (HPLC) chromatogram obtained for liquid solvent extract of wood shavings; (b) HPLC chromatogram obtained for liquid solvent extract of wood shavings spiked with chlorinated phenols; (c) on-line supercritical fluid extraction (SFE–HPLC) analysis of wood shavings spiked with chlorinated phenols. Reproduced with permission from ref. 58.

The SFE–HPLC recoveries of individual phenols obtained from spiked wood samples relative to spiked glass beads is shown in Table 8.15. Extraction efficiencies of less than 90% were obtained for the phenols with more than two chlorine atoms as a result of their increased polarity.

**Table 8.15** Extraction and determination of phenols from wood samples by supercritical fluid extraction–high-performance liquid chromatography. Source: ref. 58

| Components | Recovery (%)[a] | Detection limit (ppm for 15 g sample) |
|---|---|---|
| phenol | 106.4 | 2.5 |
| 2-chlorophenol | 93.3 | 1.0 |
| 4-chlorophenol | 94.6 | 1.0 |
| 3-chlorophenol | 96.1 | 1.0 |
| 2,4-dichlorophenol | 100.7 | 1.0 |
| 2,6-dichlorophenol | 101.1 | 1.0 |
| trichlorophenol | 85.2 | 1.0 |
| tetrachlorophenol | 85.0 | 2.5 |
| pentachlorophenol | 84.7 | 3.3 |

[a] Based on the reference standard spiked on the glass beads for SFE.

Stalling *et al.* [59] used on-line SFE–GPC in the analysis of a standard US Environmental Protection Agency (EPA) calibration mixture of corn oil. The author's system provided a facility to perform on-line and off-line SFE–GPC fractionation of the target analytes. In the on-line mode the extracted analytes were collected on an interface column with the spent carbon dioxide discharged to waste via a system of valves. After the extraction was completed the valves were switched and the analytes washed onto the Envirosep®-ABC column. The actual fractionation process was then monitored by UV. The EPA CLA calibration mixture chosen to test the efficacy of the SFE–GPC process contained corn oil (25 mg/ml), di(2-ethylhexyl)phthalate (DEHP, 1.0 mg/ml), methoxychlor (MOC, 0.2 mg/ml), perylene (PRL, 0.02 mg/ml) and sulphur (0.08 mg/ml). This was loaded onto Hydromatrix prior to on-line SFE–GPC (5000 psi, 45°C, followed by elution with dichloromethane). The recoveries of all analytes were reported to range from 86% to 107%.

Off-line SFE involving the analysis of soils and related matrices [17–22] has presented difficulties because of co-extractives. Mougin *et al.* [60] have developed an on-line SFE–HPLC procedure for the analysis of atrazine and its metabolites in soil. As is common in on-line studies the extraction/trapping and the analytical system are isolated from each other by a valve arrangement. This particular system is set up in such a way as to allow simultaneous extraction of one sample while a second is being assayed by HPLC. Method development involved the use of a spiked inert matrix (hyFlo-supercel, 700 mg) or loamy soil of 1 mm particle diameter. As a result of these studies on-line SFE–HPLC was performed using supercritical fluid carbon dioxide at 0.8 g/ml density (3335 psi, 50°C). The extracted analytes were trapped onto an MCH-10 $C_{18}$ HPLC column (30 mm × 4 mm i.d.). After the extraction/trapping was completed a valve was switched to isolate the trap column, the excess carbon dioxide flushed from the system with a solvent wash and finally the analytes were eluted to the analytical column. The HPLC system used to assay the analytes was a Varian® ODS-80TM 250 mm × 4.6 mm i.d. column with a linear gradient (water with 0.05 vol/vol phosphoric acid to acetonitrile with 0.05% vol/vol phosphoric acid). With these SFE–HPLC conditions the radiolabelled atrazine, desethylatrazine and deisopropylatrazine were extracted from hyFlo-supercel with efficiencies greater than 99.5% after 15 min; however, hydroxyatrazine could not be extracted even in the presence of methanol. Spiked soil samples yielded efficiencies greater than 99.5% after extraction for 30 min for atrazine and desethylatrazine, with deisopropylatrazine being significantly lower, at 90.6%. In a practical sense this inefficiency proved not to be an issue since desethylatrazine was the only metabolite produced in an experiment where a soil sample spiked with $^{14}C$ labelled atrazine was incubated for two weeks, in the dark, at 25°C. The experiment effectively mimicked what would actually happen to atrazine under normal use. Since dissolved

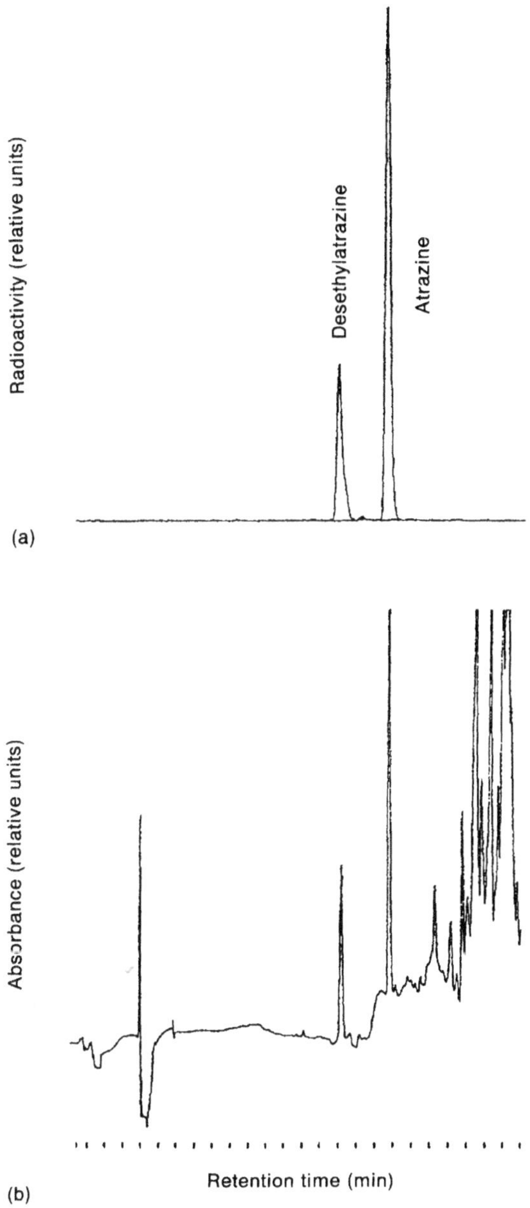


**Figure 8.39** On-line supercritical fluid extraction–high-performance liquid chromatography analysis of triazine compounds extracted from soil: (a) radiochromatogram; (b) ultraviolet chromatogram. Reproduced with permission from ref. 60.

carbon dioxide and non-labelled extractives do not interfere with radioactive detection, the quality of the radiochromatogram, shown in Figure 8.39(a), is, as expected, good. The fact that all the carbon dioxide was eliminated from the SFE–HPLC system using a 5 min pre-wash is confirmed by the UV/VIS chromatogram shown in Figure 8.39(b), which shows no baseline instability. Figure 8.39(b) shows additional extracted 'analytes' which are well resolved from those of interest but may affect the longevity of the HPLC column.

A direct aqueous SFE vessel has been interfaced to an LC–MS system by Ramsey *et al.* [61]. A coupled ODS–amino column system (Techsphere® 5ODS 250 mm × 4.6 mm i.d.–Nucleosil® 5 μm amino 150 mm × 4.6 mm i.d.) was used for analyte trapping and analysis using the ten-port valve arrangement shown in Figure 9.24. With this system three drugs (estrone, hexestrol and zeranol) were each analysed in water at the 200 ppb level using SFE–LC–MS with simultaneous UV/VIS diode-array detection. Following SFE, the trapped analytes were eluted from the ODS column onto the top of the amino column using pure heptane. The normal phase gradient programme was developed to purge the coupled column system of any residual carbon dioxide prior to the onset of chromatography. Although further details of this study are given in section 9.7 it is worth noting that the band-focusing effect of the coupled HPLC column system was such that it provided potential for the addition of an internal standard post-SFE using an off-line sample injection valve. This capability would help facilitate on-line quantitative SFE–HPLC studies.

## 8.6 Conclusions

Many of the off-line applications of SFE followed by HPLC analysis described within this chapter serve to dispel the notion that this sample preparation technique is limited to analytes which are amenable to GC and SFC. As illustrated in several of the HPLC chromatograms shown, the selectivity of SFE often results in the provision of clean extracts for HPLC analysis. Consequently, in appropriate circumstances this can provide a means of eliminating complications arising from possible interferences from co-extracted species. In situations where high quantities of extracted endogenous compounds can adveresely affect the final HPLC analysis (e.g. lipids extracted from tissue samples being assayed for drug residues) it is possible to use in-line cleanup procedures.

The development of on-line SFE–HPLC techniques is still in its infancy. Given the preceding comments, this situation can perhaps be attributed to the more technically challenging requirements of interfacing SFE to HPLC compared with GC or SFC. However, several of the applications described, most notably the analyses of PCBs in biological samples [51] and pesticides in vegetable oils [52, 56], serve to demonstrate the potential of SFE–HPLC for the rapid processing of complex samples for the isolation of clean extracts for a final off-line analysis stage. The few examples of direct on-line SFE–HPLC analytical procedures serve to demonstrate the feasibility of this approach for rapid analysis with minimum sample handling. It can be safely concluded that at present the full potential of on-line SFE–HPLC remains largely unexplored.

## References

1. Howard, A.L. and Taylor, L.T. (1993) In *Supercritical Fluid Extraction and its Use in Chromatographic Sample Preparation* (ed. S.A. Westwood), Blackie Academic & Professional, London, pp. 145–68.
2. Vannoort, W., Chervet, J.P., Lingeman, H. *et al.* (1990) Coupling of supercritical fluid extraction with chromatographic techniques. *Journal of Chromatography*, **505**, 45–77.
3. Jinno, K. and Saito, M. (1991) Coupling supercritical fluid extraction with chromatography. *Analytical Sciences*, **7**, 361–9.
4. Chester, T.L., Pinkston, J.D. and Raynie, D.E. (1994) Supercritical chromatography and extraction. *Analytical Chemistry*, **66**, 106R–130R.
5. Stuart, I.A., MacLachlan, J. and McNaughtan, A. (1996) Compounds of agricultural significance using environmental analytical supercritical fluid extraction. *Analyst*, **121**, 11R–28R.
6. Verschuere, M., Sandra, P. and David, F. (1992) Fractionation by SFE and microcolumn analysis of the essential oil and the bitter principles of hops. *Journal of Chromatographic Science*, **30**, 388–91.
7. Ashraf-Khorassani, M., Barzegar, M. and Yamini, Y. (1995) On-line coupling of supercritical fluid extraction with high performance liquid chromatography. *Journal of High Resolution Chromatography*, **18**, 472–6.
8. Unger, K. K. and Roumeliotis, P. (1983) On-line high pressure extraction–high performance liquid chromatography. *Journal of Chromatography*, **282**, 519.
9. Li, K., Ong, C.P. and Li, S.F.Y. (1994) Systematic multivariate optimisation of supercritical fluid extraction. *Journal of Chromatographic Science*, **32**, 53–6.
10. Ashraf-Khorassani, M., Gidanian, S. and Yamini, Y. (1995) Effect of pressure, temperature, modifier concentration and sample matrix on supercritical fluid extraction efficiency of different phenolic compounds. *Journal of Chromatographic Science*, **33**, 658–62.
11. Thompson, P.G., Taylor, L.T., Richter, B.E. *et al.* (1993) Trapping efficiencies of various collection solvents after supercritical fluid extraction. *Journal of High Resolution Chromatography*, **16**, 713–16.
12. Gere, D.R., Knipe, C.R., Castelli, P. *et al.* (1993) Bridging the automation gap between sample preparation and analysis: an overview of SFE, GC, GC–MS and HPLC applied to environmental samples. *Journal of Chromatographic Science*, **31**, 246–58.
13. Scalia, S. and Games, D. (1992) Determination of parabens in cosmetic products by supercritical fluid extraction and high performance liquid chromatography. *Analyst*, **117**, 839–41.
14. Scalia, S., Guarneri, M. and Menegatti, E. (1994) Assay of triclosan in deodorant sticks and soaps by supercritical fluid extraction and HPLC. *Journal of the Society of Cosmetic Chemists*, **45**, 35–42.
15. Scalia, S., Renda, A., Ruberto, G. *et al.* (1995) Assay of vitamin A palmitate and vitamin E acetate in cosmetic creams and lotions by supercritical fluid extraction and HPLC. *Journal of Pharmaceutical and Biomedical Analysis*, **13**(3), 273–7.
16. Scalia, S., Ruberto, G. and Bonina, F. (1995) Determination of vitamin A, vitamin E and their esters in tablet preparations using supercritical fluid extraction and HPLC. *Journal of Pharmaceutical Sciences*, **84**(4), 433–6.
17. D'Odorico, A.D., Gilmore, I.T., Morris, A. *et al.* (1993) Isolation and identification of putative carcinogens and anticarcinogens from dietary and faecal samples by supercritical fluid extraction and HPLC. *Food and Cancer Prevention: Chemical and Biological Aspects*, **4**, 397–400.

18. Wenclawiak, B., Rathmann, C. and Teuber, A. (1992) Supercritical fluid extraction of soil samples and determination of polycyclic aromatic hydrocarbons (PAHs) by HPLC. *Fresenius Journal of Analytical Chemistry*, **344**, 497–500.
19. Meyer, A., Kleibohmer, W. and Cammann, K. (1993) SFE of PAHs from soils with a high carbon content and analyte collection *via* combined liquid/solid trapping. *Journal of High Resolution Chromatography*, **16**, 491–4.
20. Fernandez, P., Alder, A.C. and Giger, W. (1993) Quantitative determination of cationic surfactants in sewage sludges and sediments by supercritical fluid extraction and HPLC applying post column ion pair extraction. *Abstract from the Meeting of the Division of Environmental Chemistry American Chemical Society*, **33**(1), 303–6.
21. Fernandez, P., Alder, A.C., Suter, M.J.-F. and Giger, W. (1996) Determination of the quaternary ammonium surfactant ditallowdimethylammonium in digested sludges by supercritical fluid extraction and HPLC with postcolumn ion-pair formation. *The 4th World Surfactants Congress*, **4**, 102–9.
22. Breen, D.G.P.A., Horner, J.M., Bartle, K.D. *et al.* (1996) Supercritical fluid extraction and off-line HPLC analysis of cationic surfactants from dried sewage sludge. *Water Research*, **30**(2), 476–80.
23. Martindale (1996) The extra pharmacopoeia. *The Royal Pharmaceutical Society*, **31**, 1429–41.
24. Howard, A.L., Braue, C. and Taylor, L.T. (1993) Feasibility of thiocarbamate pesticide analysis in apples by supercritical fluid extraction and high performance liquid chromatography. *Journal of Chromatographic Science*, **31**, 323–9.
25. Atienza, J., Jimenez, J.J., Bernal, J.L. and Martin, M.T. (1993) Supercritical fluid extraction of fluvalinate residues in honey. Determination by high performance liquid chromatography. *Journal of Chromatography A*, **655**, 95–9.
26. Aharonson, N., Lehotay, S.J. and Ibrahim, M.A. (1994) Supercritical fluid extraction and HPLC analysis of benzimidazole fungicides in potato, apple and banana. *Journal of Agricultural Food Chemistry*, **42**, 2817–23.
27. Shun Huang, A., Robinson, L.R., Gursky, L.G. *et al.* (1994) Identification and quantification of SALATRIM 23CA in foods by the combination of supercritical fluid extraction, particle beam LC–mass spectrometry and HPLC with light scattering detector. *Journal of Agricultural Food Chemistry*, **42**, 468–73.
28. Argauer, R.J., Eller, K.I., Ibrahim, M.A. and Brown, R.T. (1995) Determining propoxur and other carbamates in meat using HPLC fluorescence and gas chromatography/ion trap mass spectrometry after supercritical fluid extraction. *Journal of Agricultural Food Chemistry*, **43**, 2774–8.
29. Järvenpää, E., Huopalhti, R. and Tapanainen, P. (1996) Use of supercritical fluid extraction high performance liquid chromatography in the determination of polynuclear aromatic hydrocarbons from smoked and broiled fish. *Journal of Liquid Chromatography and Related Technology*, **19**(9), 1473–82.
30. Din, N., Bartle, K.D., Clifford, A.A. and Castle, L. (1996) An investigation of supercritical fluid extraction of trenbolone from beef. *Journal of High Resolution Chromatography*, **19**, 465–9.
31. Huopalahti, R.P. and Henion, J.D. (1996) Application of supercritical fluid extraction and high performance liquid chromatography/mass spectrometry for the determination of some anabolic agents directly from bovine tissue samples. *Journal of Liquid Chromatography and Related Technology*, **19**(1), 69–87.
32. Cross, R.F., Ezzell, J.L. and Richter, B.E. (1993) The supercritical fluid extraction of polar drugs from inert matrices and meat animal products. *Journal of Chromatographic Science*, **31**, 162–9.
33. Parks, O.W. and Maxwell, R.J. (1994) Isolation of sulphonamides from fortified chicken tissues with supercritical $CO_2$ and in-line adsorption. *Journal of Chromatographic Science*, **32**, 290–3.
34. Pensabene, J., Fiddler, W. and Parks, O.W. (1997) Isolation of sulphonamides from whole egg by supercritical fluid extraction. *Journal of Chromatographic Science*, **35**, 270–4.
35. Parks, O.W., Lightfield, A.R. and Maxwell, R.J. (1995) Effect of matrix dehydration during supercritical fluid extraction on the recoveries of drug residues from fortified chicken liver. *Journal of Chromatographic Science*, **33**, 654–7.

36. Din, N., Bartle, K.D., Clifford, A.A. *et al.* (1997) Supercritical fluid extraction of sulphamethazine and its metabolites from meat tissue. *Journal of Chromatographic Science*, **35**, 31–7.
37. Hedrick, J.L. and Taylor, L.T. (1992) Direct supercritical fluid extraction of nitrogenous bases from aqueous solution. *Journal of High Resolution Chromatography*, **15**, 151–4.
38. Engelhardt, H. and Haas, P. (1993) Possibilities and limitations of SFE in the extraction of aflatoxin B1 from food matrices. *Journal of Chromatographic Science*, **31**, 13–19.
39. Yao, W.X., Cheng, V.M. and Zhang, Y.Q. (1995) Determination of flavonoid compounds in *Ginko biloba* leaves supercritical fluid extraction and high performance liquid chromatography (SFE/HPLC). *Chinese Chemical Letters*, **6**(7), 589–92.
40. Cocks, S., Wrigley, S.K., Ines Chicarelli-Robinson, M. and Smith, R.M. (1995) High performance liquid chromatography comparison of supercritical fluid extraction and solvent extraction of microbial fermentation products. *Journal of Chromatography A*, **697**, 115–22.
41. Gawdzik, J., Kawka, S., Mardarowicz, M. *et al.* (1995) Supercritical fluid extraction of furanocoumarins from the fruits of *Archangelica off.* Hoffm. *Journal of High Resolution Chromatography*, **18**, 781–3.
42. Marsili, R. and Callahan, D. (1993) Comparison of a solvent extraction technique and supercritical fluid extraction for the determination of $\alpha$- and $\beta$-carotene in vegetables. *Journal of Chromatographic Science*, **31**, 422–7.
43. Bicking, M.K.L. (1992) A simplified design approach to optimisation of SFE conditions for extraction of an amine hydrochloride. *Journal of Chromatographic Science*, **30**, 358–60.
44. Messer, D.C. and Taylor, L.T. (1992) Development of analytical SFE of a polar drug from an animal food matrix. *Journal of High Resolution Chromatography*, **15**, 238–41.
45. Sauvage, E., Rocca, J.L. and Toussaint, G. (1993) The use of nitrous oxide for supercritical fluid extraction of pharmaceutical compounds from animal feed. *Journal of High Resolution Chromatography*, **16**, 234–8.
46. Barnabas, I.J., Dean, J.R., Hitchen, S.M. and Owen, S.P. (1994) Selective supercritical fluid extraction of organochlorine pesticides and herbicides from aqueous samples. *Journal of Chromatographic Science*, **32**, 547–51.
47. Howard, A.L. and Taylor, L.T. (1992) Quantitative supercritical fluid extraction of sulfonyl urea herbicides from aqueous matrices via solid phase extraction disks. *Journal of Chromatographic Science*, **30**, 374–82.
48. Venema, A., van der Ven Henk, J.F.M, David, F. and Sandra, P. (1993) Supercritical fluid extraction of Nylon 6: an investigation into the factors affecting the efficiency of extraction of caprolactam and its oligomers. *Journal of High Resolution Chromatography*, **16**, 522–4.
49. Cortes, H.J., Green, L.S. and Campbell, R.M. (1991) On-line coupling of supercritical fluid extraction with multidimensional microcolumn liquid chromatography/gas chromatography. *Analytical Chemistry*, **63**, 2719–24.
50. Bally, R.W. and Cramers, C.A. (1986) Tapered *versus* constant diameter post-column restrictors in capillary SFC. *Journal of High Resolution Chromatography, Chromatographic Communications*, **9**, 626–32.
51. Johansen, H.R., Becher, G. and Greibrokk, T. (1994) Determination of planar PCBs by combining on-line SFE–HPLC and GC–ECD or GC/MS. *Analytical Chemistry*, **66**, 4068–73.
52. Ahmad, M. (1993) *Pesticide Residue Analysis using Supercritical Fluid Extraction*, PhD Thesis, School of Applied Sciences, University of Glamorgan, Pontypridd.
53. King, J.W. (1989) Fundamentals and applications of supercritical fluid extraction in chromatographic science. *Journal of Chromatographic Science*, **27**, 355.
54. Nam, K.S., Kapila, S., Yanders, A.F. and Puri, R.K. (1991) A multiple sample extraction and on-line system for the analysis of chlorinated compounds. *Chemosphere*, **23**(8–10), 1109–15.
55. Murugaverl, B. and Voorhees, K.J. (1991) On-line supercritical fluid extraction/chromatography system for trace analysis of pesticides in soybean oil and rendered fats. *Journal of Microcolumn Separations*, **3**, 11–16.
56. Ahmad, M., Berry, A.J. and Ramsey, E.D. (1993) In *The 2nd European Symposium on Analytical Supercritical Chromatography and Extraction, Riva Garda* (ed. P. Sandra and K. Markides), Huethig, Heidelberg, p. 282.

57. Ischi, F. and Haerdi, W. (1995) Supercritical fluid extraction–high performance liquid chromatography on-line coupling: extraction of some model aromatic compounds. *Chromatographia*, **41**(3/4), 238–42.
58. Liu, M.H., Kapila, S., Nam, K.S. and Elseewi, A.A. (1993) Tandem supercritical fluid extraction and liquid chromatography system for determination of chlorinated phenols in solid matrices. *Journal of Chromatography*, **639**, 151–7.
59. Stalling, D.L., Saim, S., Kuo, K.C. and Stunkel, J.J. (1992) Integrated sample processing by on-line supercritical fluid extraction–gel permeation chromatography. *Journal of Chromatographic Science*, **30**, 486–90.
60. Mougin, C., Dubroca, J. and Barriuso, E. (1996) On-line supercritical fluid extraction and high performance liquid chromatography for determination of triazine compounds in soil. *Journal of High Resolution Chromatography*, **19**, 700–2.
61. Ramsey, E.D., Minty, B. and Rees, A.T. (1997) Drugs in water: analysis at the ppb level using direct supercritical fluid extraction of aqueous samples coupled on-line with diode array UV/VIS liquid chromatography–mass spectrometry. *Analytical Communications*, **34**, 261–4.

# 9 Supercritical fluid extraction coupled on-line with mass spectrometry and spectroscopic techniques

B. MINTY, E.D. RAMSEY, A.T. REES, D.I. JAMES, P.M. O'BRIEN AND M.I. LITTLEWOOD

## 9.1 Introduction

Previous chapters have described the on-line coupling of supercritical fluid extraction (SFE) with a variety of chromatographic techniques, including: gas chromatography (GC), capillary supercritical fluid chromatography (cSFC), packed column supercritical fluid chromatography (SFC) and high-performance liquid chromatography (HPLC). Three major advantages of coupling SFE with chromatography are that:

- all of the extracted analytes can be transferred (if required) to the chromatographic system, this being particularly valuable in the case of ultra-trace-level studies;
- provided robust on-line SFE procedures can be developed, considerable savings in analyses times can be achieved;
- potential errors associated with off-line sample preparation manipulations can be eliminated.

Integration of SFE (either directly or via an intermediate on-line chromatographic inlet) to spectroscopic techniques such as Fourier transform infra-red spectroscopy (FTIR), nuclear magnetic resonance spectroscopy (NMR) or mass spectrometry (MS) offers further potential advantages:

- confirmation that target analytes have been extracted;
- structural determination of unknown extracted species;
- in the case of MS, improved detection limits using the wide range of ionisation techniques possible with SFE–GC–MS, SFE–SFC–MS and SFE–LC–MS.

Of course, the on-line coupling of other sample preparation techniques with FTIR, NMR and MS offers similar potential advantages. For example, Slobodnik *et al.* [1] and Puig *et al.* [2] have combined solid phase extraction (SPE) on-line with LC–MS for the determination of organic pollutants in aqueous samples. Also, Ollers *et al.* [3] have described the use of SPE–GC–MS for the analysis of caffeine in urine and triazine in drinking water.

The purpose of this chapter is to illustrate, with detailed reference to a variety of case studies, how the on-line combination of SFE with FTIR, NMR and MS has potential to provide solutions for a range of analytical problems. In focusing on such combinations we have purposefully made only limited reference to the wide range of other chromatographic detection systems which can also be integrated into combined systems; for example, flame ionisation detection (FID), electron capture detection (ECD) and ultraviolet/visible detection (UV/VIS), especially with diode-array capability. Previous chapters have described the application of these detection systems.

## 9.2 On-line supercritical fluid extraction–Fourier transform infra-red spectroscopy

Taylor and Jordan [4] have produced an excellent review concerning the underlying theory and principles of SFE–FTIR and the reader is referred to that article for a more detailed and extended treatment of the points raised in this section.

At low pressures $CO_2$ absorbs in the IR region at 3504–3822 $cm^{-1}$, 2137–2551 $cm^{-1}$ and below 900 $cm^{-1}$. Since few organic compounds absorb in these regions, windows of opportunity exist for the development of SFE–FTIR analytical procedures. The appearance of Fermi resonance bands at 2070 $cm^{-1}$, 1944 $cm^{-1}$, 1387 $cm^{-1}$ and 1282 $cm^{-1}$, at increased $CO_2$ pressures, can provide a source of interference. Although these symmetry-forbidden transitions may affect the quality of SFE–FTIR analyses, standard software background-subtraction packages can minimise their contribution to the final FTIR spectrum. Raynor *et al.* [5] published an early report describing the potential of on-line cSFC–FTIR for the analysis of organic compounds. However, since then a steady increase in favour of applications of SFE–FTIR have appeared in the literature.

The use of SFE–FTIR offers several advantages.

1. Since FTIR is a non-destructive technique, potential exists for the incorporation of a second detector in series into a fully integrated system (e.g. MS or FID). Also, a variety of chromatographic inlets including SFC (packed and capillary) and GC can also be successfully interfaced to FTIR. Examples of such instrumentation are SFE–FTIR–FID, SFE–SFC–FTIR–MS, SFE–GC–FTIR–MS, etc. Such hyphenated systems require the use of a high-pressure FTIR flow cell.
2. Selective detection is possible. In the event of co-extracted species, acquisition of SFE–FTIR spectra over a narrow absorbance range specific to a functional group (or groups) associated with an analyte (or analytes) may obviate the need for chromatographic separation. The

analyte concentration can then be determined by integrating its signal through the selected absorbance range.

3. There is a reduction in the use of restricted solvents. The Montreal Protocol treaty [6] set out a series of deadlines for the phasing out of production of ozone-depleting solvents in Europe and North America. Amongst these solvents are those which have been widely used in IR spectroscopy (e.g. chlorofluorocarbons and carbon tetrachloride). The fact that $CO_2$ is cheap, non-toxic, non-flammable, readily available in high purity with no-disposal costs means that development of replacement SFE–FTIR technology offers great potential to become cost effective.

The success of SFE–FTIR procedures involving the direct use of flow cells is largely determined by the extent of the analytes' SF-$CO_2$ solubility. Figure 9.1 shows the theoretical extraction profiles of a highly soluble analyte (profiles A) and a solubility-limited analyte (profiles B). In the former case, dynamic SFE will result in the extraction of the analyte over a narrow time interval (promoting detectability). In the latter case a high-volume plug, the maximum

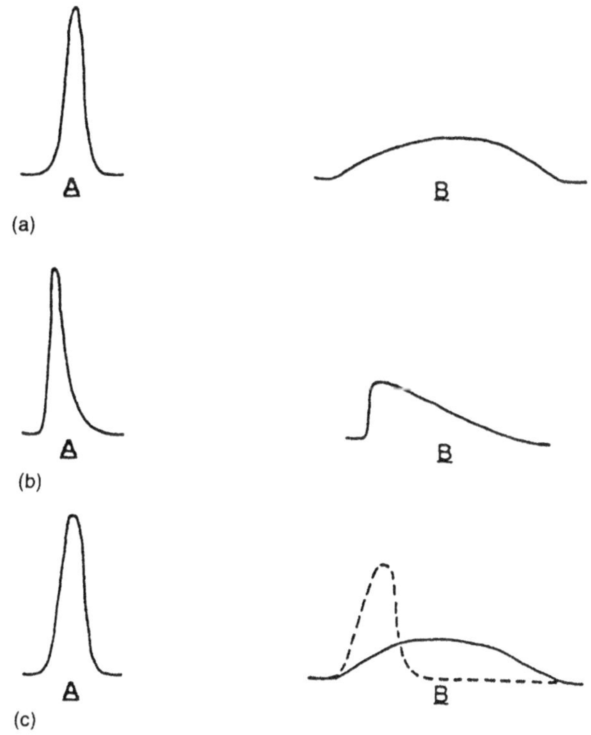


**Figure 9.1** Extraction profiles for supercritical-fluid (SF) soluble (A) compared with SF-solubility-limited (B) analytes: (a) dynamic extraction; (b) static–dynamic extraction; (c) dynamic extraction with cryotrapping and recovery. Reproduced with permission from ref. 4.

concentration of which may not be indicative of the amount of analyte present within the sample, is obtained. In both instances, static–dynamic SFE can lead to increased analyte concentration during the onset of the dynamic SFE stage. The use of an intervening cryotrapping stage during dynamic SFE can provide improved detection limits, particularly for analytes which are strongly bound within a matrix, assuming that the trapped analytes can be rapidly desorbed for FTIR analysis. Multicomponent extractants can further complicate these considerations, for example a composite extraction profile may appear broad despite the presence of components possessing high to medium SF solubilities. In these circumstances cryotrapping with efficient sample desorption can help minimise FTIR sample discrimination effects.

The applications described in the following sections illustrate the potential of SFE–FTIR procedures for the analysis of a range of complex mixtures.

### *9.2.1 Flow cell supercritical fluid extraction–Fourier transform infra-red spectroscopy*

Ikushima *et al.* [7] have studied the extraction dynamics of stearic acid methyl ester, linolenic acid methyl ester and DL-$\alpha$-tocopherol using dynamic SFE–FTIR with UV/VIS detection. A schematic of the system is shown in Figure 9.2. Selection of the appropriate valve configuration enables either

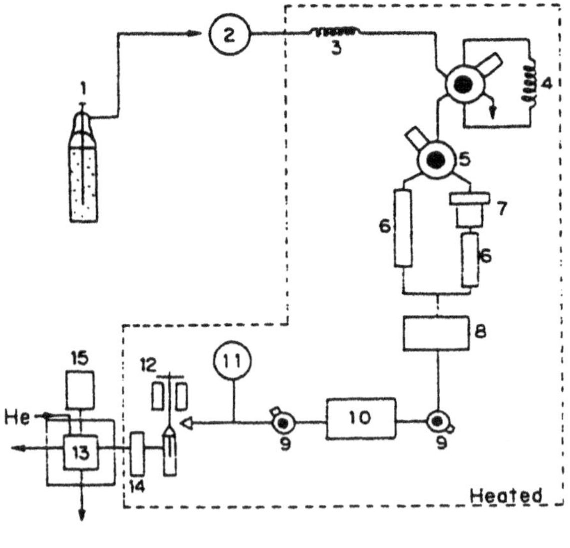


**Figure 9.2** Schematics of system used to study extraction dynamics of lipids. 1 = $CO_2$ cylinder; 2 = pump; 3 = preheater; 4 = injection valve; 5 = three-way valve; 6 = separation column; 7 = supercritical fluid extraction vessel; 8 = ultraviolet detector; 9 = on–off valve; 10 = Fourier transform infra-red spectrometer; 11 = pressure transducer; 12 = restrictor; 13 = flame ionisation detector; 14 = gas sampler; 15 = recorder. Reproduced with permission from ref. 7.

SFE–UV/VIS–FTIR or SFC–UV/VIS–FTIR. The extraction of DL-$\alpha$-tocopherol and stearic methyl ester at various SF-$CO_2$ densities was determined by monitoring $\nu$(C–O–C) stretching at 1053 $cm^{-1}$ and $\nu$(C=O) stretching at 1744 $cm^{-1}$, respectively. For these initial SFE–FTIR studies an empty column was placed between the extraction vessel and FTIR spectrometer. With this approach it was demonstrated that a mixture of these two compounds could be separated because of their different solubilities in SF-$CO_2$. However, DL-$\alpha$-tocopherol and linolenic acid methyl ester could not be separated using SFE alone. Following these studies, SFC–UV/VIS–FTIR using a silica gel packed column [150 mm × 4.6 mm inner diameter (i.d.)] with and without a silver nitrate coating (0.1% m/m loading onto the silica stationary phase) was used to separate mixtures of the lipids. Mathematical equations were developed to model separation efficiency using several aspects of SF-$CO_2$ solvation power and forces of interaction between the lipids with individual stationary phases and the SF-$CO_2$ phase.

A versatile SFE–FTIR system has been developed by Kirschner and Taylor [8] which enables either static, dynamic, static–dynamic or dynamic–cryotrap modes of operation with minor alterations to a commercial SFE–SFC system (Suprex, Pittsburgh, PA). A 1 m × 0.26 mm i.d. length of stainless-steel tubing connected to 50 µm i.d. fused-silica capillary tubing served as a transfer line between the SFE vessel (0.167 ml) and a commercial high-pressure FTIR flow cell (Nicolet, Madison, WI). The back pressure within the FTIR cell was maintained using a 60 µm i.d. fused-silica tapered restrictor. A $C_{24}$ hydrocarbon served as a test analyte for dynamic–cryotrap SFE–FTIR. Dichloromethane solutions of the $C_{24}$ standard were loaded onto Celite® placed within the SFE vessel. Dynamic SFE ($CO_2$ at 300 atm, 75°C; extraction time 10 min) was performed using a stainless-steel cryotrap at −20°C. In order to desorb the $C_{24}$ for FTIR analysis, the trap was heated to 180°C and flushed with SF-$CO_2$ (300 atm). Quantitation was performed using four replicate injections of five $C_{24}$ standards (ranging from 0.216 to 10.8 µg), integrating the absorbance 2951–2916 $cm^{-1}$. For masses 1–10 µg, the relative standard deviations (RSDs) were 3%–5%, whereas 0.216 µg had an RSD of 12%. A linear calibration graph was obtained with a correlation coefficient of 0.99998 for this analyte range. Limits of detection and quantification for $C_{24}$ were 74 ng and 250 ng, respectively, using propagation of errors statistical calculations. Variations in restrictor transmission characteristics and cryotrap desorption dynamics were cited as factors which could adversely affect this technique. Although an increased SF-$CO_2$ flow rate promotes analyte detectability by producing a sharper extraction profile, in practice this was limited because of the production of a noisy FTIR background. A mixture of $C_{24}$ and caffeine loaded onto Celite® were partially separated following dynamic SFE by flushing the analyte cryotrap with SF-$CO_2$ at 300 atm. Caffeine could be detected in the presence of $C_{24}$

by extracting the absorbance chromatogram at 1685–1665 $cm^{-1}$ for the carbonyl group.

Using the same instrumentation as that used in the above study, Kirschner *et al.* [9] analysed the finishes applied to three textile fibres (polyurethane, polyamide and aramid). In order to improve the reliability of the system, a manually adjusted high-pressure needle valve linked the FTIR flow cell outlet to a 50 μm i.d. linear fused-silica capillary restrictor that was inserted into an FID heating block assembly. The feasibility of on-line SFE–FTIR was first evaluated using off-line dynamic SFE. The procedure involved loading neat fibre finishes (50–75 mg) onto Celite® (25 mg) and collecting the SFE extract into dichloromethane. The total SFE recoveries for the fibre finishes were determined gravimetrically, with off-line SFC being used to determine the recoveries of individual components. A 100 mm × 1 mm i.d. cyanopropyl packed column was used for all SFC studies.

Certain components within the polyamide finish (six components) and aramid finish (four components) were not highly soluble in SF-$CO_2$ (350 atm, 75°C; extraction time 20 min) resulting in lower total recoveries than that obtained for the polyurethane finish [one component, poly(dimethylsiloxane) oil]. On-line SFE–SFC–FTIR analysis indicated that low molecular weight polyurethane species could also be extracted from the polyurethane sample itself. These co-extractants did not present a problem to the development of SFE–FTIR for the determination of polyurethane fibre finish, since poly(dimethylsiloxane) oil could be differentiated by monitoring $\nu(Si-CH_3)$ at 820–790 $cm^{-1}$. In order to compensate for different extraction profiles, a static–dynamic SFE–FTIR procedure was developed. SFE–FTIR calibration graphs were constructed by loading fibre finish standards onto pre-extracted fibre matrices. The $\nu$(C–O) stretch between 1120–1100 $cm^{-1}$ and the $\nu$(C–H) stretch at 2940–3920 $cm^{-1}$ were used, respectively, to analyse the polyamide and aramid fibre finishes by SFE–FTIR. The SFE–FTIR procedure was used to analyse 2–10 mg real-world polyurethane, polyamide and aramid finished fibre samples for the percentage finish on yarn (%FOY) values. The results compared favourably with conventional organic solvent extraction followed by off-line FTIR analyses.

In a continuation of the above studies, Jordan *et al.* [10] developed an SFE–FTIR procedure for determining the fibre finish applied to Dacron® polyester fibres. This was a more demanding study than that described above for the three other textile fibres. The Dacron® finish studied had seven components, several of which possessed only limited SF-$CO_2$ solubility. The individual components were fatty acids, fatty acid esters, phosphorus-containing antioxidants, sorbitan derivatives and two ethoxylate–propoxylate copolymers. The dynamic SF-$CO_2$ solubilities (350 atm, 75°C; extraction time 20 min) of each individual finish component was determined off-line using SFE with gravimetric analysis. The composite value of these results correlated with the SFE recovery for a neat mixture of the fibre finish. This indicated that no

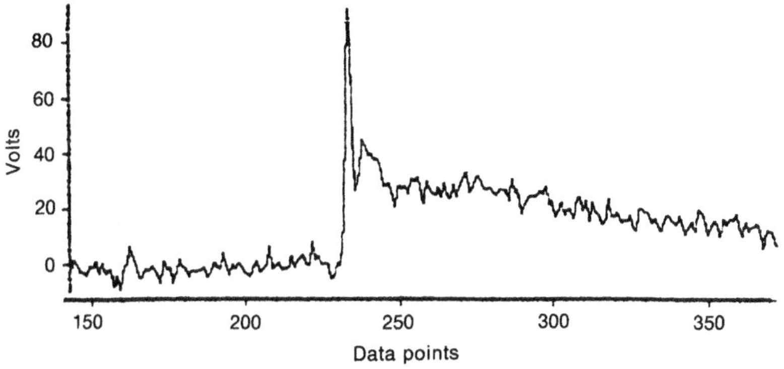


**Figure 9.3** Integrated transmittance plot for the 2940–2920 cm$^{-1}$ region using the direct-static supercritical fluid extraction–Fourier transform infra-red scheme. Reproduced with permission from ref. 10.

synergistic effects existed during SFE. The Dacron® fibre finish static–dynamic extraction profile exhibited a high-level tailing for several minutes, as shown in Figure 9.3, that was attributed to the low SF-$CO_2$ solubilities of the two sorbitan derivatives. Although a calibration curve could be constructed using the $\nu$(C–H) stretch at 2940–2920 cm$^{-1}$ by spiking the finish onto Celite®, the finish could not be quantified on the yarn itself using static–dynamic SFE–FTIR. It was suggested that this failure could be due to extensive diffusion of the finish into the matrix and/or aging, giving rise to the formation of additional analyte–matrix bonds. In order to develop an SFE–FTIR procedure, an intermediate analyte cryotrapping stage was introduced that effectively eliminated the problems associated with individual finish component extraction profiles. A range of trapping materials were evaluated based upon analyte recoveries and standard deviations. A stainless-steel trap operated at $-10°C$ was selected in preference to deactivated fused silica and poly(methylsiloxane)-coated fused silica. With this approach, 5–8 mg Dacron® samples were subjected to dynamic–cryotrap SFE ($CO_2$ at 350 atm, 75°C). The analytes were recovered from the trap at 60°C with SF-$CO_2$ at 350 atm. Calibration curves were constructed by spiking fibre finish standards onto pre-extracted finish-free Dacron® fibres. Two different absorbance ranges were used to compare the results of SFE–FTIR against a conventional organic solvent extraction, off-line FTIR procedure. The %FOY results were as follows: conventional procedure, $0.31\% \pm 0.035\%$; SFE–FTIR, $0.18\% \pm 0.015\%$ using $\nu$(C–H) 2940–2920 cm$^{-1}$; and SFE–FTIR, $0.27\% \pm 0.031\%$ using an absorbance at 1120–1100 cm$^{-1}$. The 1120–1100 cm$^{-1}$ region was more efficient for quantitative purposes because of the large absorbance of a single highly extractable finish component.

An SFE–FTIR method has been developed by Liescheski [11] for the determination of the iodine number in vegetable oils. A 500 atm rated high-pressure flow cell (Foxboro Co., Foxboro, MA) was connected to a 2.5 ml SFE vessel by 1 m × 100 μm i.d. stainless-steel tubing that was ohmically heated by the passage of an electric current. The pressure in the flow cell was maintained by the connection of a 1.5 ml/min coaxially heated capillary restrictor (Isco, Lincoln, NE; see Figure 2.8). In order to prevent the FTIR flow cell windows from becoming 'misted' as a result of the deposition of extracted oil, the temperature of the transfer line and flow cell was maintained at 80°C. The extraction profile of triolein was determined by SFE–FTIR. The procedure involved loading 0.8 mg of this triglyceride onto diatomite support and monitoring the vinyllic $\nu$(H–C=C) and symmetric $\nu$(H–C) stretches at 3011 $cm^{-1}$ and 2930 $cm^{-1}$, respectively. These experiments indicated that triolein was almost completely extracted within 5 min dynamic SFE using SF-$CO_2$ (400 atm, 80°C). Confirmation of these results was obtained by gravimetric analysis of 25 mg sesame seed oil extracted using the same SFE conditions. In order to demonstrate the correlation between iodine number (defined as the number of grams of elemental iodine absorbed per 100 g of fat, oil or wax) and the $\nu$(H–C=C) stretch in the region 3010–3020 $cm^{-1}$, four triglyceride standards and eight refined vegetable oils were analysed by dynamic SFE–FTIR. Although a relationship between iodine number and intensity of SFE–FTIR $\nu$(H–C=C) was readily demonstrated, some of the data points had non-linear coordinates. In order to compensate for instrumental parameters including optical pathlength, a dimensionless ratio $\varepsilon$(H–C=C)/$\varepsilon$(sym $CH_2$) was plotted against iodine number for the 12 standards, resulting in a linear relationship. The validity of this approach was demonstrated by analysing the vegetable oil of corn tortilla chips, the experimental SFE–FTIR iodine number of which was determined as $102 \pm 8$. This value approximates to the iodine number reported for corn oil (109–133) as reported by the 11th edition of the *Merck Index* [12].

In a further study using the same instrumentation, Liescheski *et al.* [13] described a direct SFE–FTIR method for the analysis of the fibre finish, poly(methylsiloxane) oil on polyurethane fibre. A series of diatomite supports that had been spiked with finish oil standards were analysed using direct SFE–FTIR ($CO_2$ at 30.4 MPa, 80°C; extraction time 5 min). Three calibration graphs were obtained using the absorbances at 1025 $cm^{-1}$ and 1097 $cm^{-1}$ for a double band, and at 1259 $cm^{-1}$ and 2966 $cm^{-1}$ for the single bands. A kinetic study indicated that almost complete recovery of the finish oil was obtained within 5 min dynamic SFE (confirmed by off-line SFE gravimetric analysis). The calibration graphs obtained for 1025 $cm^{-1}$ and 1097 $cm^{-1}$ and for 1259 $cm^{-1}$ indicated some curvature. This was attributed to the strong absorbance of these bands and the pathlength of the FTIR high-pressure flow cell. The weaker absorbance at 2966 $cm^{-1}$ was used in further studies to exploit the fact that a linear calibration graph could be constructed through the expected range for the analysis of finish oil in real samples. In order to test the procedure, real-world polyurethane fibres were analysed (in triplicate) using SFE–FTIR, off-line SFE gravimetric analysis and conventional solvent extraction with off-line FTIR. The results for the fibre finish on the polyurethane samples were as follows: SFE–FTIR, 4.7% ± 0.3%; SFE gravimetric analysis, 4.5% ± 0.1%; and solvent extraction, 4.7% ± 0.2%. These results further demonstrate the potential of SFE–FTIR for the analysis of fibre finishes, previously reported by Kirschner *et al.* [9] and Jordan *et al.* [10].

### 9.2.2 *Stop-flow supercritical fluid extraction–Fourier transform infra-red spectroscopy*

In appropriate circumstances on-line FTIR analysis of SFE extracts can be performed without using a flow cell arrangement. Advantages of this approach include the elimination of using a restrictor with potential plugging problems and the possible loss of volatile components associated with solvent/cryotrap analyte collection.

The construction of an inexpensive fibre optic high pressure FTIR cell has been described in detail by Heglund *et al.* [14]. The interface was designed to replace flow cells, whose windows are typically constructed from thick, optically transparent materials such as ZnSe, IRTRAN (ZnS) and $CaF_2$. The SFE–FTIR interface was fabricated from two lengths of chalcogenide (AsSeTe) glass-clad fibre optics housed opposite each other in a stainless-steel union cross. One arm of the cross delivered the SFE medium, whereas the other was connected to a vent valve which enabled cleaning the cell *in situ* with SF-$CO_2$ between analyses. The working spectral range of the interface was 5000–800 $cm^{-1}$ and was rugged to 400 atm. The interface was connected to a 6.94 ml SFE vessel (Keystone Scientific, Bellefonte, PA) and was used to obtain the SFE–FTIR spectrum of caffeine by subtracting the SFE–FTIR spectra of caffeinated and decaffeinated coffee from each other. A quantitative study was also performed involving the determination of the total petroleum hydrocarbon (TPH) content in diesel-contaminated and gasoline-contaminated soil samples. A calibration graph was constructed by injecting 1–20 μl of a US Environmental Protection Agency (EPA) calibration standard (2:3:3 vol/vol/vol mixture of chlorobenzene, *n*-hexadecane and isooctane) into the extraction vessel and monitoring $\nu$(C–H) absorbance at 2932 $cm^{-1}$, in triplicate, of the SF-$CO_2$ solvated (340 atm, 77°C–80°C) hydrocarbons. The FTIR results of off-line SFE and Soxhlet extraction (both techniques involving the use of freon 113) and on-line SFE–FTIR for the TPH analyses of contaminated soil samples are shown in Table 9.1. Further qualitative studies demonstrating the utility of this SFE–FTIR fibre optic interface have been reported by Tilotta *et al.* [15]. These include the analysis of orange peel, and of vegetable fat from potato chips.

Various spectroscopic procedures, including SFE–FTIR, have been used for the determination of catalyst residues in polymers [16]. A 0.5 ml FTIR

**Table 9.1** Comparison of on-line supercritical fluid extraction–Fourier transform infra-red (SFE–FTIR) with off-line SFE–IR and Soxhlet extraction determinations of total petroleum hydrocarbons for selected soil samples. Source: ref. 14

| Sample | TPH (ppm)[a] | | |
|---|---|---|---|
| | Off-line SFE–IR[b] | On-line SFE–IR | Soxhlet extraction[b] |
| Farm diesel | 31 000 (3) | 38 041 (24) | 38 500 (11) |
| Farm gasoline | 6 240 (5) | 6 538 (17) | 5 840 (3) |

[a] Total petroleum hydrocarbon (TPH) results are the average values obtained from triplicate determinations. Relative standard deviations (%) are given in parentheses.
[b] Off-line SFE was performed for 30 min with collection in 5 ml of freon 113. The Soxhlet extraction was performed at 4 h using 150 ml freon 113.

high-pressure cell loaded with polyethylene film was used to monitor the impregnation and extraction of $CpMn(CO)_3$ over periods up to 18 h. The high-frequency shift of the two $\nu$(C–O) absorptions of the organometallic complex in SF-$CO_2$ solution, relative to those observed when this complex is entrained in the polymer, were used to study the extraction/impregnation kinetics.

The feasibility of interfacing a direct aqueous SFE vessel (150 ml capacity) to an FTIR spectrometer equipped with a high-pressure cell has been demonstrated [17]. With this system, the potential of analysing oil in water was described. Further improvements in the system have resulted in detection limits and linearity of response appropriate for determining the hydrocarbon content in real-world effluents and process waters destined for discharge. Figure 9.4 shows the analysis of hexadecane standards in water, ranging from 7.7 to 38.5 ppm.

### 9.2.3 *On-line supercritical fluid extraction–supercritical fluid chromatography–Fourier transform infra-red spectroscopy and supercritical fluid extraction–capillary supercritical fluid chromatography–Fourier transform infra-red spectroscopy*

Hedrick and Taylor [18] described the analysis of diisopropyl methylphosphonate (DIMP) in water using an on-line SFE–SFC–FTIR system. Details of the direct aqueous SFE vessel and multivalve configuration used in these studies are given in sections 4.2.1 and 4.8, respectively. A recirculating pump was used to deliver SF-$CO_2$ (350 atm, 50°C) to 3 ml aqueous DIMP samples with 20 µl aliquots of the SF phase being introduced onto either 1.0 mm or 4.6 mm i.d. × 25 cm cyanopropyl-packed columns at specific time intervals. With this arrangement, aqueous SFE kinetics were studied using SFE–SFC/FID. On-line FTIR was used to confirm the extraction of DIMP. Addition of 0.1 g sodium chloride reduced the DIMP equilibrium distribution time from 1.5 h to less than 5 min.

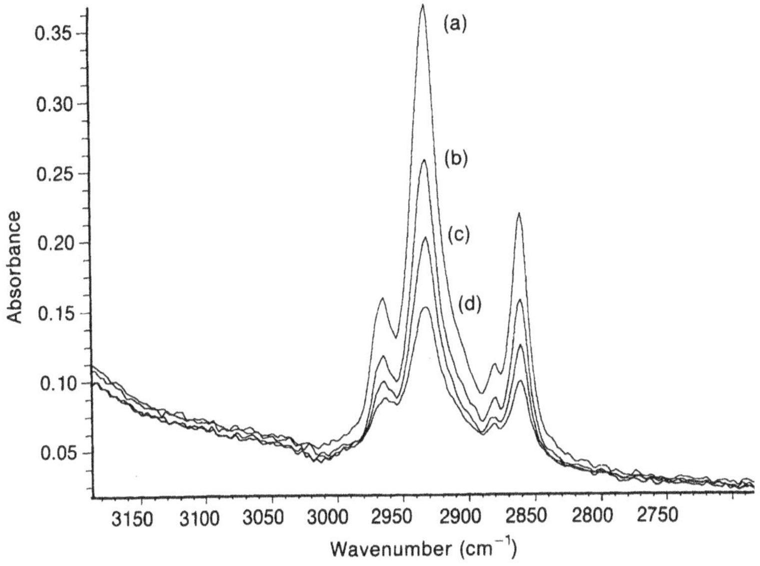


**Figure 9.4** Analysis of hexadecane standards in water by direct aqueous supercritical fluid extraction–Fourier transform infra-red spectroscopy: (a) 38.5 ppm; (b) 23.1 ppm; (c) 15.4 ppm; (d) 7.7 ppm.

Wiebold *et al.* [19] have characterised a range of antioxidants in polyethylene using SFE–cSFC–FTIR. Approximately 75 mg polyethylene were subjected to 5 min dynamic SFE (SF-$CO_2$, 100–400 atm at 150 atm/min, 50°C) with extracted analytes being cryofocused onto a capillary column (10 m × 100 μm i.d. SB-Biphenyl-30) at −30°C. A Nicolet (Madison, WI) SFC–FTIR interface was used for on-line FTIR data acquisition. Approximately 15 g of polyethylene were subjected to Soxhlet-type extraction using dichloromethane with the concentrated extract being analysed using cSFC–FTIR. The Gram–Schmidt reconstructed (GSR) chromatograms obtained for the off-line and on-line analyses are shown in Figures 9.5(a) and 9.5(b), respectively. The most apparent difference between the two sets of results corresponds to the retention time period 30–40 min, where the polymer additives elute. The Soxhlet-type extract provided a regular spaced series of peaks through this region of the chromatogram corresponding to a homologous series of low molecular weight polymers, confirmed by FTIR. The SFE–cSFC–FTIR GSR chromatograms provided less interference, thereby facilitating more rapid characterisation of polyethylene additives using standard FTIR data-based library search procedures. Figure 9.6 shows the reconstructed absorbance chromatogram through the region 1088–1080 cm$^{-1}$. This was used to confirm the presence of the antioxidant tris (2,4-di-*tert*-butylphosphate) phosphite, trade name Irgafos 168, at retention time 34.08 min by SFE–cSFC–FTIR. Rapid identification of oxidised Irgafos 168 was possible

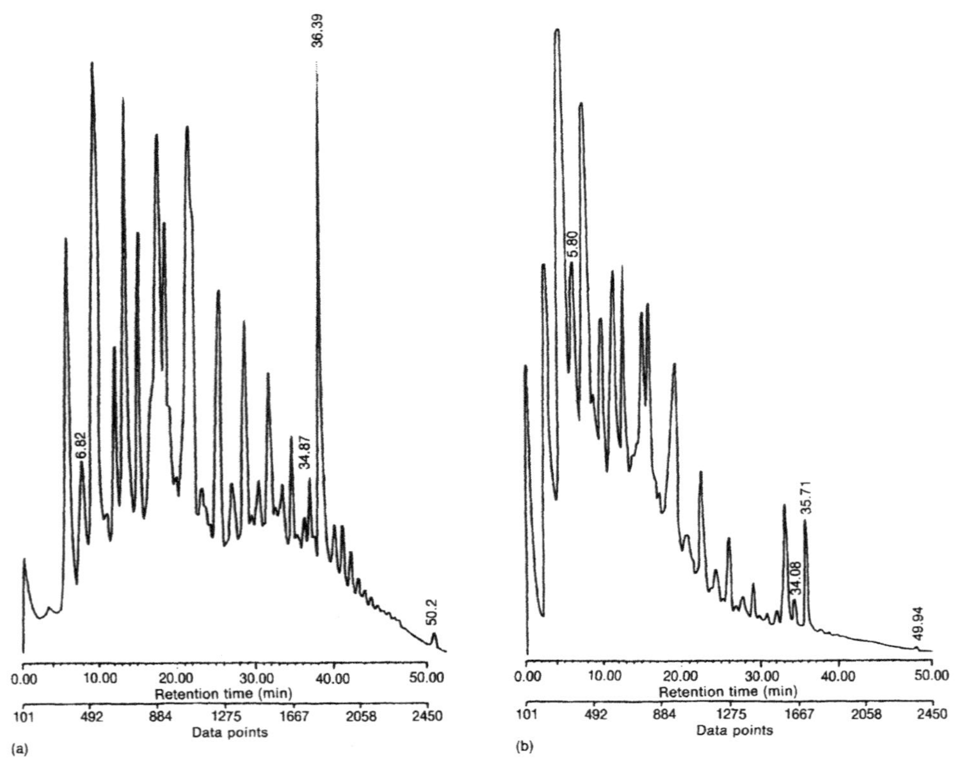


**Figure 9.5** (a) Capillary supercritical fluid chromatography–Fourier transform infra-red spectroscopy (cSFC–FTIR) Gram–Schmidt reconstructed (GSR) chromatogram obtained for a Soxhlet type extraction of polyethylene; (b) the supercritical fluid extraction–capillary supercritical fluid chromatography–Fourier transform infra-red spectroscopy (SFE–cSFC–FTIR) GSR chromatogram obtained for polyethylene. Reproduced with permission from ref. 19.

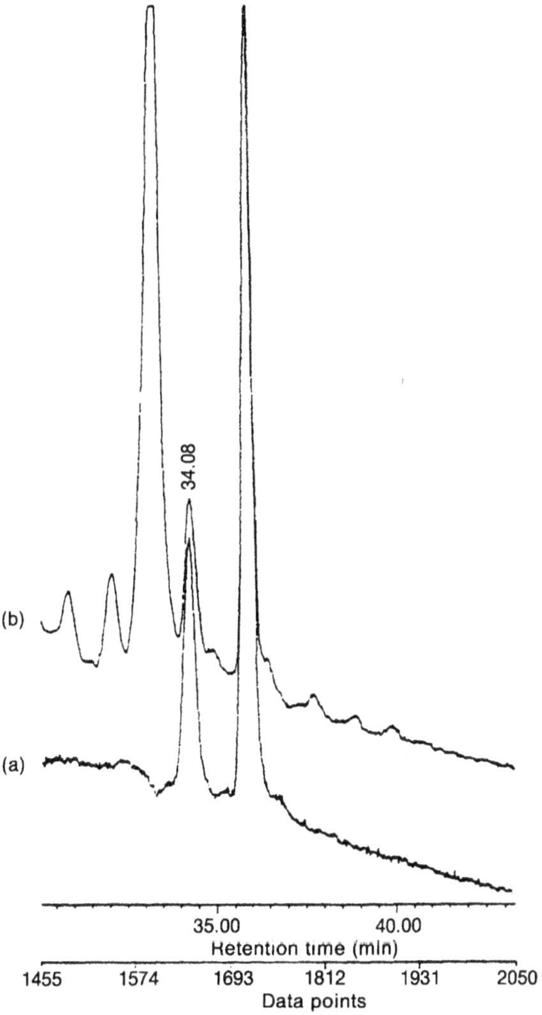


**Figure 9.6** (a) Expanded portion of the selective 1088–1080 cm$^{-1}$ reconstructed chromatogram for the supercritical fluid extraction–capillary supercritical fluid chromatography–Fourier transform infra-red spectroscopy (SFE–cSFC–FTIR) analysis of polyethylene, confirming the extraction of Irgafos 168 at retention time 34.08 min; (b) a section of the Gram–Schmidt reconstructed chromatogram taken from Figure 9.5(b) superimposed for comparison. Reproduced with permission from ref. 19.

as a result of the difference in IR absorption bands for the P–O–C stretch observed for the two oxidation states of phosphorus: 995–915 cm$^{-1}$ for pentavalent phosphorus, and 875–855 cm$^{-1}$ for trivalent phosphorus.

Jordan *et al.* [20] have analysed various additives and monomers in nylon and polystyrene using SFE–SFC–FTIR procedures. For both polymers a cryotrap was used to accumulate SFE extractables prior to SFC–FTIR analysis. In order to promote SFE efficiency, both polymers were milled under liquid nitrogen followed by particle sieving. A Suprex (Pittsburgh, PA) SFE–SFC system was interfaced to an FTIR spectrometer via a Nicolet (Madison, WI) high-pressure flow cell interface whose back pressure was maintained using a 50 µm i.d. fused-silica, tapered restrictor. The feasibility of determining the levels of caprolactam in a copolymer of nylon 6,6 (whose monomers are hexamethylenediamine and adipic acid) and nylon 6 (produced from caprolactam) was demonstrated using off-line SFE (with solvent trapping) followed by SFC–FTIR analysis. These studies confirmed that without initially milling the copolymer pellets residual caprolactam could not be recovered in sufficient quantity using static–dynamic SFE ($CO_2$ at 350 atm, 75°C; total extraction time 75 min) to enable its detection. A 100 mm × 1 mm i.d. cyanopropyl-packed column was used for off-line and on-line studies. The results of on-line SFE–SFC–FTIR indicated that four major components (water, adipic acid, caprolactam and a later eluting component that was not identified by FTIR) could be extracted from the ground copolymer. Following these studies, the levels of caprolactam in three different nylon copolymer samples was determined using SFE–SFC/FID. Prior to these studies, four different analyte traps were evaluated, of which a 100 µm cyano-coated silica capillary provided the best performance for the caprolactam quantitative studies. As shown in Table 9.2 the addition of methanol to the samples prior to dynamic SFE ($CO_2$ at 250 atm, 75°C; extraction time 45 min) aided the recovery of caprolactam. Off-line studies indicated that these improvements were associated with methanol swelling the copolymer (making it more permeable to SF-$CO_2$) rather than improved caprolactam solubility in modified SF-$CO_2$. Low molecular weight extractables in polystyrene, including styrene dimers and trimers, were characterised using SFE–SFC–FTIR–FID. Analyses were performed using a 200 mm × 1 mm

**Table 9.2** Results for quantification of caprolactam: (a) without modifier; (b) with 25 µl methanol added to the matrix. Analyses were performed in triplicate. Source: ref. 20

| Polymer number[a] and nylon 6 content | Percentage caprolactam in the bulk polymer[b] | Percentage caprolactam in nylon 6 component |
|---|---|---|
| (a) | | |
| 268, 1.5% | 0.0044 (4) | 0.31 |
| 262, 5.0% | 0.0154 (14) | 0.3 |
| 188, 1.5% | 0.0043 (6) | 0.29 |
| (b) | | |
| 268, 1.5% | 0.011 (6) | 0.73 |
| 262, 5.0% | 0.031 (7) | 0.61 |
| 188, 1.5% | 0.011 (2) | 0.7 |

[a] This number designates a specific nylon polymer. The 188 and 268 polymers differ only in the amount of $TiO_2$.
[b] Percentage relative standard deviations are given in parentheses.

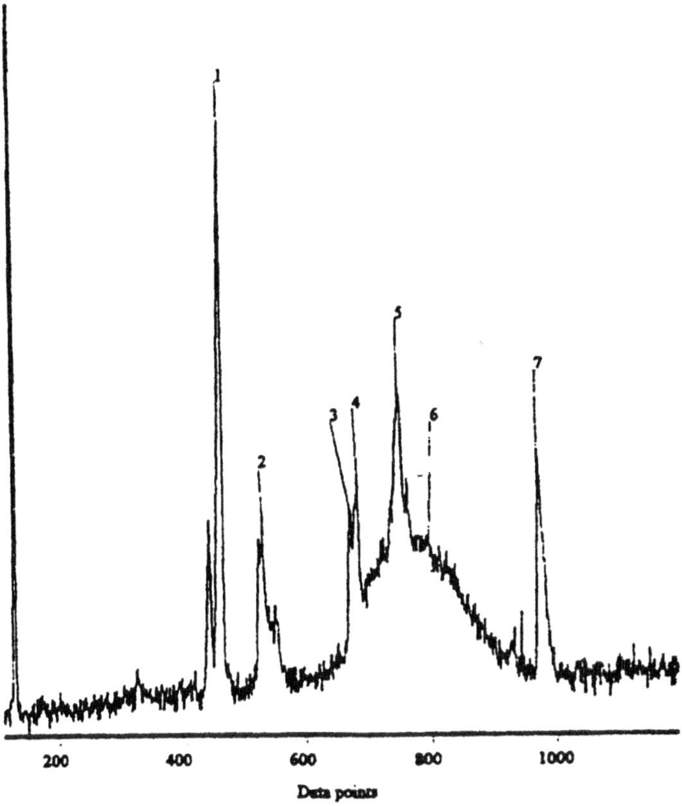


**Figure 9.7** Supercritical fluid extraction–supercritical fluid chromatography–Fourier transform infra-red spectroscopy (SFE–SFC–FTIR) Gram–Schmidt reconstructed chromatogram for polystyrene extract. Peak assignments: 1 = dimer; 2 = phosphite additive; 3 = stearic acid; 4 = linear trimers; 5 = cyclic trimers; 6 = mineral oil; 7 = Irganox 1076. Reproduced with permission from ref. 20.

i.d. octadecylsilyl-bonded silica ($C_{18}$) packed column coupled to a 100 mm × 1 mm i.d. cyanopropyl-packed column. The polystyrene samples were mixed with Celite® (1:1) to prevent the polymer, when it expanded under the SFE conditions employed, from forming a solid impermeable plug in the SFE vessel. In contrast to the nylon 6,6 copolymer (few extractables using high SFE pressures and methanol as modifier), polystyrene proved to be highly extractable. This was attributed to relatively low molecular weight range and lack of polymer crystallinity. The extent of polystrene swelling was related to SFE temperature. Figure 9.7 shows the on-line SFE–SFC–FTIR GSR chromatogram obtained for the analysis of 17.5 mg polystyrene sample subjected to dynamic SFE ($CO_2$ at 107 atm, 120°C; extraction time 30 min). The identities of the seven major peaks were assigned using standard condensed and vapour phase library searches, spectral comparison to standards run using the SFE–SFC–FTIR system, retention times and knowledge of melt flow additives.

## 9.3 On-line supercritical fluid extraction–nuclear magnetic resonance spectroscopy

Braumann *et al.* [21] have described on-line monitoring of the SFE of roasted coffee and black pepper using $^{1}H$ NMR. These studies were conducted using an SFC–NMR probe that had been developed by Albert *et al.* [22]. During extraction the probe could also be isolated by means of a switching valve for stopped-flow measurements. The SFE–NMR contour plot obtained for 820 mg roast coffee is shown in Figure 9.8. This plot was obtained using dynamic SFE ($CO_2$ linear pressure programme from 71 to 294 bar in 55 min, 45°C). The extraction profiles recorded for caffeine and the triglycerides were obtained by summation of the NMR intensities obtained for these compounds. The maximum extraction obtained for caffeine was at 105 bar whereas maximum extraction of the triglycerides was at approximately 200 bar. The extraction profile obtained for the triglycerides indicated they had similar chemical structures, verified by examination of the on-line $^{1}H$ NMR spectra. Figure 9.9 shows the SFE–NMR spectrum obtained for caffeine following data acquisition using 2D software. It was reported that increasing the SF-$CO_2$ density gave rise to upfield changes of chemical

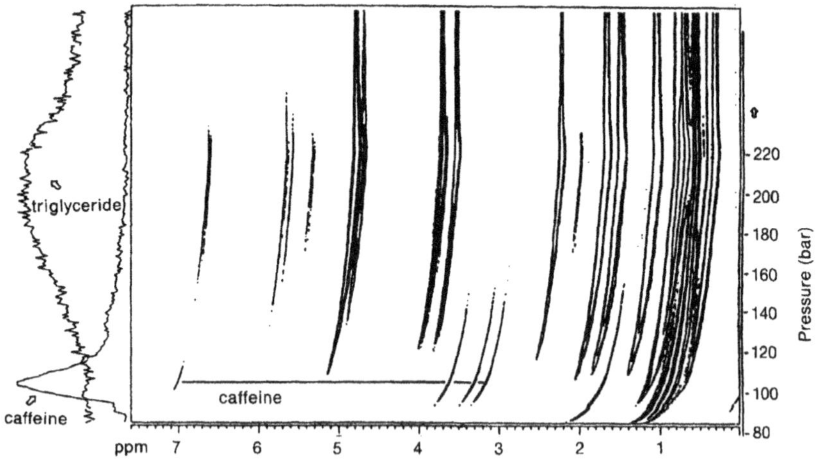


**Figure 9.8** Supercritical fluid extraction of roasted coffee using a linear pressure programme. The projection was reconstructed from intensities of $^{1}H$ NMR signals, 3.4–2.9 ppm caffeine, 4.5–5.3 ppm for the triglycerides. Reproduced with permission from ref. 21.

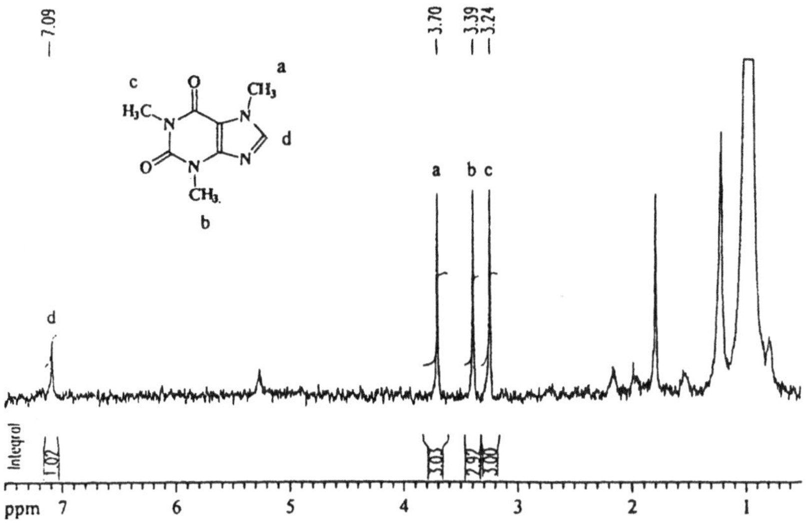


**Figure 9.9** Supercritical fluid extraction–nuclear magnetic resonance spectrum obtained for caffeine collected on-line during coffee extraction. Reproduced with permission from ref. 21.

shift values and that a highfield-shifted impurity of $CO_2$ was used as the internal standard set to 0 ppm. Piperine was identified as an extracted component from black pepper using a stop-flow measurement procedure. It was concluded that SFE–NMR analysis of more polar components should be possible using deutrated organic modifiers. These modifiers could also serve as an internal lock for the spectrometer, since pressure-dependent changes of chemical shift values for the sample and background (modifier) NMR signals should be equally affected.

## 9.4 On-line supercritical fluid extraction–gas chromatography–mass spectrometry

Techniques for coupling SFE to GC were reviewed in Chapter 5. In principle the extension of a successful SFE–GC procedure to SFE–GC–MS presents no significant difficulties other than the retention of a high vacuum within the mass spectrometer and avoiding rapid contamination of the mass spectrometer ion source resulting from the extraction of high levels of compounds.

Hawthorne and Miller [23] described an SFE–GC–MS procedure for the identification of phenolics, nicotine and related compounds in cigarette ash. The fused-silica capillary restrictor (15 cm × 25 μm i.d.) from the SFE vessel (containing 10 mg ash sample) was inserted into the GC column through the on-column injection port. The GC oven was operated at 5°C to cryofocus the extracted components onto the top of the column. Extractions were performed using SF-$N_2O$ (300 atm, 45°C; extraction time 10 min). At the end of the SFE period the restrictor was withdrawn from the injection port and the column was purged with helium to remove traces of $N_2O$ gas. After this period, electron ionisation (EI) GC–MS was performed using a temperature programme. With this approach good chromatographic peak shape was obtained, although some peak broadening was detected for the more volatile components. Extraction was considered to be complete within 10 min since a second SFE–GC–MS analysis of the ash sample failed to enable the detection of any residual organic species. The study was extended to determine whether the SFE–GC–MS procedure could perform quantitative analysis of polycyclic aromatic hydrocarbons (PAHs) within a dust sample (National Institute of Standards and Technology, USA, formerly National Bureau of Standards, urban dust reference material, SRM 1649). The PAHs were separated using a wide-bore column (30 m × 0.32 mm i.d.). In order to maintain the high vacuum of the benchtop mass spectrometer used in these studies, a 150 μm i.d. fused-silica capillary was used to connect the wide-bore column to the ion source. This arrangement proved effective in limiting the flow of $N_2O$ gas into the mass spectrometer during SFE. Quantitative GC–MS of five PAHs was performed using selected ion monitoring (SIM), with calibration standards being prepared from SRM 1647 dust which contained certified concentrations of PAHs in acetonitrile. An internal standard, 4,4′-dichlorophenyl, was added to each 2 mg dust sample prior to SFE ($N_2O$ at 350 atm, 45°C). Using an extraction time of 15 min, it was demonstrated that the entire SFE–GC–MS procedure allowed the determination of PAH concentrations, with good general agreement with certified values, within 1 h.

Using a similar approach to that above, Hawthorne *et al.* [24] conducted a further series of SFE–GC–MS studies using standard instrumentation. Several different sample types were subjected to dynamic SFE ($CO_2$ at 300 atm, 45°C; extraction time 10 min). On-column and split/splitless modes of sample introduction were performed, depending upon the quantity of material extracted. A narrow-bore column (60 m × 0.25 mm i.d.) directly coupled to the mass spectrometer ion source was used for SFE–GC–MS studies. A standard solution of alkanes ($C_7$–$C_{24}$), spiked onto 10 mg clean river sediment was used to evaluate SFE–GC/FID with on-column cryogenic trapping at −30°C. Careful selection of the fused-silica capillary restrictor i.d. resulted in retention of chromatographic fidelity using SFE–GC compared with off-line GC analyses (with split/splitless or on-column injections of the standard alkane mixture). The identities of diesel exhaust pollutants trapped onto a polyruethane foam sorbent were rapidly determined using SFE–GC–MS. Other applications reported in this study included SFE–GC–MS identification of components extracted from oregano, cloves, basil and allspice and from cigarette particulates trapped onto a quartz fibre. Efficient on-column cryogenic trapping meant that 1 mg samples of herbs and spices were sufficient to enable the analysis of flavour and fragrance compounds using SFE–GC–MS.

Huston and Ji [25] have evaluated on-column SFE–GC–MS for the analysis of volatile and semivolatile compounds extracted from cloves. The technique was compared with an off-line SFE procedure involving the use of propan-2-ol as trapping solvent prior to GC–MS analysis. A glass liner placed within the GC injection port served as an extension to the column since a seal was made between the column's polyimide coating and the tapered inner surface of the glass liner. The SFE fused-silica capillary restrictor was introduced into the glass liner (packed with glass wool) through a syringe needle that was used to pierce the injector septum. This arrangement provided a pseudo on-column injection facility for SFE–GC–MS studies. Off-line SFE analyte collection was performed by inserting the SFE restrictor through a syringe needle piercing the septum of the recovery vial containing propan-2-ol. A range of different SFE experiments were performed using different SF-$CO_2$ densities and temperatures. A valve situated between the SFE vessel and restrictor enabled short on-column SFE–GC–MS injection times, after which the restrictor was withdrawn from the injection port. Off-line GC–MS analyses were performed by injecting propan-2-ol SFE extracts into the 'on-column' interface. For any given SF-$CO_2$ density, the on-column SFE–GC–MS total ion chromatogram (TIC) enabled the detection of a greater number of extracted clove components than did the off-line analyte collection procedure. Typical results are shown in Figures 9.10(a) and 9.10(b). This was attributed to lower flow splitting and a reduction in the possibility for thermal decomposition during the injection stage. Furthermore, the on-column procedure was reported to enable the detection of more extracted components than the previous SFE–GC–MS technique described by Hawthorne *et al.* [24]. This on-column procedure allowed a flexible approach for the utilisation of SF-$CO_2$ selectivity factors, including fluid density and extraction time, for the analysis of clove constituents.

Snyder *et al.* [26] have developed an SFE–GC–MS procedure to identify and quantify volatile and semivolatile marker compounds in fire-exposed meat products. Approximately 0.5 g of meat sample, held in place by glass wool plugs, was placed into a 5 ml SFE vessel. A heated transfer line was used to house a 50 μm frit restrictor which was inserted into the GC injection port. Static SFE was performed ($CO_2$ at 100 atm, 60°C) for 5 min, after which the volatiles and semivolatiles were transferred onto the capillary column (held at −50°C to facilitate cryogenic trapping) using 5 min dynamic SFE. The flow rate of expanded gaseous $CO_2$ during dynamic SFE was 40 ml/min. The extracted compounds were then analysed using EI GC–MS. Increased levels of volatile carbonyl and aromatic compounds were found to be present in the fire-exposed meat samples. Naphthalene was selected as a marker compound for fire-damaged meat. Quantitative SIM GC–MS studies were

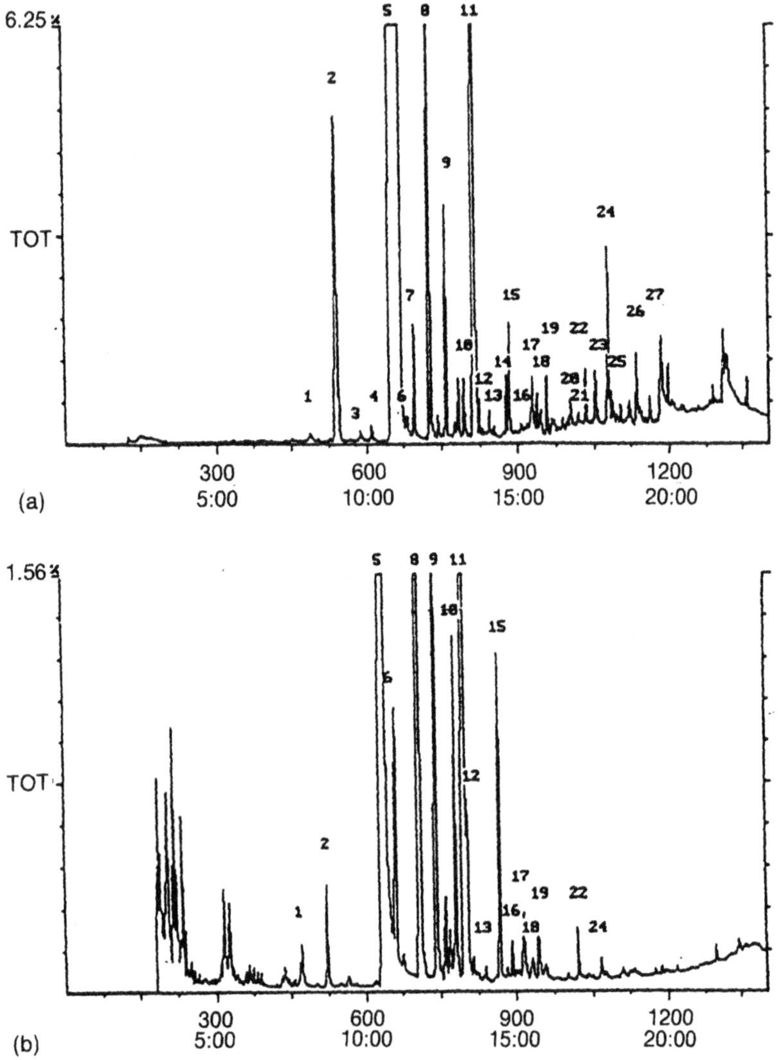


**Figure 9.10** Supercritical fluid extraction–gas chromatography–mass spectrometry total ion chromatograms obtained for cloves with (a) on-column interface; (b) solvent trapping. Reproduced with permission from ref. 25.

performed using the molecular ion of naphthalene [mass to charge ratio ($m/z$) = 128] with hexadecane as internal standard. The SFE recovery of naphthalene was determined by the addition of [$^2H_8$]-naphthalene to the meat samples prior to extraction. The SFE–GC–MS TIC of a fire-exposed chicken product is shown in Figure 9.11. A 1 ppb limit of quantitation for naphthalene was reported using this procedure.

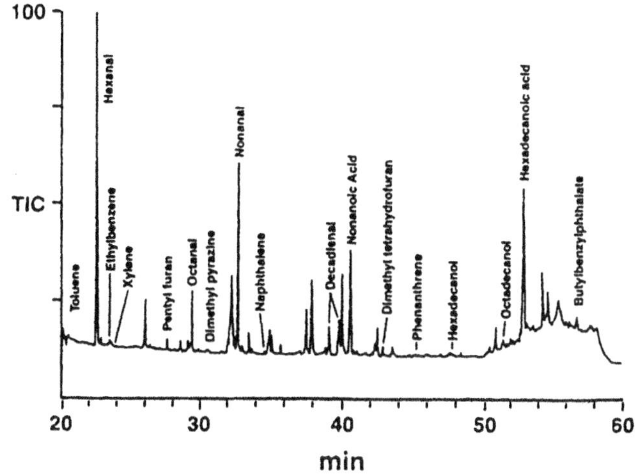


**Figure 9.11** Supercritical fluid extraction–gas chromatography–mass spectrometry total ion chromatogram (TIC) of smoked chicken product that was exposed to fire. Reproduced with permission from ref. 26.

## 9.5 On-line supercritcal fluid extraction–capillary supercritical fluid chromatography–mass spectrometry

Murugaverl *et al.* [27] investigated the use of various sorbents to provide an *in situ* cleanup stage for the SFE–cSFC–MS analysis of three carbamate pesticides spiked into samples of bovine muscle. In an earlier study (section 4.2.4) Murugaverl and Voorhees [28] demonstrated an indirect liquid SFE–cSFC procedure for the analysis of pesticides in soybean oil loaded onto $C_{18}$ sorbent. In that study seven sorbents, including divinylbenzene-styrene, alumina, silica, $C_{18}$, cyanopropyl, aminopropyl and diol, were evaluated to provide an on-line cleanup stage. This was necessary to prevent lipids (including fatty acids, glycerides and sterols) which are readily soluble in SF-$CO_2$ from adversely affecting the chromatographic performance and from causing rapid mass spectrometer ion source contamination. The bovine tissue samples were ground with Hydromatrix (30% m/m) before being loaded onto sorbents contained within a 40 mm × 3 mm i.d. SFE vessel. Dynamic SFE was performed ($CO_2$ at 219 atm, 90°C; extraction time 30 min) such that the extracted material was cryogenically trapped at the top of a cSFC column (1.5 m × 50 μm i.d.). Following this, cSFC–MS was performed in EI mode using a benchtop mass spectrometer. Optimum *in situ* fat cleanup was obtained using a diol:$C_{18}$ (7:93) mixed sorbent. Typical results from these preliminary investigations are shown in Figures 9.12(a) and 9.12(b). Using this SFE–cSFC–MS procedure, bovine tissue samples ranging from 50 mg to 130 mg loaded onto 100 mg mixed sorbent gave a 200 ppb SIM detection limit for bendiocarb.

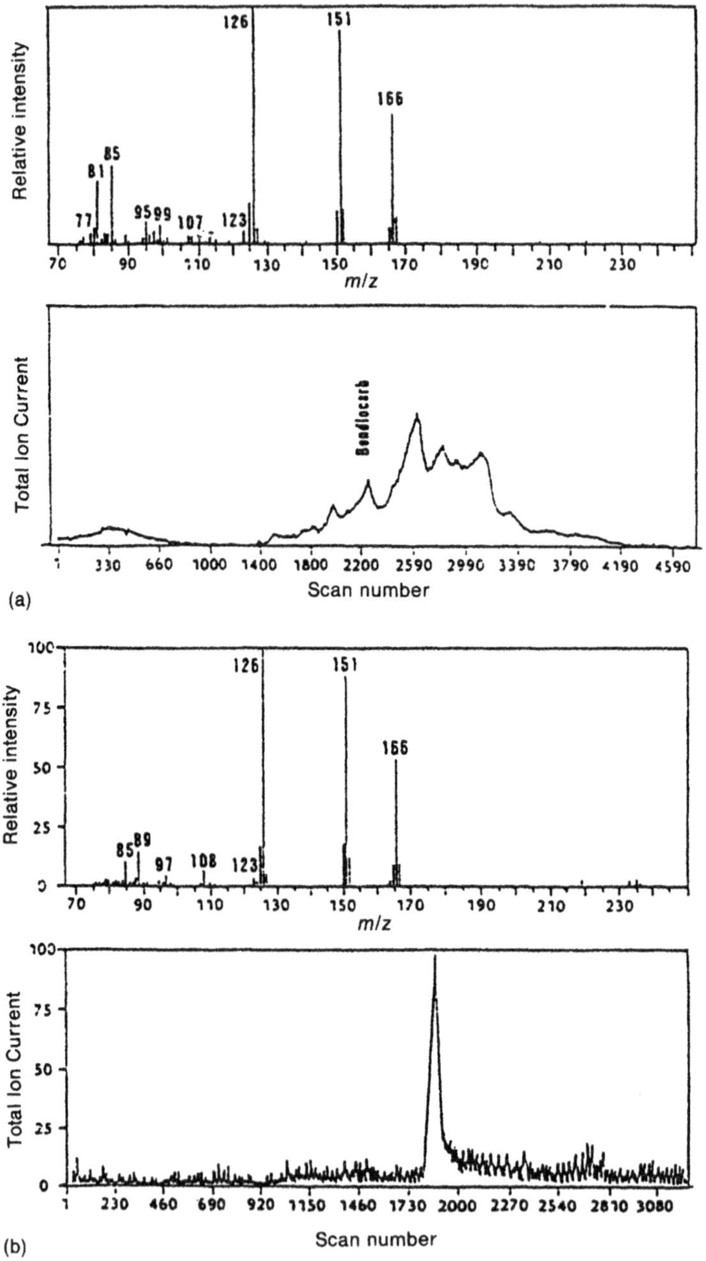


**Figure 9.12** Supercritical fluid extraction–capillary supercritical fluid chromatography–mass spectrometry total ion chromatograms (TICs) and electron ionisation (EI) mass spectra obtained for bendiocarb using (a) $C_{18}$ cleanup only; (b) diol : $C_{18}$ (7 : 93) cleanup. Reproduced with permission from ref. 27.

An SFE–cSFC–MS procedure has been described by MacKay and Smith [29] as a rapid means for the qualitative determination of four chlorinated organophosphate flame retardants in polyurethane foams. Members of this class of polymer additives are difficult to analyse using GC since they are thermally labile. Furthermore, their total water immiscibility renders them unsuitable for reverse phase HPLC whereas their rapid elution in organic phase solvent fronts was reported to hinder the use of normal phase HPLC. Since flame retardants are present in foam at 1%–3% m/m, approximately 0.2 mg of sample loaded into a 0.5 ml SFE vessel was used. Dynamic SFE ($CO_2$ at 300 atm, 60°C; extraction time 5 min) was performed, with extracted material being cryogenically trapped onto the top of a cSFC column (10 m × 50 μm i.d.). A length of 50 μm i.d. fused silica (housed within a stainless-steel transfer tube at room temperature) was used to connect the column with an integral frit restrictor linked to a magnetic sector mass spectrometer EI ion source The ion source temperature was maintained at 350°C to help prevent $CO_2$ cluster ion formation. All four fire retardants could be rapidly determined by this SFE–cSFC–MS technique despite a reduction in chromatographic performance attributed to imprecise analyte trapping. The on-line EI mass spectra were characterised by the higher mass ions being present at lower relative abundances compared with the direct insertion probe (DIP) mass spectra obtained for these compounds. This distortion was thought to arise from the higher source pressures associated with cSFC–MS. In general, this effect becomes more pronounced for strongly retained components, the cSFC separation of which requires the use of a pressure programme; that is, such components will be introduced into the mass spectrometer ion source in the presence of higher levels of $CO_2$.

In a further study using the same instrumentation as described above, MacKay and Smith [30] extended their SFE–cSFC–MS studies of polymers, characterising a number of additives associated with polyurethane thermoplastics. Dynamic SFE ($CO_2$ at 300 atm, 60°C; extraction time 10 min) with on-column (10 m × 50 μm i.d.) trapping was performed. The TIC obtained following the EI SFE–cSFC–MS analysis of 90-N is shown in Figure 9.13. A number of extracted polymer additives were identified using a standard mass spectral library. Examples include butylated hydroxytoluene (peak 3), triphenyl phosphate (peak 7) and a homologous series of ethers (labelled as peaks 1). Peak 9 was identified as the antioxidant Irganox 1010 ($M_r = 1178$) on the basis of retention time. Other polymer additives, some of which are not amenable to GC–MS, associated with pellethane and estane were also rapidly identified using this SFE–cSFC–MS procedure.

Bucherl *et al.* [31] have also developed a rapid SFE–cSFC–MS procedure for the rapid identification of a variety of polymer additives. Polymer samples ranging from 2 mg to 5 mg or liquid standards (2–10 μl) were loaded into an extraction chamber. Figure 9.14 shows a schematic of the system used. Dynamic SFE was performed using an SF-$CO_2$ pressure

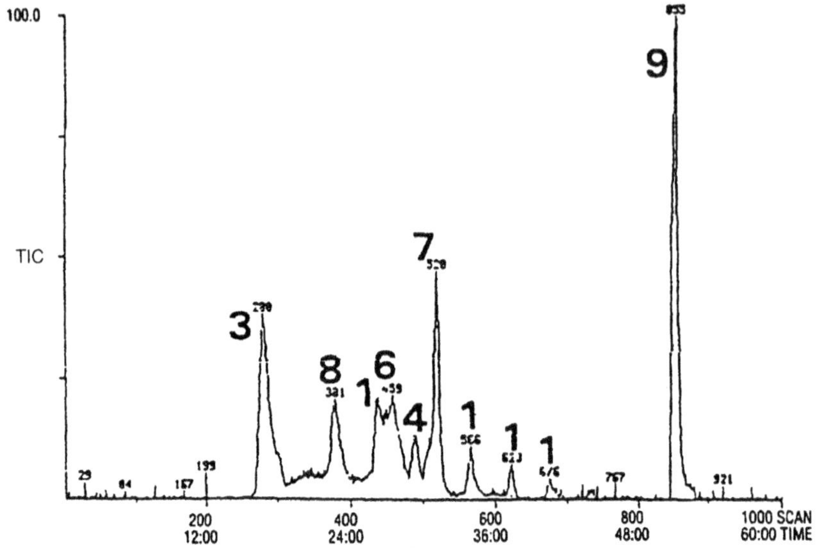


**Figure 9.13** Supercritical fluid extraction–capillary supercritical fluid chromatography–mass spectrometry total ion chromatogram (TIC) obtained for 90-N. Peak assignments: 1 = homologous series of ethers; 3 = butylated hydroxytoluene (BHT); 4 = naphthalene-based compounds; 6 = an adipate; 7 = triphenyl phosphine; 8 = tentatively assigned as a sterically hindered phenol, related to BHT; 9 = Irganox 1010. Reproduced with permission from ref. 30.

programme (60°C; extraction time 30 min) with the extracted components being concentrated in a cryotrap. After extraction, the cryotrap was heated to 200°C and the components were separated using a multiramp cSFC density programme. The use of a column eluent flow splitter, equipped with two tapered restrictors, enabled simultaneous SFC/FID and SFC–MS detection. A 1:1 split resulted in 2 ml/min decompressed $CO_2$ being supplied to each detector. A range of polymer additives were analysed including: light stabilisers, antioxidants, processing stabilisers and an antistatic agent. The EI TIC obtained for 2 mg polyolefin film analysed using SFE–cSFC–MS is shown in Figure 9.15. The levels of the three additives (Tinuvin 326, Irgafos 168 and Irganox 1330) were estimated to range from 10 ng to 100 ng, based upon the responses obtained for standards. Detailed interpretation of EI fragmentation patterns led to the identification and classification of seven components in Irgafos P-EPQ (a processing stabiliser) analysed using SFE–cSFC–MS. Various polymer additives in samples of polypropylene film and lacquers were also rapidly identified.

In a separate report, Bucherl *et al.* [32] described the construction of the cSFC–MS interface used in the above studies. In order to minimise the amount of gaseous $CO_2$ entering the ion source, the integral restrictor interface was equipped with a jet separator, as shown in Figure 9.16.

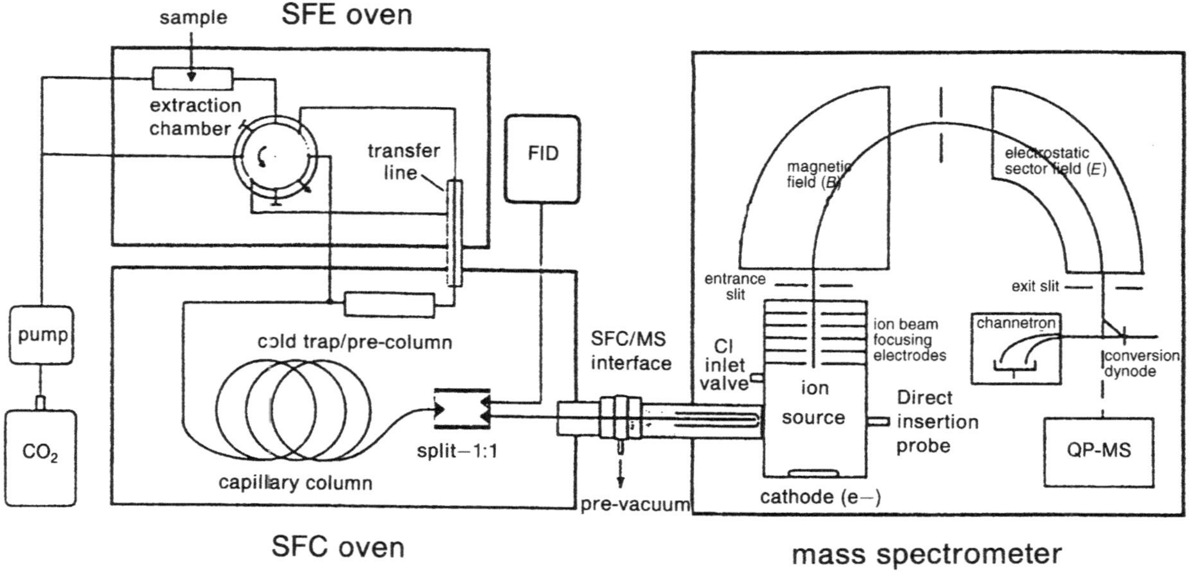


**Figure 9.14** Schematic of a supercritical fluid extraction–capillary supercritical fluid chromatography–mass spectrometry system. CI = chemical ionisation; FID = flame ionisation detection; QP = quadrupole. Reproduced with permission from ref. 31.

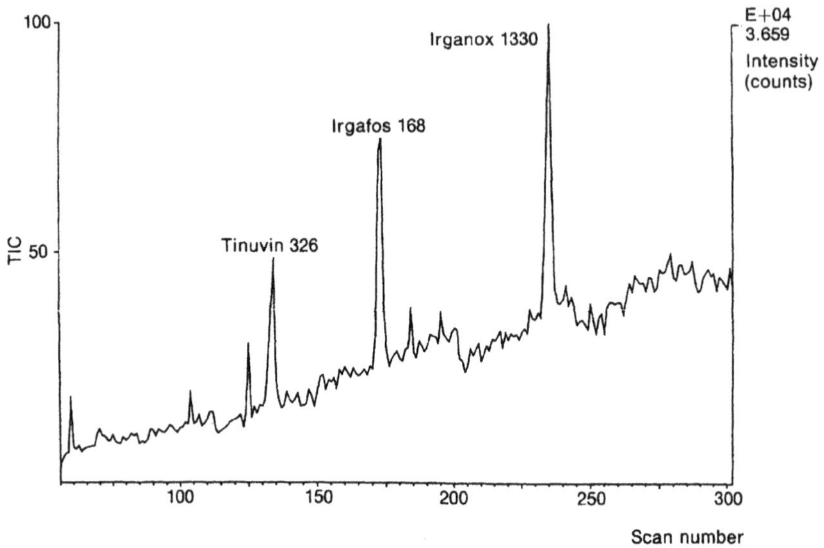


**Figure 9.15** Supercritical fluid extraction–capillary supercritical fluid chromatography–mass spectrometry (SFE–cSFC–MS) total ion chromatogram (TIC) obtained for polyolefin film. E = ion count, expressed exponentially. Reproduced with permission from ref. 31.

Fractionation of sample vapour from gaseous $CO_2$ occurs in the expanding gas stream. The heavier analyte species have sufficient forward momentum to escape the influence of the 'initial vacuum' whereas the lighter $CO_2$ molecules do not. The interface is also equipped with its own independent heaters which prevent deposition of analytes in the interface and plugging due to the formation of solid $CO_2$. Different interface operating conditions, including temperature, restrictor position, flow rate and sample transfer conditions, were optimised. A standard solution of polymer additives was analysed using SFE–cSFC–MS. The on-line dynamic SFE and cSFC stages both utilised

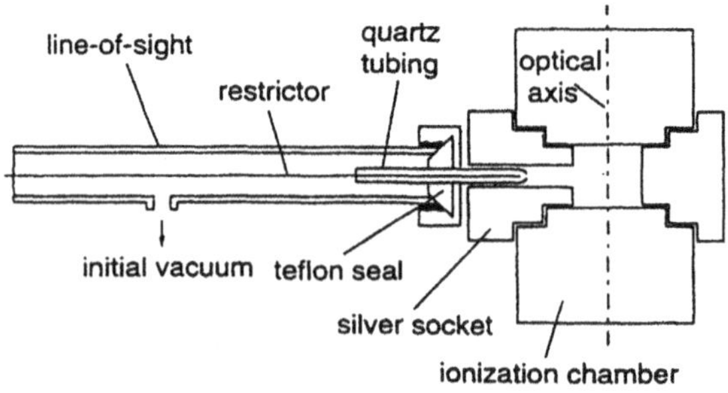


**Figure 9.16** Diagram of a capillary supercritical fluid chromatography–mass spectrometry interface. Reproduced with permission from ref. 32.

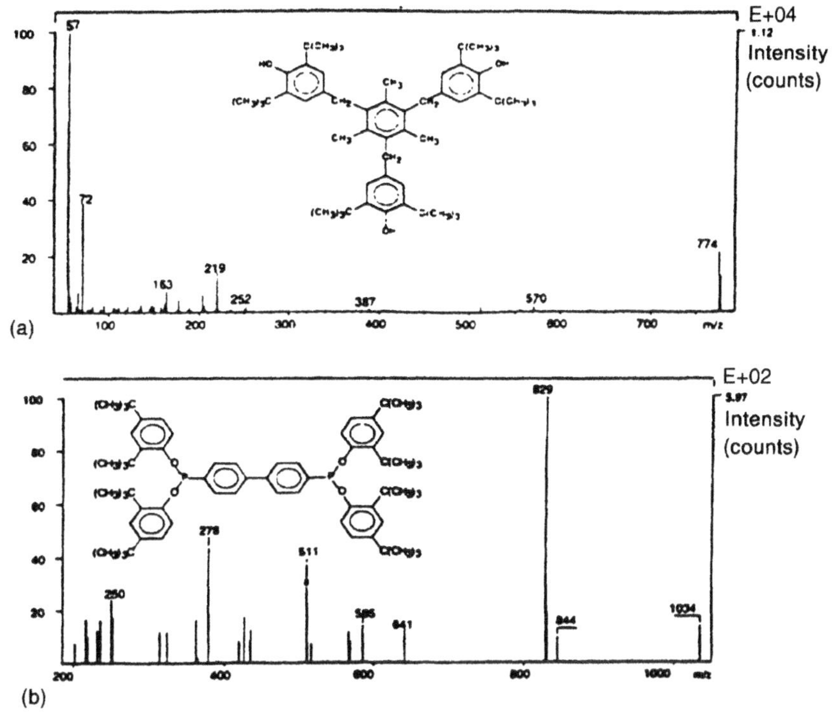


**Figure 9.17** Supercritical fluid extraction–capillary supercritical fluid chromatography–mass spectrometry mass spectra obtained for: (a) Iganox 1330 ($M_r = 774$) and (b) Irgafos P-EPQ ($M_r = 1034$). E = ion count, expressed exponentially. Reproduced with permission from ref. 32.

pressure programmes. DIP reference EI spectra were also obtained. Using this interface it was demonstrated that the on-line EI SFE–cSFC–MS mass spectra obtained for all the polymer additives were identical to those obtained using DIP. Figure 9.17 shows examples of these EI SFE–cSFC–MS mass spectra.

## 9.6 On-line supercritical fluid extraction–packed column supercritical fluid chromatography–mass spectrometry

The use of SFE–SFC–MS–MS has been evaluated for the detection of a small group of veterinary drugs (trimethoprim, hexestrol, diethylstilbestrol and dienestrol) in freeze-dried pig's kidney [33]. A hybrid tandem mass spectrometer (a double focusing magnetic sector instrument linked to two quadrupole analysers) equipped with a moving belt interface was used for these studies. An HPLC guard column served as the SFE vessel which had an average loading of 28.4 mg freeze dried tissue. Off-line dynamic SFE experiments indicated that approximately 20% m/m of the sample (predominantly lipids) were extracted in 8.5 min ($CO_2$ at 302 atm, 45°C). The use of direct SFE–MS–MS would have been impractical since rapid contamination of the mass spectrometer would have occurred. In order to minimise this problem an SFE–SFC cleanup stage was developed (initial studies were conducted with a standard solution of the four drugs) using a packed aminopropyl column (100 mm × 4.6 mm i.d.). During the SFE of spiked

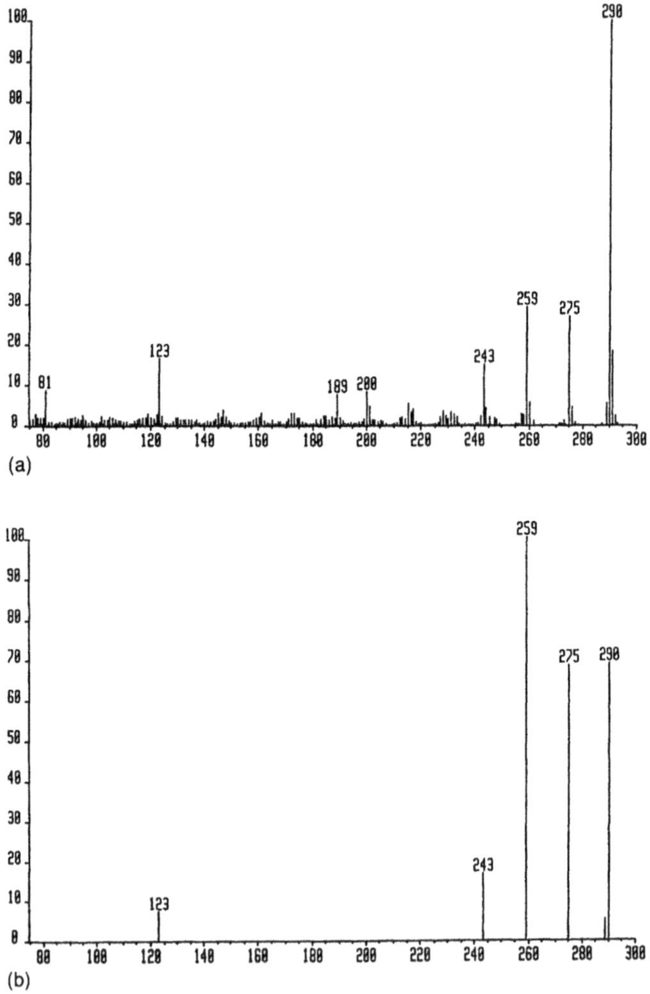


**Figure 9.18** (a) Direct insertion probe electron ionisation mass spectrum obtained for trimethoprim ($M_r = 290$), the molecular ion being the most abundant species was selected as the precursor for collision-induced dissociation (CID) product-ion studies; (b) CID product-ion mass spectrum obtained for trimethoprim's molecular ion. Reproduced with permission from ref. 33.

tissue samples, using unmodified SF-$CO_2$, the more polar drugs were strongly retained at the top of the column whereas the majority of non-polar lipids eluted to waste. Following SFE, a methanol-modified SF-$CO_2$ gradient programme was used to elute the drugs. Only the eluant associated with the drugs' retention time peroid was analysed using mass spectrometry. Despite this it was estimated that approximately 100–200 μg of endogenous material was applied to the moving belt interface per analysis. The EI full scan and MS–MS collision-induced dissociation (CID) product ion mass spectra recorded for pure trimethoprim ($M_r = 290$) and trimethoprim's molecular ion are shown in Figures 9.18(a) and 9.18(b), respectively. Figure 9.19 shows the SFE–SFC chromatogram, UV detection at 215 nm, obtained for a blank kidney sample, with the retention time interval for elution of the drugs being indicated. Confirmation of the high levels of co-extracted lipids is provided by Figure 9.20(a) which shows the SFE–SFC–MS EI mass spectrum obtained at the retention time of trimethoprim, spiked at the 10 ppm level. As shown in Figure 9.20(b), resorting to CID product-ion scanning, SFE–SFC–MS–MS enabled the detection of trimethoprim at the 10 ppm level. The two CID product ions observed at $m/z$ 127 and 133, shown in Figure 9.20(b), were not produced in the corresponding

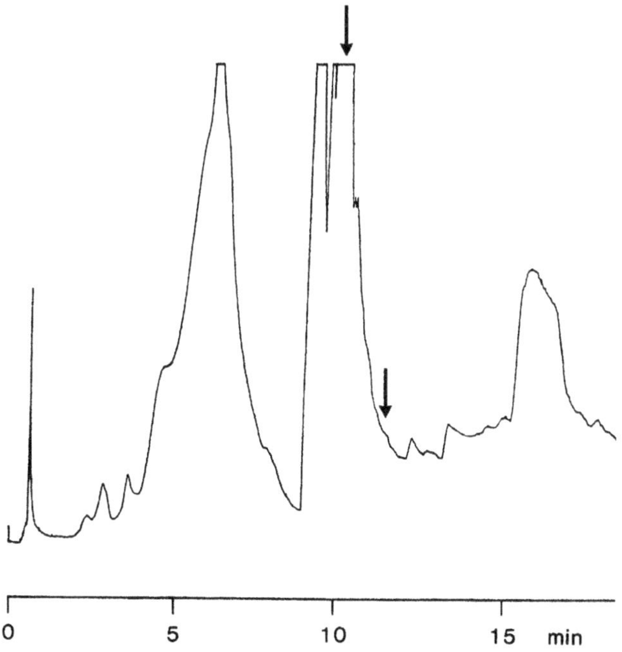


**Figure 9.19** Supercritical fluid extraction–supercritical fluid chromatography analysis of a blank porcine kidney sample. Ultraviolet detection at 215 nm. The elution range of the drugs is indicated by the arrows. Reproduced with permission from ref. 33.

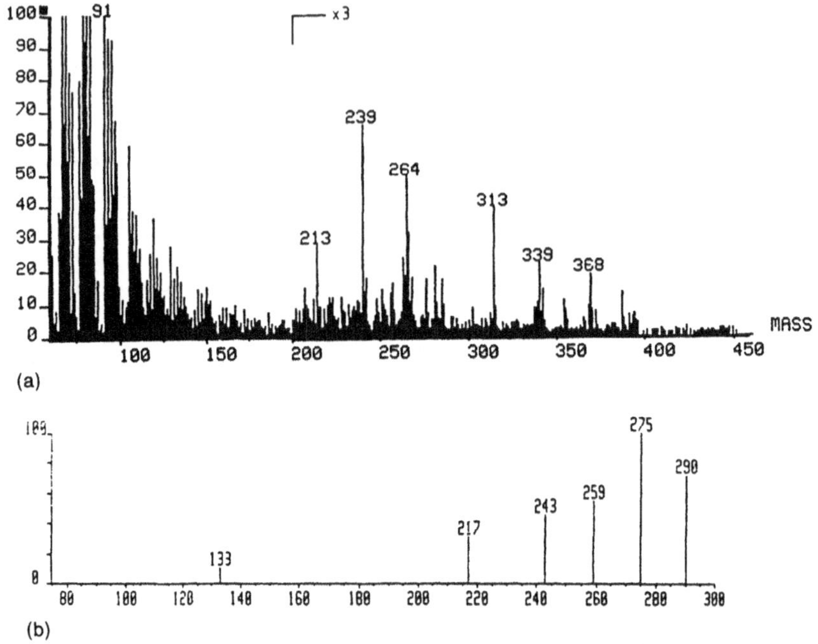


**Figure 9.20** (a) Electron ionisation mass spectrum obtained at supercritical fluid extraction–supercritical fluid chromatography (SFE–SFC) retention time of trimethoprim; the result of SFE–SFC–mass spectrometry (MS) analysis of kidney sample spiked at 10 ppm with each drug; (b) SFE–SFC–MS–MS collision-induced dissociation product-ion mass spectrum obtained at retention time of trimethoprim. Reproduced with permission from ref. 33.

CID product-ion mass spectrum obtained for pure trimethoprim, shown in Figure 9.18(b). Presumably these two extraneous ions originated from co-chromatographing species, with trimethoprim also providing an $m/z$ 290 precursor ion. Hexestrol, diethylstilbestrol and dienestrol were also each detected at the 10 ppm level using SFE–SFC–MS–MS with CID product-ion scanning. Using selected reaction monitoring (an MS–MS procedure which involves two-stage SIM) each of the drugs could be easily detected at the 1 ppm level. In order to improve the sensitivity of this technique it would be necessary to reduce the co-elution of lipids with the drugs. As shown in Figure 9.20(a), the EI mass spectrum recorded at the retention time of trimethoprim is completely dominated by ions produced from other species. Consequently, the primary ionisation efficiency of trimethoprim is low (the ionising electron beam being effectively 'mopped up' by lipids), which results in poor MS–MS sensitivity for its analysis. A possible rapid solution to this problem (which would also help reduce mass spectrometer contamination) involves an *in situ* cleanup procedure. France *et al.* [34], Parks and Maxwell [35] and Murugaverl *et al.* [27] (section 9.5) have described the use of solid phase sorbents for *in situ* cleanup of lipids from

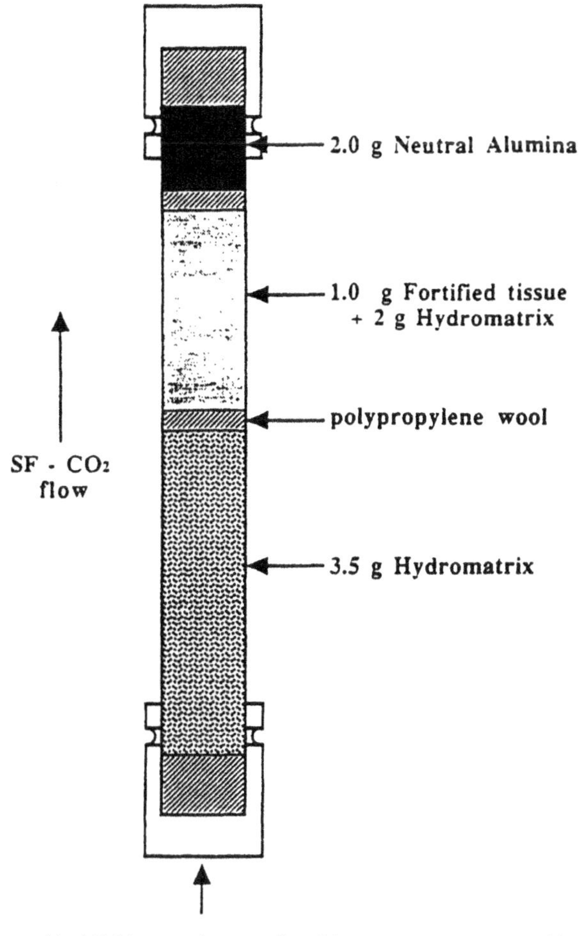


**Figure 9.21** Supercritical fluid extraction vessel packing arrangement to provide *in situ* cleanup of extracted lipids. Reproduced with permission from ref. 35.

fat and tissue samples. Figure 9.21 summarises the procedure developed by Parks and Maxwell, with the neutral alumina retaining the lipids extracted from chicken tissues. Also, see section 3.1.3 for further details concerning lipid cleanup procedures.

A direct aqueous SFE vessel has been interfaced to a packed column SFC–MS system with use of a high-pressure multivalve system for the analysis of phenols in water [36]. Schematics of the mutivalve system are provided in section 7.4.2. Analyses were performed using 250 ml water samples spiked with three phenols (2,4-dibromophenol, 2,4,5-trichlorophenol and 4-nitrophenol) each at the 40 ppb level. A 5 μm aminopropyl column (150 mm × 4.6 mm i.d.) was used as the analyte trap during direct aqueous SFE ($CO_2$ at 32.33 MPa,

65°C; extraction time 45 min). After extraction, the valve configuration was switched such that a second 5 μm aminopropyl column (150 mm × 4.6 mm i.d.) became linked in series with the first column. This arrangement was necessary to compensate for degradation of chromatographic performance arising from imprecise analyte trapping on the first column. All SFE–SFC–MS analyses were performed using negative ion atmospheric pressure chemical ionisation (APCI), since each phenol provided an intense response

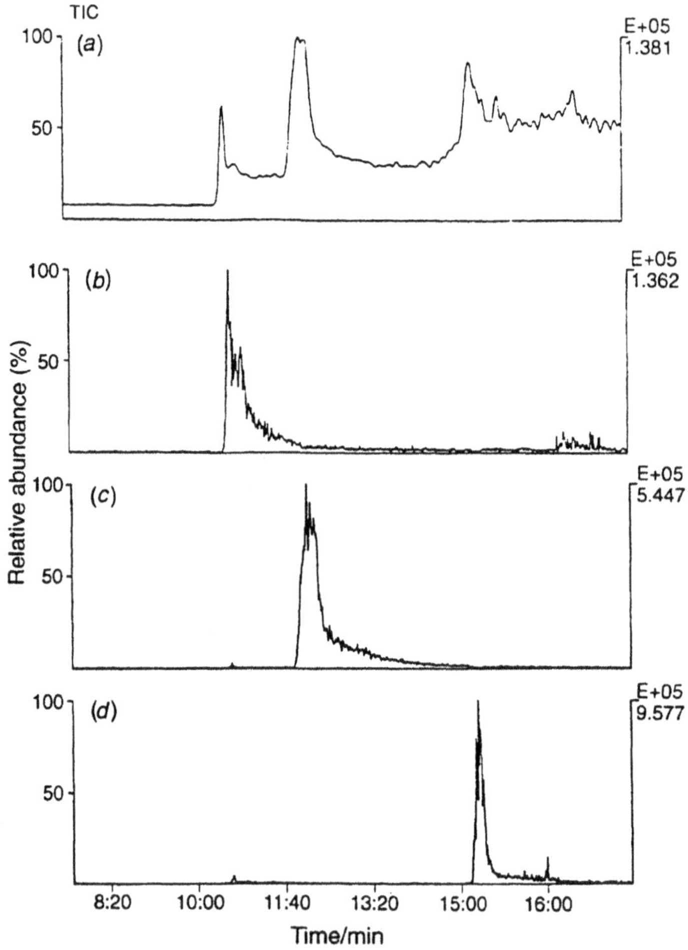


**Figure 9.22** (a) Supercritical fluid extraction–supercritical fluid chromatography–mass spectrometry total ion chromatogram (TIC) obtained for 250 ml water sample, spiked with three phenols, each at the 40 ppb level. Ion profile responses were obtained for the $[M-H]^-$ species of individual phenols; (b) $m/z = 251$ for 2,4-dibromophenol; (c) $m/z = 195$ for 2,4,5-trichlorophenol; (d) $m/z = 138$ for 4-nitrophenol. E = ion count, expressed exponentionally. Reproduced with permission from ref. 36.

using this ionisation mode. Each phenol's negative ion APCI mass spectrum was dominated by the presence of an $[M-H]^-$ ion. The SFE–SFC–MS negative ion APCI TIC obtained for the analysis of the three phenols at the 40 ppb level is shown in Figure 9.22(a). Although the full-scan response from all the phenols provided a poor signal to noise ratio, each could be readily detected by inspection of the respective $[M-H]^-$ ion profiles, as shown in Figures 9.22(b) and 9.22(d). The SFE–SFC–MS full-scan negative ion APCI mass spectrum obtained for 2,4,5-trichlorophenol is shown in Figure 9.23. The results of quantitative GC–MS following liquid–liquid extraction of the SFE extracted water sample indicated the following SFE phenol recoveries: 2,4-dibromophenol, 88%; 2,4,5-trichlorophenol, 81%; 4-nitrophenol, 67%. The entire SFE–SFC–MS analysis, leading to confirmation of the identities of the target analytes, was achieved in a period of approximately 1 h with minimal sample handling. It was concluded that given the intensity

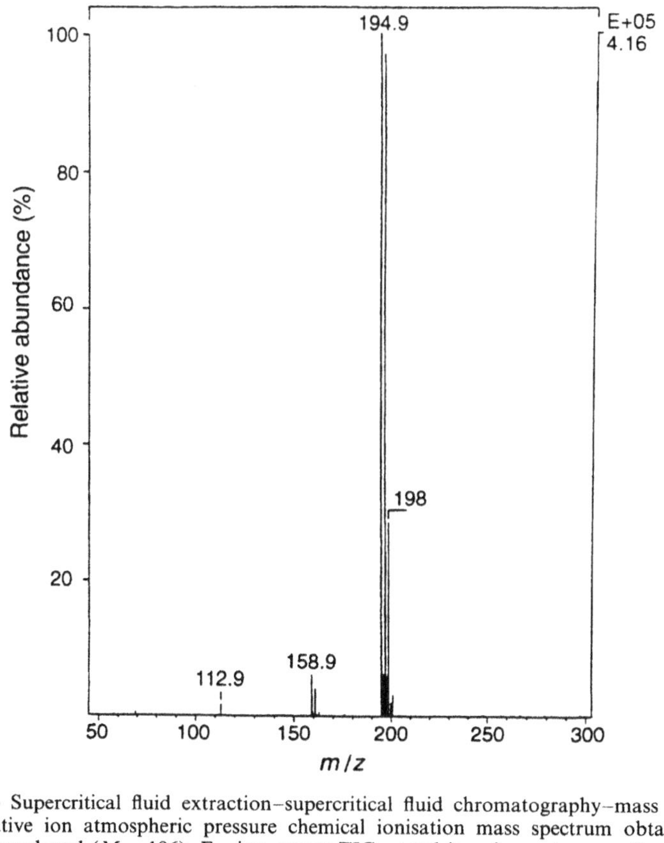


**Figure 9.23** Supercritical fluid extraction–supercritical fluid chromatography–mass spectrometry negative ion atmospheric pressure chemical ionisation mass spectrum obtained for 2,4,5-trichlorophenol ($M_r = 196$). E = ion count; TIC = total ion chromatogram. Reproduced with permission from ref. 36.

of the mass spectrum obtained for each phenol, the determination of these compounds at levels lower than 40 ppb should be readily achievable in full-scan mode. The possibility of sub-parts-per-trillion detection limits for these phenols seems realistic by resorting to the use of SIM SFE–SFC–MS procedures.

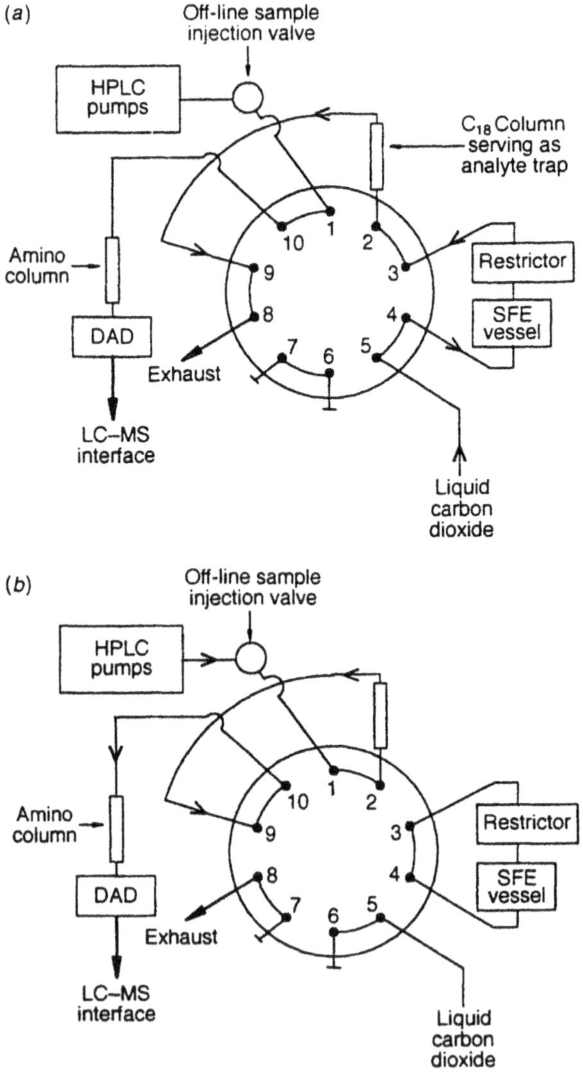


**Figure 9.24** Configuration of a ten-port high-pressure valve used for (a) direct aqueous supercritical fluid extraction (SFE); (b) liquid chromatography–mass spectrometry (LC–MS). DAD = photodiode array detector; HPLC = high-performance liquid chromatography. Reproduced with permission from ref. 37.

## 9.7 On-line supercritical fluid extraction–liquid chromatography–mass spectrometry

The use of a single ten-port switching valve to interface a direct aqueous SFE vessel to an LC–MS instrument equipped with a UV/VIS diode array detector has been reported [37]. Figure 9.24 shows the valve configurations for the SFE and LC–MS stages of the analysis. Analyses were performed using 250 ml water samples spiked with three drugs (estrone, hexestrol and zeranol), each at the 200 ppb level. A 5 μm $C_{18}$ column (250 mm × 4.6 mm i.d.) acted as the analyte trap during direct aqueous SFE ($CO_2$ at 24.1 MPa, 55°C; extraction time 30 min). After SFE, the trapped drugs were eluted from the $C_{18}$ column and band focused onto the top of a 5 μm 150 mm × 4.6 mm i.d. aminopropyl column. All SFE–LC–MS analyses were performed using a moving belt interface with ammonia chemical ionisation (CI). After band focusing, the drugs were eluted from the aminopropyl column using a heptane–ethanol gradient programme (section 8.5). The direct aqueous SFE–LC–MS ammonia CI TIC obtained for the drugs at the 200 ppb level is shown in Figure 9.25(a). The protonated molecular ion profiles obtained for estrone and zeranol are shown in Figures 9.25(b) and 9.25(d), respectively, along with the *m/z* 135 fragment ion profile

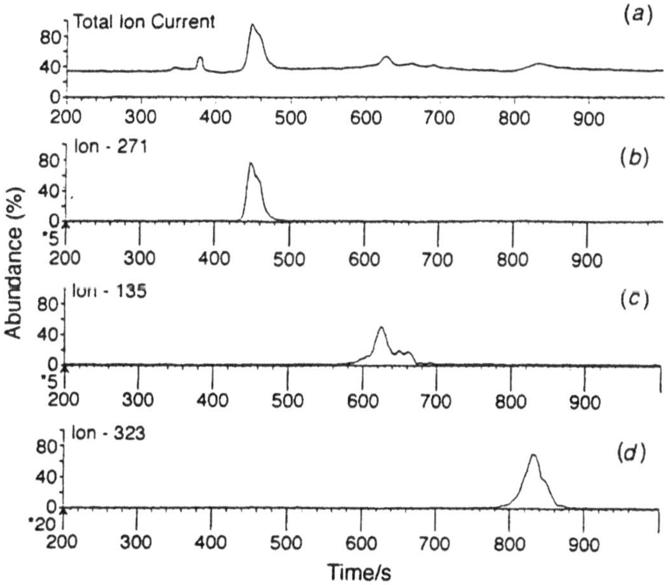


**Figure 9.25** (a) Full-scan ammonia chemical ionisation supercritical fluid extraction–liquid chromatography–mass spectrometry total ion chromatogram obtained for 250 ml water, spiked with three drugs, each at the 200 ppb level. Ion profiles were obtained for: (b) estrone $[M+H]^+$ species at $m/z = 271$; (c) hexestrol fragment ion at $m/z = 135$; (d) zeranol $[M+H]^+$ species at $m/z = 323$. Reproduced with permission from ref. 37.

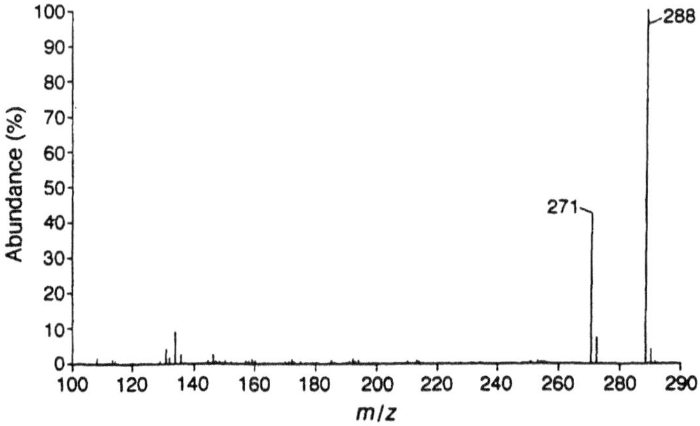


**Figure 9.26** Supercritical fluid extraction–liquid chromatography–mass spectrometry ammonia chemical ionisation mass spectrum obtained for estrone ($M_r = 270$). The ions at $m/z = 272$ and $m/z = 288$ correspond to the protonated and ammonium adduct molecular species, respectively. Reproduced with permission from ref. 37.

characteristic of hexestrol in Figure 9.25(c). The full-scan ammonia CI SFE–LC–MS mass spectrum obtained for estrone is shown in Figure 9.26. Although not investigated, it was surmised that by using this SFE–LC–MS procedure sub-parts-per-billion detection limits for these drugs using SIM should be possible. Selection of appropriate HPLC columns should enable a range of other LC–MS techniques, including particle beam, APCI, electrospray and dynamic fast atom bombardment, to be interfaced with SFE systems with the same valve arrangement.

## 9.8 Conclusions

At present, on-line combinations of SFE with FTIR, MS and NMR are in the early stages of development. Despite this, the applications described within this chapter serve to demonstrate the potential of these on-line techniques for solving a wide range of analytical problems.

As pointed out by Taylor and Jordan [4], a general purpose SFE–FTIR instrument would need to be quite versatile in terms of its ability to provide static, dynamic and static–dynamic modes of SFE–FTIR operation, with the option of cryotrapping. These facilities are required to prevent different analyte extraction profiles giving rise to sample discrimination effects and/or poor quantitation. It is anticipated that the continued phase-out of the production of solvents typically used in IR spectroscopy (e.g. carbon tetrachloride and freon 113) offers real potential for the development of both off-line and on-line SFE sample preparation procedures for FTIR.

The interfacing of SFE with GC–MS and cSFC–MS is relatively straightforward, with split/splitless and on-column modes of extract introduction being possible. However, care is needed to avoid the possibility of column overload and/or rapid mass spectrometer contamination. For this reason off-line SFE pilot studies should be performed on unknown samples to determine the levels of extracted compounds.

The development of integrated *in situ* cleanup procedures, in particular with packed column SFE–SFC–MS and SFE–LC–MS, should afford a means of analysing complex samples where high levels of endogenous compounds other than target analytes are also extracted (e.g. lipids from tissue samples being screened for drug residues).

With the development of appropriate valve systems, direct aqueous SFE–SFC–MS and SFE–LC–MS procedures have shown their potential to achieve detection limits associated with more established on-line aqueous sample preparation procedures (e.g. SPE). Further developments in this area, particularly automation and the use of indirect aqueous on-line SFE procedures, can be anticipated.

## References

1. Slobodnik, J., Hoekstra-Oussoren, S.J.F., Jager, M.E. *et al.* (1996) On-line solid-phase extraction–liquid chromatography–particle beam mass spectrometry and gas chromatography–mass spectrometry of carbamate pesticides. *Analyst*, **121**, 1327–34.
2. Puig, D., Silgoner, I., Grasserbauer, M. and Barcelo, D. (1997) Part-per-trillion level determination of priority methyl-, nitro-, and chlorophenols in river water samples automated on-line liquid/solid extraction followed by liquid chromatography/mass spectrometry using atmospheric pressure chemical ionization and ion spray interfaces. *Analytical Chemistry*, **69**, 2756–61.
3. Ollers, S., van Lieshout, M., Janssen, H.-G. and Cramers, C. A. (1997) Development of an interface for directly coupled solid phase extraction and GC–MS analysis. *LC–GC International*, **10**(7), 435–9.
4. Taylor, L.T. and Jordan, S.L. (1995) Supercritical fluid extraction coupled directly with Fourier transform infrared spectrometry. *Journal of Chromatography A*, **703**, 537–48.
5. Raynor, M.W., Davies, I.D., Bartle, K.D. *et al.* (1987) Interfacing supercritical fluid chromatography with Fourier transform infrared spectrometry. *European Chromatography News*, **1**, 18–22.
6. Noble, D. (1993) Here today, gone tomorrow: halogenated solvents in analytical chemistry. *Analytical Chemistry*, **65**, 693A–695A.
7. Ikushima, Y., Saito, N., Hatakeda, K. *et al.* (1992) In situ monitoring of extraction and separation behavior of lipids with supercritical carbon dioxide. *Industrial Engineering and Chemical Research*, **31**, 568–74.
8. Kirschner, C.H. and Taylor, L.T. (1993) Quantitative analysis by on-line supercritical fluid extraction/Fourier transform infrared spectrometry. *Analytical Chemistry*, **65**, 78–83.
9. Kirschner, C.H., Jordan, S.L., Taylor, L.T. and Seemuth, P.D. (1994) Feasibility of extraction and quantification of fiber finishes via on-line SFE/FT-IR. *Analytical Chemistry*, **66**, 882–7.
10. Jordan, S.L., Taylor, L.T. and Seemuth, P.D. (1995) Extracting and quantifying a polyester fiber finish using on-line SFE/FT-IR with intermediate trapping. *Textile Research Journal*, **65**, 230–5.

11. Liescheski, P.B. (1996) Supercritical fluid extraction coupled to infrared spectroscopy for iodine number analysis of edible oils. *Journal of Agricultural Food Chemistry*, **44**, 823–8.
12. Budavari, S. (ed.) (1989) *Merck Index*, 11th edn, Merck & Co., Rahway, NJ.
13. Liescheski, P.B., Seemuth, P.D. and Trask, T.O. (1996) Analysis of fiber finishes using SFE coupled to a simple, universal IR flow cell. *Textile Research Journal*, **66**, 436–41.
14. Heglund, D.L., Tilotta, D.C., Hawthorne, S.B. and Miller, D.J. (1994) Simple fiber-optic interface for on-line supercritical fluid extraction Fourier transform infrared spectrometry. *Analytical Chemistry*, **66**, 3543–51.
15. Tilotta, D.C., Heglund, D.L. and Hawthorne, S.B. (1996) On-line SFE-FTIR spectrometry with a fiber optic transmission cell. *American Laboratory*, **28**, 36R–36T.
16. Howdle, S.M., Ramsay, J.M. and Cooper, A.I. (1994) Spectroscopic analysis and in situ monitoring of impregnation and extraction of polymer films and powders using supercritical fluids. *Journal of Polymer Science: Part B: Polymer Physics*, **32**, 541–49.
17. Minty, B., Ramsey, E.D. and Lewis, R. (1996) Hydrocarbons in water: analysis using on-line aqueous supercritical fluid extraction–Fourier transform infrared spectroscopy. *Analytical Communications*, **33**, 203–4.
18. Hedrick, J. and Taylor, L.T. (1989) Quantitative supercritical fluid extraction/supercritical fluid chromatography of a phosphonate from aqueous media. *Analytical Chemistry*, **61**, 1986–8.
19. Wieboldt, R.C., Kempfert, K.D. and Dalrymple, D.L. (1990) Analysis of antioxidants in polyethylene using supercritical fluid extraction/supercritical fluid chromatography and infrared detection. *Applied Spectroscopy*, **44**, 1028–34.
20. Jordan, S.L., Taylor, L.T. and Miller, R.J. (1997) Analysis of additives and monomers in nylon and polystyrene. *Textile Chemist and Colorist*, **29**, 25–32.
21. Braumann, U., Handel, H., Albert, K. *et al.* (1995) On-line monitoring of the supercritical fluid extraction process with proton nuclear magnetic resonance spectroscopy. *Analytical Chemistry*, **67**, 930–5.
22. Albert, K., Braumann, U., Tseng, L. *et al.* (1994) On-line coupling of supercritical fluid chromatography and proton high-field nuclear magnetic resonance spectroscopy. *Analytical Chemistry*, **66**, 3042–6.
23. Hawthorne, S.B. and Miller, D.J. (1987) Directly coupled supercritical fluid extraction–gas chromatographic analysis of polycyclic aromatic hydrocarbons and polychlorinated biphenyls from environmental solids. *Journal of Chromatography*, **404**, 63–76.
24. Hawthorne, S.B., Miller, D.J. and Krieger, M.S. (1989) Rapid and quantitative extraction and analysis of trace organics using direcly coupled SFE–GC. *Journal of High Resolution Chromatography*, **12**, 714–20.
25. Huston, C.K. and Ji, H. (1991) Optimization of the analytical supercritical fluid extraction of cloves via an on-column interface to an ion trap GC/MS system. *Journal of Agricultural Food Chemistry*, **39**, 1229–33.
26. Snyder, J.M., King, J.W. and Nam, K.S. (1996) Determination of volatile and semivolatile contaminants in meat by supercritical fluid extraction/gas chromatography/mass spectrometry. *Journal of the Science of Food and Agriculture*, **72**, 25–30.
27. Murugaverl, B., Gharaibeh, A. and Voorhees, K.J. (1993) Mixed adsorbent for cleanup during supercritical fluid extraction of three carbamate pesticides in tissues. *Journal of Chromatography A*, **657**, 223–6.
28. Murugaverl, B. and Voorhees, K.J. (1991) On-line supercritical fluid extraction/chromatography system for trace analysis of pesticides in soybean oil and rendered fats. *Journal of Microcolumn Separations*, **3**, 11–16.
29. MacKay, G.A. and Smith, R.M. (1993) Supercritical fluid extraction and chromatography–mass spectrometry of flame retardants from polyurethane foams. *Analyst*, **118**, 741–5.
30. MacKay, G.A. and Smith, R.M. (1994) Supercritical fluid extraction–supercritical fluid chromatography–mass spectrometry for the analysis of additives in polyurethanes. *Journal of Chromatographic Science*, **32**, 455–60.
31. Bucherl, T., Gruner, A. and Palibroda, N. (1994) Rapid analysis of polymer homologues and additives with SFE/SFC–MS coupling. *Packaging Techology and Science*, **7**, 139–54.
32. Bucherl, T., Eschler, M., Gruner, A. *et al.* (1994) Direct coupling of capillary supercritical fluid chromatography with double-focusing high resolution mass spectrometry. *Journal of High Resolution Chromatography*, **17**, 765–9.

33. Ramsey, E.D., Perkins, J.R., Games, D.E. and Startin, J.R. (1989) Analysis of drug residues in tissues by combined supercritical-fluid extraction–supercritical-fluid chromatography–mass spectrometry–mass spectrometry. *Journal of Chromatography*, **464**, 353–64.
34. France, J.E., King, J.E. and Snyder, J.M. (1991) Supercritical fluid-based cleanup technique for the separation of organochlorine pesticides from fats. *Journal of Agricultural Food Chemistry*, **39**, 1871–74.
35. Parks, O.W. and Maxwell, R.J. (1994) Isolation of sulphonamides from fortified chicken tissues with supercritical $CO_2$ and in-line adsorption. *Journal of Chromatographic Science*, **32**, 290–3.
36. Ramsey, E.D., Minty, B., McCullagh, M.A. *et al.* (1997) Analysis of phenols in water at the ppb level using direct supercritical fluid extraction of aqueous samples combined on-line with supercritical fluid chromatography–mass spectrometry. *Analytical Communications*, **34**, 3–6.
37. Ramsey, E.D., Minty, B. and Rees, A.T. (1997) Drugs in water: analysis at the part-per-billion level using direct supercritical fluid extraction of aqueous samples coupled on-line with ultraviolet–visible diode-array liquid chromatography–mass spectrometry. *Analytical Communications*, **34**, 261–4.

# 10 Modern alternatives to supercritical fluid extraction

J.R. DEAN AND N. SAIM

## 10.1 Introduction

The use of alternative methods to supercritical fluid extraction (SFE) has been necessary because of some limitations (economic, lack of acceptance of SFE and arrival of new technologies) that have been identified. Economically, SFE equipment is expensive when compared with the traditional approach of Soxhlet extraction. Environmental concern over the use of hazardous and chlorinated solvents, and the cost of solvent disposal, should mean that the use of Soxhlet should be dwindling. However, this decline is not in strong evidence. Inevitably, laboratories (commercial and industrial) are slower to respond to newer technology unless strong economic motives are in evidence (extractions can be done cheaper, faster with automated equipment) or that government legislation forces the reduction in solvent usage. Another aspect to the lack of acceptance of SFE is the unavailability of robust methods. For a particular technique to be used extensively requires 'approved methods'. In this context the US Environmental Protection Agency (EPA) is regarded worldwide as a main producer of official methods. So far, only two EPA methods for SFE have been produced (methods 3561 and 3560, for polycyclic aromatic hydrocarbons and total petroleum hydrocarbons from soils, respectively). Furthermore, the arrival of alternative extraction techniques has particularly highlighted these inadequacies of SFE. It should be considered that the investment in terms of SFE since its inception in the analytical laboratory in 1982 must be enormous. The vast amount of literature relating to methods for SFE in the environmental, polymer, food, pharmaceutical and biomedical areas has resulted in numerous texts, reviews and journal papers.

Of the emerging technologies for the extraction of organic pollutants from solid matrices two particular methods are discussed in this chapter: microwave-assisted extraction and accelerated solvent extraction, although the use of solid phase extraction and solid phase microextraction for the analysis of aqueous samples should not be neglected. However, this particular chapter focuses on the extraction of analytes from solid matrices.

## 10.2 Microwave-assisted extraction

The use of microwave irradiation in organic extraction has been successfully conducted on various types of matrices. Microwave-assisted extraction (MAE) has the potential to be a feasible alternative to Soxhlet extraction since it yields quantitative recovery of target analytes with less organic solvent consumption and reduction in extraction time. Analytical applications of MAE include the extraction of organic pollutants, for example polycyclic aromatic hydrocarbons (PAHs), organochlorine pesticides (OCPs), organophosphorus pesticides (OPPs) and polychlorinated biphenyls (PCBs) from soil and sediment [1–5], crude fat and antinutritives from soya bean and cottonseed [6], organotin species from sediments [7], methylmercury from aquatic sediments [8], butyltin compounds from sediments [9] and ergosterol from fungal spores and mushrooms [10]. In addition, but not the subject of this review, has been the use of MAE for the extraction of pollutants from aqueous samples directly [11, 12], by elution from solid phase extraction disks [13] or the headspace above the aqueous sample [14].

### *10.2.1 Theory of microwave heating*

Microwaves are a form of electromagnetic energy which extends from 300 to 300 000 MHz. Within this range 2450 MHz is the frequency commonly utilised in industrial microwave systems. The basis of microwave heating includes orientation polarisation and interfacial polarisation arising from an alternating electric field, together with ionic conduction [15]. Orientation polarisation or dipolar polarisation occurs as a result of the alignment of molecules in the solvent or sample that have permanent or induced dipole moments under the influence of a changing electric field. At high frequency, the molecule is unable to restore the induced disorder, which results in thermal energy being released. At 2450 MHz the alignment of the molecules followed by their return to disorder occurs $4.9 \times 10^9$ times per second, which results in rapid heating [16]. Therefore, organic solvents with high dipole moments are generally used in MAE. Table 10.1 lists the dipole moment for commonly used organic solvents [17]. Interfacial or Maxwell–Wagner polarisation relates to the build up of charged particles at interfaces in heterogeneous dielectrics [15] whereas ionic conduction is the conductive migration of dissolved ions in the applied field [16]. The ability of the solvent to convert electromagnetic energy to heat can be expressed as a dissipation factor or loss tangent ($\tan \delta$) which is a ratio of the sample's dielectric loss ($\varepsilon''$) to its dielectric constant ($\varepsilon'$): $(\tan \delta) = (\varepsilon''/\varepsilon')$. The dissipation factor or loss tangent is related to the solvent's ability to be penetrated by the microwave energy and dissipate the energy as heat. The dielectric constant for commonly used organic solvents is shown in Table 10.1.

**Table 10.1** Dielectric constants and dipole moment of commonly used solvents. Source: ref. 17

| Solvent | Dielectric constant | Dipole moment |
|---|---|---|
| Acetone | 2.77 | 20.7 |
| Benzene | 0 | 2.27 |
| Cyclohexane | 0 | 2.02 |
| Ethanol | 1.71 | 25.00 |
| Acetonitrile | 3.47 | 37.5 |
| Hexane | 0 | 1.89 |
| Methanol | 1.68 | 33.62 |
| Methylethyl ketone | 2.76 | 18.51 |
| Methylene chloride | 1.14 | 8.93 |

For closed-vessel MAE, in addition to the rapid heating of the solvent, solvents can be heated to approximately 100°C above their normal boiling point [17], thereby leading to more efficient extraction. Table 10.2 shows the solvent boiling points at atmospheric pressure and their measured boiling points in the closed MAE vessel.

### 10.2.2 *Instrumentation*

For microwave ovens, a fixed-tuned magnetron is used to generate microwave energy at a frequency of 2450 MHz. A magnetron is a cylindrical diode with a cathode where electrons are emitted and an anode where the microwave signal is propagated [18]. For a closed-vessel MAE the microwaves generated by the magnetron are conducted by the waveguide to the microwave cavity where the sample vessels are placed. Waveguides, constructed of reflective material such as sheet metal, are assigned to direct microwaves to the cavity without reflectance. A mode stirrer is generally

**Table 10.2** Solvent boiling point comparison. Source: ref. 17

| Solvent | Boiling point (°C) | Boiling point in closed vessel (°C)[a] |
|---|---|---|
| Methylene chloride | 39.8 | 140 |
| Acetone | 56.2 | 164 |
| Methanol | 64.7 | 151 |
| Hexane | 68.7 | – |
| Cyclohexane | 80.7 | 164 |
| Acetonitrile | 81.6 | – |
| 2-propanol | 82.4 | 194 |
| Petroleum ether | 35–80 | – |
| Acetone: hexane 1:1 vol/vol | 52 | 156 |
| Acetone: cyclohexane 70:70 vol/vol | 52 | 160 |
| Acetone: petroleum ether 1:1 vol/vol | 39 | 147 |

[a] Measured at 175 psi.

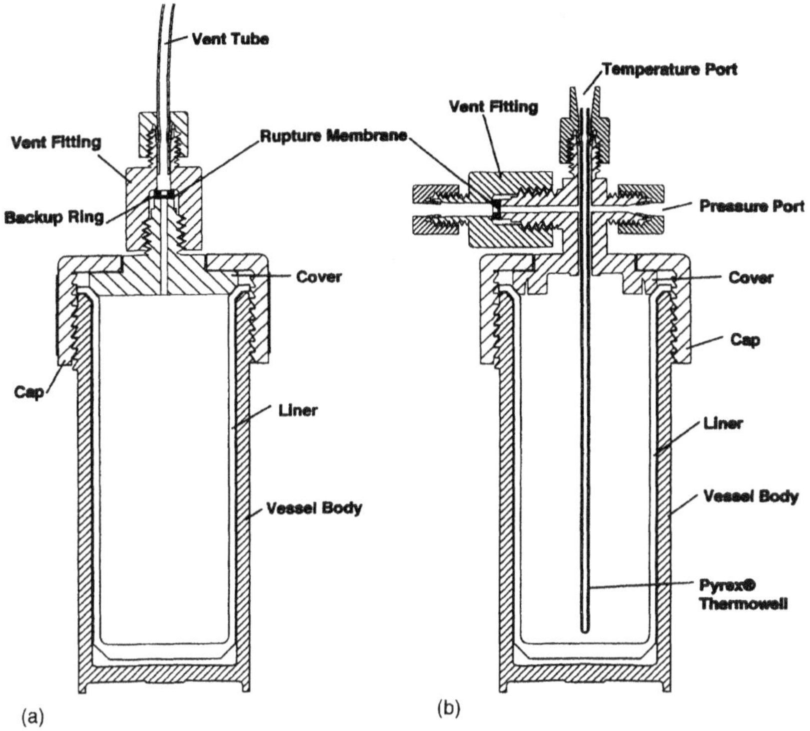


**Figure 10.1** Diagram of lined extraction vessels for microwave-assisted extraction: (a) standard lined extraction vessel; (b) lined extraction vessel for pressure and temperature control. Reproduced with kind permission from CEM Microwave Technology.

used to reflect and distribute the energy evenly within the cavity so that the heating of the sample will be independent of position.

A CEM model MES 1000 microwave system (CEM Corporation, UK) consists of a turntable holding up to 12 vessels, which enables 12 samples to be extracted simultaneously. The sample vessels (Figure 10.1) are designed for heating under pressure. Extraction vessels are made of Ultem poly(ether imide), lined with perfluoroalkoxy or teflon. They are constructed from materials that are transparent to microwave energy so that the sample mixture can absorb the maximum amount of microwave energy and are lined to withstand various organic solvents. Both temperature and pressure probes are usually placed in one of the vessels. Since the extraction involves the use of flammable solvent at high temperatures and pressures the microwave extraction system is equipped with several safety features. A solvent detector is usually placed in the system air exhaust, which turns off the magnetron or alerts the operator if solvent vapours are detected in the microwave cavity. For overpressurisation protection, each MAE system is equipped with

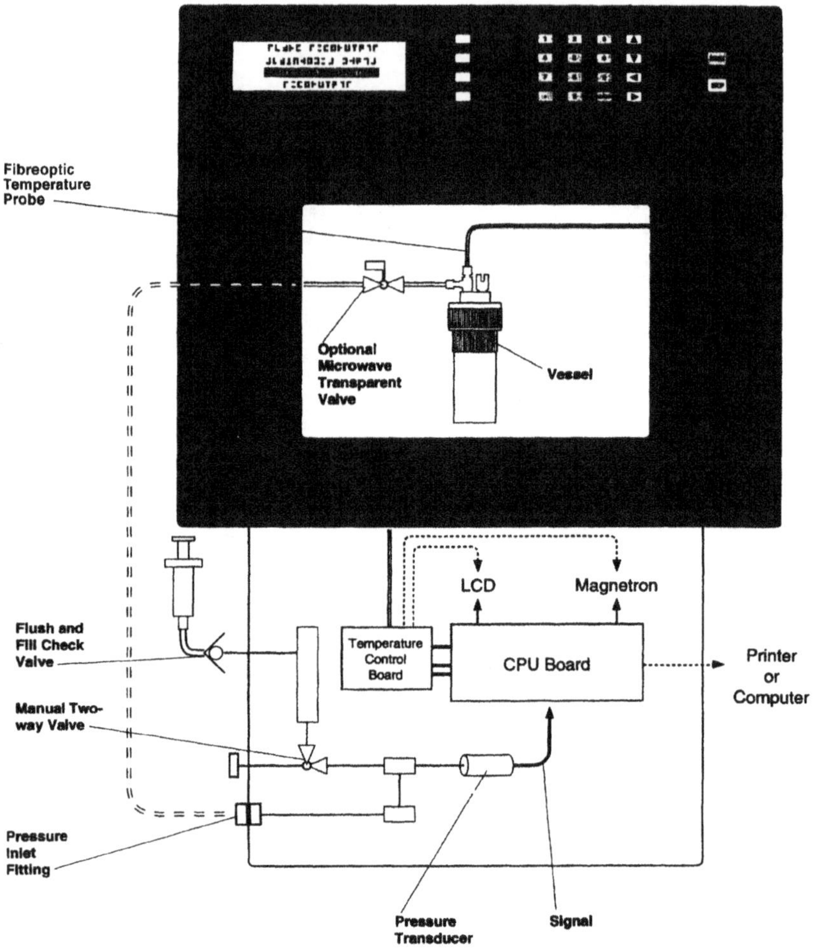


**Figure 10.2** Schematic diagram of a microwave-assisted extraction system. Reproduced with kind permission from CEM Microwave Technology.

pressure relief valves or a membrane rupture which will break at high pressure. A schematic diagram of a typical MAE system is shown in Figure 10.2.

For open-vessel MAE, individual sample vessels are heated sequentially. A typical commercial system is the Soxwave (Prolabo Ltd., France) which operates at percentage power increments from 0% to 100%, corresponding to a maximum of 300 W. The sample is introduced into a glass container, along with the appropriate solvent, which is then located within a protective glass sheath. The sample-containing glass vessel, which has the appearance of a large boiling or test tube, is connected to an air or water (according to the preferred option) condenser. This prevents loss of volatile analyte and solvent. The sample-containing glass vessel is subjected to the microwave radiation.

In closed-vessel MAE, typically a 1–5 g sample is extracted with 30–50 ml of solvent and irradiated for a short time period (10–30 min). Following extraction, the vessels are cooled to room temperature (approximately 30 mins) and the contents of the vessels are filtered or centrifuged and concentrated as necessary prior to analysis. Open-vessel extractions permit extraction of bigger samples, up to 40 g, with 30–50 ml of solvent, with extraction time ranges between 10 and 30 min. No significant post-extraction cooling is required.

### *10.2.3 Selection of solvent and extraction conditions*

In MAE an organic solvent which has a relatively high dipole moment and a high dielectric constant is generally used. Non-polar solvents, such as hexane, that do not have a permanent dipole moment can only be used with the addition of a polar solvents such as acetone. A minimum percentage (10%) of polar compound is required in MAE [3]. The majority of published MAE analyses of organic pollutants (i.e. PAHs, PCBs) have utilised hexane:acetone (1:1) as the solvent system [5, 17]. Other solvent systems include 100% acetone [1] and a 1:1 mixture of isooctane and acetonitrile [4].

Other than the choice of solvent systems there are several other parameters that are of key importance in conducting MAE, including the temperature, extraction time and solvent volume. For a closed-system, pressure is dependent on temperature. Temperature ranges from 115°C to 150°C, with extraction times between 5 min and 30 min. Solvent volume varies between 30 ml and 50 ml. The amount of solvent required will be dependent on the mass of the sample. Optimisation of these operating parameters should lead to the quantitative recovery of analytes.

### *10.2.4 Applications of microwave-assisted extraction*

#### *Polycyclic aromatic hydrocarbons*

The applicability of MAE in the extraction of PAHs from environmental matrices, such as soil and sediments, has been extensively investigated. PAHs from standard reference marine sediments (HS-3, HS-4, HS-5 and SRM 1941) and two certified soils (SRS 103-100 and ERA lot no. 321) were extracted quantitatively with hexane:acetone (1:1) at a temperature of 115°C for 10 min [3]. Optimisation of the MAE conditions (solvent system, extraction temperature, extraction time, solvent volume) for the extraction of PAHs from certified reference materials (HS-4 and HS-6) was investigated by Chee *et al.* [2]. The optimum MAE conditions were as follows: 30 ml of acetone–hexane (1:1), an optimum microwave heating temperature of 115°C and a 5 min duration of microwave heating. However, Barnabas *et al.* [1] investigated the extraction efficiency for PAHs from soil using a range of hexane:acetone compositions and found that increasing the polarity of the solvent system resulted in greater amounts of PAHs being recovered. To ascertain the repeatability of the extraction, multiple extractions were performed using acetone. The results obtained gave an average recovery of 422.9 $mg\,kg^{-1}$ total PAHs with an average relative standard deviation (RSD) of 2.4% (based on six determinations). They concluded that the most effective solvent for the extraction of PAHs from native contaminated soil was 100% acetone. Using this information they investigated the influence of MAE operating variables (temperature, 40°C–120°C; extraction time, 5 min–20 min; solvent volume, 30–50 ml) using an experimental design approach based on a central composite design. The results showed that MAE using 100% acetone was insensitive to the operating variables investigated.

*Organochlorine pesticides and compounds*

Onuska and Terry [4] investigated the solvent suitable for OCPs using open-vessel MAE. An isooctane–acetonitrile (1 : 1) solvent system gave recoveries better than those achieved with 6 h Soxhlet extraction. In the study it was found that sample moisture level is a significant parameter for good recovery in MAE. A moisture level of 15% was found to give the best recovery.

The effect of sample ageing on the recovery of 45 OCPs from spiked topsoil was investigated by Lopez-Avila *et al.* [5]. The recoveries were assessed in terms of freshly spiked topsoil stored for 24 h at 4°C, and topsoil aged for 14 days at 4°C. On the freshly spiked topsoil, the MAE recoveries using a 1 : 1 hexane:acetone mixture were greater than 70%, with the exception of 1,2-dibromo-3-chloropropane (DBCP). Ageing, however, had more pronounced effects with 20 OCPs having less than 70% recovery after 14 days. Dichlone was severely retained by the soil (less than 10% recovery) after even the shortest ageing process. They then went on to compare Soxhlet extraction, sonication and MAE [5]. In this situation, soils were spiked and their average recoveries determined. Although the recovery for MAE was in almost all cases superior to that of the other two extraction techniques it should be noted that the spike level for MAE was also 400 times higher. Sonication was the least effective extraction method.

The MAE of OCPs from pork fat and seal blubber with *n*-hexane as the solvent has been described [19]. Normally, in MAE the choice of solvent is restricted to those solvents with a dielectric constant. However, in this paper a microwave transformer (Weflon®, an inert material based on carbon-containing teflon) was used to transfer energy to the sample. To prevent solvent and analyte leakage, the MAE system was operated in short time durations, typically 30 s at full power followed by 5 min cooling. As might be expected, no certified reference materials were available for this work so the blubber of a grey seal stranded on Rugen, Baltic Sea, Germany, was used. To provide a value (reference point) for the MAE work the naturally contaminated seal blubber was initially melted, subjected to matrix cleanup with deactivated silica and analysed. For all organochlorine compounds studied [PCB 153, PCB 138, PCB 180, *p,p'*-DDE (*p,p'*-dichlorodiphenyldichloroethylene), α-HCH (hexachlorocyclohexane), γ-HCH and HCB (hexachlorobenzene)] the best recoveries were obtained using seven MAE extraction cycles; average recoveries for the seven compounds were 96.9% ± 0.5%. Spiking experiments on pork fat of the organochlorine compounds gave excellent recoveries (88.5% ± 1.3% to 98.6% ± 2.3%) based on seven MAE cycles and $n = 5$.

*Organophosphorus pesticides*

Lopez-Avila *et al.* [5] report the recoveries of 43 OPPs after spiking onto topsoil and extracting immediately, after ageing for 24 h at 4°C or after 21 days at 4°C. Recoveries of more than 80% were noted for 33 OPPs from freshly spiked (0.5–2 mg $kg^{-1}$) topsoil. Recoveries of less than 10% were noted for 4 OPPs [monocrotophos, naled, tetraethylenepyrophosphate (TEPP) and trichlorfon]. Ageing for 21 days did not significantly influence the recovery of OPPs, except for demeton-O, demeton-S and phosmet, whose recoveries fell from approximately 100% to 33%, 5% and 45%, respectively after ageing for 21 days.

*Herbicides*

Steinheimer [20] investigated the extraction of atrazine and its metabolites (deethylatrazine and deisopropylatrazine) from spiked soil (Nashua and Treynor sites). Each soil type was characterised as follows:

- Nashua soil
  - organic carbon content 4.48%,
  - cation exchange capacity 36.1 mequiv per 100 g,
  - pH 6.7,
  - sand content 42.5%,
  - silt content 47.5%,
  - clay 10.0%;
- Treynor soil
  - organic carbon content 1.81%,
  - cation exchange capacity 32.3 mequiv per 100 g,
  - pH 6.1,
  - sand content 10.0%,
  - silt content 77.5%,
  - clay content 12.5%.

Each soil (10 g) was extracted using 25 ml of organic-free water a'
turc of 95°C–98°C. This process was repeated using organic-frcc

followed by three times with 0.35 M hydrochloric acid. The combined extracts (*c.* 125 ml), were adjusted to pH 7.0, centrifuged, filtered and preconcentrated using solid phase extraction (SPE) prior to high-performance liquid chromatography (HPLC) analysis. The results indicated that the soil type did not significantly affect recovery. Average ($n=4$) recoveries of approximately 62% for atrazine, 95% for deethylatrazine and 88% for deisopropylatrazine were obtained from the Nashua soil whereas the recoveries of atrazine, deethylatrazine and deisopropylatrazine from the Treynor soil were approximately 58%, 83% and 118%, respectively. It was noted that the microwave extracts done in water and HCl resulted in cleaner extracts and simpler chromatograms than those extracts produced using the more traditional solid–liquid extraction methods using methanol–water and acetonitrile–water.

As previously described, Lopez-Avila *et al.* [5] evaluated the recovery of atrazine and simazine from freshly spiked soils, spiked topsoil aged for 24 h at 4°C and spiked topsoil aged for 21 days at 4°C. At a spike level of 1 mg $kg^{-1}$ the recoveries in all cases ranged from 85.6% to 123% with good precision (less than 1.6% deviation), with the exception of simazine from topsoil which had been aged for 24 h, where the average recovery was 85.6% with an RSD of 40%.

MAE has been evaluated for the extraction of a new class of low-use-rate, reduced-environmental-risk herbicides, the imidazolinones [21]. As the use of the imidazolinones is at a low-use-rate the level of determination in soil is at the low parts-per-billion level. Current methodology uses 50 g of soil with 200–300 ml of 0.5 M NaOH for 1 h followed by extensive cleanup using precipitation and centrifugation followed by partitioning with 200–300 ml of dichloromethane. After evaporation the final extract is cleaned up using SPE, with both a strong cation exchange column and a $C_{18}$ column. This allows a level of quantitation using HPLC with ultraviolet (UV) detection of 5 ppb. The sample preparation time is estimated to be *c.* 12 h for a set of six samples. The authors considered the use of MAE as an alternative to this approach. After suitable experimentation and keeping the MAE parameters constant (temperature 125°C and time 3 min) it was concluded that the optimum extraction solvent was 0.1 M $NH_4OAc$ at pH 10.0. Incorporation of the MAE procedure, followed by SPE preconcentration and cleanup using a $C_{18}$ cartridge, allowed quantitation at the 1 ppb level using gas chromatography (GC) with electron capture negative chemical ionisation mass spectrometry. The sample preparation time (extraction and cleanup) was estimated to be *c.* 10 min per sample, as 12 samples could be extracted in parallel using MAE. The recoveries from the aged soil samples ranged from 85% to 104%.

*Polychlorinated biphenyls*

Recovery of PCBs from spiked soils (clay, topsoil and sand) was reported [22]. Good recoveries at a spike level of 100 ng $g^{-1}$ of Arochlor 1016 and

1260 were obtained from sand. However, a noticeable decrease in recovery was noted for recovery from topsoil (76% for Arochlor 1260 and 84% for Arochlor 1016) and clay (62% for Arochlor 1260 and 78% for Arochlor 1016). The same workers [5] also report the recoveries of Arochlor 1254 and 1260 from certified reference samples. Acceptable recoveries were obtained from the National Research Council of Canada marine sediments (for HS-1, the recovery of Arochlor 1254 was 93.2% with an RSD of 8.1% whereas for HS-2 the recovery of Arochlor 1254 was 76.7% with an RSD of 6.0%) and for ERA soil (lot no. 9801) the recovery of Arochlor 1260 was 89.9% with an RSD of 2.6%.

MAE has also been compared with two different types of detection systems [GC with electron capture detection (ECD) and an immunoassay method, enzyme-linked immunosorbent assay (ELISA)] for assessing the concentration of Arochlors in standard reference materials (marine sediment and soil) [22]. Comparable results, at the level of approximately 400 000 mg $kg^{-1}$, were obtained for Arochlor 1260 in ERA soil (lot no. 9801) by both detection systems. However, for the marine sediment SRMs, at the levels of approximately 20 mg $kg^{-1}$ and 110 mg $kg^{-1}$ the immunoassay approach gave consistently poorer results.

*Phenols*

Lopez-Avila *et al.* [3] evaluated the recoveries of four phenols (2-methylphenol, 2,4-dichlorophenol, 2,4,6-trichlorophenol and pentachlorophenol) from an ERA standard reference soil (lot no. 321) using either room temperature solvent extraction or MAE conditions for time periods of 5 min, 10 min and 20 min using a 1:1 mixture of hexane and acetone [3]. The MAE temperature and pressure conditions investigated were 80°C and 22 psi, 115°C and 72 psi, and 145°C and 150 psi. Acceptable recoveries for 2,4-dichlorophenol and 2,4,6-trichlorophenol were obtained (typical recoveries ranged from 87.1% to 112%) using MAE once conditions of 115°C and 72 psi were achieved. Higher MAE conditions (temperature 145°C and a pressure of 150 psi) were required to obtain acceptable recoveries (76%–83.6%) for pentachlorophenol. However, under all conditions investigated, the recovery of 2-methylphenol was poor, never exceeding 28%.

Further work by the same group [5] investigated the recovery of 18 phenols from spiked soils (at the 5 mg $kg^{-1}$ level) and extraction using MAE at 115°C for 10 min. Acceptable recoveries for all phenols were obtained, ranging from 85.3% to 107% (6%–12% RSD). The exceptions were 4,6-dinitro-2-methylphenol (57.6% recovery and 9.3% RSD) and 2,4-dinitrophenol (17.2% recovery and 39% RSD). After ageing of the soil (24 h at 4°C) the recoveries for most phenols decreased. However, recoveries of more than 90% were obtained for 4-chloro-3-methylphenol, 4-nitrophenol, pentachlorophenol, 2,3,4,6-tetrachlorophenol, 2,4,5-trichlorophenol, 2,4,6-trichlorophenol and 2,4,6-tribromophenol. Acceptable recoveries of seven phenols from a certified reference soil (ERA soil, lot no. 324) were also reported [5]. The recoveries ranged from 71.1% to 99.5%, with RSDs ranging from 1.3% to 15.7%.

A chemometric approach, based on a central composite design, has been applied to the MAE of phenol and methylphenol isomers from spiked soils that have been aged for 25 days [23]. The variables investigated were as follows: temperature, 70–130°C; proportion of acetone, 20%–80%; and solvent volume, 15–50 ml, at a fixed extraction time of 10 min and for 5 g of sample. Temperature was found to be significant for all four phenols studied and the proportion of acetone to hexane and the solvent volume were found to be significant in some cases. It was concluded that 10 ml of acetone–hexane (80:20) at 130°C was the most appropriate volume, producing good recoveries.

*Base and neutral compounds*

Recoveries of nine base and neutral compounds [dibenzofuran, 1,2-dichlorobenzene, *N*-nitroso-di-*n*-propylamine, nitrobenzene, 1,2,4-trichlorobenzene, 2,4-dinitrobenzene, 9H-carbazole, di-*n*-butyl phthalate and bis(2-ethylhexyl) phthalate] from an ERA standard reference soil (lot no. 321) were investigated [3]. The extraction conditions evaluated were a room-temperature solvent extraction and MAE. Both approaches used different time periods for solvent–matrix interactions, that is, 5 min, 10 min and 20 min using a 1:1 mixture of hexane and acetone. The MAE temperature and pressure conditions investigated were 80°C and 22 psi, 115°C and 72 psi and 45°C and 150 psi. It was found that four compounds – *N*-nitroso-di-*n*-propylamine, nitrobenzene, 2,4-dinitrotoluene and di-*n*-butyl phthalate – did not require MAE and could be quantitatively extracted (93.7%, 90.2%, 139% and 129% recovery, respectively) under room-temperature conditions in 5 min. Also, 1,2-dichlorobenzene and 1,2,4-trichlorobenzene could not be quantitatively extracted using room-temperature or MAE conditions. The highest reported recovery was approximately 38% for 1,2-dichlorobenzene and 67% for 1,2,4-trichlorobenzene using MAE. For the three remaining compounds [dibenzofuran, 9H-carbazole and bis(2-ethylhexyl) phthalate] a moderate increase in recovery was noted when going from room-temperature to MAE conditions. Acceptable recoveries (ranging from 82.2% to 136%) were achieved at 115°C and 72 psi.

As for previous analytes, Lopez-Avila *et al.* [5] evaluated the recoveries of 54 compounds from freshly spiked (5 mg $kg^{-1}$) soils and from soils aged for 24 h at 4°C, using 30 ml of a 1:1 mixture of hexane and acetone at 115°C. In general it could be concluded that the effect of soil ageing could be classified as having no significant effect on the recovery of compounds or that some influence was noted. Perhaps more importantly it is important to consider the recoveries of base and neutral compounds from certified reference soils [5]. Seven base and neutral compounds were extracted and the amounts compared with the certificate values. In all cases recoveries ranged from 83% to 150%, with RSDs of 1.9% to 15.8%.

*Phthalate esters*
Phthalate esters [dimethyl phthalate, diethyl phthalate, diallyl phthalate, dibutyl phthalate, benzyl *n*-butyl phthalate and bis(2-ethylhexyl) phthalate] have been extracted using MAE from spiked and native sediment samples [24]. The MAE operating parameters were evaluated using spiked sediment samples. It was found that no statistical difference was observed between a 1:1 acetone:hexane mixture and dichloromethane (DCM). The former was, however, preferred as it required no solvent exchange prior to GC/ECD analysis. The optimum conditions used for the native contaminated sediment samples located in the Tuas–Jurong industrial area of Singapore were 30 ml of a 1:1 solvent mixture (hexane:acetone) at a temperature of 115°C for 10 min. The results obtained indicated that dibutyl phthalate and bis(2-ethylhexyl) phthalate were present in the ranges 0.68–1.60 mg $kg^{-1}$ and 0.16–2.79 mg $kg^{-1}$, respectively.

## 10.3 Accelerated solvent extraction

Accelerated solvent extraction (ASE) utilises organic solvents at high temperature and pressure to extract pollutants from environmental matrices. It was first proposed as a method (method 3545) in Update III of the 1995 USEPA SW-846 Methods [25].

### *10.3.1 Theoretical considerations*

It has been proposed [26] that the two main reasons why the use of organic solvents may offer enhanced performance at elevated temperature and pressure are solubility and mass transfer effects and the disruption of surface equilibria. Both of these will now be discussed.

*Solubility and mass transfer effects*
Three factors are considered to be important [26]:

- higher temperatures increase the capacity of solvents to solubilise analytes;
- faster diffusion rates occur as a result of the increased temperature;
- improved mass transfer and hence increased extraction rates occur when fresh solvent is introduced.

*Disruption of surface equilibria*
As both temperature and pressure are involved both need to be considered [26]:

- temperature effects:
  - increased temperatures can interfere with the strong solute–matrix interactions,
  - the viscosity of organic solvents decreases as the temperature increases, resulting in improved penetration of the matrix by the organic solvent and hence improved extraction efficiency;
- pressure effects:
  - the use of high pressures allows solvents to remain liquefied above their boiling points;
  - high pressures also allow the solvent to penetrate the sample matrix with obvious improved extraction efficiencies.

### *10.3.2 Instrumentation*

The technique of accelerated solvent extraction (ASE) was launched in the spring of 1995 by Dionex Corp. (Salt Lake City, UT, USA) as a fully automated sequential extraction system (Figure 10.3). The system (ASE 200) can operate with up to 24 sample-containing extraction vessels and up to 26 collection vials plus an additional 4 vial positions for rinse or waste collection. Three sample extraction vessel sizes are available: 11 ml, 22 ml and 33 ml, with an internal diameter of 19 mm. The sample vessel has removable end caps that allow for ease of cleaning and sample filling. Each sample vessel is fitted with two finger-tight caps with compression seals for high-pressure closure. To fill a sample vessel, one end cap is introduced and screwed on to finger tightness, then a filter paper is introduced into the sample vessel, followed by the sample itself. The other end cap is then screwed on to finger tightness and the sample vessel placed in the carousel. Through computer control the carousel introduces selected extraction vessels consecutively.

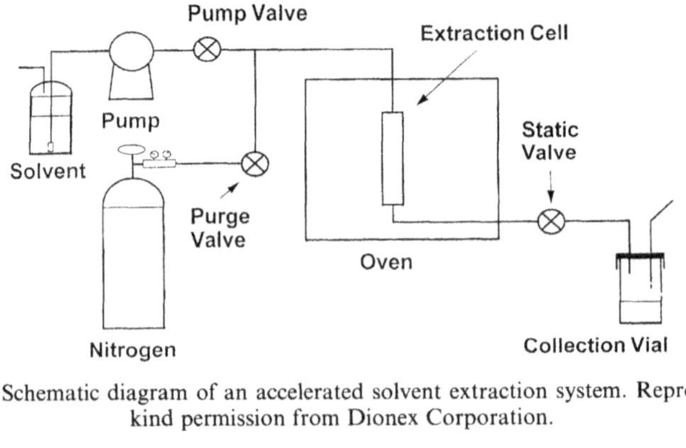


**Figure 10.3** Schematic diagram of an accelerated solvent extraction system. Reproduced with kind permission from Dionex Corporation.

An autoseal actuator places the extraction vessel into the system and then returns the vessel to the carousel after extraction. The system is operated as follows: the sample cell, positioned vertically, is filled from top to bottom with the selected solvent or solvent mixture from a pump and then heated to a designated temperature (up to 200°C) and pressure (up to 20 MPa or 3000 psi). These operating conditions are maintained for a pre-specified time using static valves. After the appropriate time (typically 5 min) the static valves are released and a few millilitres of clean solvent (or solvent mixture) is passed through the sample cell to expel the existing solvent(s) and extracted analytes. This rinsing is enhanced by the passage of $N_2$ gas to purge both the sample cell and the stainless steel transfer lines. After gas purging, all extracted analytes and solvent(s) are passed through stainless steel tubing that punctures a septa [solvent-resistant; coated with poly(tetrafluoroethylene) (PTFE) on the solvent side] located on top of the glass collection vials (40 ml or 60 ml capacity). If required, multiple extractions can be performed per extraction vessel. The arrival and level of solvent in the collection vial is monitored using an infra-red (IR) sensor. In the event of system failure, an automatic shut-off procedure is instigated.

As ASE has only been commercially available since Spring 1995 the literature at the present time is sparse. However, it is anticipated that this will not be the case in the near future. Some papers have been published and these will now be reviewed.

### *10.3.3 Applications: environmental matrices*

*Polycyclic aromatic hydrocarbons*

Richter *et al.* [26] were able to demonstrate that PAHs could be extracted from urban dust and marine sediment using the following conditions: pressure, 2000 psi; temperature, 100°C; extraction time, 5 min equilibriation plus 5 min static extraction; and an extraction solvent of 1:1 vol/vol DCM: acetone. The results in each case indicated that agreement was achievable (average recoveries ranged from 88.5% to 125% from urban dust, SRM 1649, and 57.7%–364.3% for marine sediment, HS-3) compared with the certificate value (or non-certified value). Subsequent work [27] compared the recoveries for 16 individual PAHs from a natively contaminated soil after a 6 h Soxhlet extraction using DCM with those obtained using 10% methanol-modified supercritical $CO_2$, MAE using acetone and ASE using DCM and acetone (1:1 vol/vol). The results obtained using the instrumental methods of extraction (SFE, MAE and ASE) compared favourably with each other (the total amount of the 16 individual PAHs recovered was: 422.9 mg kg$^{-1}$ by MAE; 458.0 mg kg$^{-1}$ by SFE; and 421.9 mg kg$^{-1}$ by ASE). However, lower recoveries were observed using Soxhlet extraction (the total amount of the 16 individual PAHs recovered was 297.4 mg $kg^{-1}$). In further work Soxhlet extraction was compared with pressurised microwave-assisted extraction, atmospheric microwave-assisted extraction, supercritical fluid extraction and accelerated solvent extraction for the extraction of PAHs from natively contaminated soil [28]. All the extraction methods were operated under standard conditions, optimised conditions, as previously determined in this laboratory for PAHs, or according to manufacturers' recommendations. In the case of the standard conditions, Soxhlet extraction was done in accordance with USEPA method 3540 using a 10 g sample, 150 ml of DCM and extracted for 24 hours. Both the SFE and the pressurised MAE were operated under experimentally optimised conditions, as determined for the extraction of PAHs from native, contaminated soil. These conditions were as follows:

- SFE
  - sample size 1 g,
  - solvent 20% methanol-modified supercritical $CO_2$, and 25 ml,
  - pressure 250 kg $cm^{-2}$,
  - temperature 70°C,
  - extraction time 35 min;
- pressurised MAE
  - sample size 2 g,
  - solvent acetone,
  - solvent volume 40 ml,
  - temperature 120°C,
  - extraction time 20 min.

Atmospheric MAE was operated under conditions as recommended by the manufacturer, that is:

- atmospheric MAE
  - sample size 2 g,
  - solvent DCM,
  - solvent volume 70 ml,
  - power 99% at 300 W,
  - extraction time 20 min.

ASE was operated according to the proposed EPA method 3545:

- ASE
  - sample size 7 g,
  - solvent DCM/acetone,
  - solvent volume 25 ml,
  - pressure 2000 psi,
  - temperature 100°C,
  - extraction time 12 min.

The results obtained, presented as the total mean ($n=6$) recovery of 16 individual PAHs, were as follows: Soxhlet, 1623 mg $kg^{-1}$ and an RSD of 5%; pressurised MAE, 1576 mg $kg^{-1}$ and an RSD of 7%; atmospheric pressure MAE, 1491 mg $kg^{-1}$ and an RSD of 9%; SFE, 1545 mg $kg^{-1}$ and an RSD of 6%; accelerated solvent extraction, 1536 mg $kg^{-1}$ and an RSD of 6%. Individual PAHs were then compared against Soxhlet extraction for statistical significance at the 95% confidence interval. It was noted that the majority of PAH determinations were not significant for both pressurised and atmospheric pressure MAE. However, statistical significance was noted for SFE and ASE. In all cases, except benz(*a*)anthracene by ASE, the statistically significant results produced extraction recoveries less than those obtained by Soxhlet. It was concluded that at the present time the method of choice still appears to be Soxhlet extraction. However, this statement should be qualified by inferring that the other extraction techniques could challenge Soxhlet in the near future.

ASE has also been evaluated as a rapid, low-solvent replacement for Soxhlet and bath sonication/shaking extraction in established soil-screening methods [29]. In this study a natively contaminated soil and a spiked uncontaminated clay loam were extracted using a 1:1 vol/vol solvent mixture of DCM and acetone. For the spiked soil (naphthalene and pyrene were the only PAHs present in the spiking solution mixture) ASE recoveries were higher than those obtained by Soxhlet extraction and generally approximated to 100% recovery at the 4 mg $kg^{-1}$ and 20 mg $kg^{-1}$ spiking level. For the natively contaminated soils, ASE recoveries were equivalent or superior to bath sonication/shaking recoveries, with ASE giving approximately double the total PAH content for matrices containing small stones (total PAH recovery, 2.2 mg $kg^{-1}$ by sonication and shake compared with 5.3 mg $kg^{-1}$ by ASE) and/or coal (total PAH recovery 1000 mg $kg^{-1}$ by sonication and shake compared with 1900 mg $kg^{-1}$ by ASE). A particular problem noted for ASE was the level of background interference observed in the GC–MSD (mass selective detector) chromatogram compared with Soxhlet extraction. This problem was not totally remedied even when operating the GC–MSD in single-ion monitoring mode. The authors concluded that the performance of ASE for the general screening of contaminated soils for semivolatile organic pollutants is at least equivalent to Soxhlet extraction and bath sonication/shaking.

*Pesticides (including fungicides)*

As part of an equivalence study, Ezzell *et al.* [30] investigated the extraction of 32 OPPs from three soil matrices (clay, loam and sand). The soils were spiked at two different levels – low-level (in the range 50.5–500 μg $kg^{-1}$) and high-level (505–5000 μg $kg^{-1}$) – depending upon the OPP. The results indicated that comparable results were obtainable with an 18 h DCM Soxhlet extraction and were independent of the soil matrix.

More recently, ASE has also been evaluated as a rapid, low-solvent replacement for Soxhlet and bath sonication/shaking extraction in established soil-screening methods by Fisher *et al.* [29]. In this study an OCP-contaminated blended soil was extracted using a 1:1 vol/vol solvent mixture of DCM and acetone. It was found that recoveries obtained by ASE ($1.35 \pm 0.38\,\mathrm{mg\,kg^{-1}}$ for dieldrin and $0.10 \pm 0.04\,\mathrm{mg\,kg^{-1}}$ for aldrin) were in agreement with those obtained by Soxhlet extraction ($1.11 \pm 0.34\,\mathrm{mg\,kg^{-1}}$ for dieldrin and $0.11 \pm 0.03\,\mathrm{mg\,kg^{-1}}$ for aldrin) but less than those obtained by sonication ($1.61\,\mathrm{mg\,kg^{-1}}$ for dieldrin and $0.20\,\mathrm{mg\,kg^{-1}}$ for aldrin). Both ASE and Soxhlet used a 1:1 dichloromethane–acetone solvent system whereas sonication was done using a 4:1 hexane–acetone solvent mixture. As noted for PAHs the level of background interference observed in the GC–MSD chromatogram for ASE was high compared with that in Soxhlet extraction. This problem was not totally remedied even when operating the GC–MSD in single-ion-monitoring mode. The authors concluded that the performance of ASE for the general screening of contaminated soils for semivolatile organic pollutants is at least equivalent to Soxhlet extraction and bath sonication/shaking.

Frost *et al.* [31] compared Soxhlet extraction with MAE, SFE and ASE for the extraction of hexaconazole, a fungicide, from weathered Canadian soils. Each soil sample had been collected at different times after hexaconazole application, air dried for 48 h, sieved through a 2–3.5 mm sieve and stored at a temperature of −18°C until analysed. Two soil types were used in this study, a sandy loam soil (5.7% organic matter, 81% sand, 8% silt, 11% clay and a pH of 7.4) and a sandy clay soil (1.5% organic matter, 53% sand, 25% silt, 22% clay and a pH of 8.3). Soxhlet extraction was done using 40 g of soil and 80 ml of acetonitrile–water (1:1) for 6 h. The extract was then partitioned into DCM, reduced to dryness and re-diluted in 1 ml of DCM. The extract was then selectively cleaned up using a silica SPE cartridge column. The final solvent composition was acetone:hexane (2:8). SFE was done using a Jasco system and operated under the following conditions: sample, 4 g; pressure, $250\,\mathrm{kg\,cm^{-2}}$; temperature, 55°C; 20% methanol modifier; and an extraction time of 20 min. The sandy loam soil extracts were subjected to a pre-chromatographic cleanup using a $C_{18}$ SPE cartridge. The final solvent composition was methanol. MAE was done using a Milestone 1200 mega unit. Soil samples (5 g) were extracted using acetone as solvent at 115°C and a power of 1000 W for 15 min. After cooling (30 min) both soil types were subjected to pre-chromatographic SPE cleanup ($C_{18}$). The final solvent composition was methanol. ASE was done using an automated system (Dionex, ASE 200). The sample (5 g) was extracted using acetone at a pressure of 2000 psi and a temperature of 100°C for 10 min. No pre-chromatographic cleanup was necessary.

The results indicate that the level of hexaconazole residue in the soil depletes over a 52-week period. On the basis of hexaconazole recovery from the sandy clay soil (lower organic matter content) SFE, MAE and

ASE give comparable results to Soxhlet extraction [Figure 10.4(a)]. However, some matrix dependency compared with Soxhlet extraction is noted for the sandy loam soil (higher organic matter content) for both SFE and MAE [Figure 10.4(b)]. No such dependency is noted for ASE. The highest precison was obtainable using the automated ASE system.

*Polychlorinated biphenyls*

PCBs have been extracted from oyster tissue and sewage sludge by ASE and their recoveries compared with a 6 h Soxhlet extraction using hexane as solvent [26]. The results for the PCB congeners (28, 52, 101, 153, 138 and 180) ranged from 86.3%–90.0% recovery for the oyster tissue and 110%–160% for the sewage sludge. The analyte concentration ranges were 50–150 $\mu g\,kg^{-1}$ per component for the oyster tissue and 160–200 $\mu g\,kg^{-1}$ per component for the sewage sludge.

*Phenols*

The extraction of seven phenols (phenol, 2,4-dichlorophenol, 4-nitrophenol, 4-chloro-3-methylphenol, 2-methylphenol, 2-nitrophenol and 2,4-dimethylphenol) from slurry-spiked soils has been investigated using a home-made ASE system [32]. Preliminary work investigated the effect of operating variables (pressure, 4–20 MPa; temperature, 30°C–70°C; and extraction time, 5–25 min) on the recovery of the phenols using acetonitrile as the extraction solvent. The variables were investigated using a central composite design approach. After statistical data treatment using multiple linear regression it was found that only 2-methylphenol was influenced by the operating variables, that is, by pressure and extraction time. It was also noted that the recovery of 2,4-dimethylphenol was low (24.5%) in all instances. The ASE approach was compared with shake-flask extraction and similiar recoveries were obtained. It was also noted that in the case of shake-flask extraction 2,4-dimethylphenol was not recovered at all.

Another interesting feature was observed when comparing the chromatograms obtained after ASE extraction (using acetonitrile as solvent) with those obtained using the shake-flask approach (using an acetonitrile–water mixture as solvent). It is clearly observed (Figure 10.5) that the use of acetonitrile only (ASE, hot solvent extraction) resulted in a cleaner chromatogram than that obtained when using a cold acetonitrile–water mixture (shake-flask extraction).

### *10.3.4 Applications: food matrices*

In contrast to others Obana *et al.* [33] extracted 19 OPPs from spiked foods using ASE. The OPPs were extracted at a pressure of 1500 psi, a temperature of 100°C and an extraction time of approximately 20 min. Wet samples were extracted after mixing with Extrelut® drying agent. The foodstuffs

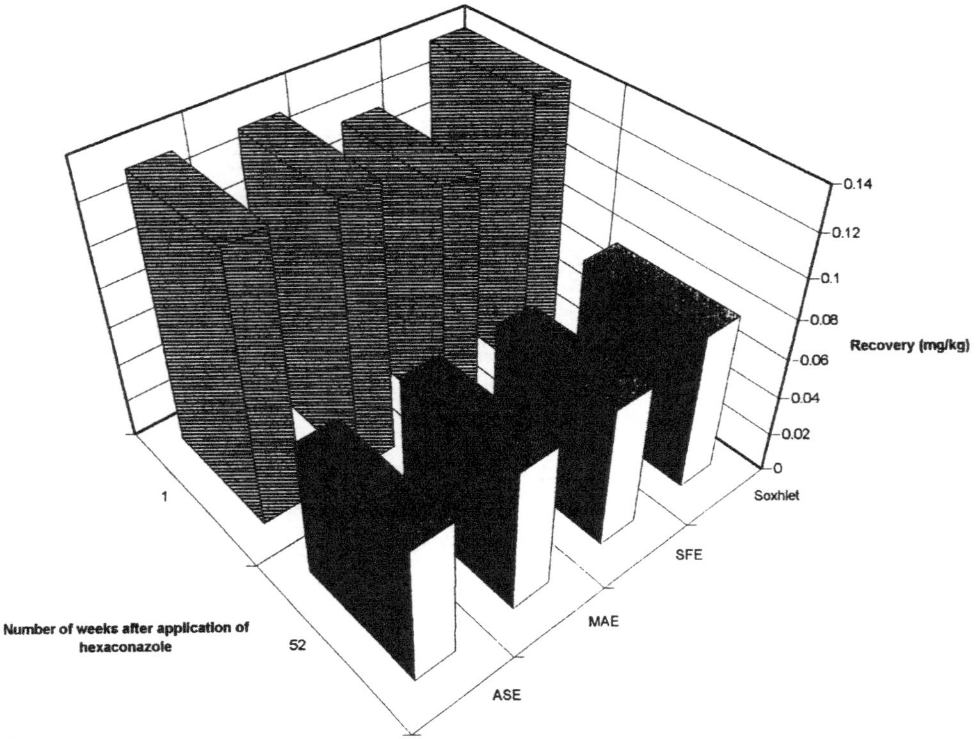
0.14
0.12
0.1
0.08
0.06
0.04
0.02
0
Recovery (mg/kg)
Soxhlet
SFE
MAE
ASE
1
52
Number of weeks after application of hexaconazole

(a)

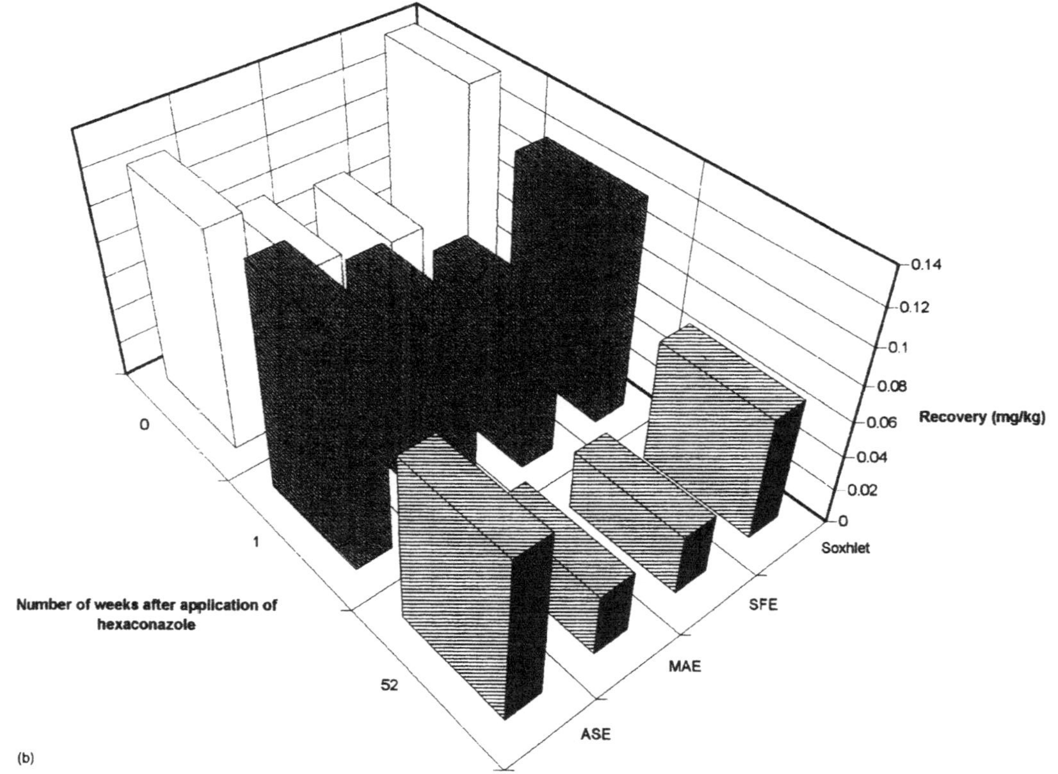


(b)

**Figure 10.4** Extraction of hexaconazole from (a) weathered Canadian sandy clay soil; (b) weathered Canadian sandy loam soil. ASE = accelerated solvent extraction; MAE = microwave-assisted extraction; SFE = supercritical fluid extraction; Soxhlet = Soxhlet extraction.

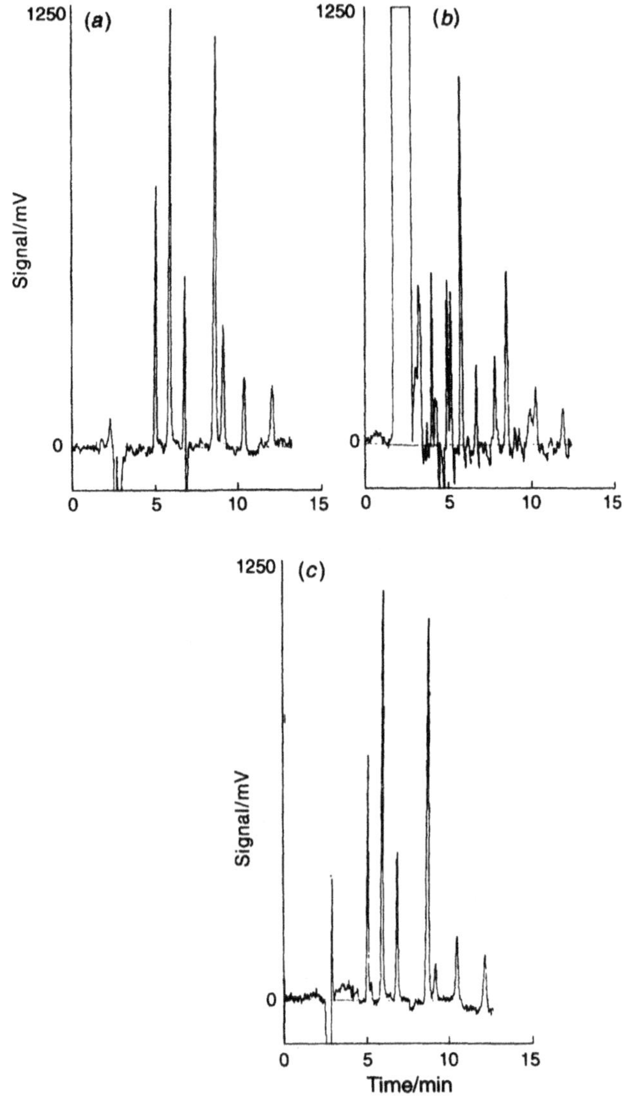


**Figure 10.5** High-performance liquid chromatograms of (a) a 100 ng ml$^{-1}$ phenol standard mixture; (b) shake-flask extract; (c) accelerated solvent extract. Reproduced with permission from ref. 32.

(flour, grapefruit juice, orange juice, broccoli) were spiked at levels between 0.05 ppm and 0.1 ppm and their recoveries assessed. Initially the effect of extraction solvent on the recovery of 17 OPPs (0.05 ppm) from flour (10 g) was done. The solvents evaluated were: cyclohexane–acetone, dichloromethane–acetone and ethylacetate–acetone (all in 1 : 1 ratios). The recoveries ranged from 85.5%

to 115.8% with RSDs ($n = 3$) of less than 12% using cyclohexane–acetone. The exception were dichlorvos and dimethoate which were only poorly recovered using this solvent system (36.6% and 49.1% for dichlorvos and dimethoate, respectively). With the other solvent systems used, that is, with dichloromethane–acetone and ethylacetate–acetone, the average recoveries obtained ranged from 83.2% to 106.0% and 95.5% to 155.4%, respectively, with the exception of dichlorvos and dimethoate. The poor recovery of dichlorvos was suggested to be a result of its volatility; no explanation was offered in the paper for the unexpected recoveries of dimethoate. Using the solvent system of cyclohexane–acetone (1:1) 19 OPPs were extracted from orange and grapefruit juice. The average recoveries ($n = 3$) ranged from 51% to 110% and 82% to 105% in orange juice and grapefruit juice, respectively. Exceptions were noted in the cases of methamidophos and acephate, which were only poorly recovered (less than 11%). The precision of the results was moderate to poor in most cases. These poor recoveries were investigated further by considering the effects of pressure, temperature and extraction time on the recovery of methamidophos and acephate in orange juice. The ASE conditions were varied as follows: temperature, 50°C, 100°C and 150°C; pressure, 1500 psi, 2000 psi and 2500 psi; and extraction time, 5 min, 10 min and 15 min. No significant enhancement in recovery was noted by adjusting the ASE operating conditions. Subsequent work investigated the effect of solvent on the recovery of both OPPs from orange juice. The extraction solvents considered were: cyclohexane–acetone, toluene–acetone, dichloromethane–acetone, ethylacetate–acetone, acetonitrile and ethylacetate. The highest recoveries were obtained using ethylacetate (56.1% and 47.8% for methamidophos and acephate, respectively). Ethylacetate was used as the extraction solvent for subsequent work. The ASE approach was then applied to the extraction of 19 OPPs from orange juice (spike level 0.1 ppm), grapefruit juice (spike level 0.1 ppm), broccoli (spike level 0.1 ppm) and flour (spike level 0.05 ppm) with average recoveries of 79.3%, 87.4%, 90.4% and 82.9%, respectively. Finally, ASE was compared with a hexane extraction for the analysis of food samples (banana, okra and sweetie) which had been found to be contaminated with pesticides. Three OPPs (chlorpyrifos, phosalone and malathion) were identified in the food samples at approximately similar levels to those obtained using hexane. The authors suggested that the use of an automated ASE system offers potential advantages over existing methods for pesticide residue analysis in foodstuffs.

### *10.3.5 Applications: polymeric matrices*

Two ground polymeric materials, nylon-6 and poly(1,4-butylene terephthalate) (PBT) have been extracted for monomers and oligomers using a home-made ASE system [34]. In particular, ASE was used to extract caprolactam from nylon-6 and the dimer and trimer from PBT. Various extraction parameters (temperature, pressure, static extraction time, flow rate and type

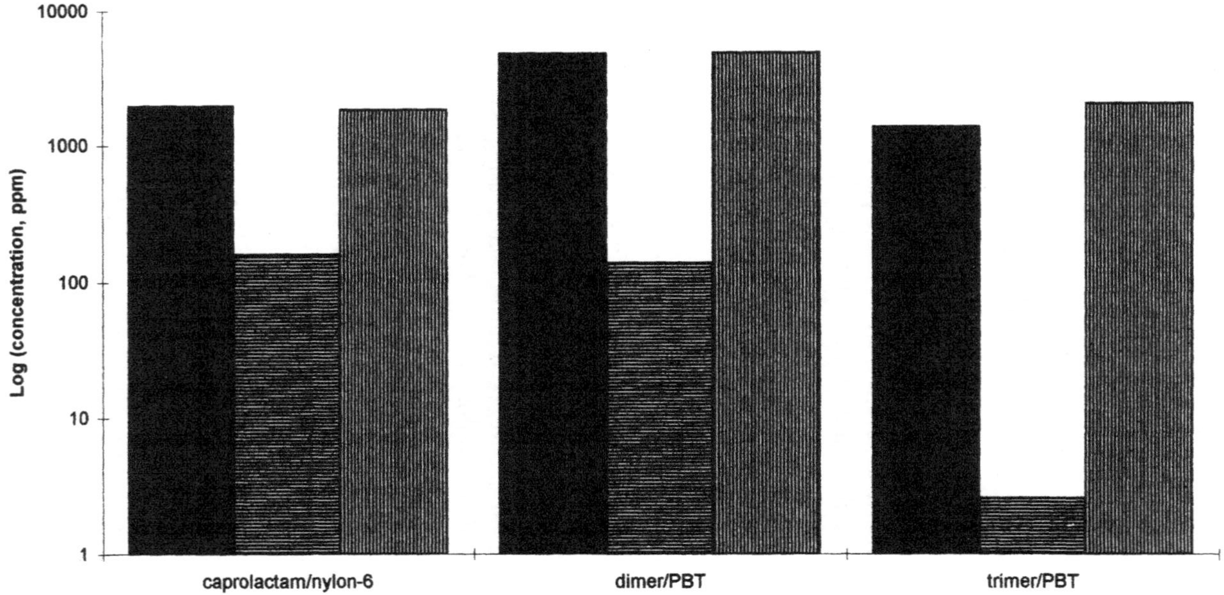


**Figure 10.6** Comparison of accelerated solvent extraction (ASE) with Soxhlet extraction for polymeric materials. PBT = poly(1,4-butylene terephthalate); Soxhlet (1) = hexane solvent for 40 h; Soxhlet (2) = 30 h extraction using methanol for nylon-6, and chloroform for PBT. ▨ = ASE; ▤ = Soxhlet (1); ▥ = Soxhlet (2).

and volume of solvent) were investigated in a systematic manner. It was suggested that for polymeric samples the normal solvents used for Soxhlet extraction are not ideal for ASE. This is because the solvents used for Soxhlet extraction of polymers are normally good swelling agents for the polymer to be extracted. Swelling of polymeric samples in ASE is an undesirable effect as it can cause blockages. For this reason hexane was selected. The recoveries of caprolactam from nylon-6 and the dimer and trimer from PBT both increased as the temperature was increased from 50°C to 170°C. No effect on recovery of the analytes was noted when the pressure increased from 100 bar to 300 bar. Additionally, the effects of static time were investigated for both nylon-6 and PBT. It was noted that although an increase of static time from 10 min to 20 min caused an increase in recovery of the three analytes no further increase was noted when the time was increased to 40 min. A static time of 20 min was therefore adopted for further work. Furthermore, the influence of solvent flow rate on the recovery of the three analytes was investigated at a temperature of 150°C and a pressure of 200 bar. No effect on recovery was noted when the flow rate was increased from 0.4 ml $min^{-1}$ to 2.0 ml $min^{-1}$. Finally, ASE was compared with Soxhlet extraction. The results can be seen in Figure 10.6. It is noted that the recoveries of the three analytes in Soxhlet extraction when using hexane is poor. However, comparable recoveries (except trimer from PBT) are obtained using hexane in ASE with Soxhlet extraction using either methanol (nylon-6) or chloroform (PBT). The effects observed with hexane can clearly be attributed to the use of higher temperature and pressure in ASE.

## 10.4 Conclusions

As Soxhlet extraction is regarded by many as the benchmark technique against which all other extraction techniques are compared it is necessary to consider why this has remained unchanged for many years. The answer you may come up with is that Soxhlet is low-cost, reliable and provides robust methods that can be repeated by even the least experienced member of the laboratory. It is against this simple yardstick all other extraction techniques have to compare. So although analytical supercritical fluid extraction has been around since 1982 it has failed to provide reliable and robust methods, in most cases, that are transferable from one laboratory to another. This is not to say that SFE cannot, in the right hands, provide reliable results, but it is this dependence on an experienced operator that continues to prevent its widespread use. The arrival of both MAE and ASE, as discussed in this chapter, merely reinforces this need to obtain extraction techniques to compete with Soxhlet extraction.

What of the future? Well, as with SFE previously, both MAE and ASE offer similar promises, that is, lower solvent consumption and faster extractions at both elevated temperature and pressure. In addition, both MAE and ASE allow a range of extraction solvents (but they must have a dipole moment for MAE) to be used. The ability to choose an appropriate solvent system may lead to either selective extractions or reactive solvents being used (e.g. *in situ* derivatisation). Beyond this is the need to perform extractions that can reliably extract native pollutants and not just spiked samples, to use 'environmentally-friendly' solvents on easy-to-use instruments that provide reliable and robust methods with some degree of automation. Perhaps all this can be done outside the laboratory, that is, in the field or in the chemical plant, with unattended operation. Some of this may seem like 'pie in the sky' but what of Baron Cagniard de la Tour who in 1822 discovered the phenomenon of the supercritical state, or Percy Spencer of Raytheon who discovered the heating effect of the microwave magnetron in 1947; could they have forseen what became of their discoveries? We await the future with interest.

## References


1. Barnabas, I.J., Dean, J.R., Fowlis, I.A. and Owen, S.P. (1995) Extraction of polycyclic aromatic hydrocarbons from highly contaminated soils using microwave energy. *Analyst*, **120**, 1897–1904.
2. Chee, K.K., Wong, M.K. and Lee, H.K. (1996) Optimization of microwave-assisted solvent extraction of polycyclic aromatic hydrocarbons in marine sediments using a microwave extraction system with high-performance liquid chromatography–fluorescence detection and gas chromatography–mass spectrometry. *Journal of Chromatography A*, **723**, 259–71.
3. Lopez-Avila, V., Young, R. and Beckert, W.F. (1994) Microwave-assisted extraction of organic compounds from standard reference soils and sediments. *Analytical Chemistry*, **66**, 1097–106.
4. Onuska, F.I. and Terry, K.A. (1993) Extraction of pesticides from sediments using a microwave technique. *Chromatographia*, **36**, 191–4.
5. Lopez-Avila, V., Young, R., Benedicto, J. *et al.* (1995) Extraction of organic pollutants from solid samples using microwave energy. *Analytical Chemistry*, **67**, 2096–102.
6. Ganzler, K., Salgo, A. and Valko, K. (1986) Microwave extraction. A novel sample preparation method for chromatography. *Journal of Chromatography*, **371**, 299–306.
7. Donard, O.F.X., Lalere, B., Martin, F. and Lobinski, R. (1995) Microwave-assisted leaching of organotin compounds from sediments for speciation analysis. *Analytical Chemistry*, **67**, 4250–4.
8. Vasquez, M.J., Carro, A.M., Lorenzo, R.A. and Cela, R. (1997) Optimization of methylmercury microwave-assisted extraction from aquatic sediments. *Analytical Chemistry*, **69**, 221–5.
9. Szpunar, J., Ceulemans, M., Schmitt, V.O. *et al.* (1996) Microwave-accelerated speciation analysis for butyltin compounds in sediments and biomaterials by large volume injection capillary gas chromatography quartz furnace atomic absorption spectrometry. *Analytica Chimica Acta*, **332**, 225–32.
10. Young, J.C. (1995) Microwave-assisted extraction of the fungal metabolite ergosterol and total fatty acids. *Journal of Agriculture and Food Chemistry*, **43**(11), 2904–10.
11. Onuska, F.I. and Terry, K.A. (1995) Microwave extraction in analytical chemistry of pollutants: polychlorinated biphenyls. *Journal of High Resolution Chromatography*, **18**, 417–21.
12. Onuska, F.I. and Terry, K.A. (1995) Microwave extraction in analytical chemistry of pollutants: chlorinated benzenes. *Journal of Microcolumn Separations*, **7**(4), 319–26.
13. Chee, K.K., Wong, M.K. and Lee, H.K. (1996) Microwave-assisted solvent elution technique for the extraction of organic pollutants in water. *Analytica Chimica Acta*, **330**, 217–27.

14. Pare, J.R.J., Belanger, J.M.R., Li, K. and Stafford, S.S. (1995) Microwave-assisted process (MAP): Application to the headspace analysis of VOCs in water. *Journal of Microcolumn Separations*, **7**, 37–40.
15. Metaxas, A.C. and Meredith, R.J. (1983) *Industrial Microwave Heating, Power Engineering Series 4*, Peter Peregrinus, London.
16. Kingston, H.M. and Jassie, L.B. (1988) *Introduction to Microwave Sample Preparation. Theory and Practice*, ACS Professional Reference Book, American Chemical Society, Washington, DC.
17. Hasty, E. and Revesz, R. (1995) Total petroleum hydrocarbon determination by microwave solvent extraction. *American Laboratory* (February), 66–73.
18. Cheung, W.S. and Levein, F.H. (1985) *Microwaves Made Simple: Principles and Applications*, Artech House, Boston, MA.
19. Hummert, K., Vetter, W. and Luckas, B. (1996) Fast and effective sample preparation for determination of organochlorine compounds in fatty tissue of marine mammals using microwave extraction. *Chromatographia*, **42**, 300–4.
20. Steinheimer, T.R. (1993) HPLC determination of atrazine and principal degradates in agricultural soils and associated surface and ground water. *Journal of Agricultural Food Chemistry*, **41**, 588–95.
21. Stout, S.J., daCunha, A.R. and Allardice, D.G. (1996) Microwave-assisted extraction coupled with gas chromatography/electron capture negative chemical ionization mass spectrometry for the simplified determination of imidazolinone herbicides in soil at the ppb level. *Analytical Chemistry*, **68**, 653–8.
22. Lopez-Avila, V., Benedicto, J., Charan, C. *et al.* (1995) Determination of PCBs in soils/sediments by microwave-assisted extraction and GC/ECD or ELISA. *Environmental Science and Technology*, **29**, 2709–12.
23. Llompart, M.P., Lorenzo, R.A. Cela, R. and Pare, J.R.J. (1997) Optimization of a microwave-assisted extraction method for phenol and methylphenol isomers in soil samples using a central composite design. *Analyst*, **122**, 133–7.
24. Chee, K.K., Wong, M.K. and Lee, H.K. (1996) Microwave extraction of phthalate esters from marine sediment and soil. *Chromatographia*, **42**, 378–84.
25. USEPA (1995), Testing methods for evaluating solid waste, Method 3545, USEPA SW-846, 3rd edn, Update III, US Environmental Protection Agency, Washington DC.
26. Richter, B.E., Jones, B.A., Ezzell, J.L. *et al.* (1996) Accelerated solvent extraction: a technique for sample preparation. *Analytical Chemistry*, **68**, 1033–9.
27. Dean, J.R. (1996) Accelerated solvent extraction of polycyclic aromatic hydrocarbons from contaminated soil. *Analytical Communications*, **33**, 191–2.
28. Saim, N., Dean, J.R., Abdullah, M. P. and Zakaria, Z. (1997) Extraction of polycyclic aromatic hydrocarbons from contaminated soil using Soxhlet extraction, pressurised and atmospheric microwave-assisted extraction, supercritical fluid extraction and accelerated solvent extraction. *Journal of Chromatography A*, **791**, 361–6.
29. Fisher, J.A., Scarlett, M.J. and Scott, A.D. (1997) Accelerated solvent extraction: an evaluation for screening of soils for selected US EPA semivolatiles organic priority pollutants. *Environmental Science and Technology*, **31**, 1120–7.
30. Ezzell, J.L., Richter, B.E., Felix, W.D. *et al.* (1995) A comparison of accelerated solvent extraction with conventional solvent extraction for organophosphorus pesticides and herbicides. *LC–GC*, **13**, 390–8.
31. Frost, S.P., Dean, J.R., Evans, K.P. *et al.* (1997) Extraction of hexaconazole from weathered soils: a comparison between Soxhlet extraction, microwave-assisted extraction, supercritical fluid extraction and accelerated solvent extraction. *Analyst*, **122**, 895–8.
32. Dean, J.R., Santamaria-Rekondo, A. and Ludkin, E. (1996) Accelerated solvent extraction of phenols from soil. *Analytical Communications*, **33**, 413–16.
33. Obana, H., Kikuchi, K., Okihashi, M. and Hori, S. (1997) Determination of organophosphorus pesticides in foods using an accelerated solvent extraction system. *Analyst*, **122**, 217–20.
34. Lou, X., Janssen, H.-G. and Cramers, C.A. (1997) Parameters affecting the accelerated solvent extraction of polymeric samples. *Analytical Chemistry*, **69**, 1598–603.

# Appendix

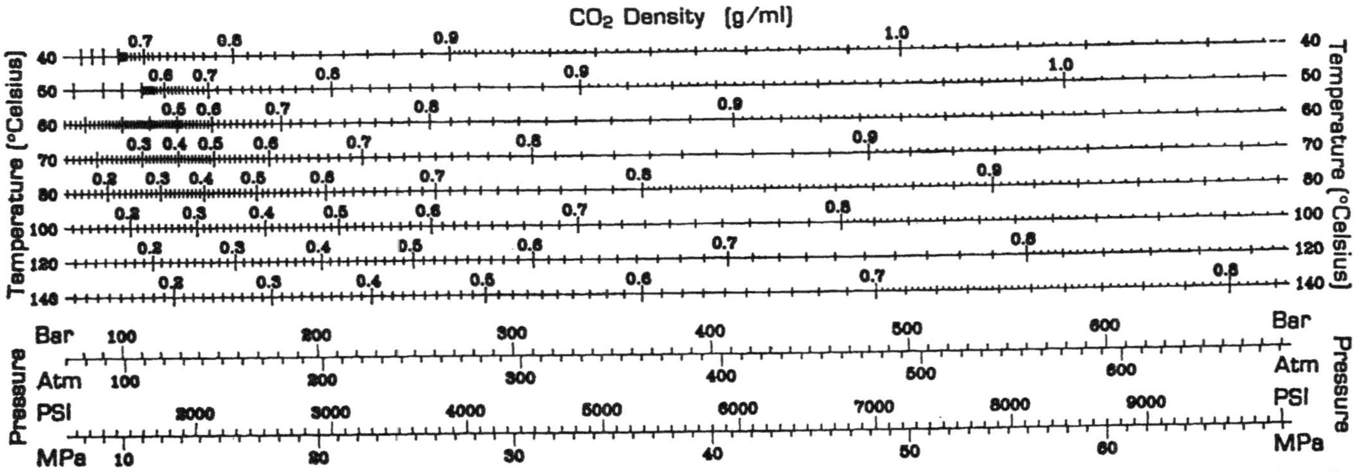


**Figure A.1** Supercritical fluid extraction conditions: density–pressure–temperature relationships for supercritical fluid $CO_2$. Find $CO_2$ pressure for a given temperature and density, select the scale for the desired temperature, set a rule on the desired density and read the pressure from the bottom scales. The pressure scales also serve for unit conversion. Reproduced with kind permission from Isco Inc.

**Table A.1** Supercritical fluid $CO_2$ densities (g/ml) at various temperatures ($T$) and pressures

| | Pressure (atm) | | | | | | | | | | | | |
|---|---|---|---|---|---|---|---|---|---|---|---|---|---|
| $T$ (°C) | 14.6 | 29.1 | 43.7 | 58.3 | 72.8 | 87.4 | 102.0 | 116.6 | 131.1 | 145.7 | 160.3 | 174.8 | 189.4 |
| −29.7 | 0.0388 | 1.0783 | 1.0717 | 1.0733 | 1.0856 | 1.1099 | 1.1126 | 1.1039 | 1.1142 | 1.1147 | 1.1222 | 1.1285 | 1.1339 |
| −14.5 | 0.0349 | 1.0094 | 1.0101 | 1.0132 | 1.0304 | 1.0461 | 1.0537 | 1.0491 | 1.0578 | 1.0648 | 1.0706 | 1.0755 | 1.0797 |
| 0.7 | 0.0320 | 0.0755 | 0.9529 | 0.9561 | 0.9674 | 0.9781 | 0.9805 | 0.9908 | 0.9951 | 1.0056 | 1.0111 | 1.0188 | 1.0225 |
| 15.9 | 0.0296 | 0.0671 | 0.1206 | 0.8418 | 0.8714 | 0.8911 | 0.9024 | 0.9227 | 0.9205 | 0.9290 | 0.9450 | 0.9531 | 0.9543 |
| 31.1 | 0.0277 | 0.0609 | 0.1028 | 0.1643 | 0.4716 | 0.7331 | 0.7915 | 0.8120 | 0.8332 | 0.8545 | 0.8668 | 0.8784 | 0.8915 |
| 46.3 | 0.0260 | 0.0561 | 0.0918 | 0.1367 | 0.1999 | 0.3116 | 0.5068 | 0.6308 | 0.6930 | 0.7337 | 0.7620 | 0.7895 | 0.8081 |
| 61.5 | 0.0246 | 0.0523 | 0.0839 | 0.1210 | 0.1664 | 0.2246 | 0.3053 | 0.4067 | 0.5066 | 0.5823 | 0.6429 | 0.6797 | 0.7098 |
| 76.7 | 0.0234 | 0.0492 | 0.0780 | 0.1100 | 0.1472 | 0.1906 | 0.2422 | 0.3018 | 0.3700 | 0.4421 | 0.4998 | 0.5517 | 0.5946 |
| 91.9 | 0.0222 | 0.0464 | 0.0727 | 0.1016 | 0.1335 | 0.1695 | 0.2088 | 0.2539 | 0.3035 | 0.3539 | 0.4007 | 0.4466 | 0.4885 |
| 107.1 | 0.0212 | 0.0441 | 0.0684 | 0.0948 | 0.1233 | 0.1544 | 0.1875 | 0.2233 | 0.2632 | 0.3035 | 0.3429 | 0.3824 | 0.4217 |
| 122.3 | 0.0203 | 0.0419 | 0.0647 | 0.8090 | 0.1149 | 0.1426 | 0.1713 | 0.2019 | 0.2350 | 0.2691 | 0.3043 | 0.3386 | 0.3710 |
| 152.7 | 0.0188 | 0.0383 | 0.0587 | 0.0797 | 0.1020 | 0.1252 | 0.1485 | 0.1725 | 0.1981 | 0.2244 | 0.2516 | 0.2786 | 0.3060 |
| 183.2 | 0.0174 | 0.0354 | 0.0538 | 0.0728 | 0.0924 | 0.1126 | 0.1329 | 0.1538 | 0.1751 | 0.1970 | 0.2186 | 0.2415 | 0.2634 |
| 213.6 | 0.0163 | 0.0329 | 0.0498 | 0.0671 | 0.0849 | 0.1028 | 0.1210 | 0.1393 | 0.1578 | 0.1768 | 0.1959 | 0.2151 | 0.2343 |
| 244.0 | 0.0153 | 0.0308 | 0.0465 | 0.0623 | 0.0784 | 0.0948 | 0.1115 | 0.1278 | 0.1443 | 0.1613 | 0.1783 | 0.1955 | 0.2127 |
| 274.4 | 0.0144 | 0.0289 | 0.0435 | 0.0583 | 0.0731 | 0.0883 | 0.1034 | 0.1184 | 0.1335 | 0.1487 | 0.1640 | 0.1795 | 0.1946 |
| 304.8 | 0.0136 | 0.0273 | 0.0410 | 0.0548 | 0.0686 | 0.0826 | 0.0965 | 0.1105 | 0.1242 | 0.1383 | 0.1524 | 0.1665 | 0.1806 |
| 335.2 | 0.0129 | 0.0259 | 0.0388 | 0.0518 | 0.0648 | 0.0778 | 0.0908 | 0.1037 | 0.1167 | 0.1297 | 0.1428 | 0.1557 | 0.1681 |
| 487.3 | 0.0103 | 0.0205 | 0.0307 | 0.0408 | 0.0508 | 0.0608 | 0.0706 | 0.0806 | 0.0903 | 0.1001 | 0.1095 | 0.1192 | 0.1284 |
| 639.3 | 0.0086 | 0.0171 | 0.0255 | 0.0338 | 0.0421 | 0.0504 | 0.0586 | 0.0666 | 0.0747 | 0.0826 | 0.0903 | 0.0980 | 0.1057 |
| 791.4 | 0.0073 | 0.0146 | 0.0218 | 0.0289 | 0.0360 | 0.0430 | 0.0500 | 0.0568 | 0.0638 | 0.0706 | 0.0773 | 0.0839 | 0.0905 |
| 943.4 | 0.0064 | 0.0128 | 0.0191 | 0.0253 | 0.0315 | 0.0376 | 0.0437 | 0.0498 | 0.0558 | 0.0618 | 0.0677 | 0.0735 | 0.0794 |

**Table A.1** *Cont.*

| | Pressure (atm) | | | | | | | | | | | | |
|---|---|---|---|---|---|---|---|---|---|---|---|---|---|
| $T$ (°C) | 204.0 | 218.5 | 233.1 | 247.7 | 262.3 | 276.8 | 291.4 | 327.8 | 364.2 | 437.1 | 509.9 | 582.8 | 655.6 |
| −29.7 | 1.1415 | 1.1482 | 1.1515 | 1.1544 | 1.1593 | 1.1615 | 1.1635 | 1.1683 | 1.1704 | 1.1808 | 1.1921 | 1.2018 | 1.2103 |
| −14.5 | 1.0861 | 1.0858 | 1.0910 | 1.0905 | 1.0971 | 1.1032 | 1.1066 | 1.1171 | 1.1241 | 1.1388 | 1.1522 | 1.1668 | 1.1762 |
| 0.7 | 1.0225 | 1.0279 | 1.0298 | 1.0365 | 1.0399 | 1.0454 | 1.0523 | 1.0651 | 1.0772 | 1.0986 | 1.1145 | 1.1253 | 1.1381 |
| 15.9 | 0.9631 | 0.9709 | 0.9729 | 0.9814 | 0.9891 | 0.9937 | 1.0002 | 1.0168 | 1.0275 | 1.0536 | 1.0731 | 1.0883 | 1.1003 |
| 31.1 | 0.8979 | 0.9104 | 0.9150 | 0.9255 | 0.9307 | 0.9431 | 0.9471 | 0.9650 | 0.9798 | 1.0068 | 1.0317 | 1.0505 | 1.0613 |
| 46.3 | 0.8204 | 0.8336 | 0.8452 | 0.8596 | 0.8728 | 0.8829 | 0.8942 | 0.9133 | 0.9265 | 0.9567 | 0.9877 | 1.0123 | 1.0244 |
| 61.5 | 0.7307 | 0.7499 | 0.7671 | 0.7849 | 0.7999 | 0.8135 | 0.8264 | 0.8535 | 0.8738 | 0.9073 | 0.9431 | 0.9726 | 0.9882 |
| 76.7 | 0.6295 | 0.6576 | 0.6801 | 0.7029 | 0.7216 | 0.7408 | 0.7577 | 0.7903 | 0.8148 | 0.8590 | 0.8956 | 0.9300 | 0.9514 |
| 91.9 | 0.5287 | 0.5649 | 0.5979 | 0.6257 | 0.6505 | 0.6758 | 0.6944 | 0.7321 | 0.7609 | 0.8128 | 0.8534 | 0.8904 | 0.9133 |
| 107.1 | 0.4589 | 0.4944 | 0.5278 | 0.5565 | 0.5851 | 0.6092 | 0.6297 | 0.6740 | 0.7093 | 0.7637 | 0.8090 | 0.8501 | 0.8755 |
| 122.3 | 0.4038 | 0.4353 | 0.4674 | 0.4977 | 0.5241 | 0.5504 | 0.5737 | 0.6218 | 0.6611 | 0.7211 | 0.7694 | 0.8116 | 0.8427 |
| 152.7 | 0.3334 | 0.3606 | 0.3868 | 0.4122 | 0.4355 | 0.4580 | 0.4791 | 0.5283 | 0.5707 | 0.6418 | 0.6922 | 0.7385 | 0.7744 |
| 183.2 | 0.2857 | 0.3075 | 0.3300 | 0.3511 | 0.3727 | 0.3925 | 0.4112 | 0.4562 | 0.4966 | 0.5669 | 0.6267 | 0.6730 | 0.7115 |
| 213.6 | 0.2537 | 0.2721 | 0.2915 | 0.3105 | 0.3282 | 0.3461 | 0.3639 | 0.4063 | 0.4435 | 0.5100 | 0.5682 | 0.6168 | 0.6545 |
| 244.0 | 0.2295 | 0.2464 | 0.2633 | 0.2794 | 0.2951 | 0.3112 | 0.3264 | 0.3632 | 0.3981 | 0.4597 | 0.5156 | 0.5626 | 0.6024 |
| 274.4 | 0.2097 | 0.2246 | 0.2393 | 0.2539 | 0.2682 | 0.2822 | 0.2960 | 0.3298 | 0.3612 | 0.4188 | 0.4701 | 0.5150 | 0.5557 |
| 304.8 | 0.1944 | 0.2080 | 0.2216 | 0.2349 | 0.2479 | 0.2606 | 0.2731 | 0.3037 | 0.3330 | 0.3855 | 0.4330 | 0.4751 | 0.5138 |
| 335.2 | 0.1806 | 0.1933 | 0.2057 | 0.2178 | 0.2299 | 0.2421 | 0.2538 | 0.2826 | 0.3096 | 0.3594 | 0.4034 | 0.4437 | 0.4799 |
| 487.3 | 0.1378 | 0.1470 | 0.1562 | 0.1654 | 0.1743 | 0.1832 | 0.1918 | 0.2130 | 0.2336 | 0.2725 | 0.3084 | 0.3415 | 0.3712 |
| 639.3 | 0.1133 | 0.1207 | 0.1283 | 0.1356 | 0.1429 | 0.1503 | 0.1571 | 0.1746 | 0.1914 | 0.2232 | 0.2534 | 0.2802 | 0.3059 |
| 791.4 | 0.0970 | 0.1034 | 0.1097 | 0.1185 | 0.1222 | 0.1284 | 0.1345 | 0.1496 | 0.1641 | 0.1921 | 0.2184 | 0.2429 | 0.2664 |
| 943.4 | 0.0850 | 0.0907 | 0.0964 | 0.1020 | 0.1074 | 0.1129 | 0.1182 | 0.1316 | 0.1446 | 0.1697 | 0.1934 | 0.2157 | 0.2371 |

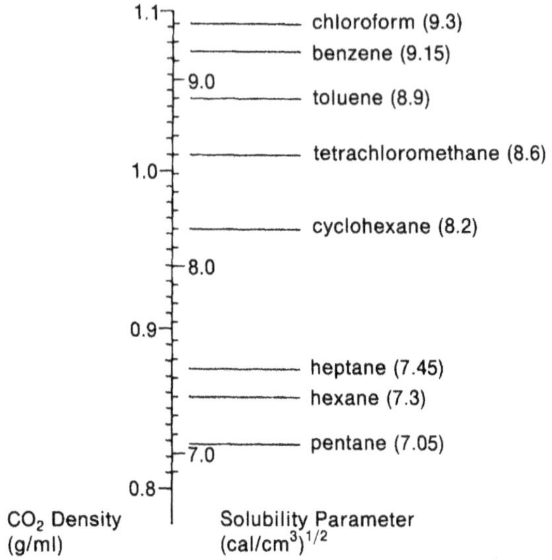


**Figure A.2** Organic solvent replacement guide: supercritical fluid $CO_2$ density and solubility parameter and equivalent common solvents. Guide to use:

1. find the organic solvent used in Soxhlet extraction to be replaced with supercritical fluid extraction (SFE);
2. note the corresponding solubility parameter and $CO_2$ density;
3. use Table A.1 or the scales in Figure A.1 to find starting conditions for SFE.

Reproduced with kind permission from Isco Inc.

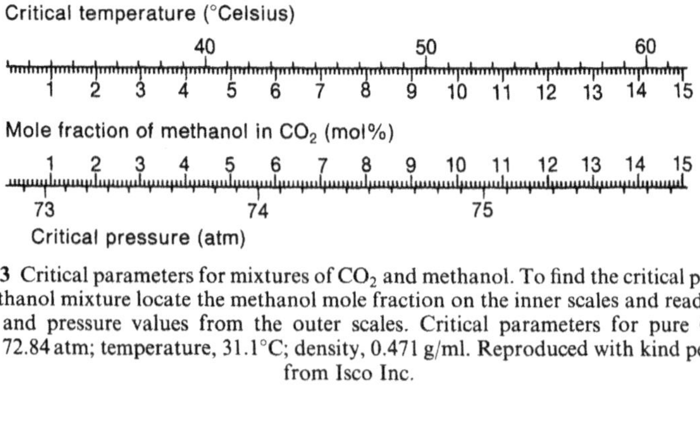


**Figure A.3** Critical parameters for mixtures of $CO_2$ and methanol. To find the critical point for a $CO_2$–methanol mixture locate the methanol mole fraction on the inner scales and read the temperature and pressure values from the outer scales. Critical parameters for pure $CO_2$ are: pressure, 72.84 atm; temperature, 31.1°C; density, 0.471 g/ml. Reproduced with kind permission from Isco Inc.

# Subject index

# Matrix index

As a very brief SFE applications guide, two additional indices are provided based upon the SFE applications cited within this text. These indices do not include information extracted from either the text tables or figures. The reader is referred to Chapter 10 to find applications of microwave-assisted extraction and accelerated solvent extraction.

# Analyte index

**Drugs and biochemicals**

**Flavour and fragrance compounds**



**Hydrocarbons**



**Metals, inorganics and organometallic compounds**



**Pesticides, herbicides and fungicides**

**Phenols**



**Polychlorinated biphenyls (PCBs) and polychlorinated dibenzo-*p*-dioxins (PCDDs)**



**Polycyclic aromatic hydrocarbons (PAHs) and BTEX compounds**

**Polymer additives**



**Polymers and oligomers**



**Miscellaneous**

9 780751 404463

GPSR Compliance

*The European Union's (EU) General Product Safety Regulation (GPSR) is a set of rules that requires consumer products to be safe and our obligations to ensure this.*

*If you have any concerns about our products, you can contact us on ProductSafety@springernature.com*

In case Publisher is established outside the EU, the EU authorized representative is:

Springer Nature Customer Service Center GmbH
Europaplatz 3
69115 Heidelberg, Germany

**Batch number: 10809439**

Printed by Printforce, the Netherlands